Quality Production Improvement and System Safety
QPI 16 - CZOTO 10

16th Quality Production Improvement, QPI 2022, Zaborze, Poland, 20-22.06.2022.
10th System Safety: Human-Technical Facility-Environment (CZOTO), Zakopane, Poland, 07-09.12.2022

Editors
Robert Ulewicz[1], Norbert Radek[2], Jacek Pietraszek[3]

[1]Częstochowa University of Technology, Faculty of Management, Poland
[2]Kielce University of Technology, Centre for Laser Technologies of Metals, Poland
[3]Cracow University of Technology, Faculty of Mechanical Engineering, Poland

Peer review statement

All papers published in this volume of "Materials Research Proceedings" have been peer reviewed. The process of peer review was initiated and overseen by the above proceedings editors. All reviews were conducted by expert referees in accordance to Materials Research Forum LLC high standards.

Published under License by **Materials Research Forum LLC**
Millersville, PA 17551, USA

Published as part of the proceedings series
Materials Research Proceedings
Volume 34 (2023)

ISSN 2474-3941 (Print)
ISSN 2474-395X (Online)

ISBN 978-1-64490-268-4 (Print)
ISBN 978-1-64490-269-1 (eBook)

This book contains information obtained from authentic and highly regarded sources. Reasonable efforts have been made to publish reliable data and information, but the author and publisher cannot assume responsibility for the validity of all materials or the consequences of their use. The authors and publishers have attempted to trace the copyright holders of all material reproduced in this publication and apologize to copyright holders if permission to publish in this form has not been obtained. If any copyright material has not been acknowledged please write and let us know so we may rectify in any future reprint.

Distributed worldwide by

Materials Research Forum LLC
105 Springdale Lane
Millersville, PA 17551
USA
https://www.mrforum.com

Manufactured in the United State of America
10 9 8 7 6 5 4 3 2 1

Table of Contents

Preface

We are delighted and excited to present to you this outstanding compilation of 55 papers that have received positive reviews. These papers were presented at two influential conferences: the 16th International Conference on Quality Production Improvement (QPI) and the 11th International Conference on Safety Systems (Human-Technical Facility-Environment) held in 2022. These conferences served as platforms for experts, researchers, practitioners, and scholars from diverse disciplines to exchange their knowledge, perspectives, and advancements in the fields of safety management, quality production, and improvement.

The papers selected for this collection encompass a wide range of topics discussed during the conferences, reflecting the diverse interests and cutting-edge research presented. While the primary focus lies on safety systems, the papers delve into various key areas, including health and safety management, technological safety, cleaner production, and the influence of human factors on occupational health and safety management systems. Additionally, they explore the safe operation of technical facilities, environmental safety management, and the safety management of products and services. These papers emphasize the importance of adopting a comprehensive approach that ensures both security and sustainability in our environment.

Furthermore, the papers presented at the QPI conference offer valuable insights into various aspects of quality engineering, production management, and improvement strategies. They provide a deeper understanding of optimizing production processes, implementing visual control and management systems, and integrating concepts from Industry 4.0 and 5.0. These papers also explore the sustainable development of production systems, the application of additive manufacturing, the implementation of cleaner production methods, and the utilization of waste materials for sustainable production practices. The research shared in these papers contributes to advancing and improving production systems in a holistic and environmentally conscious manner.

Within this collection, you will discover papers that address critical topics, including total productive maintenance, Six Sigma, lean manufacturing, and the enhancement of work safety. These papers also delve into the operation and management of technical and construction facilities, the effective handling of product and process innovations, and the integration of renewable and alternative energy sources within production systems. Furthermore, they discuss the significance of human resource management in quality improvement, the pivotal role of human capital in production engineering, and the importance of fostering active learning in engineering education. The insights shared in these papers contribute to the understanding and advancement of these areas, providing valuable knowledge for researchers, practitioners, and scholars alike.

The extensive range of topics covered in this collection underscores the multidisciplinary nature of these conferences. The contributions made by the authors have significantly enriched our understanding and offered practical solutions to intricate challenges in safety management, quality production, and improvement. These papers not only contribute to academic knowledge but also hold practical implications for industries, organizations, and policymakers who are dedicated to establishing safe, sustainable, and high-quality environments. The insights shared within these papers can inform and guide real-world practices, fostering positive changes and advancements in various sectors.

We express our heartfelt appreciation to the authors of these exceptional papers for their valuable contributions, as well as to the reviewers who diligently assessed each submission.

Their expertise, dedication, and commitment to excellence have ensured the remarkable quality of this collection.

It is our sincere hope that this compilation will serve as a wellspring of inspiration and knowledge for researchers, professionals, and students alike. May it ignite further research endeavors, foster collaboration among various stakeholders, and spark innovative ideas that pave the way toward a safer, more productive, and sustainable future.

The organizers would also like to express their gratitude to all who have updated and reviewed the papers submitted to the conference, the scientific secretary for her editing work, and Materials Research Forum LLC for producing the volume.

On behalf of the organizing committee of the conferences:

Robert Ulewicz

Quality and Production Managers Association

and

Production Engineering and Safety Department, Faculty of Management
Czestochowa University of Technology

Conference Organizer

Production Engineering and Safety Department
Faculty of Management
Częstochowa University of Technology

Scientific Committee

CHAIRMAN
Prof. Robert Ulewicz (PL), Częstochowa University of Technology

DEPUTY CHAIRMAN
Prof. Branislav Hadzima (SK), University of Žilina
Prof. Ružica R. Nikolić (SRB), University of Kragujevac
Prof. Dorota Klimecka-Tatar (PL), Częstochowa University of Technology

MEMBERS
Prof. Witold Biały (PL)
Prof. Martina Blašková (CZ)
Prof. Tatianá Čorejova (SK)
Prof. Nicoletto Gianni (I)
Prof. Radosław Wolniak (PL)
Prof. Otakar Bokůvka (SK)
Prof. Miloš Hitka (SK)
Prof. Eva Schmidova (CZ)
Prof. František Holešovskỳ (CZ)
Prof. Jozef Hrubec (SK)
Prof. Denis Jelačić (HR)
Prof. Dorota Jelonek (PL)
Prof. Jiří Kliber (CZ)
Prof. Ludvík Kunz (CZ)
Prof. Tomasz Lipiński (PL)
Prof. František Nový (SK)
Prof. Andrzej Pacana (PL)
Prof. Jacek Pietraszek (PL)
Prof. Norbert Radek (PL)
Prof. Anna Šatanova, (SK)
Prof. Jacek Selejdak (PL)
Prof. Tamer Rizaoglu (TR)
Prof. Agnieszka Dardzinska-Głębocka (PL)
Prof. Augustín Sládek (SK)
Prof. János Takács (H)
Prof. Richard Vlosky (USA)
Prof. Wieslaw „Wes" Grebski (USA)
Prof. Pisut Koomsap (THA)
Atul B. Borade, PhD. (IN)
Juhani Anttila, M.Sc. (FI)
Prof. José Carlos Sá (PT)

Prof. Henryk Noga (PL)
Prof. Roman Khmil (UA)
Agnieszka Deja, PhD (PL)
Mariusz Urbański, PhD (PL)

Organizing committee CZOTO

Robert Ulewicz, Chairman
Dorota Klimecka-Tatar, Deputy Chairman
Marta Jagusiak-Kocik, Secretary
Adam Idzikowski
Manuela Ingaldi
Marta Niciejewska
Magdalena Mazur
Marek Krynke
Krzysztof Knop
Renata Stasiak-Betlejewska
Monika Osyra
Krzysztof Mielczarek

Organizing committee QPI

Robert Ulewicz, Chairman
Magdalena Mazur, Deputy Chairman
Manuela Ingaldi
Renata Stasiak-Betlejewska
Adam Idzikowski
Monika Osyra
Paweł Smolnik
Piotr Czaja

Quality Production Improvement and System Safety: QPI 16 - CZOTO 10   Materials Research Forum LLC
Materials Research Proceedings 34 (2023) 1-5   https://doi.org/10.21741/9781644902691-1

Effect of Welding Methods on Mechanical Properties of Domex 700MC Steel

NOVÝ František [1,a], MEDVECKÁ Denisa [1,b]*, TRŠKO Libor [2,c],
BOKŮVKA Otakar [1,d], DRÍMALOVÁ Petra [1,e]

[1]Department of Materials Engineering, Faculty of Mechanical Engineering, University of Žilina, Univerzitná 8215/1, 010 26 Žilina, Slovakia

[2]Research Centre of the University of Žilina, University of Žilina, Univerzitná 8215/1, 010 26 Žilina, Slovakia

[a]frantisek.novy@fstroj.uniza.sk, [b]*denisa.medvecka@fstroj.uniza.sk, [c]libor.trsko@rc.uniza.sk, [d]otakar.bokuvka@fstroj.uniza.sk, [e]petra.drimalova@fstroj.uniza.sk

Keywords: Fatigue, Welded Joints, HSLA Steel, Rotating Bending Test

Abstract. Recent years have seen a lot of research on modern steels with higher yield strengths. The study investigates the microstructure and mechanical properties of two different welding techniques - gas metal arc (GMA) and laser beam (LB). The result of this study showed a significant softening effect in the heat-affected zone (HAZ) for both welding techniques, the GMA and LB, resulting from formation of a coarse-grained microstructure during welding. The findings of this study increased understanding and added to the body of knowledge in the rapidly growing field of GMA and LB welding processes of HSLA steels.

Introduction
High mechanical properties and fine-grained microstructure of high-strength low-alloyed (HSLA) steels resulting from low carbon content and microalloying up to 0.15 wt.% by niobium, titanium, and vanadium are achieved by thermomechanical controlled treatment [1,2]. Domex 700MC steel is a high-strength, low-alloy steel that is widely used in bridge and crane construction. The welding procedure employed during fabrication can have marked impact on the mechanical properties. Gas metal arc welding (GMA) and laser beam welding (LB) are two of the most frequent welding processes used on Domex 700MC steel. Each of these techniques can generate different rates of heating and cooling, heat input and distortion, all of which can alter the mechanical properties of the steel [2,3]. The core problem is the formation of the coarse-grained microstructure in the HAZ during welding, after the rapid cooling from high temperatures. It is essential to conduct the behavior of these steels in the heat-affected zone during the welding process, especially in terms of safety.

Experimental Material and Methods
Experimental material. The 10 mm thick sheet of hot-rolled structural steel Domex 700MC produced by SSAB steelwork company was used to perform experimental part of the paper. An optical emission spectrometer (OES) was used to define the chemical composition of experimental material, and the obtained result is shown in Table 1.

Table 1. Chemical composition of Domex 700MC [wt.%]

C	Si	Mn	S	P	Al	Nb	V	Ti
0.12	0.09	1.99	0.07	0.014	0.02	0.05	0.104	0.07

For microhardness identification of base metal and the weld joint samples, the standard hardness testing method, and the procedure, specifically Vickers line microhardness testing (HV1), was used.

Welding procedures. Two welding techniques, GMA and the LB, have been applied to provide the experimental part. In accordance with the recommendations of the EN 10146-2 standard, low-alloy wire OK AristoRod 69 with a diameter of 1.2 mm was used as filler material for GMA welding. Welding conditions for GMA welding are described in Table 2. For the laser welding regime, the parameters are shown in Table 3.

Table 2. Specifications for the GMA welding procedure

	Electrode	Diameter [mm]	Voltage [V]	Current [A]	Welding speed [mm/s]
Root layer	OK AristoRod 69	1.2	23.3	164	4.8
Top layer			28.2	230	5

Table 3. Specifications for the LB welding procedure

Power [kW]	Defocusing distance [mm]	Welding speed [mm/s]
5	-4	7.5

The welding conditions for both methods were optimized to ensure complete material flow, proper formation of the weld root and weld bead, and also the absence of external or internal defects.

Microstructure. The microstructure was observed with an optical microscope. Experimental samples of base metal and HAZ were prepared and analyzed. Etching by 2% Nital ensured the reveal of ferrite grain boundaries.

Results and Discussion

Microhardness. The line measurements in the cross-sections of welded samples revealed distinctive changes in the final microhardness profiles, starting in the base metal, continuing in HAZ, and ending in the weld metal (mirrored results). The results of the GMA weld joint sample are shown in the left column of Fig. 1, and those of the LB weld joint sample in the right column.

Since GMA welding uses a wire electrode with a similar chemical composition to the base metal, the hardness results of base and weld metal are comparable. The laser beam is directed at the point of the weld, creating a small, highly concentrated heat-affected zone that melts the metal, creating a weld. The HAZ and weld metal width in the outcomes of the GMA and LB weld joints varied significantly, which is related to the heat input. The GMA welding process is a relatively low heat input process, while laser beam welding is a high heat input process which results in narrower welds as well as the HAZ. The width of the weld metal also varies depending on the measurement position of the individual measure lines A, B, and C. Sharp decrease in microhardness (softening phenomenon) is a result of the rapid coling rates from high temperatures and is associated with the formation of a coarse-grained microstructure. The formation of coarser grains occurs in a region referred to the coarse-grained HAZ (CGHAZ). In comparison to base metal, the microhardness of the CGHAZ decreased in GMA weld joint samples by 28.6% and in LB weld joint samples by 15.4%.

Quality Production Improvement and System Safety: QPI 16 - CZOTO 10   Materials Research Forum LLC
Materials Research Proceedings 34 (2023) 1-5   https://doi.org/10.21741/9781644902691-1

Microstructure. The microstructure of the base metal (Fig. 2) represents a very fine ferritic microstructure with uniformly dispersed very fine cementite particles and locally dispersed coarse TiC-based carbides.

The coarse-grained heat-affected zone (CGHAZ) and the fine-grained heat-affected zone (FGHAZ) were identified as the two primary regions of the HAZ in each weld (Fig. 3). The CGHAZ suffers a higher temperature during the welding process than the FGHAZ since it is situated nearer the weld metal.

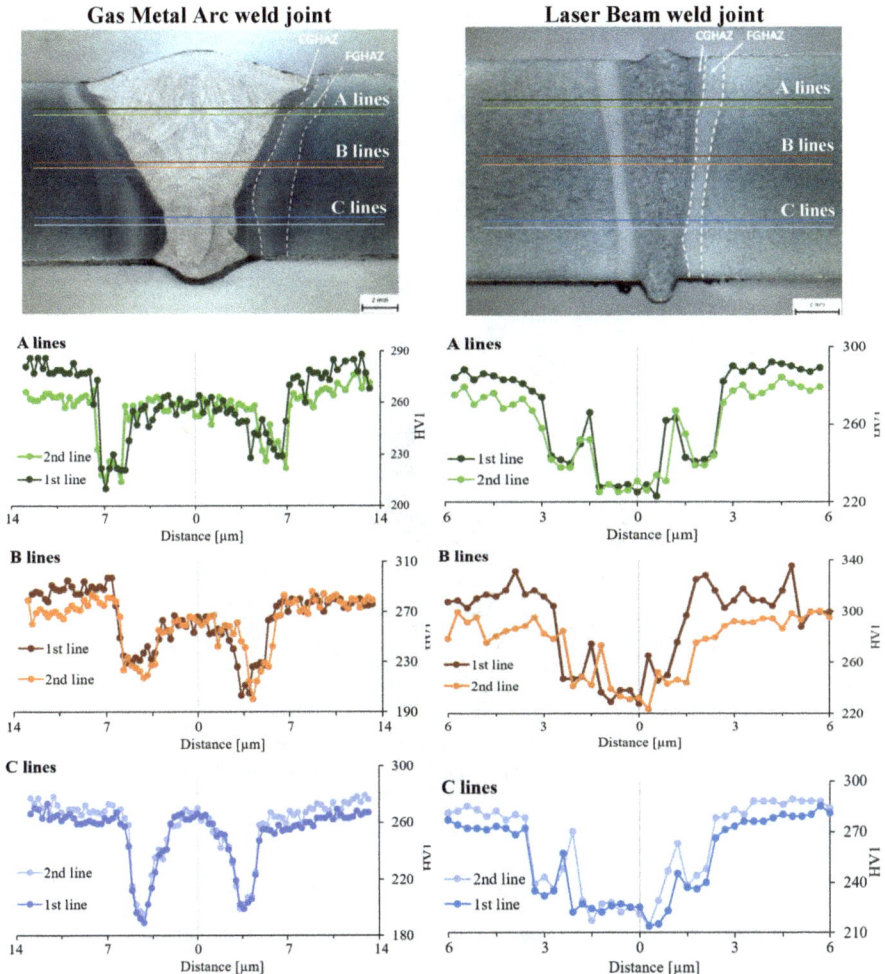

Fig.1. Microhardness line measurements of: GMA weld joint sample (left column), and LB weld joint sample (right column).

Quality Production Improvement and System Safety: QPI 16 - CZOTO 10   Materials Research Forum LLC
Materials Research Proceedings 34 (2023) 1-5   https://doi.org/10.21741/9781644902691-1

The high cooling rate and the large grain size of the CGHAZ result in a coarser microstructure. The grain size increases as a result of the recrystallization process, giving the material a coarser microstructure. On the contrary, the FGHAZ has a finer microstructure since it is farther away from the weld metal and receives a lower temperature during the welding process.

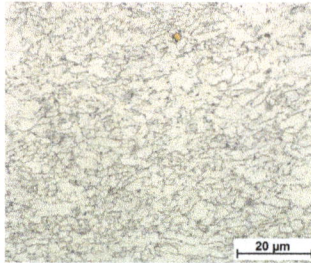

Fig.2. *Microstructure of the base metal Domex 700MC.*

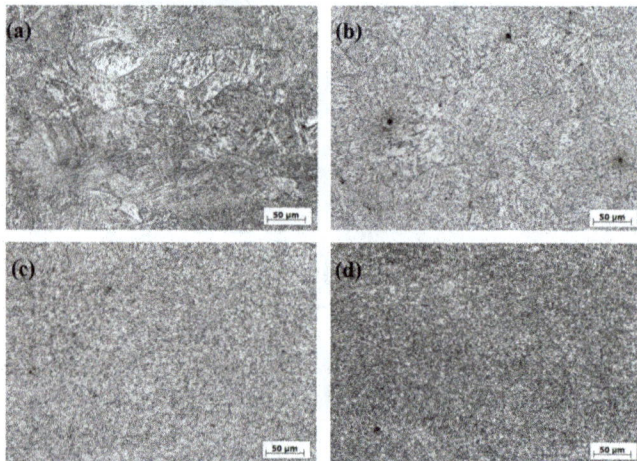

Fig.3. *The HAZ microstructure analysis:*
(a) CGHAZ of the GMA weld joint sample; (b) CGHAZ of the LB weld joint sample;
(c) FGHAZ of the GMA weld joint sample; (d) FGHAZ of the LB weld joint sample.

Conclusions
The following can be concluded from the testing and analysis that were conducted:
- The character of the microhardness profiles varies as a result of the application of specific welding techniques.
- In both cases, despite the difference in the welding methods used, softening occurs in the HAZ.
- GMA welded joints promoted a larger decrease in microhardness in the HAZ than LB welded joints. Microhardness results showed a decrease of 15.4% in the HAZ for samples welded using the LB method and a decrease of 28.6% for samples welded using the GMA method.

- The LB welding method has greater impact on the microstructural changes and mechanical properties than GMA welding method.
- The base metal has a very fine-grained ferritic microstructure with uniformly dispersed fine cementite particles. Two primary regions in the HAZ were identified in samples of both welding methods, CGHAZ and FGHAZ.
- A problematic CGHAZ region is represented by microstructure with coarse grains resulting from the recrystallization process.

It may be concluded that the most critical part of the weld join is the HAZ, especially the CGHAZ region, indicating the degradation of mechanical properties due to the formation of a coarse-grained microstructure.

Acknowledgement

The research was funded by the Slovak Ministry of Education, Science, Research, and Sport's Scientific Grant Agency under contract VEGA No. 1/0741/21 and the Slovak Research and Development Agency under contract No. APVV-20-0427.

References

[1] P. Kopas et al. Comparison of the mechanical properties and microstructural evolution in the HAZ of HSLA DOMEX 700MC welded by gas metal arc welding and electron beam welding, MATEC Web of Conf. 244 (2018) art.01009. https://doi.org/10.1051/matecconf/201824401009

[2] M. Mazur et al. The Structure and Mechanical Properties of Domex 700 MC Steel, Commun. – Sci. Lett. Univ. Zilina 15 (2013) 54-57. https://doi.org/10.26552/com.C.2013.4.54-57

[3] Y.A. Takahashi et al. Influence of Roughness on Corrosion Resistance of Carbon Steel SAE 1020 and DOMEX 700 MC Steel, Mater. Sci. Forum 1012 (2020) 441-446. https://doi.org/10.4028/www.scientific.net/MSF.1012.441

Quality Production Improvement and System Safety: QPI 16 - CZOTO 10   Materials Research Forum LLC
Materials Research Proceedings 34 (2023) 6-13   https://doi.org/10.21741/9781644902691-2

Experimental Determination of the Mechanical Properties of Onyx Material

ČUCHOR Dávid [1, a *], KOPÁS Peter [1, b], BRONČEK Jozef [1, c], BRUMERČÍK František [1, d]

[1] University of Žilina, Faculty of Mechanical Engineering, Depertment of Mechanical Design, Univerzitná 8215/1, 010 26 Žilina, Slovakia

[a]david.cuchor@fstroj.uniza.sk, [b]peter.kopas@fstroj.uniza.sk, [c]jozef.broncek@fstroj.uniza.sk, [d]frantisek.brumercik@fstroj.uniza.sk

Keywords: Composite Material, 3D Printing, Onyx, Mechanical Properties

Abstract. The main objective of the study is to determine the mechanical properties of the composite material produced by 3D printing. Experimental verification of the influence of the orientation and thickness of the layers was carried out. The test specimens for the experiment are fabricated by 3D printing from Onyx material using the Fused filament fabrication (FFF) method. The investigation of the mechanical properties is carried out using static tensile testing on the Inova FU-O-160-1260-V1 machine. The shape and dimensions of the test specimens are defined by the ASTM D638-14 standard. The study aims to determine the mechanical properties of Onyx material and specify the influence of the orientation of the layers and the thickness of the specimens on these mechanical properties.

Introduction

The issue of research and development of advanced transport vehicles that use unconventional lightweight construction materials, alternative drives, modular construction, and intelligent assistance systems is essential for the further development of intelligent transport in the future. The development of vehicles that do not produce any emissions due to their operation is the current trend in the automotive industry. This is visible in the development of special vehicles for rescue services, agriculture, and forestry. The driving parameter for the new design solutions being developed is to increase the proportion of energy-saving materials that are preferably recycled or recyclable while maintaining the same or better properties. The use of fiber-reinforced polymer composites in the design of a special vehicle is essential for zero emissions achievement. This trend of new solutions leads to the development of lightweight material structures in order to reduce the environmental impact [1]. The main benefits of composite materials are their stiffness, strength, lightweight, resistance to corrosion, stability, and reliability [2]. The reinforcing function performs reinforcement in the form of fibers, flakes, or dispersed particles [3,4]. With the expansion of plastics into the market, fiber-reinforced polymer composites have begun to be applied in practice [5]. Depending on the fiber length, the polymers can be reinforced with long or short fibers [4,6], however, composites reinforced with long fibers achieve better mechanical properties [7]. Fiber-reinforced polymer composites are produced by combining resin and fibers. Commonly, fibers are made from materials such as glass, carbon, aramid or basalt, and others [1, 8, 9]. More frequent use of glass fiber-reinforced polymer composites is due to their excellent ratio between price and mechanical properties. The carbon composite showed higher tensile and compressive modulus. In-plane shear properties of both the composites were comparable and interlaminar shear properties of glass composites were observed to be better than the carbon composite, because of the better nesting between the glass fabric layers [10, 11].

Quality Production Improvement and System Safety: QPI 16 - CZOTO 10   Materials Research Forum LLC
Materials Research Proceedings 34 (2023) 6-13   https://doi.org/10.21741/9781644902691-2

Although 3D printing has been undergoing a significant increase in interest only recently, the current state is the result of long-term development that began in the 1980s. The technology of 3D printing has advanced significantly in recent years and is now crucial in the technical field. It enables the design of complex components, reduces the cost of manufacturing them, and saves prototyping and development time. Various polymers with different mechanical and thermal properties are currently used as filament materials for the FFF printing method. Low strength and excessive thermal expansion are polymers' drawbacks, which restrict their wider application. Reinforcing fibers or particles are included in the polymer matrix to alleviate such a disadvantage. As a result, the mechanical and thermal properties of the newly formed composite material are improved [12].

This study aims to experimentally measure and evaluate the selected mechanical properties of test specimens fabricated using the FFF method (Fig. 1) from Onyx material. The specimens will be subjected to static tensile testing. Subsequently, the influence of the thickness and orientation of the layers concerning the loading direction will be evaluated.

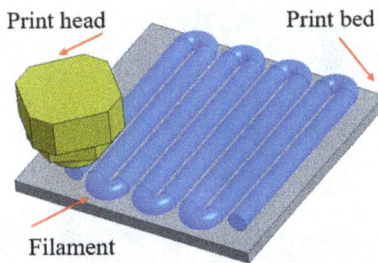

Fig. 1. *Fundamental scheme of FFF method.*

Materials and Methods

The test sample is made by 3D printing technology according to ASTM D638-14 standard from Onyx material. The dimensions of the test specimen are designed according to the Type I specimen in this standard (Fig. 2). The experiment was conducted with specimen thicknesses of 0.5 mm or 1 mm.

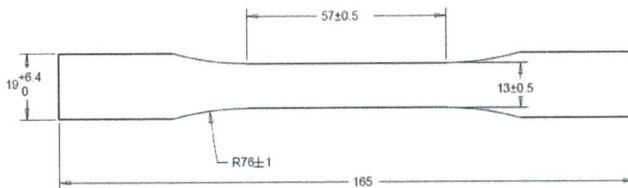

Fig.2. *Test specimen according to ASTM D638-14 standard.*

The Onyx material is a blend of nylon and chopped short randomly oriented carbon fibers. The material components have a very high surface quality and the fiber reinforcements add rigidity. These fibers form a micro-carbon reinforcement and make the components strong, rigid, and achieve very precise dimensions with digital design. The combination of nylon and carbon fiber has achieved both strength and high heat resistance. The material can withstand temperatures up to 145° C. The material can be utilized either on its own, as previously noted, or it can be further reinforced with high-strength fibers for even superior mechanical qualities [13].

The bi-axial experimental testing device INOVA FU-O-160-1260-V1 located in the laboratory of the Department of Applied Mechanics, SJF Žilina (Fig. 3) was used for the experimental measurements. To achieve accurate results from the static tensile test, the equipment was calibrated prior to the actual test. The device loads the test specimen by pulling away the jaws of the machine until it breaks. The specimen is placed in the jaws of the machine so that the long axis of the specimen is coincident with the imaginary axis of the jaws of the machine in the load direction. The jaws grip the specimen evenly and firmly so that the specimen does not slip during the test, but care must be taken to ensure that the specimen is not damaged by too strong of a grip.

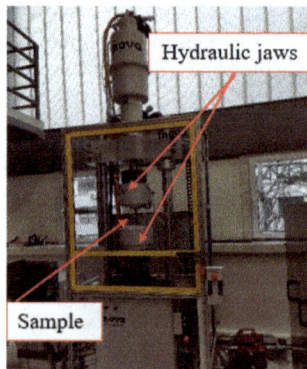

Fig. 3. *Testing device INOVA FU-O-160-1260-V1.*

A total of 4 sets of test specimens were produced, each set containing a different layer arrangement as well as different layer thicknesses. In terms of their identification, the sample sets are labeled A, B, C, and D. Each set consists of five test specimens numbered 1 to 5. All samples are printed with the face lying on the x-axis. Figures 4 and 5 illustrate the arrangement and direction of the layer stacking of printed samples, as well as the direction of loading.

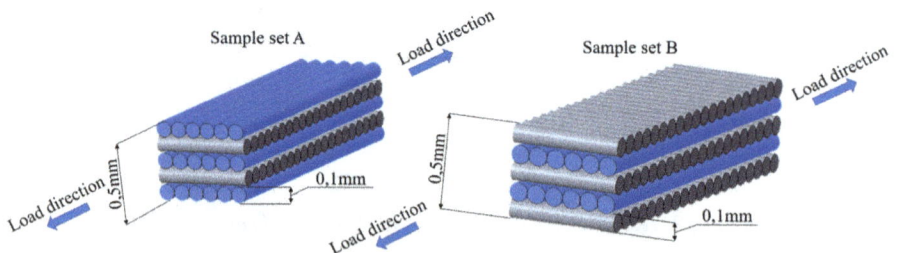

Fig. 4. *Arrangement of layers of sample sets A and B.*

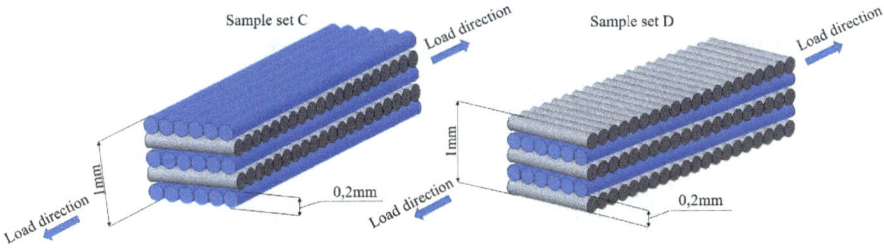

Fig. 5. Arrangement of layers of sample sets C and D.

Results and Discussion

Based on the tensile test performed, information about the test progress was obtained from the software of the experimental device. The test specimens after the static tensile test are shown in Figures 6 and 7 together with the tensile diagrams that were created.

To summarize and evaluate the experimentally verified mechanical properties of the Onyx composite material under static loading, a summary Table 1 was created. The average values from the experimental measurements for the tested sets A to D are recorded in the table. Figures 8 and 9 show the average values of ultimate strength and ductility.

Cross-sections of the ruptured specimen from sample set A were examined using a scanning electron microscope (SEM) to determine the behavior of the material during loading. The presence of voids between each filament within the single lamina is shown in the illustration (Fig. 10).

Fig. 6. Test specimen No 2 of the sample set A after the static tensile test.

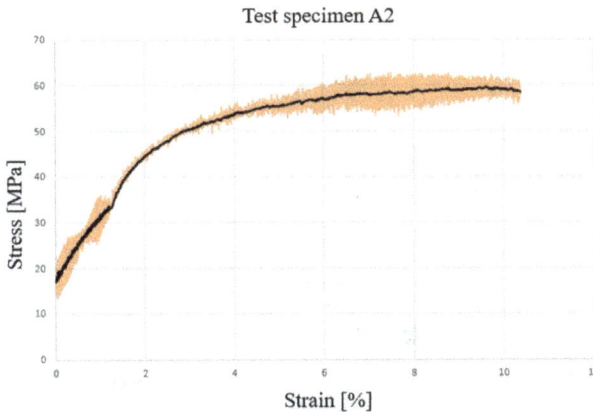

Fig. 7. Tensile test diagram of test specimen A2.

Quality Production Improvement and System Safety: QPI 16 - CZOTO 10   Materials Research Forum LLC
Materials Research Proceedings 34 (2023) 6-13   https://doi.org/10.21741/9781644902691-2

Table 1. Summary results of mechanical properties of Onyx composite material

Sample set	R_m [MPa]	$R_{p0.2}$ [MPa]	F_m [N]	A_t [%]
A	56.14	17.30	364.89	9.46
B	43.93	12.32	284.28	14.84
C	39.78	14.40	517.19	15.56
D	38.22	14.56	496.92	16.01

Average ultimate tensile strength values

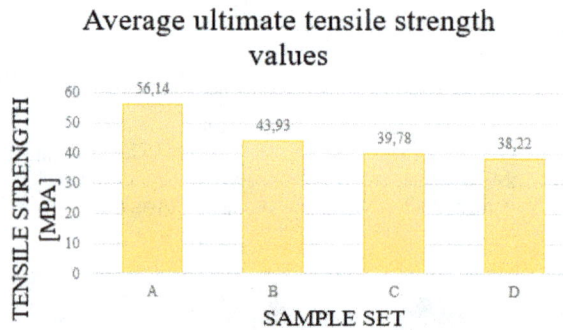

Fig. 8. Average ultimate tensile strength values.

Average strain values

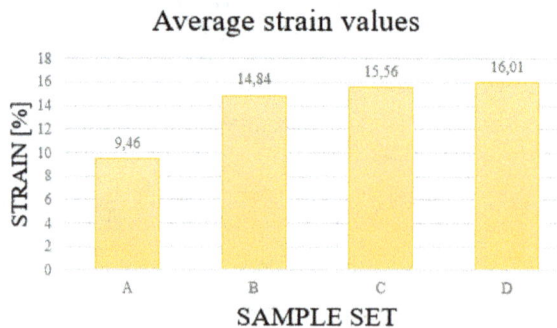

Fig. 9. Average strain values.

The presence of voids contributes to a limited connection between the filaments and laminas. The impact of tensile loading on the various composite phases is shown in images with a higher magnification - nylon filaments that are overstretched and broken carbon fibers (Fig. 11). The

Quality Production Improvement and System Safety: QPI 16 - CZOTO 10   Materials Research Forum LLC
Materials Research Proceedings 34 (2023) 6-13   https://doi.org/10.21741/9781644902691-2

illustrations also demonstrate a laminate failure process. The broken carbon fibers may be observed in the laminas that are oriented in the load direction. The rupture propagated between the filaments in the case of laminas aligned perpendicular to the load direction, showing a lower strength between them. Only laminas aligned with the load direction effectively transfer tensile loading [4].

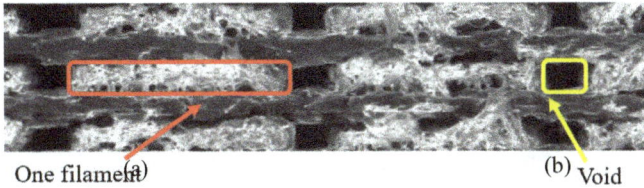

One filament (a)   (b) Void

Fig. 10. *Cross section of ruptured specimen (100x magnification, detector LVSD) [4].*

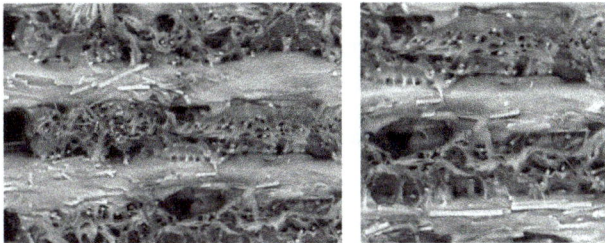

Fig. 11. *Cross section of ruptured specimen (a) 300x magnification, detector BSE; (b) 500x magnification, detector BSE [4].*

Conclusions

From the evaluation of the experimental results, it can be concluded that the strength and stiffness of the composite material depend critically on the strength and stiffness of the fibers themselves. The orientation of the fibers itself is also very important as it significantly influences the other mechanical properties of the composites. At the same time, it is evident that the very different mechanical properties of the fibers and the matrix give rise to a complex state of stress in the composite structure. This is mainly a matter of the bonding between the fibers and the matrix itself and, at the same time, the behavior of the particular arrangement of these fibers in the volume. For a given composite material, i. e. with randomly oriented fibers, changes in the mechanical properties can most likely only be altered by changing the content or volume of these fibers in the individual layers. In this particular case, the reinforcement content of the material tested did not change. The only parameters that changed were the orientation and stacking of the layers of the composite. And it was this factor that proved to be the reason for the changes in the experimental measurement results. The explanation for the decrease or increase in the individual mechanical properties of the composite used could be explained by the quality of the bond between the fiber and the matrix - the material thickness (number of layers) vs. ultimate strength. An explanation can be sought among the infinite number of interfaces, which usually exhibit less strength than each component individually. These interfacial areas and their volume fraction in the composite are where failure is expected to occur and propagate. Thus, with their enormous number, a large

Quality Production Improvement and System Safety: QPI 16 - CZOTO 10   Materials Research Forum LLC
Materials Research Proceedings 34 (2023) 6-13   https://doi.org/10.21741/9781644902691-2

amount of energy is converted in a small area into the work required to create the failures or cracks. The more of these layers were 3D printed on top of each other, the lower the strength values were - confirming the assumption.

In this work, the mechanical properties of composite test specimens made of Onyx material were experimentally verified. The mechanical properties of the test sets of specimens differed from each other based on the thickness of the specimens and the orientation of the layers during 3D printing. This work opened up possibilities for further investigation of the properties of composite materials fabricated by 3D printing. There are many ways to investigate composites and thus achieve a significantly more comprehensive view of their properties. Further research could also focus on other types of stresses whether static, dynamic or hardness testing.

Acknowledgement
This article was elaborated in the framework of the project titled Implementation of the language of geometric product specification in the field of coordinate 3D metrology KEGA 033 ŽU-4/2022.

References
[1] M. F. Ashby. Materials Selection in Mechanical Design, Elsevier Butterworth Heinemann. Burlington, 2005.

[2] R.M. Jones. Mechanics of Composite Materials, 2nd Ed., Taylor and Francis, New York, 1999.

[3] M.Y. Khalid et al. Natural fiber reinforced composites: Sustainable materials for emerging applications. Results Eng. 11 (2021) art.100263. https://doi.org/10.1016/j.rineng.2021.100263

[4] J. Majko et al. Tensile Properties of Additively Manufactured Thermoplastic Composites Reinforced with Chopped Carbon Fibre, Materials 15 (2022) art.4224. https://doi.org/10.3390/ma15124224

[5] F. Van der Klif et al. 3D Printing of Continuous Carbon Fibre Reinforced Thermo-Plastic (CFRTP) Tensile Test Specimens. Open J. Compos. Mater (2016), 6, 18–27. http://dx.doi.org/10.4236/ojcm.2016.61003

[6] G.W. Melenka et al. Evaluation and prediction of the tensile properties of continuous fiber-reinforced 3D printed structures. Compos. Struct 153 (2016) 866-875. https://doi.org/10.1016/j.compstruct.2016.07.018

[7] F. Ning et al. Additive manufacturing of carbon fiber reinforced thermoplastic composites using fused deposition modeling. Compos. Part B Eng. 80 (2015) 369-378. https://doi.org/10.1016/j.compositesb.2015.06.013

[8] Fiber Reinforced Polymer (FRP) Composites Market Analysis By Fiber Type (Glass, Carbon, Basalt, Aramid), By Application (Automotive, Construction, Electronic, Defense), By Region, And Segment Forecasts, 2018-2025, Grand View Research (2017). Report ID: GVR-2-68038-006-4. [online]. 2017. [viewed: 2023-01-31] Available from: https://www.grandviewresearch.com/industry-analysis/fiber-reinforced-polymer-frp-composites-market

[9] T. Liptáková, P. Alexy, E. Gondár, V. Khunová, Polymérne konštrukčné materiály, EDIS, Žilina, 2012. ISBN 978-80-554-0505-6

[10] G.D. Goh et al. Recent progress in additive manufacturing of fiber reinforced polymer composite, Adv. Mater. Technol. 4 (2019) art.1800271. https://doi.org/10.1002/admt.201800271

[11] G.D. Goh et al. Characterization of mechanical properties and fracture mode of additively manufactured carbon fiber and glass fiber reinforced thermoplastics, Mater. Des. 137 (2018) 79-89. https://doi.org/10.1016/j.matdes.2017.10.021

[12] K. Kandananond. Surface Roughness Reduction in A Fused Filament Fabrication (FFF) Process using Central Composite Design Method, Prod. Eng. Arch. 28 (2022) 157-163. https://doi.org/10.30657/pea.2022.28.18

[13] Introducing Our New Markforged Material: Onyx. [online]. 2023. [viewed: 2023-01-31]. Available from: http://markforged.com/resources/blog/introducing-our-new-markforged-material-onyx

Quality Production Improvement and System Safety: QPI 16 - CZOTO 10
Materials Research Proceedings 34 (2023) 14-23

Materials Research Forum LLC
https://doi.org/10.21741/9781644902691-3

Experimental Results of Helical Metal Expansion Joints Fabrication

KURP Piotr[1, a *], DANIELEWSKI Hubert[1, b] and CEDRO Leszek[1, c]

[1] Faculty of Mechatronics and Mechanical Engineering, Kielce University of Technology, Aleja Tysiąclecia Państwa Polskiego 7, 25-314 Kielce, Poland

[a]pkurp@tu.kielce.pl, [b]hubert_danielewski@o2.pl, [c]lcedro@tu.kielce.pl

Keywords: Laser Forming, Metal Expansion Joints, Laser Treatment, Pipeline Compensation

Abstract. This paper discusses the technological assumptions for the production of helical metal expansion joint as a new type of expansion joints used to eliminate torsional deformations of industrial pipelines. The method of mechanically assisted laser forming, which was used as a manufacturing technology, was presented as well. Furthermore, technological parameters and experimental results obtained during the fabrication of helical metal expansion joints were presented. Mainly presented and discussed are the results such as: obtained geometry, forces necessary to produce the product, processing temperature, strain rate and others. Satisfactory treatment results were obtained, which are illustrated below.

Introduction

Metal expansion joints are pipeline components designed to compensate for installation deformations related to changes in operating parameters such as temperature and pressure. Without compensating for this type of deformation, the piping installation would fail quickly. In addition, compensators eliminate the assembly inaccuracies of such an installation. The deformations that are compensated are [1-3]:

– lateral deformation,
– axial deformation,
– angular deformation.

There are two main types of metal expansion joints described above [4, 5]:

– bellows expansion joints,
– lens expansion joints.

Examples of metal expansion joints produced by ENERGOMET Wrocław are shown in Fig.1.

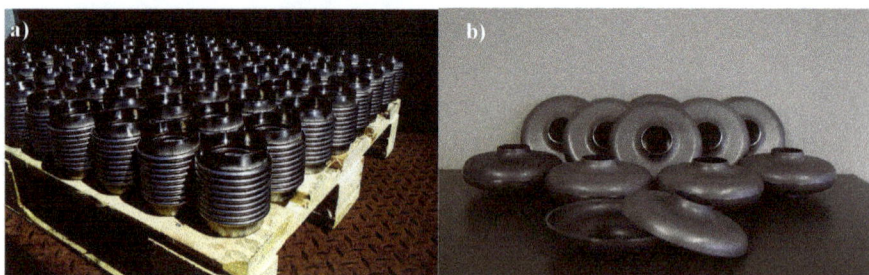

Fig. 1. Typical metal expansion joint: a) bellows expansion joints, b) lens expansion joints [6].

The mentioned components do not compensate for deformations generated by torques, which are the result of the operation of fittings, main pumps, etc. installed in the pipeline. To come up

Quality Production Improvement and System Safety: QPI 16 - CZOTO 10   Materials Research Forum LLC
Materials Research Proceedings 34 (2023) 14-23   https://doi.org/10.21741/9781644902691-3

with the idea of manufacturing helical metal expansion joints, a task should be used to compensate for this type of deformation.

The fabrication of this type of components is carried out mainly by various types of cold plastic forming methods [1]. The authors of this paper will present the results of tests on the fabrication of metal expansion joints using the mechanically assisted laser forming method. Laser treatment is now a widely used fabrication method. It covers a whole range of various types of technologies such as: cutting, welding, surfacing, surface treatment with and without remelting, micromachining, additive technology and many others [7-9].

Laser forming uses induced internal stresses caused by the temperature difference between the areas of the component heated by the laser beam and the rest of the material. This method uses the phenomenon of thermal expansion, specifically the differences in expansion between "warm" and "cold" areas. In the literature, there are three basic mechanisms of laser forming, described in the mid-1990s, they are [10]:

- Temperature Gradient Mechanism (TGM),
- Upsetting Mechanism (UM),
- Buckling Mechanism (BM).

However, the laser freeform forming method (as we can call it) is a very time-consuming process. Therefore, the authors will present the results of research into the fabrication of helical metal expansion joints using a hybrid method of mechanically assisted laser forming.

Production and use of compensators are of significant importance for the reliability of pressure systems, which are widely utilized in industries as well as in many commonly used devices and machinery. This influences the quality [11-13] and safety of the functioning of the equipment and machinery that utilize them [14-16]. Material and thermomechanical equations allow for the solution of a straightforward issue: calculating deformations based on a given energy stream. However, the appropriate shaping of a compensator using a focused laser energy beam [17] requires solving a highly complex inverse problem, which is time-consuming and usually involves the method of successive approximations [18].

The installation of a compensator reduces the risks arising from the geometric imperfections of piping systems and their deformation under the influence of pressure and temperature changes. However, it also introduces new risks associated with corrosion [19,20] and biocorrosion [21], as well as joint fatigue [22,23] and wear [24,25]. The remedy for these problems lies in the proper selection of materials, both commonly used [26-28] and special alloys [29]. Their technological properties can be adjusted towards desired values by applying appropriate protective coatings [30-32] or modifying the surface layer [33]. Here, increasingly popular techniques such as electrospark deposition [34-36] and diamond-like carbon coatings [37-39] can be mentioned.

The benefits derived from such shaping techniques for compensators include energy consumption reduction [40] and less environmental pollution through waste separation [41]. The reliability of machinery [42] equipped with them, including railway rolling stock [43], significantly increases. This increased reliability does not go unnoticed by a customer known for their exceptionally high-quality standards: the military [44].

Sophisticated technological procedures often necessitate a decrease in the quantity of examined variables [45,46], enabling subsequent process optimization and stabilization through the utilization of statistical methodologies. These methodologies may take the form of classical approaches [47-49] such as factorial design, response surface methodology (RSM), and Taguchi methods, or nonparametric techniques [50-52], even with resampling support [53].

Technology and Materials
Ideas and assumptions about technology were presented in the article [54]. The technology will be briefly described below to introduce the reader to the subject. The helical metal expansion joint

has bellows in the form of a spiral on its circumference. This spiral is analogous in appearance to a thread. It can be both right and left-handed. In short: in order to make this type of metal expansion joint using the hybrid method of mechanically assisted laser forming, it is necessary to bring part of the pipe to the plasticizing temperature by heating it around the perimeter in a spiral, and then compressing the pipe with the appropriate force. The diagram of the technology is shown in Fig.2 and Fig.3.

Fig. 2. *Diagram of the method of heating the element (for better illustration, the pipe has been presented as transparent)*

The output material for the experiment was a pipe made of X5CrNi18-10 grade stainless steel with dimensions ϕ50x1.5 mm. The chemical composition and selected material properties are presented in Table 1 [55]. The research stand and the results of the experiment are presented below.

a)

b)

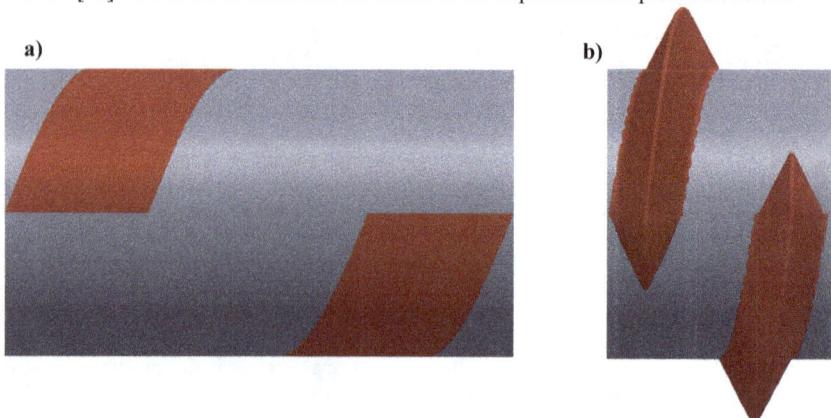

Fig. 3. *Scheme: a) output pipe heated in a helix, b) component after compression.*

Table 1. *Chemical composition and main physical properties of X5CrNi18-10 austenitic steel [12].*

Chemical composition							
C	Cr	Ni	Mn	Si	P	S	N
<0.07	17.5÷19.5	8.0÷10.5	<2.0	<1.0	<0.045	<0.015	<0.11
Density ρ, kg/m^3		Thermal expansion coefficient α, 1/K		Heat capacity C_{p20}, J/kgK		Thermal conductivity λ, W/mK	
8020		14.2x10^{-6}		480		15	

Experiment and Results

The test stand presented in Fig. 4 consisted of a TRUMPF TruFlow 6000 CO_2 laser generating a rectangular laser beam with a wavelength of λ=10.6 μm.

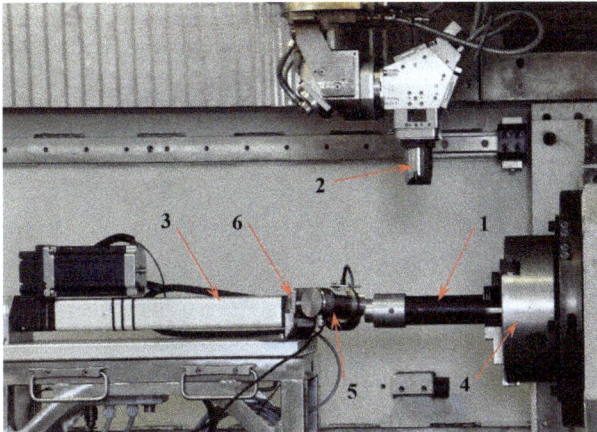

Fig. 4. *View of the test stand: 1 - pipe, 2 - laser head, 3 - actuator, 4 - swivel chuck, 5 - force sensor, 6 - pyrometer.*

A tubular pipe was installed between the actuator and the swivel chuck. The process temperature was controlled by a monochromatic pyrometer, the laser power was controlled by feedback to the pyrometer readings to keep the zone temperature constant. The surface of the sample was covered with a special absorber (matt black enamel) in order to increase the laser radiation absorption coefficient. The laser processing parameters are listed below:

– process temperature: approx. T = 1100°C,
– laser power: depending on element temperature P ∈ <900, 2500> W
– compressive length (total): s = 15 mm,
– initial beam pitch: p_1 = 140 mm,
– pipe compressive speed: v = 10 mm/s.
– initial beam pitch: p = 80 mm,
– number of coils: i = 3.

Quality Production Improvement and System Safety: QPI 16 - CZOTO 10   Materials Research Forum LLC
Materials Research Proceedings 34 (2023) 14-23   https://doi.org/10.21741/9781644902691-3

Due to the upsetting of the helix during the process, the pitch of the helix was corrected six times. Each time the correction was by 10 mm. The results of the experiment and their discussion are presented in the paragraph below.

Results

The result of the experiment was the DN50 helical metal expansion joint fabrication shown in Fig. 5. As can be observed in Fig. 5, upsets (below) were formed around the circumference of the pipe. Force needed to create a helix around the pipe circumference is shown in Fig.6. The results of the experiment are discussed in the paragraph below.

Discussion and Conclusions

Experimental studies confirmed the possibilities of the new technology in the aspect of helical metal expansion joints fabrication. On the basis of previously performed experimental studies, it was possible to select the parameters of the laser treatment, which allowed for the formation of a helix on the surface of the pipe element. Macroscopic examinations allow the assessment of the obtained helical metal expansion joint at the correct level. No cracks, kinks, unfavorable corrugations, etc. were noticed on the surface. The helix obtained is symmetrical and repeatable, no defects in the upset geometry were noticed.

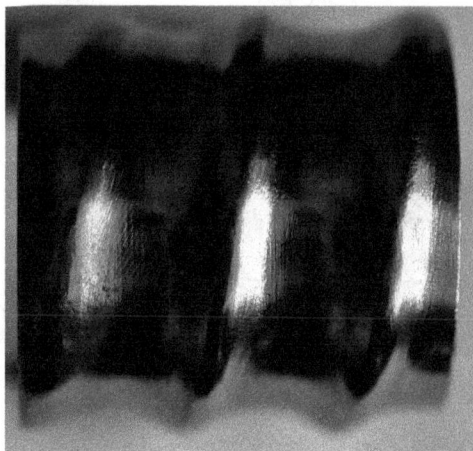

Fig. 5. View of the DN50 helical metal expansion joint surface.

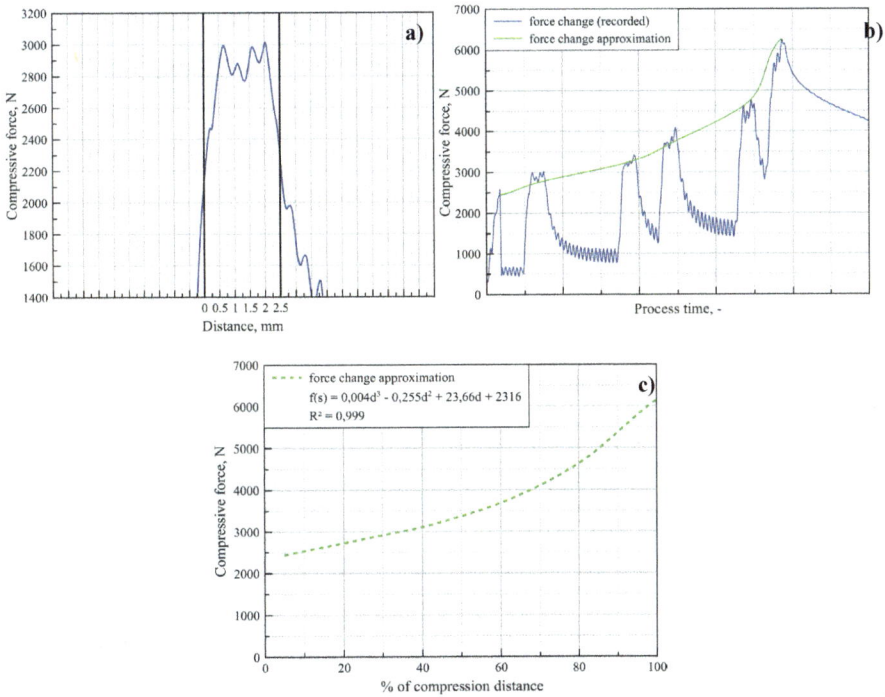

Fig. 6. Force needed to create a helix around the pipe circumference: a) for a single pass, b) for six passes, c) approximated force needed to produce a complete metal expansion joint for given technological parameters.

Recorded force show an increase in the force needed to make a hel;ix with each successive pass. Most likely, this is due to the distribution of forces in the emerging upset, which must be overcome in order to increase the resulting upset. As noted from the graphs (Fig.6), the increase in force can be described by an exponential function in the form (for the correlation coefficient R^2=0.999):

$$f(s) = 0.004\ d^3 + 0.255\ d^2 + 23.66\ d + 2316, \tag{1}$$

where: d - compressive distance.

Of course, the values of the *d* term in the formula (1) will depend on the values of the forces, which will be different for different diameters and thicknesses. The results of the experiment may be helpful in planning and developing technology for making this type of metal expansion joints. In order to check the possibility of compensating the torsional torques by the expansion joint, additional tests should be carried out. Such studies are planned for the future.

Acknowledgement
The research reported herein was supported by a grant from the National Centre for Research and Development: grant *Lider XI*, title: "*Development of new type metal expansion joints and their manufacturing technology*", contract number: LIDER/44/0164/L-11/19/NCBR/2020.

References
[1] Standards of the Expansion Joint Manufacturers Association, Tenth Edition, Expansion Joints Manufacturer Association Inc., 2020.

[2] S. Sharma, A.J. Obaid. Design and Analysis of Metal Expansion bellows under Axial and Transverse loads using CATIA V5 R21 software, IOP Conf. Ser.: Mater. Sci. Eng. (2022), 1-9. https://doi.org/10.1088/1757-899X/1145/1/012054

[3] A. Vinoth, B. Vignesh, L. Khurana, A. Rai. A Review on Application of Bellows Expansion Joints and Effect of Design Parameters on System Characteristics, Indian J. Sci. Technol. 9, (2016) 1-8. https://doi.org/10.17485/ijst/2016/v9i32/94320

[4] C. Soares. Accessory Systems, in: Gas Turbines (Second Edition) A Handbook of Air, Land and Sea Applications, 2nd ed., Butterworth-Heinemann, Oxford, 2015, 413-484. https://doi.org/10.1016/B978-0-12-410461-7.00008-0

[5] K. Sotoodeh, Piping special items, in: A Practical Guide to Piping and Valves for the Oil and Gas Industry, Gulf Professional Publishing, 2021, 721-798. https://doi.org/10.1016/B978-0-12-823796-0.00012-X

[6] Metal Expansion Joint Manufacturer: ENERGOMET Wrocław, Poland. Available from: http://www.energomet.com.pl [viewed: 2023-01-26]

[7] S. Tofil, M. Manoharan, A. Natarajan. Surface Laser Micropatterning of Polyethylene Terephthalate (PET) to Increase the Shearing Strength of Adhesive Joints, Materials Research Proceedings 24 (2022) 27-33. https://doi.org/10.21741/9781644902059-5

[8] N. Radek, A. Kalinowski, J. Pietraszek, L, Orman, M. Szczepaniak, A. Januszko, J. Kamiński, J. Bronček, O. Paraska. Formation of coatings with technologies using concentrated energy stream, Production Engineering Archives 28 (2022) 117-122. https://doi.org/10.30657/pea.2022.28.13

[9] P. Sęk. Experimental studies on the possibility of using a pulsed laser for spot welding of thin metallic foils, Open Engineering 10 (2020) 674-680. https://doi.org/10.1515/eng-2020-0076

[10] F. Vollertsen. Mechanisms and models for laser forming, in: M. Geiger and F. Vollertsen (Eds.), Laser Assisted Net Shape Engineering, Proceedings of the LANE'94, Meisenbach Bamberg, Germany, 1994, Vol. I, pp. 345–360.

[11] R. Ulewicz, F. Nový. Quality management systems in special processes, Transp. Res. Procedia 40 (2019) 113-118. https://doi.org/10.1016/j.trpro.2019.07.019

[12] D. Siwiec et al. Improving the non-destructive test by initiating the quality management techniques on an example of the turbine nozzle outlet, Materials Research Proceedings 17 (2020) 16-22. https://doi.org/10.21741/9781644901038-3

[13] R. Ulewicz et al. Logistic controlling processes and quality issues in a cast iron foundry, Mater. Res. Proc. 17 (2020) 65-71. https://doi.org/10.21741/9781644901038-10

[14] G. Filo, E. Lisowski, M. Domagała, J. Fabiś-Domagała, H. Momeni. Modelling of pressure pulse generator with the use of a flow control valve and a fuzzy logic controller, AIP Conf. Proc. 2029 (2018) art.20015. https://doi.org/10.1063/1.5066477

[15] E. Lisowski et al. Flow Analysis of a 2URED6C Cartridge Valve, Lecture Notes in Mechanical Engineering 24 (2021) 40-49. https://doi.org/10.1007/978-3-030-59509-8_4

[16] M. Domagala et al. CFD Estimation of a Resistance Coefficient for an Egg-Shaped Geometric Dome, Appl. Sci. 12 (2022) art.10780. https://doi.org/10.3390/app122110780

[17] N. Radek et al. The WC-Co electrospark alloying coatings modified by laser treatment, Powder Metall. Met. Ceram. 47 (2008) 197-201. https://doi.org/10.1007/s11106-008-9005-7

[18] L. Cedro. Model parameter on-line identification with nonlinear parametrization – manipulator model, Technical Transactions 119 (2022) art. e2022007. https://doi.org/10.37705/TechTrans/e2022007

[19] M. Scendo et al. Influence of laser treatment on the corrosive resistance of WC-Cu coating produced by electrospark deposition, Int. J. Electrochem. Sci. 8 (2013) 9264-9277.

[20] T. Lipiński, J. Pietraszek. Corrosion of the S235JR Carbon Steel after Normalizing and Overheating Annealing in 2.5% Sulphuric Acid at Room Temperature, Mater. Res. Proc. 24 (2022) 102-108.

[21] E. Skrzypczak-Pietraszek et al. Enhanced accumulation of harpagide and 8-O-acetyl-harpagide in *Melittis melissophyllum* L. agitated shoot cultures analyzed by UPLC-MS/MS. PLoS ONE 13 (2018) art. e0202556. https://doi.org/10.1371/journal.pone.0202556

[22] N. Radek et al. The impact of laser welding parameters on the mechanical properties of the weld, AIP Conf. Proc. 2017 (2018) art.20025. https://doi.org/10.1063/1.5056288

[23] N. Radek et al. Properties of Steel Welded with CO2 Laser, Lecture Notes in Mechanical Engineering (2020) 571-580. https://doi.org/10.1007/978-3-030-33146-7_65

[24] S. Marković et al. Exploitation characteristics of teeth flanks of gears regenerated by three hard-facing procedures, Materials 14 (20210 art. 4203. https://doi.org/10.3390/ma14154203

[25] M. Krynke et al. Maintenance management of large-size rolling bearings in heavy-duty machinery, Acta Montan. Slovaca 27 (2022) 327-341. https://doi.org/10.46544/AMS.v27i2.04

[26] D. Klimecka-Tatar, R. Dwornicka. The assembly process stability assessment based on the strength parameters statistical control of complex metal products, METAL 2019 28th Int. Conf. Metall. Mater. (2019) 709-714. ISBN 978-808729492-5

[27] D. Siwiec et al. Improving the process of achieving required microstructure and mechanical properties of 38mnvs6 steel, METAL 2020 29th Int. Conf. Metall. Mater. (2020) 591-596. https://doi.org/10.37904/metal.2020.3525

[28] P. Jonšta et al. The effect of rare earth metals alloying on the internal quality of industrially produced heavy steel forgings, Materials 14 (2021) art.5160. https://doi.org/10.3390/ma14185160

[29] A. Dudek, B. Lisiecka, R. Ulewicz. The effect of alloying method on the structure and properties of sintered stainless steel, Archives of Metallurgy and Materials 62 (2017) 281-287. https://doi.org/10.1515/amm-2017-0042

[30] N. Radek et al. Technology and application of anti-graffiti coating systems for rolling stock, METAL 2019 28th Int. Conf. Metall. Mater. (2019) 1127-1132. ISBN 978-8087294925

[31] N. Radek et al. The effect of laser beam processing on the properties of WC-Co coatings deposited on steel. Materials 14 (2021) art. 538. https://doi.org/10.3390/ma14030538

[32] N. Radek et al. Formation of coatings with technologies using concentrated energy stream, Prod. Eng. Arch. 28 (2022) 117-122. https://doi.org/10.30657/pea.2022.28.13

[33] N. Radek et al. The influence of plasma cutting parameters on the geometric structure of cut surfaces, Mater. Res. Proc. 17 (2020) 132-137. https://doi.org/10.21741/9781644901038-20

[34] N. Radek et al. Microstructure and tribological properties of DLC coatings, Mater. Res. Proc. 17 (2020) 171-176. https://doi.org/10.21741/9781644901038-26

[35] N. Radek et al. Influence of laser texturing on tribological properties of DLC coatings, Prod. Eng. Arch. 27 (2021) 119-123. https://doi.org/10.30657/pea.2021.27.15

[36] N. Radek et al. Operational properties of DLC coatings and their potential application, METAL 2022 31st Int. Conf. Metall. Mater. (2022) 531-536. https://doi.org/10.37904/metal.2022.4491

[37] N. Radek, J. Konstanty. Cermet ESD coatings modified by laser treatment, Arch. Metall. Mater. 57 (2012) 665-670. https://doi.org/10.2478/v10172-012-0071-y

[38] N. Radek, K. Bartkowiak. Laser Treatment of Electro-Spark Coatings Deposited in the Carbon Steel Substrate with using Nanostructured WC-Cu Electrodes, Physics Procedia 39 (2012) 295-301. https://doi.org/10.1016/j.phpro.2012.10.041

[39] N. Radek et al. The effect of laser treatment on operational properties of ESD coatings, METAL 2021 30th Ann. Int. Conf. Metall. Mater. (2021) 876-882. https://doi.org/10.37904/metal.2021.4212

[40] Ł.J. Orman. Enhancement of pool boiling heat transfer with pin-fin microstructures, J. Enhanc. Heat Transf. 23 (2016) 137-153. https://doi.org/10.1615/JEnhHeatTransf.2017019452

[41] M. Zenkiewicz et al. Electrostatic separation of binary mixtures of some biodegradable polymers and poly(vinyl chloride) or poly(ethylene terephthalate), Polimery/Polymers 61 (2016) 835-843. https://doi.org/10.14314/polimery.2016.835

[42] R. Ulewicz, M. Mazur. Economic aspects of robotization of production processes by example of a car semi-trailers manufacturer, Manuf. Technol. 19 (2019) 1054-1059. https://doi.org/10.21062/ujep/408.2019/a/1213-2489/MT/19/6/1054

[43] N. Radek, R. Dwornicka. Fire properties of intumescent coating systems for the rolling stock, Commun. – Sci. Lett. Univ. Zilina 22 (2020) 90-96. https://doi.org/10.26552/com.C.2020.4.90-96

[44] W. Przybył et al. Virtual Methods of Testing Automatically Generated Camouflage Patterns Created Using Cellular Automata, Mater. Res. Proc. 24 (2022) 66-74. https://doi.org/10.21741/9781644902059-11

[45] J. Pietraszek et al. The principal component analysis of tribological tests of surface layers modified with IF-WS2 nanoparticles, Solid State Phenom. 235 (2015) 9-15. https://doi.org/10.4028/www.scientific.net/SSP.235.9

[46] J. Pietraszek, E. Skrzypczak-Pietraszek. The uncertainty and robustness of the principal component analysis as a tool for the dimensionality reduction. Solid State Phenom. 235 (2015) 1-8. https://doi.org/10.4028/www.scientific.net/SSP.235.1

[47] J. Pietraszek et al. The fixed-effects analysis of the relation between SDAS and carbides for the airfoil blade traces. Archives of Metallurgy and Materials 62 (2017) 235-239. https://doi.org/10.1515/amm-2017-0035

[48] R. Dwornicka, J. Pietraszek. The outline of the expert system for the design of experiment, Prod. Eng. Arch. 20 (2018) 43-48. https://doi.org/10.30657/pea.2018.20.09

[49] J. Pietraszek et al. Challenges for the DOE methodology related to the introduction of Industry 4.0. Prod. Eng. Arch. 26 (2020) 190-194. https://doi.org/10.30657/pea.2020.26.33

[50] J. Pietraszek. The modified sequential-binary approach for fuzzy operations on correlated assessments, LNAI 7894 (2013) 353-364. https://doi.org/10.1007/978-3-642-38658-9_32

[51] J. Pietraszek et al. Non-parametric assessment of the uncertainty in the analysis of the airfoil blade traces, METAL 2017 26th Int. Conf. Metall. Mater. (2017) 1412-1418. ISBN 978-8087294796

[52] J. Pietraszek et al. The non-parametric approach to the quantification of the uncertainty in the design of experiments modelling, UNCECOMP 2017 Proc. 2nd Int. Conf. Uncert. Quant. Comput. Sci. Eng. (2017) 598-604. https://doi.org/10.7712/120217.5395.17225

[53] J. Pietraszek, L. Wojnar. The bootstrap approach to the statistical significance of parameters in RSM model, ECCOMAS Congress 2016 Proc. 7th Europ. Congr. Comput. Methods in Appl. Sci. Eng. 1 (2016) 2003-2009. https://doi.org/10.7712/100016.1937.9138

[54] P. Kurp. Ideas and Assumptions of a New Kind Helical Metal Expansion Joints, Materials Research Proceedings 24 (2022) 233-239. https://doi.org/10.21741/9781644902059-34

[55] K.C. Mills. Recommended values of thermophysical properties for selected commercial alloys, 1st Ed., Woodhead, Abington, England, 2002.

Materials Research Forum LLC
https://doi.org/10.21741/9781644902691-4

High-Temperature Oxidation of High-Entropy FeNiCoCrAl Alloys

GUMEN Olena [1, a *], KARPETS Myroslav [2, b], SMAKOVSKA Ganna [3, c] and YAKUBIV Mykola [4, d]

[1] National Technical University of Ukraine "Igor Sikorsky Kyiv Polytechnic Institute", 37 Peremohy Avenue, Kyiv, 03056, Ukraine, ORCID: 0000-0003-3992-895X

[2] National Technical University of Ukraine "Igor Sikorsky Kyiv Polytechnic Institute", Kyiv, Ukraine, ORCID: 0000-0001-9528-1850

[3] National Technical University of Ukraine "Igor Sikorsky Kyiv Polytechnic Institute", Kyiv, Ukraine, ORCID: 0000-0003-3900-4431

[4] National Technical University of Ukraine "Igor Sikorsky Kyiv Polytechnic Institute", Kyiv, Ukraine

[a] gumens@ukr.net, [b] mkarpets@ukr.net, [c] anna-07@ukr.net, [d] fmf_ikg@ukr.net

Keywords: High-Entropy Alloy, Valence Electron Concentration, Oxidation, Solid Solution, Ordering, Automated Indentation, Hardness, Microstructure

Abstract. Phase composition and mechanical properties and the formation of oxide layers on $Fe_{40-x}NiCoCrAl_x$ (x = 5 and 10 at.%) alloys in long-term oxidation at 900 and 1000°C were studied. In the initial cast state, depending on the aluminum content and valence electron concentration, the alloys contain only an fcc solid solution (VEC = 8 e/a) or a mixture of fcc and bcc phases (VEC = 7.75 e/a). Thin continuous oxide scales containing Cr_2O_3 and $NiCr_2O$ spinel formed on the surface of both alloys oxidized at 900°C for 50 h. A further increase in the annealing time to 100 h leads to the formation of aluminum oxide Al_2O_3 in the scale on the $Fe_{30}Ni_{25}Co_{15}Cr_{20}Al_{10}$ alloy, having high protective properties. An increase in the oxidation temperature to 1000°C results in partial failure of the protective layer on the alloy with 10 at.% Al. Long-term holding at 900°C (100 h) + 1000°C (50 h) does not change the phase composition of the $Fe_{35}Ni_{25}Co_{15}Cr_{20}Al_5$ alloy matrix, being indicative of its high thermal stability. In the two-phase $Fe_{30}Ni_{25}Co_{15}Cr_{20}Al_{10}$ alloy, the quantitative ratio of solid solutions sharply changes: the amount of the bcc phase increases from 4 to 54 wt.% and its B2-type ordering is observed. The mechanical characteristics of the starting alloys and those after long-term high-temperature annealing were determined by automated indentation. The hardness (HIT) and elastic modulus (E) of the cast $Fe_{35}Ni_{25}Co_{15}Cr_{20}Al_5$ alloy are equal to 2 and 147 GPa, respectively, and decrease to 1.8 and 106 GPa after a series of long-term annealing operations. The $Fe_{30}Ni_{25}Co_{15}Cr_{20}Al_{10}$ alloy shows the opposite dependence: HIT increases from 2.5 in the initial state to 3.1 GPa after annealing and E decreases from 152 to 134 GPa. This indicates that the $Fe_{30}Ni_{25}Co_{15}Cr_{20}Al_{10}$ alloy is promising as a high-temperature oxidation-resistant and creep-resistant material.

Introduction

A new class of alloys, called high-entropy alloys (HEAs), possessing low free energy and high mixing entropy has been of research focus recently. They contain more than four elements in an equiatomic or close ratio (Yeh et al., 2004). The HEAs are peculiar in that they form simple substitutional solid fcc or bcc solutions or a mixture of bcc + fcc solid solutions (Gao et al., 2013; Sheng et al., 2011). Differently doped alloys in which simple bcc and fcc solid solutions form are among HEAs that have been studied most extensively. Numerous papers focus on the optimization of alloy structures and mechanical properties (Firstov et al., 2013ab; Firstov et al., 2016a; Gumen et al., 2019ab; Wang et al., 2012). There are currently very limited data on the high-temperature

Quality Production Improvement and System Safety: QPI 16 - CZOTO 10   Materials Research Forum LLC
Materials Research Proceedings 34 (2023) 24-34   https://doi.org/10.21741/9781644902691-4

oxidation resistance of these materials and the stability of their phase composition and mechanical properties in long-term high-temperature holding (Butler et al., 2016ab; Holcomb et al., 2015; Kim et al., 2018; Wu et al., 2017). Our objective is to examine the effect of aluminum on the stability of phase composition, structure, and mechanical properties and on the formation of oxide layers in long-term temperature oxidation of the $Fe_{40-x}NiCoCrAl_x$ alloys (x = 5 and 10 at.% Al) at 900 and 1000°C.

Experimental part

The $Fe_{40-x}NiCoCrAl_x$ alloys (x = 5 and 10 at.% Al) were produced by arc melting in a high-purity argon atmosphere. The ingots were melted six to seven times to homogenize their composition. The starting components were high-purity materials (at least 99.98% purity).

The oxidation resistance of the alloys was examined in an electric arc furnace at 900 and 1000°C in air. The samples were periodically weighed in 5, 10, 25, 50, and 100 h. To measure the weight change of the samples, we used a Radwag precision balance (±0.0001 g). The oxidation resistance was assessed from the specific change in weight of the samples q ($mg/cm^2 \cdot h$).

The phase composition of the starting alloys and those after high-temperature oxidation was analyzed with an Ultima IV diffractometer in monochromatic Cu-K_α radiation. The X-ray diffraction data were processed using the PowderCell 2.4 software for full-profile analysis of X-ray spectra for a mixture of polycrystalline phase components. The accuracy of the measured lattice period values is ±0.0001 nm. The morphology and microstructure of scales on the starting and annealing alloys were examined with a Jeol Superprobe 733 scanning electron microscope (SEM) and with a MIM-8 optical microscope. The mechanical properties were determined by automated indentation with a Berkovich pyramid under 3N load employing a Micron Gamma unit.

Results and Discussion

The fcc solid-solution phase is known (Butler et al., 2017; Firstov et al., 2016b) to form at elevated valence electron concentrations, VEC ≥ 8 e/a, in high-entropy alloys in most cases, while the bcc solid-solution phases are stable at VEC ≤ 6.87 e/a. If the valence electron concentration is 7.2 ≤ VEC ≤ 8 e/a, a mixture of two, bcc and fcc, solid solutions forms. According to our calculations, VEC = 8 e/a for the $Fe_{35}Ni_{25}Co_{15}Cr_{20}Al_5$ alloy and 7.75 e/a for the $Fe_{30}Ni_{25}Co_{15}Cr_{20}Al_{10}$ alloy (Table 1).

Table 1.   *Phase Constituents of the Cast Alloys, Lattice Parameters of the Phases, and Valence Electron Concentrations*

Alloy	Phase constituents, wt.%	Lattice parameter a, nm	VEC, e/a
$Fe_{35}Ni_{25}Co_{15}Cr_{20}Al_5$	100 FCC	0,3571	8
$Fe_{30}Ni_{25}Co_{15}Cr_{20}Al_{10}$	96 FCC 4 BCC	0,3603 0,2862	7,75

Metallographic analysis of the high-entropy alloys in starting (cast) state showed that the $Fe_{35}Ni_{25}Co_{15}Cr_{20}Al_5$ microstructure (Fig. 1b) included grains elongated in the crystallization direction and subgrains at grain boundaries and partially inside them. With higher aluminum content of the $Fe_{30}Ni_{25}Co_{15}Cr_{20}Al_{10}$ alloy, dendritic crystallization leading to light dendrites and a darker phase in the space between the dendrites was observed (Fig. 1d).

Quality Production Improvement and System Safety: QPI 16 - CZOTO 10   Materials Research Forum LLC
Materials Research Proceedings 34 (2023) 24-34   https://doi.org/10.21741/9781644902691-4

A greater aluminum content of the alloy reduces VEC, in turn indicating that stability of the fcc solid solution reduces and the bcc phase forms through solid-phase decomposition mechanism.

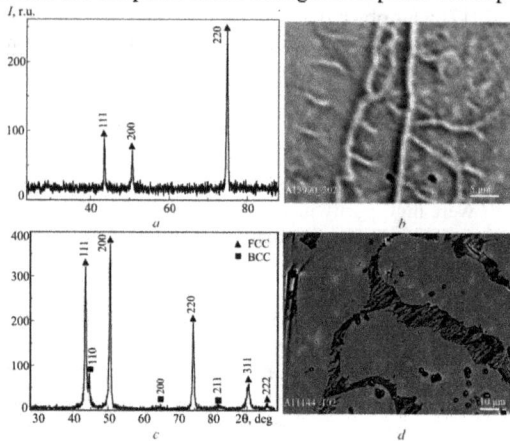

Fig. 1. *X-ray diffraction patterns (a, c) and microstructures (b, d) for the Fe$_{35}$Ni$_{25}$Co$_{15}$Cr$_{20}$Al$_5$ (a, b) and Fe$_{30}$Ni$_{25}$Co$_{15}$Cr$_{20}$Al$_{10}$ (c, d) high-entropy alloys in starting state (SEM).*

Figure 2 shows variation in the specific weight of the samples q (mg/cm^2) after oxidation of the Fe$_{35}$Ni$_{25}$Cr$_{20}$Co$_{20}$Al$_5$ and Fe$_{30}$Ni$_{25}$Cr$_{20}$Co$_{15}$Al$_{10}$ high-entropy alloys at 900°C for 100 h and then at 1000°C for 50 h. After short holding for 5 h at 900°C, the specific weight of the alloy samples increased differently. The alloy with 10 at.% Al oxidized much more slowly for the first 5 h than the alloy containing 5 at.% Al.

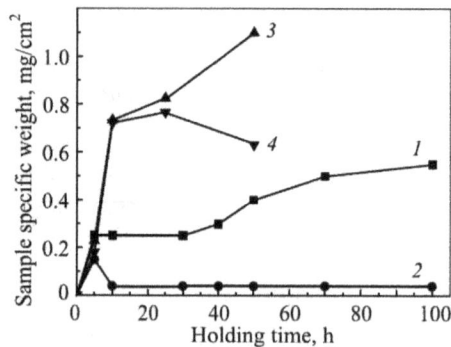

Fig. 2. *Variation in the specific weight of alloy samples after oxidation at 900°C (100 h) and further at 1000°C (50 h).*

The specific weight of the Fe$_{35}$Ni$_{25}$Cr$_{20}$Co$_{20}$Al$_5$ alloy showed a stable gain for 30 h at 900°C and naturally increased further with longer holding as oxide film began to develop and grow. The Fe$_{30}$Ni$_{25}$Cr$_{20}$Co$_{15}$Al$_{10}$ alloy behaved differently. The sample significantly gained weight after the

Quality Production Improvement and System Safety: QPI 16 - CZOTO 10   Materials Research Forum LLC
Materials Research Proceedings 34 (2023) 24-34   https://doi.org/10.21741/9781644902691-4

first 5 h and then partially lost its weight, which was accompanied by the formation of a thin brittle scale that spalled in some places. A dense scale formed further and oxidation proceeded not only without weight losses but also without weight gain, being indicative of high protective properties of the oxide film. However, a further increase of the oxidation temperature to 1000°C led to a sharp gain in the specific weight of the samples and growth of the scale thickness (Fig. 2).

The $Fe_{35}Ni_{25}Cr_{20}Co_{20}Al_5$ alloy showed a continuous increase in the specific weight within entire holding at 1000°C; q was minimum for the first 5 h and the oxidation rate sharply increased in a period between 5 to 10 h. During the next exposure in the interval from 10 to 25 hours, a significant slowdown in the oxidation process is observed, and then there is a significant increase in the increase in the mass of the samples, this may indicate that the scale formed on the surface of the alloy does not have high protective properties at a given temperature.

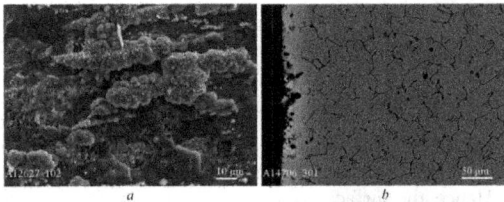

Fig. 3. *Morphology of the scale on the $Fe_{30}Ni_{25}Cr_{20}Co_{15}Al_{10}$ alloy (a) after oxidation at 900°C (100 h) + 1000°C (50 h) and cross-sectional microstructure (b).*

The situation was different for the $Fe_{30}Ni_{25}Cr_{20}Co_{15}Al_{10}$ alloy that contained more aluminum and less iron. After the first 5 h, the sample weight gain was much lower than that of the $Fe_{35}Ni_{25}Cr_{20}Co_{20}Al_5$ alloy. In a period between 5 and 10 h, q substantially increased and then (10–25 h) the process sharply slowed down. In a period between 25 and 50 h, the curve showing variation in the specific weight changed its direction. When the sample cooled down, the scale partially spalled since the protective film failed under a cyclic change in temperatures because of different thermal expansion coefficients (TECs) of the scale and alloy matrix (Fig. 3).

A two-phase oxide film containing Cr_2O_3 and $NiCr_2O_4$ spinel formed on both alloys at 900°C for 50 h. The scales were very thin (3–5 μm thick on the $Fe_{30}Ni_{25}Cr_{20}Co_{15}Al_{10}$ alloy and 7–8 μm thick on the $Fe_{35}Ni_{25}Cr_{20}Co_{20}Al_5$ alloy) and the X-ray diffraction patterns of both alloys had also reflections from the matrix phase of the fcc solid solution (Fig. 4).

The lattice parameter of the fcc solid solutions in the $Fe_{35}Ni_{25}Cr_{20}Co_{20}Al_5$ alloy increased compared to the starting state and became a = 0.3577 nm (since the near-surface metal layer was saturated with oxygen) and that in the $Fe_{30}Ni_{25}Cr_{20}Co_{15}Al_{10}$ alloy substantially decreased and became a = 0.3585 nm (Table 2), indicating that metal ions diffused toward the oxidation front. Some diffusion processes occurred under the scale. For example, the aluminum diffusion rate increased with temperature. Hence, the aluminum concentration toward the metal surface became higher to promote the formation of Al_2O_3. In turn, this indicates that the fcc phase became depleted of aluminum and its lattice parameter decreased.

The amount of the fcc phase in the 5 at.% Al alloy was 7 wt.% and that in the 10 at.% Al alloy was 41 wt.%. This substantial difference can be attributed to varying thickness of the scales; as a result, X-rays penetrate to different depths of the sample. The main phase of the scales is presented by oxides of Cr_2O_3 structure: their amount reaches 83 wt.% for $Fe_{35}Ni_{25}Cr_{20}Co_{20}Al_5$ and 42 wt.% for $Fe_{30}Ni_{25}Cr_{20}Co_{15}Al_{10}$.

The amount of $NiCr_2O_4$ spinel in the scale of the $Fe_{35}Ni_{25}Cr_{20}Co_{20}Al_5$ alloy was 10 wt.% and that in the $Fe_{30}Ni_{25}Cr_{20}Co_{15}Al_{10}$ alloy was 17 wt.%. Further holding changed the scale phase composition. The alloy with lower aluminum content had a scale after 100 h at 900°C containing two phases: $Cr_{13}Fe_7O_{30}$ and $NiCr_2O_4$ (Fig. 4b).

The presence of $Cr_{13}Fe_7O_{30}$ can be explained by the fact that iron oxidized in long-term holding to form FeO that dissolved in Cr_2O_3 to create the $Cr_{13}Fe_7O_{30}$ phase with lattice parameters a = 0.4986 nm and c = 1.3646 nm, which is isostructural to Cr_2O_3.

The $NiCr_2O_4$ lattice parameter increased after oxidation at 900°C and became 0.8350 nm (Table 2). In oxidation at 900°C for 100 h, Al_2O_3 formed on the $Fe_{30}Ni_{25}Cr_{20}Co_{15}Al_{10}$ alloy besides $Cr_{13}Fe_7O_{30}$. The X-ray diffraction pattern for the scale surface had no $NiCr_2O_4$ reflections since spinel separated from the scale surface after the thin continuous Al_2O_3 layer formed (being typical of the oxidation of Ni–Cr–Al alloys), so reflections from the fcc phase were predominant (Fig. 4d). After further oxidation at 1000°C for 50 h, the phase composition of the scales was similar to that of the oxide films after annealing for the same time at 900°C. In both cases, the X-ray diffraction patterns had reflections of $NiCr_2O_4$, Cr_2O_3, and fcc phase (Fig. 4e, f), but the amounts of phase constituents somewhat differed: the amount of Cr_2O_3 decreased and that of $NiCr_2O_4$ increased (Table 2). This will further improve the high-temperature oxidation resistance of the alloys since diffusion processes slow down significantly in the layer with $NiCr_2O_4$.

Cross-sectional studies of the $Fe_{30}Ni_{25}Cr_{20}Co_{15}Al_{10}$ alloy samples oxidized at 900°C + 1000°C revealed a wide continuous region of a light phase, identified by X-rays as the fcc solid solution, under the oxide film layer. This explains why there are no reflections of the bcc phase in the X-ray diffraction pattern. Hence, the surface layers are structured as follows: scale, fcc solid-solution region, and matrix with bcc + fcc phases.

Since the papers (Lin et al., 2011; Kao et al., 2011; Wang et al., 2014) found differences in the phase composition of the FeNiCoCrAl$_x$ alloys (in particular, for x = 5 and 10 at.% Al) after annealing at 900, 950, and 1100°C, we conducted additional X-ray diffraction and microstructural analyses of the samples' end surfaces after complete grinding of the oxide layers to specify changes in the structure and phase composition of the alloy matrices (Fig. 5).

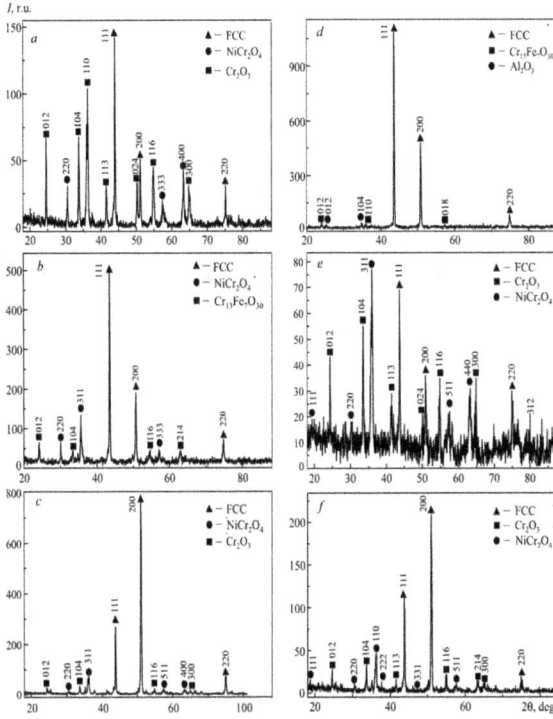

Fig. 4. *X-ray diffraction patterns for the $Fe_{35}Ni_{25}Cr_{20}Co_{20}Al_5$ (a, b, e) and $Fe_{30}Ni_{25}Cr_{20}Co_{15}Al_{10}$ (c, d, f) alloys after oxidation at 900°C for 50 (a, c) and 100 h (b, d) and after oxidation at 1000°C for 50 h (e, f).*

Table 2. *Phase Compositions and Lattice Parameters of Scale Phases after Annealing of the Alloys*

Alloy	Annealing conditions		Phase compositions. wt.%	Lattice parameters. nm	
	$T.\,°C$	$\tau.\,h$		A	C
$Fe_{35}Ni_{25}Co_{15}Cr_{20}Al_5$	900	50	7 FCC	0.3577	
			10 $NiCr_2O_4$	0.8306	
			83 $(FeCr)_2O_3$	0.4975	1.3715
	900	100	51 ГЦК	0.3590	
			29 $Cr_{13}Fe_7O_{30}$	0.4986	
			20 $NiCr_2O_4$	0.8350	1.3646
	1000	50	15 FCC	0.3583	
			33 $NiCr_2O_4$	0.8304	
			52 Cr_2O_3	0.4984	1.3597
$Fe_{30}Ni_{25}Co_{15}Cr_{20}Al_{10}$	900	50	41 FCC	0.3585	
			17 $NiCr_2O_4$	0.8299	
			42 Cr_2O_3	0.4958	1.3597
	900	100	92 FCC	0.3580	
			3 $Cr_{13}Fe_7O_{30}$	0.4985	1.3621
			5 Al_2O_3	0.4790	1.3006
	1000	50	43 FCC	0.3597	
			23 $NiCr_2O_4$	0.8318	
			34 Cr_2O_3	0.4990	1.3659

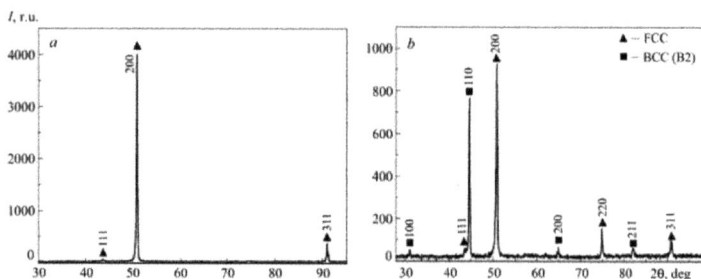

Fig. 5. *X-ray diffraction patterns for the $Fe_{35}Ni_{25}Cr_{20}Co_{20}Al_5$ (a) and $Fe_{30}Ni_{25}Cr_{20}Co_{15}Al_{10}$ (b) alloy matrices after annealing at 900°C (100 h) and 1000°C (50 h) and complete removal of the oxide.*

Quality Production Improvement and System Safety: QPI 16 - CZOTO 10   Materials Research Forum LLC
Materials Research Proceedings 34 (2023) 24-34   https://doi.org/10.21741/9781644902691-4

Table 3. *Phase Compositions and Lattice Parameters of Phases in the Alloy Matrix after*
Annealing at 900°C for 100h+1000°C for 50h

Alloy	Phase constituents	Phase content, %	Lattice parameter a, nm
$Fe_{35}Ni_{25}Cr_{20}Co_{20}Al_5$	FCC	100	0.3586
$Fe_{30}Ni_{25}Cr_{20}Co_{15}Al_{10}$	FCC	46	0.3587
	BCC (B2)	54	0.2870

After annealing at 900°C for 24 h, two phases were found to form in the $FeNiCoCrAl_5$ alloy (Lin et al., 2011): fcc and B2-ordered bcc; at 1100°C, there was only fcc phase, the phase composition of the $Fe_{35}Ni_{25}Cr_{20}Co_{20}Al_5$ alloy remained unchanged after long-term high-temperature annealing in our case; there was only the fcc solid solution. This indicates that this alloy has high thermal stability since variations in line intensities in the X-ray diffraction pattern are not associated with structural changes but result from significant texture confirmed by the respective texture coefficient along the {200} crystallographic direction. For example, this coefficient τ was $0.2775_{(200)}$ after annealing for the alloy with 5 at.% Al and $0.2553_{(200)}$ for the alloy with 10 at.% Al (note that $\tau = 1$ in the absence of texture) (Fig. 5a).

The phase composition of the $Fe_{30}Ni_{25}Cr_{20}Co_{15}Al_{10}$ alloy somewhat changed (Fig. 5b): the amount of the fcc solid solution sharply decreased from 96 to 46 wt.% and the B2-ordered bcc phase constituted 54 wt.%. Compared to the starting state, the lattice parameter of the fcc phase in the $FeNiCoCrAl_5$ alloy increased from 0.3571 to 0.3586 nm and that of the $FeNiCoCrAl_{10}$ alloy conversely decreased from 0.3603 to 0.3587 nm after long-term high-temperature annealing. The lattice parameter of the bcc solid solution increased to 0.2870 nm. These changes in lattice parameters occur because of diffusion-controlled redistribution of the alloy components (Table 3) among the phase compositions. Microstructural analysis of the annealed $Fe_{35}Ni_{25}Cr_{20}Co_{20}Al_5$ and $Fe_{30}Ni_{25}Cr_{20}Co_{15}Al_{10}$ alloy matrices by scanning electron microscopy (SEM) showed that long-term annealing substantially influenced their morphology. The alloy with 5 at.% Al became homogeneous after annealing, and thus the structure of even a preliminary etched sample was not practically revealed (Fig. 6a). The size of dendrites increased and their shape changed in the $Fe_{30}Ni_{25}Cr_{20}Co_{15}Al_{10}$ alloy, while a dark fine phase precipitated within light grains (Fig. 6b). The annealed alloy matrix and the distribution of elements in characteristic radiation were studied by SEM, which showed that the dendritic phase was enriched with chromium and iron, cobalt was distributed uniformly between the two phases, and nickel and aluminum were mainly in the space between dendrites and fine phase in the $Fe_{30}Ni_{25}Cr_{20}Co_{15}Al_{10}$ alloy.

The mechanical characteristics of the alloys in cast state and after annealing were determined by automated indentation (Table 4). After annealing, hardness of the $Fe_{35}Ni_{25}Cr_{20}Co_{20}Al_5$ alloy somewhat decreased to 1.8 GPa, which can be attributed to its high homogenization. The elastic modulus sharply reduced to 106 GPa since the lattice parameter of the fcc solid solution increased substantially, which in turn weakened interatomic forces and decreased E.

Table 4. *Mechanical Properties of the Alloys in Starting State and after Long-Term Oxidation at 900°C for 100 h + 1000°C for 50 h*

Alloy	Cast state		Matrix after annealing	
	H_{IT}, GPa	E, GPa	H_{IT}, GPa	E, GPa
$Fe_{35}Ni_{25}Cr_{20}Co_{20}Al_5$	2	147	1.8	106
$Fe_{30}Ni_{25}Cr_{20}Co_{15}Al_{10}$	2.5	152	3.1	134

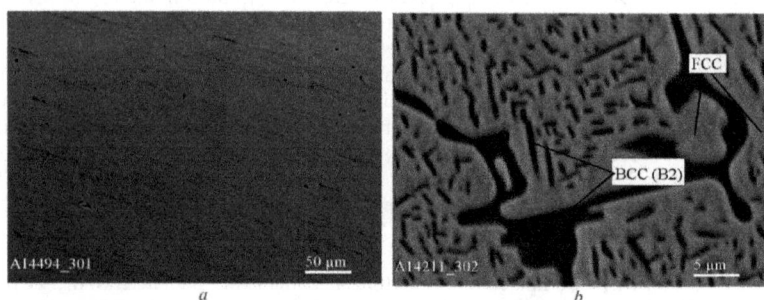

a   *b*

Fig. 6. *Structure of the $Fe_{35}Ni_{25}Cr_{20}Co_{20}Al_5$ (a) and $Fe_{30}Ni_{25}Cr_{20}Co_{15}Al_{10}$ (b) alloy matrices after subsequent annealing at 900°C + 1000°C.*

The hardness increased to 3.1 GPa in the $Fe_{30}Ni_{25}Cr_{20}Co_{15}Al_{10}$ alloy as the B2-ordered bcc phase formed. This indicates that this alloy is a promising high-temperature oxidation-resistant material.

Summary and conclusion

A series of studies focusing on the cast and annealed alloys in the $Fe_{40-x}NiCoCrAl_x$ system (x = 5 and 10 at.%) has established the following. The starting $Fe_{35}Ni_{25}Cr_{20}Co_{20}Al_5$ alloy is a single-phase fcc solid solution, whose hardness is 2 GPa and elastic modulus 147 GPa. The $Fe_{30}Ni_{25}Cr_{20}Co_{15}Al_{10}$ alloy is two-phase and contains fcc (96%) and bcc (4%) solid solutions. For this reason, it has higher hardness (2.5 GPa) and Young's modulus (152 GPa).

In oxidation at 900°C for 100 h, thin two-phase oxide films that contain $NiCr_2O_4$ spinel and $Cr_{13}Fe_7O_{30}$ oxide form on the $Fe_{35}Ni_{25}Cr_{20}Co_{20}Al_5$ alloy and Al_2O_3 and $Cr_{13}Fe_7O_{30}$ oxides on the $Fe_{30}Ni_{25}Cr_{20}Co_{15}Al_{10}$ alloy. Aluminum oxide in the scale promotes effective protection of the alloy against oxidation. However, when scale annealing temperature increases, Al_2O_3 separates and oxidation proceeds with the formation of $NiCr_2O_4$ and Cr_2O_3 on the surface. Despite this, the high-temperature oxidation resistance of the alloys is at the level of some hightemperature oxidation-resistant alloys in the Ni–Cr–Al system. A wide continuous region of the fcc solid solution under the scale on the $Fe_{30}Ni_{25}Cr_{20}Co_{15}Al_{10}$ alloy has been established to form for the first time.

Long-term high-temperature annealing substantially changes the phase composition and mechanical characteristics of the $Fe_{30}Ni_{25}Cr_{20}Co_{15}Al_{10}$ alloy: the amount of the fcc solid solution sharply decreases to 46 wt.% and an ordered bcc (B2) phase forms. This substantially increases hardness: to 3.1 GPa. This indicates that the $Fe_{30}Ni_{25}Cr_{20}Co_{15}Al_{10}$ alloy is a high-temperature oxidation-resistant material.

Reference

[1] T.M. Butler, M.L. Weaver. Investigation of the phase stabilities in AlNiCoCrFe high-entropy alloys, J. Alloys Compd. 691 (2017) 119-129. https://doi.org/10.1016/j.jallcom.2016.08.121

[2] T.M. Butler, M.L. Weaver. Oxidation behavior of arc-melted AlCoCrFeNi multi-component high-entropy alloys, J. Alloys Compd. 674 (2016) 229-244. https://doi.org/10.1016/j.jallcom.2016.02.257

[3] T.M. Butler, M.L. Weaver, Influence of annealing on the microstructures and oxidation behaviors of Al8(CoCrFeNi)92, Al15(CoCrFeNi)85, and Al30(CoCrFeNi)70 high-entropy alloys, Metals 6 (2016) art.222. https://doi.org/10.3390/met6090222

[4] S.A. Firstov et al. Structural features and solid-solution hardening of the high-entropy CrMnFeCoNi alloy, Powder Metall. Met. Ceram. 55 (2016) 225-235. https://doi.org/10.1007/s11106-016-9797-9

[5] S.A. Firstov et al. Effect of electron density on phase composition of high-entropy equiatomic alloys, Powder Metall. Met. Ceram. 54 (2016) 607-613. https://doi.org/10.1007/s11106-016-9754-7

[6] S.A. Firstov et al. New class of materials – high-entropy alloys and coatings, Vestn. Tomsk. Gos. Univ. 18(4) (2013) 1938-1940.

[7] S.A. Firstov et al. Effect of the crystallization rate on the structure, phase composition, and hardness of the high-entropy AlTiVCrNbMo alloy, Deform. Razrush. Mater. 10 (2013) 8-15.

[8] M.C. Gao, D.E. Alman. Searching for next single-phase high-entropy alloy compositions, Entropy 15 (2013) 4504-4519. https://doi.org/10.3390/e15104504

[9] O. Gumen, I. Selina, R. Selin. Projection of phase composition of lowcost titanium alloy welded joints by finite element mathematical modelling method, Construction of Optimized Energy Potential 12, (2019), 51-56. https://doi.org/10.17512/bozpe.2021.1.07

[10] O. Gumen, I. Bilyk, M. Kruzhkova. Geometrical simulation of optimized vacuum-condensation spraying technology for titanium nitride on structural steel, LNCE 47 (2020) 103-110. https://doi.org/10.1007/978-3-030-27011-7_13

[11] G.R. Holcomb, J. Tylczak, C. Carney. Oxidation of CoCrFeMnNi high-entropy alloys, JOM 67 (2015) 2326-2339. https://doi.org/10.1007/s11837-015-1517-2

[12] Y.F. Kao, S.K. Chen, T.J. Chen. Electrical, magnetic, and hall properties of Al_xCoCrFeNi high-entropy alloys, J. Alloys Compd. 509 (2011) 1607-1614. https://doi.org/10.1016/j.jallcom.2010.10.210

[13] Y.-K. Kim et al. High-temperature oxidation behavior of Cr–Mn–Fe–Co–Ni high-entropy alloy, Intermetallics 98 (2018) 45-53. https://doi.org/10.1016/j.intermet.2018.04.006

[14] C.M. Lin, H.L. Tsai. Evolution of microstructure, hardness, and corrosion properties of high-entropy Al0.5CoCrFeNi alloy, Intermetallics 19 (2011) 288-294. https://doi.org/10.1016/j.intermet.2010.10.008

[15] S. Guo, C.T. Liu. Phase stability in high-entropy alloys: formation of solid-solution phase or amorphous phase, Prog. Nat. Sci. Mater. Int. 6 (2011) 433-446. https://doi.org/10.1016/S1002-0071(12)60080-X

[16] F.J. Wang et al. Cooling rate and size effect on the microstructure and mechanical properties of AlCoCrFeNi high-entropy alloy, J. Eng. Mater. Technol. 131 (2009) 345011-345013. https://doi.org/10.1115/1.3120387]

[17] W.R. Wang et al. Effects of Al addition on the microstructure and mechanical property of Al_xCoCrFeNi high-entropy alloys, Intermetallics 26 (2012) 44-51. https://doi.org/10.1016/j.intermet.2012.03.005

[18] W.R. Wang, W.L Wang, J.W. Yeh, Phases, microstructure and mechanical properties of AlCoCrFeNi high-entropy alloys at elevated temperatures, J. Alloys Compd. 589 (2014) 143-152. https://doi.org/10.1016/j.jallcom.2013.11.084

[19] W. Kai et al. Air-oxidation of FeCoNiCr-based quinary high-entropy alloys at 700–900°C, Corr. Sci. 121 (2017) 116–125. https://doi.org/10.1016/j.corsci.2017.02.008

[20] J.W. Yeh et al. Nanostructured high-entropy alloys with multiple principal elements: novel alloy design concepts and outcomes, Adv. Eng. Mater. 6 (2004) 299-303. https://doi.org/10.1002/adem.200300567

Quality Production Improvement and System Safety: QPI 16 - CZOTO 10
Materials Research Proceedings 34 (2023) 35-42

Materials Research Forum LLC
https://doi.org/10.21741/9781644902691-5

Impact of Increased Iron Content and Manganese Addition on Intermetallic Phases and Fatigue Resistance of AlSi7Mg0.6 Secondary Alloy

MIKOLAJČÍK Martin[1, a *], TILLOVÁ Eva[1, b], KUCHARIKOVÁ Lenka[1, c] and CHALUPOVÁ Mária[1, d]

[1]University of Žilina, Faculty of Mechanical Engineering, Department of Materials Engineering, Univerzitná 8215/1, 010 26 Žilina, Slovak Republic

[a]martin.mikolajcik@fstroj.uniza.sk*, [b]eva.tillova@fstroj.uniza.sk, [c]lenka.kucharikova@fstroj.uniza.sk, [d]maria.chalupova@fstroj.uniza.sk

Keywords: AlSi7Mg0.6, A357, Aluminium, Fatigue Properties, Effect of Fe and Mn

Abstract. One of the most often utilized metals in a variety of industries is aluminium alloy. Secondary aluminium alloys have drawn a lot of interest in recent years. Scrap aluminium may be recycled, which is good for the environment. In addition, compared to primary aluminium, it produces at a lower cost due to its lower energy requirement. Because it adversely affects their characteristics, secondary aluminium alloys' higher iron concentration presents the most frequent challenge. Manganese can be added to reduce its impact to some level. The objective of this research is to improve our knowledge of how iron and manganese affect the fatigue resistance of secondary aluminium alloys. Four alloys with various iron and manganese concentrations were tested to determine this impact.

Introduction

Secondary aluminium alloys are a class of materials that are derived from recycled aluminium and are widely used in various industries due to their high strength-to-weight ratio, excellent corrosion resistance, and cost-effectiveness. These alloys have been designed to meet the demanding requirements of various applications, including aerospace, transportation, construction, and packaging [1 - 3]. Although the addition of iron to aluminium alloys can provide benefits in terms of strength and fatigue resistance, it is most often considered to be an impurity. It is because iron also has a negative impact on the properties of the material. Iron can cause a reduction in ductility, which can make the material more brittle and prone to cracking. Additionally, the presence of iron leads to the formation of Al_5FeSi intermetallic compounds, which can act as stress concentrators and weaken the material. Iron can also increase the sensitivity of the material to corrosion, especially in aggressive environments. Therefore, it is important to carefully control the amount of iron added to aluminium alloys to ensure that its positive effects are maximized, and its negative effects are minimized [4 - 6]. Manganese can play a critical role in correcting the negative effects of iron in aluminium alloys. By forming stable intermetallic compounds with iron, manganese can prevent the formation of other, potentially harmful intermetallic compounds Al_5FeSi. Manganese can also help to increase the ductility of the material and improve its toughness, reducing the risk of cracking and failure. Finally, the combination of iron and manganese in aluminium alloys can provide a beneficial balance between strength, ductility, and corrosion resistance, making these alloys suitable for a wide range of applications [7 - 11]. The fatigue properties of aluminium alloys may also be impacted by the modification in the morphology of the intermetallic phases. Due to their sharp form, the plate-like Al_5FeSi particles are stress concentration points. These phases are also brittle, making them the ideal places for the development of fatigue cracks. Large Al_5FeSi particles also block the liquid flow channels as they solidify. Castings become more porous as a result. Fatigue crack initiation is most frequently caused by casting defects beneath the casting surface. It can be expected that manganese should improve the experimental material's fatigue performance since its presence should reduce the number of shrinkage cavities [12 - 14]. According to research, more than 90% of engineering components that break are fractures caused by fatigue material. Fatigue fractures in transportation, such as those in rails, tire components, plane wings, and hulls of ships, are extremely dangerous since they are frequently associated with fatalities [15]. Therefore, the fatigue properties of materials need to be further investigated. This paper investigated how secondary AlSi7Mg0.6 cast aluminium alloy with a higher iron content responds to changes in iron and manganese levels in terms of its fatigue properties. The work is part of a project aimed at the study of secondary alloys with higher iron content.

Materials and Methods

We used secondary AlSi7Mg0.6 aluminium alloy, also known as A357 alloy, as an experimental material. The alloy had a higher iron content and was supplied in the form of rods (diameter of 20mm, length of 300mm) by UNEKO Zátor, a.s. These rods were manufactured by gravity casting into sand moulds at a temperature of 750°C and refined using ECOSAL Al 113S refining salt. We tested four alloys differing in the amount of iron and manganese in the chemical composition. We investigated the effect of manganese addition on the properties of materials at two different iron contents (0.75 wt.% and 1.25 wt.%). To increase iron and manganese content, we used pre-alloys AlFe75 and AlMn75. To purely study the impact of iron and manganese, we intentionally chose materials that were not heat treated. The supplier verified the complete chemical composition according to EN 10204 3.1, which is documented in Table 1.

Table 1. Chemical composition [wt.%]

Alloy	Al	Si	Mg	Fe	Mn	Ti	Cu
A	91.17	7.374	0.477	**0.75**	0.007	0.121	0.017
B	90.91	7.252	0.5	**0.728**	**0.402**	0.12	0.04
C	90.68	7.276	0.548	**1.264**	0.008	0.117	0.012
D	90.28	7.047	0.546	**1.245**	**0.661**	0.115	0.013

To prepare samples for metallographic and quantitative analysis, standard methods were applied. A Struers-CitoPress-1 was used to press the specimens into dentacryl after they had been cut from rods using an ATM Brillant 240 saw. Using a Struers TegraPol-15 automatic machine, the samples were ground and polished in five steps with varying sandpaper roughness, duration, and chemical environments. After that, 0.5% hydrofluoric acid was used to etch the samples. Metallographic and quantitative analyses were carried out using an optical microscope NEOPHOT-32 with NIS-Element 5.20 software. The aim of the quantitative analysis was to uncover how iron and manganese content affected the Fe-rich intermetallic phases' size and morphology. We also investigated the effect of chemical composition (addition of manganese) on the porosity of the castings.

Standard chip machining techniques were used to create the test rods for the fatigue tests. They measured 150mm in length, 12mm in width, and 8mm in diameter at their thinnest point. We used ROTOFLEX to analyze the alloys under test regarding their fatigue resistance. The test rods were bent while rotating, and the system kept track of how many cycles there were until failure. Each alloy was evaluated using a total of seven test rods at room temperature. The loads were 68, 78, and 88 MPa, and the test frequency was 32.1 Hz. The Department of Materials Engineering of UNIZA carried out all measurements.

Quality Production Improvement and System Safety: QPI 16 - CZOTO 10   Materials Research Forum LLC
Materials Research Proceedings 34 (2023) 35-42   https://doi.org/10.21741/9781644902691-5

Results

Fig.1. displays the outcomes of the metallographic analysis. We have determined the main structural elements: α-phase (substitutional solid solution of Si in Al) and eutectic (eutectic silicon crystals in the α-phase). We also observed ferrous intermetallic phases in the form of needles/plates (Al_5FeSi) and skeletons/Chinese scripts ($Al_{15}(FeMnMg)_3Si_2$). The microstructure also contained Mg_2Si intermetallic phases, which do not have a significant effect in Al-Si alloys without heat treatment. Shrinkage cavities and pores were visible in the experimental material as casting defects.

Through quantitative analysis, the impact of iron and manganese on the porosity of castings was investigated. The analyses' findings are shown in Fig.2, Fig.3 and in Fig.4. The highest area proportion of casting defects was found in alloy D - 6.2%. In this material, the defects also have the largest dimensions (21 753.96 μm^2). The iron content does not affect the areal proportion or size of defects. The smallest casting defects are found in alloy B. Compared to alloys A and C, which are not alloyed with manganese, the pores in this alloy form a larger area fraction. It follows that manganese, especially at higher iron contents, has a negative effect on the area fraction of the formed casting defects. The effect of manganese on the size of the formed casting defects is also negative at higher iron contents.

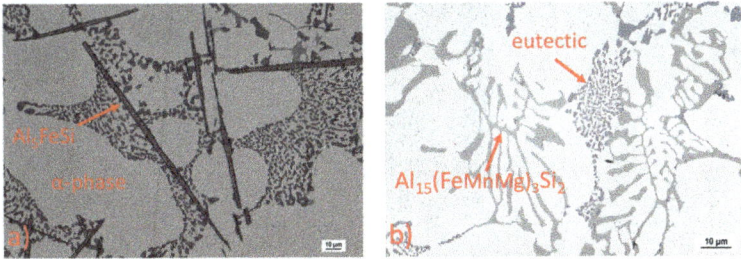

Fig.1. Identification of basic structural components in AlSi7Mg0.6 alloy.

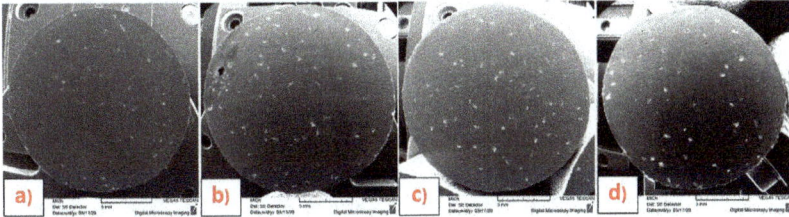

Fig.2. Effect of Mn on microporosity in AlSi7Mg0.6 alloy, SEM.

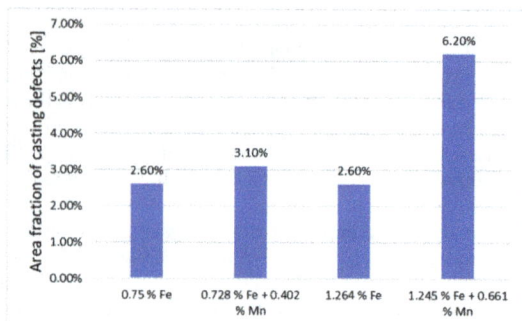

Fig.3. *Effect of Fe and Mn on the area fraction of casting defects.*

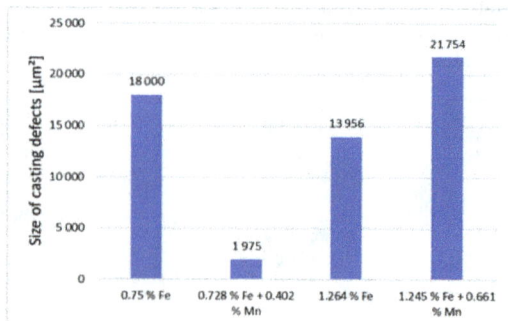

Fig.4. *Effect of Fe and Mn on the size of casting defects.*

The ferrous phases in AlSi7Mg0.6 alloys can take the form of thin plates (Al_5FeSi), which are visible as sharp needles on the cutting plane, skeletal structures ($Al_{15}(FeMnMg)_3Si_2$), or Chinese characters (also $Al_{15}(FeMnMg)_3Si_2$), depending on the chemical composition. The presence of manganese has a major impact on how the Fe-rich phases are shaped (Table. 2). Both the shape and the size of these phases undergo change. Al_5FeSi needle phases are formed in AlSi7Mg0.6 alloys, especially in the absence of Mn. In alloys B and D, which have been alloyed with manganese to correct the effect of iron, the ferrous phases in the form of plates are precipitated in much smaller amounts. At the same time, in the absence of manganese, the area fraction of needle phases in the structure increases with higher iron content. Therefore, the highest area fraction of Al_5FeSi phases in the structure is found in alloy C. The area fraction of Al_5FeSi needle phase grew from 2.2% to 3.5% with an increase in iron concentration from 0.75% to 1.25%. It significantly dropped from 2.2% to 0.7% and from 3.5% to 0.9% after manganese alloying.

In AlSi7Mg0.6 alloys, the presence or absence of manganese affects the size of the Al_5FeSi phases as well (Table 3). Shorter needles are created by adding manganese while keeping the iron amount the same. The length fell by about 55% (from 45.93 μm to 20.28 μm) at an iron concentration of 0.75%, and by about 20% (from 46.39 μm to 37.25 μm) with an iron content of about 1.25%. Longer Al_5FeSi needles occur because manganese alloyed alloys have a larger iron concentration. The influence of iron concentration on needle length is minimal in the case of alloys without manganese.

Quality Production Improvement and System Safety: QPI 16 - CZOTO 10   Materials Research Forum LLC
Materials Research Proceedings 34 (2023) 35-42   https://doi.org/10.21741/9781644902691-5

Table 2. Effect of Fe and Mn on the area fraction of Al₅FeSi phase

Alloy	A	B	C	D
Area fraction of Al₅FeSi phase	2.2 %	0.7 %	3.5 %	0.9 %

Table 3. Effect of Fe and Mn on length of Al₅FeSi phases

Alloy	A	B	C	D
Length of Al₅FeSi phases [μm]	45.83	20.28	46.39	37.25

When manganese is present, skeletal iron phases and Chinese script-shaped phases ($Al_{15}(FeMnMg)_3Si_2$) are found in the microstructure. Phases with the stated morphology were not found in the microstructure of alloys A and C because manganese is only in minimal amounts in these materials. In alloys with a higher manganese concentration (melts B and D), the area percentage of skeleton phases has a similar value (Fig.5). Alloy D, which contains more iron, had a slightly larger proportion of these phases.

The fatigue properties of the experimental alloys were studied using cyclic bending loads under rotation. Figures 6 - 8 show a comparison of the Wöhler curves for the experimental alloys (stress amplitude (σ) vs the number of cycles to fracture (N)). All the curves are decreasing in character. The number of cycles to fracture rises as the load amplitude lowers from 88 MPa to 68 MPa, showing an exponential decrease with an asymptotic approach to zero.

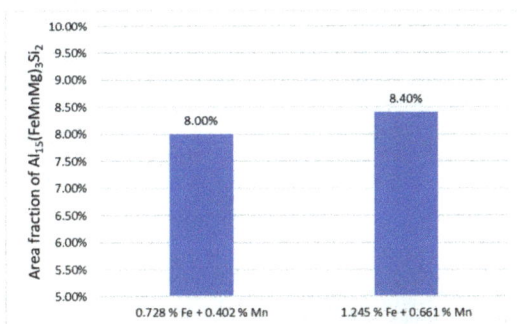

Fig.5. Effect of Fe and Mn on area fraction of $Al_{15}(FeMnMg)_3Si_2$ phases.

We discovered that the effect of iron changes with the load (Fig.6). Iron has a negative effect at higher loads. The test bars resisted cyclic loads of fewer than 10^5 cycles, independent of their iron content. This is because the fatigue crack's initiation and spread are accelerated by the fracture of brittle Al₅FeSi particles. In contrast, at lower loads, the higher iron concentration has a beneficial effect on fatigue resistance. This is likely caused by the Al₅FeSi plate-like phases in the matrix of alloy C, which have a different orientation, are much longer than in alloy A, and slow down fatigue fracture propagation. It would be necessary to do more experiments with more test rods to determine the impact of iron more precisely.

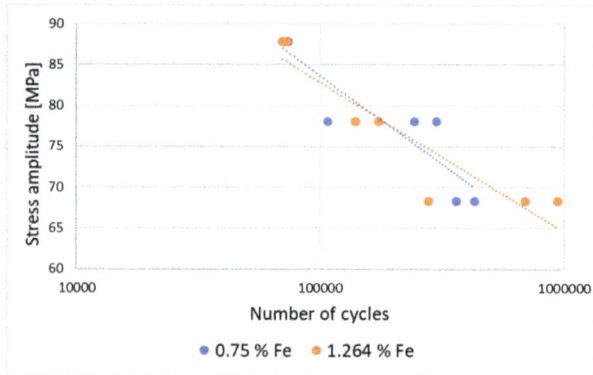

Fig.6. *Effect of Fe on the fatigue resistance of AlSi7Mg0.6 alloy.*

Manganese has a beneficial impact on the fatigue resistance of the AlSi7Mg0.6 alloy, as seen in Fig.7. and Fig.8. The fatigue curves shifted to the right as a result. This demonstrates that alloys with Mn can resist more cycles with the same amplitude of loading. Manganese was found to have a stronger impact on alloys with less iron. The beneficial effect of manganese was especially evident at greater loads at high iron contents. Samples withstand a higher number of cycles (Fig.7). In alloys with higher iron content (Fig.8), the effect of manganese was negative at a load of 68 MPa, which is probably related to an increase in porosity (Fig.3). The samples withstand a significantly lower number of cycles.

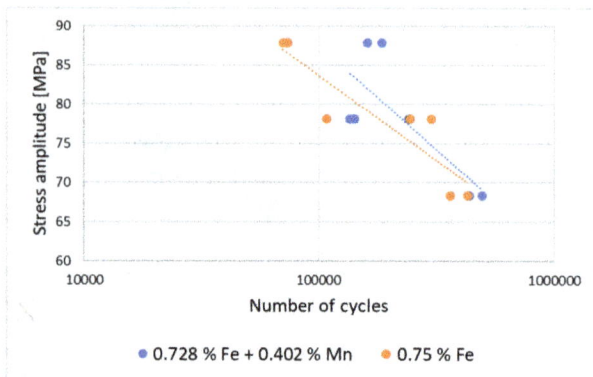

Fig.7. *Effect of Mn on fatigue resistance of AlSi7Mg0.6 alloy.*

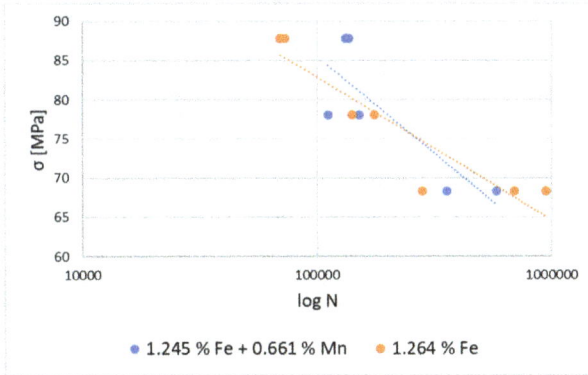

Fig.8. *Effect of Mn on the fatigue resistance of AlSi7Mg0.6 alloy.*

Summary

The following conclusions can be drawn from the experimental results:

- Pore and shrinkage cavities are more likely to occur when manganese levels are higher. With higher iron content and manganese doping, the casting defects were of the largest size.
- The plate/needle-like phases of Al_5FeSi are more widespread in alloys without significant manganese content. With an increase in Fe content, these phases become more common.
- Manganese reduces the number and dimensions (especially the length) of Al_5FeSi needles and contributes to the formation of skeleton-shaped $Al_{15}(FeMnMg)_3Si_2$ phases or Chinese scripts. The amount of $Al_{15}(FeMnMg)_3Si_2$ phase slightly increases as the percentage of Fe rises.
- Higher manganese content has a positive effect on the fatigue properties of the experimental material, which was manifested by a shift of the fatigue curves to the right. It means that Mn doped alloys resist a higher number of cycles at the same loading amplitude. Manganese was found to have a stronger impact on alloys with less iron. In higher iron content alloys, the effect of manganese was negative at lower loads. The effect of iron on fatigue resistance depends on the magnitude of the load. It would be necessary to complete the measurements with more test bars in order to assess the impact of Fe and Mn on the fatigue parameters of the AlSi7Mg0.6 alloy more precisely.

Acknowledgements

The research was supported by Grand Agency of Ministry of Education of Slovak Republic and Slovak Academy of Sciences, project KEGA 004ŽU-4/2023, project to support young researchers at UNIZA, ID project 12715 (Kucharíková) and project 313011ASY4 "Strategic implementation of additive technologies to strengthen the intervention capacities of emergencies caused by the COVID-19 pandemic".

References

[1] Y-Ch. Tzeng et al. Effect of Trace Be and Sc Additions on the Mechanical Properties of A357 Alloys, Metals 8 (2018) art.194. https://doi.org/10.3390/met8030194

[2] H. Nunes et al. Adding Value to Secondary Aluminium Casting Alloys: A Review on Trends and Achievements, Materials 16 (2023) art.895. https://doi.org/10.3390/ma16030895

[3] I. Švecová et al. Improving the quality of Al-Si castings by using ceramic filters, Prod. Eng. Arch. 26 (2020) 19-24. https://doi.org/10.30657/pea.2020.26.05

[4] J. Kasińska et al. The Influence of Remelting on the Properties of AlSi9Cu3 Alloy with Higher Iron Content, *Materials* 13 (2020) art.575. https://doi.org/10.3390/ma13030575

[5] B. Chen et al. The Effect of Cu Addition on the Precipitation Sequence in the Al-Si-Mg-Cr Alloy, Materials 15 (2022) art.8221. https://doi.org/10.3390/ma15228221

[6] L. Pastierovičová et al. Quality of automotive sand casting with different wall thickness from progressive secondary alloy, Prod. Eng. Arch. 28 (2022) 172-177. https://doi.org/10.30657/pea.2022.28.20

[7] A. Bakedano et al. Comparative Study of the Metallurgical Quality of Primary and Secondary AlSi10MnMg Aluminium Alloys, Metals 11 (2021) art.1147. https://doi.org/10.3390/met11071147

[8] M. Gnatko et al. Purification of Aluminium Cast Alloy Melts through Precipitation of Fe-Containing Intermetallic Compounds, Metals 8 (2018) art.796. https://doi.org/10.3390/met8100796

[9] L. Ceschini et al. Influence of Sludge Particles on the Fatigue Behaviour of Al-Si-Cu Secondary Aluminium Casting Alloys, Metals 8 (2018) art.268. https://doi.org/10.3390/met8040268

[10] L. Stanček et al. Structure and properties of silumin castings solidified under pressure after heat treatment, Met. Sci. Heat Treat. 56 (2014) 197-202. https://doi.org/10.1007/s11041-014-9730-0

[11] B. Vanko, L. Stanček. Utilization of heat treatment aimed to spheroidization of eutectic silicon for silumin castings produced by squeeze casting, Arch. Foundry Eng. 12 (2012) 111-114. https://doi.org/10.2478/v10266-012-0021-1

[12] M. Brochu et al. High cycle fatigue strength of permanent mold and rheocast aluminium 357 alloy. In: Int. J. Fatigue 32 (2010) 1233-1242. https://doi.org/10.1016/j.ijfatigue.2010.01.001

[13] D. Závodská et al. The Effect of Iron Content on Microstructure and Porosity of Secondary AlSi7Mg0.3 Cast Alloy, Period. Polytech. Transp. Eng. 47 (2018) 283-289. https://doi.org/10.3311/PPtr.12101

[14] M. Oberreiter et al. A Probabilistic Fatigue Strength Assessment in AlSi-Cast Material by a Layer-Based Approach, Metal 12 (2022) art.784. https://doi.org/10.3390/met12050784

[15] L. Kucharíková et al. High-cycles Fatigue of Different Casted Secondary Aluminium Alloy, Manuf. Technol. 17 (2017) 756-761. https://doi.org/10.21062/ujep/x.2017/a/1213-2489/MT/17/5/756

Quality Production Improvement and System Safety: QPI 16 - CZOTO 10   Materials Research Forum LLC
Materials Research Proceedings 34 (2023) 43-52   https://doi.org/10.21741/9781644902691-6

Microstructural Analysis of Wound Composites with Considerations on the Fiber Winding Force

KRYSIAK Piotr

Military Institute of Engineer Technology, Obornicka 136 Street, 50-961 Wrocław, Poland

krysiak@witi.wroc.pl

Keywords: Resin, Carbon Fiber, Glass Fiber, Composite, Microstructural Analysis

Abstract. The paper concerns issues related to the fabrication and microstructural analysis of carbon and glass fiber-reinforced epoxy composites. For the purposes of the tests, ring samples of carbon fiber and glass fiber were fabricated, each sample being made of the same amount of material. The rings were produced by circumferentially wrapping fibers onto a rigid core. During winding, the fiber tension was changed for each sample to investigate whether the amount and distribution of fibers in the composite depending on the force applied. The applied tension forces ranged from 18 to 138 N. In the next stage of tests, small fragments were cut out of the rings, cutting perpendicularly to the arrangement of fiber bundles. The cut elements were then embedded in epoxy resin. The microsections prepared in this were ground using a grinding disc with a gradation of 80 to 4000 and then polished, washed and dried. After these treatments, microscopic photos of the tested surfaces were taken. Observations were made using a NEOPHOT 32 microscope with an integrated camera and a KEYENCE VHX microscope. The volume fraction of voids and discontinuities in the matrix, as well as the volume fraction of fibers, were determined by measuring the percentage of the surface areas occupied by the appropriate components of the composite. On each cross-section of individual samples, photographs were taken in three planes parallel to the composite layers and three planes perpendicular to the fibers – in the matrix system. The image analysis method was used for verification, in which it is assumed that the assessment of fiber distribution in a two-dimensional section is representative of its volumetric distribution. This method is mainly used to analyze the distribution of fibers with a constant cross-section. The most important conclusions from the conducted analyses allow the authors to state that in both composites the fibers in the matrix are highly packed and that there was no significant and noticeable effect of fiber tension during winding on the "packing" density of fibers in the composites.

Introduction

A significant group of products made of reinforced plastics are pipes and high-pressure tanks, made by winding. They effectively replace products made of traditional construction materials in many technical fields. An essential factor for the increase in the use of these materials in wound products is the knowledge of the impact of manufacturing technology on the performance properties, as these materials are characterized by high anisotropy of mechanical properties. This anisotropy is created in the manufacturing process and can be freely adjusted depending on the intended use of the structure and the type of loads. In composite fibrous structures, the main load is carried by the fibers, while the binder (resin) binds the composite and ensures the fixation/solidification of the given shape of the product [1, 2].

Unidirectionally reinforced composites are the basic composites and, at the same time, the easiest ones to describe. As a result of the reinforcement, a macroscopically isotropic composite is formed in a plane perpendicular to the direction of the reinforcement, i.e. a composite with transversely isotropic symmetry. Since these groups of composites have a wide practical

Quality Production Improvement and System Safety: QPI 16 - CZOTO 10   Materials Research Forum LLC
Materials Research Proceedings 34 (2023) 43-52   https://doi.org/10.21741/9781644902691-6

application, the problem of predicting the elastic properties of a unidirectionally reinforced composite based on the properties of its components becomes vital. There are many models describing these kinds of materials. Typically, the authors of these models adopted the linear-elastic properties of the fibers and matrix, and the hypothesis of flat cross-sections is commonly accepted in the strength of materials [4, 6, 7, 8].

Material Preparation for the Tests
In order to determine the actual mechanical properties of the materials (composites) used for the tests, samples with ring geometry were produced by the winding method, each sample being made of the same amount of fibers. During the winding process, the roving tension was changed for each sample.

In the presented studies, to produce test samples, "continuous" fibers in the form of roving, wound on a suitable core, were used. The exact properties of glass fibers are given in Table 1 [9], and carbon fibers in Table 2 [10].

Table 1. Physical and mechanical properties of ER 3005 glass fiber (Krosglass)

Property	Value
Longitudinal modulus of elasticity [GPa]	73
Tensile strength [MPa]	3400
Poisson's ratio [–]	0.21
Elongation at break [%]	3.5
Density [g/cm3]	2.55
Linear density [tex]	1200±7%
Monofilament diameter [μm]	10÷15

Table 2. Properties of UTS 5631 12K carbon fiber (TohoTenax)

Property	Value
Longitudinal modulus of elasticity [GPa]	240
Tensile strength [MPa]	4800
Poisson's ratio [–]	0.285
Elongation at break [%]	$-0.1 \cdot 10^{-6}$
Density [g/cm3]	1.8
Linear density [tex]	1.79
Monofilament diameter [μm]	800
Number of monofilaments	12000
Monofilament diameter [μm]	6.9
Rowing width [mm]	2.99

Epolam 5015 epoxy resin and AXSON Epolam 2016 hardener were used in the composite matrix. The properties of the cured resin are given in Table 3 [11].

Using this method, three rings were made of carbon fiber (epoxy/carbon) and glass fiber (epoxy-glass) with the tensile strength of the fibers of 18 N; 78N; 138 N, (Fig. 1).

Table 3. *Properties of Epolam 5015 (Axson) resin*

Property	Value
Flexural modulus [GPa]	2.9
Tensile strength [MPa]	73
Elongation at break [%]	7
Poisson ratio [–]	0.35
Hardness [Shore D15]	84
Mixing ratio [by weight]	32
Density at 25°C [g/cm3]	1.12÷1.16
Pot life (on 500 g) at 25°C [min.]	360÷450
Glass transition temperature [°C]	81
Brookfield viscosity at 25°C [mPa·s]	400÷500

a)   b)

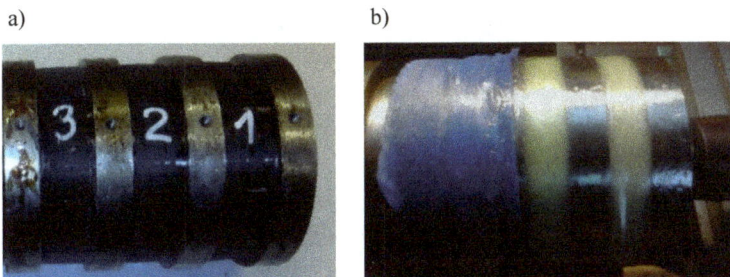

Fig.1. *Samples during the winding process; a –carbon samples, b – glass samples*

Fig.2. *The samples prepared for mechanical processing*

In the next stage, small fragments were cut out of the rings; it was done by cutting perpendicularly to the arrangement of fiber bundles. The cut elements were then embedded in epoxy resin. The microsections prepared in this way were ground on discs with a gradation of 80 to 4000 and then polished, washed and dried. Fig. 2 shows the preparation of sample sections for grinding.

Afterwards, the microscopic structure of the produced composites was examined to determine the volume fraction of fibers in the matrix. This is a key parameter affecting the material constants because the strength of the composite is mainly determined by the fibers it contains.

45

Quality Production Improvement and System Safety: QPI 16 - CZOTO 10   Materials Research Forum LLC
Materials Research Proceedings 34 (2023) 43-52   https://doi.org/10.21741/9781644902691-6

The observations were made with a NEOPHOT 32 microscope with an integrated camera and a KEYENCE VHX microscope.

The obtained microscopic images of the structures were analyzed in particular in terms of two factors:
– the number and distribution of defects in the form of discontinuities in the matrix filling;
– surface distribution and volume content of fibers in the composite.

Determination of the volume of fibers, matrix and voids in the analyzed structures was performed by image analysis.

The basic assumption of this method is that the assessment of fiber distribution in a two-dimensional section is representative of the volumetric distribution. This method is mainly used to analyze the distribution of fibers characterized by constant cross-section, as for the structures analyzed in this work. The method is described in more detail in [3].

The purpose of the method is to distinguish the color border between the fibers and the matrix, so in the first stage, the image should be reduced to a grey scale. The threshold value of individual colors can be determined based on histogram analysis (fig. 3).

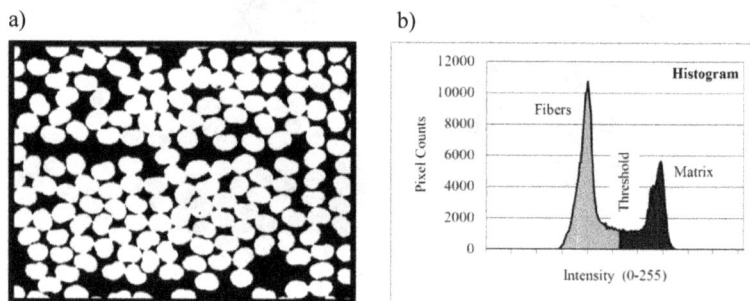

Fig.3. *Elements of image analysis; a – image of the cross-section of the sample in greyscale, b – typical histogram of fiber and resin distribution in greyscale*

The histogram shows the proportion of white and black colors (fiber and resin respectively, or vice versa). The computer then counts the number of pixels for the corresponding colors, and the ratio of these values to the total number of pixels determines the percentage of each component in the composite.

Microscopic Analysis of the Composites

The prepared surfaces of the samples were analyzed under a microscope at the following magnifications:
– ×50, to determine the distribution of roving strips and the volume fraction of voids in the composite;
– ×200, to illustrate the size of the structure discontinuity;
– ×500 or 1000, to determine the distribution of fibers in the matrix and their percentage share in the composite.

The volume fraction of voids and discontinuities in the matrix as well as the volume fraction of fibers were determined by measuring the percentage content of the surface areas occupied by the appropriate components of the composite. On each cross-section of individual samples, photographs were taken in three planes parallel to the layers and in three perpendicular ones – in the matrix system (Fig. 5).

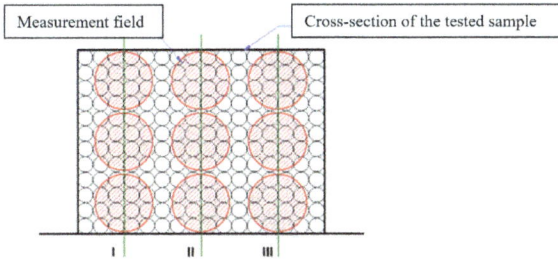

Fig. 4. Distribution of the tested fields on the cross-section of the samples

Analysis of the microstructure of the epoxy/carbon fiber-reinforced composite. In the first stage, the share of voids and structure discontinuities was analyzed at ×50 magnification (Fig. 5). After determining the content of voids, the content of the remaining components of the composite, i.e. fibers and resin, was analyzed at ×200 and ×500 magnification (Fig. 6 and 7, Table 4). In addition, photos of the topography of the sample surface were taken (example photograph: Fig. 8).

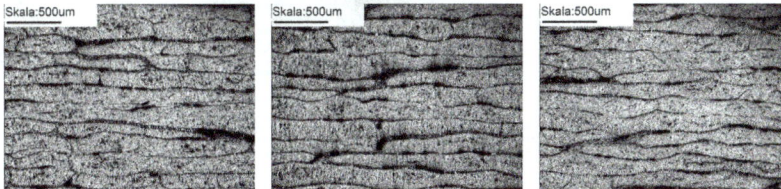

Fig. 5. Photographs of the carbon fiber composite microstructure
at ×50 magnification for samples 1÷3

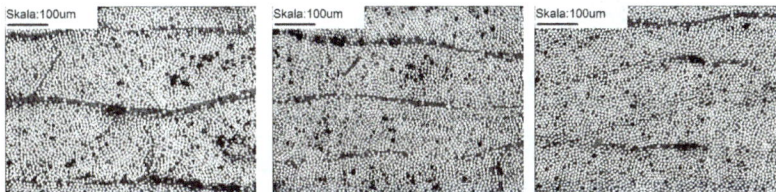

Fig. 6. Photographs of the carbon fiber composite microstructure
at ×200 magnification for samples 1÷3

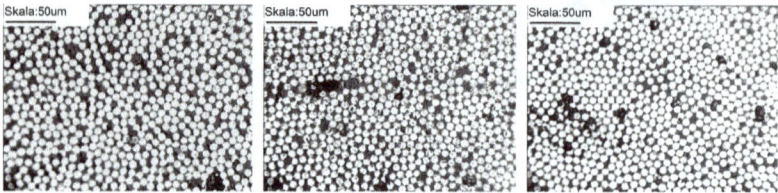

Fig. 7. *Photographs of the carbon fiber composite microstructure*
at ×500 magnification for samples 1÷3

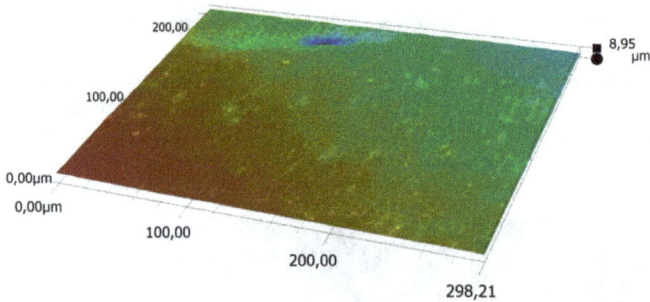

Fig. 8. *Topography of the epoxy/carbon fiber composite*

Table 4. *Tabulation of composite macrostructure surface measurement results.*

	EW composite	
	Volume fraction of fibers V_w [%]	**Volume fraction of voids V_p [%]**
Average for sample 1	70.3	10.0
Average for sample 2	64.4	8.5
Average for sample 3	65.3	6.0
Average for all the samples	66.7	8.2
Standard deviation	3.5	2.0
Volume fraction of resin $(100 - V_w - V_p)$	25.2	

Analysis of the microstructure of the epoxy/glass fiber-reinforced composite. Afterwards, tests were carried out for rings made of epoxy/glass fiber composite. In the first stage, the volume fraction of voids and discontinuities of the structure was analyzed at ×50 magnification (Fig. 9). After determining the void content, the content of the remaining components of the composite, i.e. fibers and resin, was studied at ×500 and ×1000 magnification (Fig. 10 and 11, Table 1). In addition, photographs of the topography of the sample surface were taken (example photograph - Fig. 12).

Quality Production Improvement and System Safety: QPI 16 - CZOTO 10   Materials Research Forum LLC
Materials Research Proceedings 34 (2023) 43-52   https://doi.org/10.21741/9781644902691-6

Fig. 9. *Photographs of the glass fiber composite microstructure*
at ×50 magnification for samples 1÷3

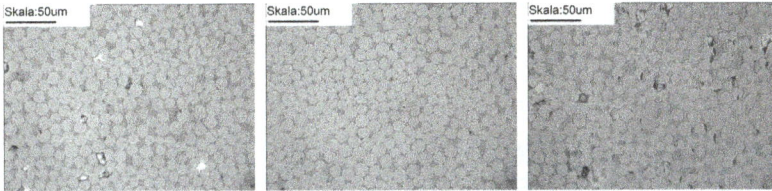

Fig. 10. *Photographs of the glass fiber composite microstructure*
at ×500 magnification for samples 1÷3

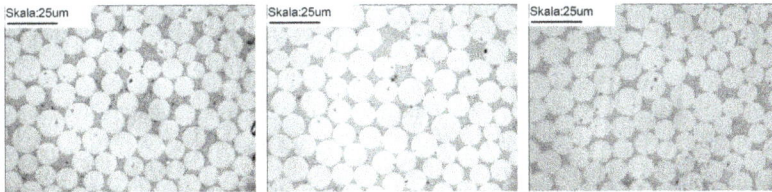

Fig. 11. *Photographs of the glass fiber composite microstructure*
at ×50 magnification for samples 1÷3

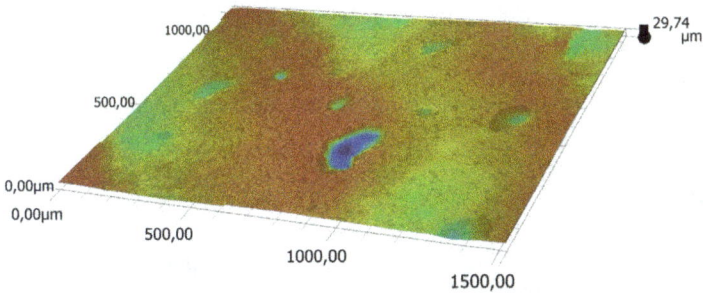

Fig. 12. *Topography of the epoxy/glass fiber-reinforced composite*

Table 5. *Tabulation of composite macrostructure surface measurement results.*

	EW composite	
	Volume fraction of fibers V_w [%]	Volume fraction of fibers V_w [%]
Average for sample 1	73.7	3.9
Average for sample 2	74.5	1.8
Average for sample 3	73.7	5.4
Average for all the samples	73.9	3.7
Standard deviation	2.1	2.5
Volume fraction of resin $(100 - V_w - V_p)$	22.4	

Conclusions

1. The structure of the distribution of roving bands in epoxy/carbon fiber composites (visible at ×50 and ×200 magnification) indicates that there is an undissolved protective layer on the outer surfaces of the roving bands, which is the reason for the lack of wettability, which in turn causes "discontinuity" of the composite. In the case of fiber composites, in which the preparation was used to cover the fibers, the structure is more homogeneous, apart from discontinuities in the form of voids;
2. Based on the statistical analysis (Tables 4 and 5), no significant and noticeable effect of fiber tension during winding on the density of "packing" fibers in composites, both epoxy/carbon fiber and epoxy/glass fiber reinforced, was found;
3. In the epoxy/carbon fiber composites, a greater "packing" was observed with increasing fiber tension, i.e. a smaller thickness of the discontinuity between the roving bands;
4. In both composites, the packing of fibers in the matrix is high; the values of 66.7% and 73.9% are considered almost borderline values (the highest tensile strength for glass fiber composites oscillates at about 72% of the fiber volume fraction in the composite [5]);
5. High content of fibers in the matrix may result in a lack of good wetting of the fibers with the resin, which may lead to reduced adhesion and delamination adhesion cracks, and as a result to "defibration" of the structure and utter destruction of the composite after the permissible load had been exceeded.

The conducted analysis focused on composites [12], although the essence of the issue falls within the scope of stereology [13-15]. The applied methods and analysis approaches can find broader application in similar cases of X-ray image observation [16], statistical analysis of material-related issues [17, 18], and process-related issues [19, 20]. Due to the projectional nature of the analyses, where 3D objects are projected onto a 2D space, nonparametric analyses [21-23] can be useful. The methodology should also be applicable in the analysis of complex surface layers such as DLC coatings [24].

References

[1] A. Błachut, P. Krysiak. Modelowanie w mikro- i makroskali nawijanej rury kompozytowej, Modelowanie Inżynierskie 35 (2018) 5-11.

[2] A. Błachut, P. Krysiak. Analysis of mechanical properties of epoxy resins used in composite high-pressure tank. Int. J. Eng. Sci. 4 (2016) 50-55.

[3] Composite Materials Handbook. Polymer matrix composites guidelines for characterization of structural materials, vol. 1, Department of Defense, Washington DC, 2002.

[4] H. Dąbrowski. Wytrzymałość polimerowych kompozytów włóknistych. Oficyna Wydawnicza Politechniki Wrocławskiej, Wrocław, 2002.

[5] W. Królikowski. Polimerowe kompozyty konstrukcyjne. PWN, Warszawa, 2012.

[6] P. Krysiak, A. Błachut, J. Kaleta. Theoretical and Experimental Analysis of Inter-Layer Stresses in Filament-Wound Cylindrical Composite Structures. Materials 14 (2021) art. 7037. https://doi.org/10.3390/ma14227037

[7] P. Krysiak, J. Kaleta, P. Gąsior, A. Błachut, R. Rybczyński. Identification of strains in a multilayer composite pipe. J. Sci. Mil. Acad. Land Forces 49 (2017) 272-282. https://doi.org/10.5604/01.3001.0010.7233

[8] P. Krysiak, R. Owczarek, W. Błażejewski, A. Błachut. Strength Testing and Ring Stiffness Testing of Underground Composite Pressure Pipes". Mater. Res. Proc. 17 (2020) 191-202. https://doi.org/10.21741/9781644901038-29

[9] Krosglass S.A. website. [online]. 2023. [viewed: 2023-01-31]. Available from: https://www.krosglass.pl

[10] Toho Tenax Europe website. [online]. 2023. [viewed: 2023-01-31]. Available from: https://www.chemeurope.com/

[11] Fatol Kunststoffen website. [online]. 2023. [viewed: 2023-01-31]. Available from: https://www.fatol.nl

[12] J. Korzekwa et al. Tribological behaviour of Al2O3/inorganic fullerene-like WS2 composite layer sliding against plastic, Int. J. Surf. Sci. Eng. 10 (2016) 570-584. https://doi.org/10.1504/IJSURFSE.2016.081035

[13] A. Gadek-Moszczak, P. Matusiewicz. Polish stereology – A historical review, Image Analysis and Stereology 36 (2017) 207-221. https://doi.org/10.5566/ias.1808

[14] J. Pietraszek, A. Szczotok, N. Radek. The fixed-effects analysis of the relation between SDAS and carbides for the airfoil blade traces. Arch. Metall. Mater. 62 (2017) 235-239. https://doi.org/10.1515/amm-2017-0035

[15] N. Radek et al. The impact of laser welding parameters on the mechanical properties of the weld, AIP Conf. Proc. 2017 (2018) art.20025. https://doi.org/10.1063/1.5056288

[16] B. Jasiewicz et al. Inter-observer and intra-observer reliability in the radiographic measurements of paediatric forefoot alignment, Foot Ankle Surg. 27 (2021) 371-376. https://doi.org/10.1016/j.fas.2020.04.015

[17] J. Pietraszek et al. The parametric RSM model with higher order terms for the meat tumbler machine process, Solid State Phenom. 235 (2015) 37-44. https://doi.org/10.4028/www.scientific.net/SSP.235.37

[18] J. Pietraszek et al. Challenges for the DOE methodology related to the introduction of Industry 4.0. Prod. Eng. Arch. 26 (2020) 190-194. https://doi.org/10.30657/pea.2020.26.33

[19] J. Pietraszek et al. Factorial approach to assessment of GPU computational efficiency in surrogate models, Adv. Mater. Res. 874 (2014) 157-162. https://doi.org/10.4028/www.scientific.net/AMR.874.157

Quality Production Improvement and System Safety: QPI 16 - CZOTO 10   Materials Research Forum LLC
Materials Research Proceedings 34 (2023) 43-52   https://doi.org/10.21741/9781644902691-6

[20] R. Dwornicka, J. Pietraszek. The outline of the expert system for the design of experiment, Prod. Eng. Arch. 20 (2018) 43-48. https://doi.org/10.30657/pea.2018.20.09

[21] J. Pietraszek. The modified sequential-binary approach for fuzzy operations on correlated assessments, LNAI 7894 (2013) 353-364. https://doi.org/10.1007/978-3-642-38658-9_32

[22] J. Pietraszek et al. Non-parametric assessment of the uncertainty in the analysis of the airfoil blade traces, METAL 2017 – 26th Int. Conf. Metall. Mater. (2017) 1412-1418. ISBN 978-8087294796

[23] J. Pietraszek et al. The non-parametric approach to the quantification of the uncertainty in the design of experiments modelling, UNCECOMP 2017 Proc. 2nd Int. Conf. Uncert. Quant. Comput. Sci. Eng. (2017) 598-604. https://doi.org/10.7712/120217.5395.17225

[24] N. Radek et al. Microstructure and tribological properties of DLC coatings, Mater. Res. Proc. 17 (2020) 171-176. https://doi.org/10.21741/9781644901038-26

Quality Production Improvement and System Safety: QPI 16 - CZOTO 10
Materials Research Proceedings 34 (2023) 53-61

Materials Research Forum LLC
https://doi.org/10.21741/9781644902691-7

Safety of Mountaineering Systems in the Industrial Area

WOŁOWCZYK Jacek [1,a] * and KLIMECKA-TATAR Dorota [1,b]

[1] Wydział Zarządzania, Politechnika Częstochowska, Al. Armii Krajowej 19B· 42-200 Częstochowa, Polska

[a] jacek.wolowczyk@pcz.pl, [b] d.klimecka-tatar@pcz.pl

Keywords: Mountaineering Systems, Industry, Work at Height, Work Safety, Industrial Safety

Abstract. This paper is devoted to the safety of mountaineering systems, which are designed to prevent falls from a height. The purpose of the paper is to investigate the safety of fall arrest systems and their effectiveness. In addition, the study is to answer the question of whether mountaineering systems are used in the industrial area. This area requires the use of different mountaineering systems than in the case of sport mountaineering. As parks at heights are included in the list of particularly dangerous works, an analysis of systems protecting employees against falling from a height was undertaken. For this purpose, surveys were conducted to determine their correct operation and the degree of employee protection. Employees working in the profession of industrial mountaineer were defined as the research group. As a result of the conducted research, it is determined that employees use individual and collective protection measures. Respondents declared that the means they use work correctly and determine high effectiveness of protection against falls from a height. The systems themselves are built in the correct way and the respondents did not clearly indicate one dominant disadvantage of the mountaineering systems they use.

Introduction

Ensuring work safety at heights relies on specialized employee protection measures. Occasionally, the creation of specific procedures for working at heights is necessary to minimize accidents and near-accidents. The profession of an industrial mountain climber and relevant legal regulations in European countries were examined to establish employer obligations towards employees. Subsequently, all protective measures designed to safeguard the well-being and safety of individuals are outlined [1]. Mountaineering involves working at heights using suspension or support systems, distinguishing it from recreational mountaineering where the primary objective is reaching a peak or specific height with a single rope. Industrial mountaineering, in addition to ascending heights, involves undertaking specific work tasks. Industrial climbing can be conducted on platforms or suspended shelves using climbing ropes. However, the most common approach involves utilizing a safety harness connected to an anchor point via a rope [1-3]. Industrial climbers perform a range of activities, including facility inspections, window cleaning, facade maintenance and washing, repair and upkeep of high-voltage and telecommunications lines, installation and removal of signage and billboards, antenna and transmitter maintenance and repair, roof snow removal, renovation and construction work, and more. The profession of an industrial mountain climber demands employees to possess high physical fitness and climbing skills. It entails significant responsibility for assigned tasks and necessitates conscientiousness and meticulous adherence to safety measures. It is not suitable for individuals with a fear of heights, as the job entails meeting various mental and physical requirements [2].

Work at height is defined as work more than one meter above ground level. Such regulation has been introduced in many European countries. Carrying out work above this height is classified as particularly dangerous work. This determines special requirements for the employer because he must ensure supervision over the work carried out in this way and the persons designated for it.

Quality Production Improvement and System Safety: QPI 16 - CZOTO 10   Materials Research Forum LLC
Materials Research Proceedings 34 (2023) 53-61   https://doi.org/10.21741/9781644902691-7

Work supervision can be carried out in a variety of ways designated by the employer or the health and safety unit. The forms of supervision include, among others: periodic work inspections, inspections and the appointment of health and safety observers. The main task of health and safety observers is to supervise the work in real time and verify its compliance with occupational safety standards. This person should have a degree in safety science and/or appropriate training. The employer is also obliged to provide protective measures. In this case, protective measures are understood as collective and individual protection measures. In addition, the employer is obliged to conduct employee training. Its purpose is to determine the order in which tasks are to be performed and to specify the requirements for performing individual activities in the areas of occupational health and safety [1-4].

Personal protective equipment against falls from a height is crucial in protecting the human factor and is mainly designed to stop a fall from a height. The choice of protection measures depends on the type of activities performed. It must be selected individually for each type of work carried out [4]. The first type of protection against falling from a height are fall arrest systems. The main idea of this type of solutions is to limit the worker to the fall hazard zone, i.e. the length of the rope determined in such a way that the worker cannot get close to the edge of the building. The elements constituting such a system are: anchoring sub-assembly, connecting component, harness.

The second type of system is used when the worker is unable to stand without additional support. Such a system is necessary when it is necessary to give positioning while working at height. The elements that make up this type of system are: anchor sub-assembly, lanyard for positioning while working, harness. The use of this type of system allows to work in conditions where a given workplace generates a high angle of inclination of the surface. Its main difference from the accident prevention system discussed earlier is that the harnesses must have two attachment points. With this system, the safety line is in constant tension. This does not allow for spontaneous loosening of the rope and the distribution of masses and forces is transferred to the entire system. The third type of mountaineer protection system is used only in cases where it is impossible to eliminate the start of falling from a height. This type of system includes: anchor sub-assembly, connecting and damping sub-assembly, safety harness [5-7]. A very important aspect that should be taken into account when planning work at height and choosing the system needed for a given job is securing the workplace. Before starting work, it is also necessary to identify places of increased risk, which may directly translate into possible accident consequences.

In addition, it should be taken into account to equip the employee with personal protective equipment in accordance with the risks that occur in a given work environment. The most commonly used personal protective equipment when working at heights are:

- head protection – they are designed to protect the employee's head against elements falling from a height,
- lower limbs protection equipment – protect the lower limbs against mechanical injuries and/or rotating elements of machines and devices that are components of a given workplace,
- upper limb protection equipment – they are designed to protect the upper limbs against threats,
- protective clothing – is designed to ensure thermal comfort and protect the employee from weather conditions.

Their use is variable depending on the type of work carried out and the place of performance. In order to make the right choice, a risk assessment should be made taking into account all hazards and variables. Changing working conditions are a key aspect determining the selection of personal protective equipment. After determining and selecting, the employee should be equipped with all necessary means for safe performance of work. This is the employer's obligation in most European countries, defined and specified in the Labor Code. Each employee should have their own personal set of protective equipment, which should be used only by a particular unit. It is a good practice for enterprises to apply work safety procedures and specify in them the use of protective measures in specific conditions and types of work. They are voluntary and are not conditioned by the legislator.

Industrial high-altitude work is a high-risk endeavor. For this reason, the equipment utilized for such tasks must be of superior quality and subject to constant supervision to identify potential flaws and malfunctions [8-10]. The primary destructive factors encompass corrosion [11-13], including biocorrosion [14], as well as wear [15-17]. These factors can be mitigated by employing appropriate materials, such as metals [18-20] and alloys [21], and by executing welded connections with precision [22-24]. Coatings [25-27] and suitable surface layers [28], including those with modified morphological and technological features [29,30], also offer effective protection. The proper design and construction of equipment for high-altitude work should also incorporate energy consumption reduction [31] throughout production and operation, as well as subsequent recycling efforts [32-34]. Accomplishing this necessitates the utilization of appropriate design methodologies [35-37] and the optimization and stabilization of processes [38-40], including nonparametric [41-43] and resampling techniques [44,45].

Experiment and Methods

According to the study, it is necessary to determine the correct operation of mountaineering systems and the degree of employee protection against falling from a height. For this purpose, a survey questionnaire was created. In this paper, the areas of the survey directly related to mountaineering safety systems used in industrial work at heights were analyzed. The analysis is based on two basic research questions (Q1 and Q2):

Q1. To what extent do mountaineering systems protect an employee against falling from a height?
Q2. Do the mountaineering systems used work properly?

The survey was developed using a web form that allowed the respondents to generate a link. After launching it, a survey opens with four questions relating to the metrics and ten survey questions. The research was conducted from March 3, 2022 to March 10, 2022. 35 industrial mountaineers participated in the study. Mountain climbers came from various industries and performed various activities. Therefore, the scope of their duties was variable, which determines the credibility of the conducted research in terms of the use of various types of fall protection systems. The first question of the survey concerned the percentage structure of respondents' gender. The results were as follows: men accounted for 94.3% and women for 5.7%. The low number of women is conditioned by the specificity of the industry.

The second question refers to the age of the respondents. This question has been divided into four possible answers within the following ranges: 18-28 years - this answer was marked by 11 people, which is 31.4%; 11 persons declared themselves to be aged 29-39 in this group, which constitutes 31.4%; 40-50 years old, this range is 7 people – 20.0%; over 50, marked by 6 people (17.1%). The next question refers to the education declared by the respondents. Most people indicated that they had secondary education, about 20%. Secondary (vocational) education was declared by 22.9%, higher education was indicated by 14.3%, primary education by 42.9% The last question concerns the seniority of the respondents in the mountaineering industry, this question is one of the key ones because they define the work experience of individual respondents. The largest number of respondents declared that their seniority is over 20 years and they constitute 28.6%. The group with 5 to 10 years of service constituted the second largest group, accounting for 20% of all answers. The distribution of responses between the individual ranges is shown in Fig.1.

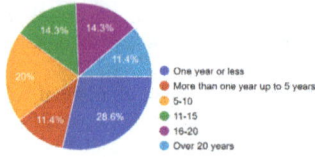

Fig.1. *Percentage structure of seniority in the industrial construction industry.*

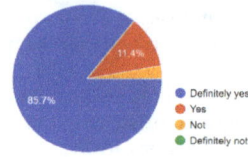

Fig.2. *The structure of the answer to the question about the effectiveness of the use of mountaineering systems.*

The next researched issue is the question of the effectiveness of mountaineering systems. The overwhelming majority of 85.7% marked the answer definitely "yes". This determines that employees consider mountaineering systems to be effective. The answers provided by the respondents confirm that the mountaineering systems are working properly. The graph is shown in Fig.2.

The question regarding the degree of protection of the employee against falling from a height was crucial in the hypothesis – mountaineering systems significantly protect the employee against falling from a height. Respondents confirmed that fall protection systems completely protected them, such an answer was selected by 74.3%. A large proportion of 22.9% believe that these systems protect the employee against falling mostly. None of the respondents marked the answer at all, as shown in Fig.3.

The question about the correct operation of the mountaineering systems showed that 77.1% rated them very well, while 20% answered well. Which means that the employees assess the correctness of operation at a high level – as shown in Fig.4. A high percentage indicates that employees trust the mountaineering protecting systems that protect their health and/or life against falling from a height. This is an important aspect of work because it determines the sense of security, which is considered fundamental in human life. The hierarchy of needs developed by A. Maslov shows it as a key enabling the satisfaction of higher human needs, but it is impossible to achieve them without satisfying lower levels [46].

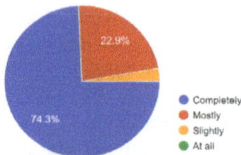

Fig.3. *Structure of answers to the question about the degree of protection of employees against falling from a height.*

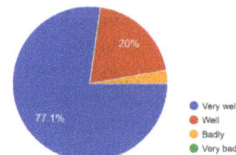

Fig.4. *Percentage structure of answers to the question on the correct operation of mountaineering systems.*

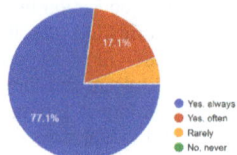

Fig.5. *Percentage structure of answers to the question on the use of the system against falling from a height by the respondents.*

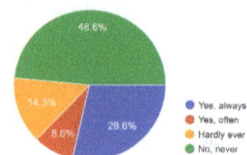

Fig.6. *Percentage structure of responses to the question on admitting an employee to work without protective equipment.*

Quality Production Improvement and System Safety: QPI 16 - CZOTO 10   Materials Research Forum LLC
Materials Research Proceedings 34 (2023) 53-61   https://doi.org/10.21741/9781644902691-7

The question relating to the use of mountaineering protecting systems by workers is a key question relating to the hypothesis – Workers use mountaineering protecting systems correctly. 77.1% answered yes, always. 17.1% answered yes, often. The mere fact that the employer provides protective measures does not determine their use by individual employees. This presupposes that employees use fall protection measures provided by the employer. The percentage structure of answers to this question is presented in Fig.5.

The next issue examined was the employer's admission without fall protection systems. This question showed that it varies depending on the approach to work and the responsibility of the employer. About 48.6% of respondents to the question: Does your employer permit you to work without any of the mountaineering protecting systems from falling down the higher answered "no, never" (Fig.6). Answer "yes", always marked 28.6%, which is a very disturbing signal, because endangering the life and health of employees is widely accepted in some workplaces. In this case, a number of actions should be taken to increase the employer's awareness of the risks of mountaineering. Particular attention should be paid to the aspects of occupational health and safety in these plants by the occupational health and safety unit. This parameter may also indirectly indicate that the cell is operating incorrectly.

The risk in the mountaineering industry, according to the respondents' answers, is at a very high level, as much as 73.5% of the respondents indicated this answer, while 20.6 indicated the answer high. Which means that employees are highly aware of the use of protective measures – Fig.7.

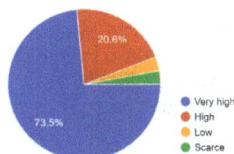

Fig.7. *Percentage structure of responses to the question regarding employees' awareness of the risk of accidents in industrial mountaineering.*

When asked about the weakest point of the whole system, the answers were as follows:

- anchoring subassembly – 14.3%
- safety cables – 77.1%
- hip harness – 11.4%
- connecting and absorbing subassembly – 8.6%
- safety harness – 22.9%

Therefore, it can be concluded that the majority of respondents notice that the weakest point in the entire fall protection system is the safety ropes. These ropes must be replaced periodically. If the equipment is assigned individually to an employee, it must be replaced after 5 years. When equipment is transferred between individual employees, its time interval is defined as one year. Ropes should be visually inspected before each use.

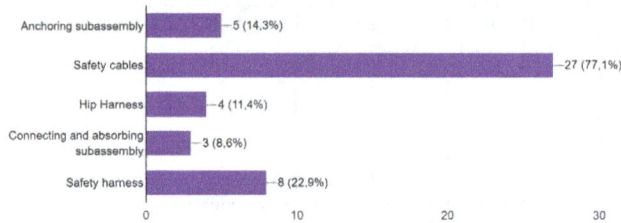

Fig.8. *The structure of the answer regarding the indication of the weakest point of the fall protection system.*

Summary

Worker fall protection systems used in the industrial mountaineering industry work properly. Their use is variable depending on the working conditions and the angle of the ground inclination. Sometimes they have to ensure the correct body position. This requires an individual approach to each work carried out and adaptation to weather conditions - if the work is carried out outside. Employers, on the other hand, are obliged to provide all means of protection, including personal protective equipment. The conducted research showed high awareness of the use of fall protection systems and the systems work properly. The respondents also noticed that these systems protect the employee to a large extent against falling from a height. Falling from a height often has serious health consequences, so remember to periodically check the condition of individual elements of the entire system. Particular attention should be paid to the safety ropes and their technical condition.

References

[1] S. Goodacre et al. Can the distance fallen predict serious injury after a fall from a height? J. Trauma 46 (1999) 1055-1058. https://doi.org/10.1097/00005373-199906000-00014

[2] F.K. Fuss, G. Niegl. Design and Mechanics of Mountaineering Equipment. In: Routledge Handbook of Sports Technology and Engineering. Routledge. 2013, 277-292. https://doi.org/10.4324/9780203851036

[3] M. Goncharova, A. Brinck, N.A. Dmitrienko. Standard legal application to works at high conditions as a method of industrial mountaineering. East European Scientific Journal 14(3) (2016) 37-41.

[4] C.F. Chi et al. Accident patterns and prevention measures for fatal occupational falls in the construction industry. App. Ergon. 36 (2005) 391-400. https://doi.org/10.1016/j.apergo.2004.09.011

[5] M. Ziemelis et al. Collective protection equipment and measures in construction during work at height. In: 19th Int. Sci. Conf. "Economic Science for Rural Development 2018", 402-416. https://doi.org/10.22616/ESRD.2018.047

[6] G.R. Mettam, L.B. Adams. How to prepare an electronic version of your article, In: B.S. Jones, R.Z. Smith (Eds.), Introduction to the Electronic Age, E-Publishing Inc., New York, 1999, 281-304.

[7] G. Pomfret. Mountaineering adventure tourists: a conceptual framework for research. Tour. Manag. 27 (2006) 113-123. https://doi.org/10.1016/j.tourman.2004.08.003

Quality Production Improvement and System Safety: QPI 16 - CZOTO 10   Materials Research Forum LLC
Materials Research Proceedings 34 (2023) 53-61   https://doi.org/10.21741/9781644902691-7

[8] G. Filo, P. Lempa. Analysis of Neural Network Structure for Implementation of the Prescriptive Maintenance Strategy, Mater. Res. Proc. 24 (2022) 273-280. https://doi.org/10.21741/9781644902059-40

[9] J. Fabis-Domagala, M. Domagala. A Concept of Risk Prioritization in FMEA of Fluid Power Components, Energies 15 (2022) art.6180. https://doi.org/10.3390/en15176180

[10] P. Lempa, G. Filo. Analysis of Neural Network Training Algorithms for Implementation of the Prescriptive Maintenance Strategy, Mater. Res. Proc. 24 (2022) 281-287. https://doi.org/10.21741/9781644902059-41

[11] K. Jagielska-Wiaderek, H. Bala, P. Wieczorek, J. Rudnicki, D. Klimecka-Tatar. Corrosion resistance depth profiles of nitrided layers on austenitic stainless steel produced at elevated temperatures, Arch. Metall. Mater. 54 (2009) 115-120.

[12] T. Lipinski, J. Pietraszek. Influence of animal slurry on carbon C35 steel with different microstructure at room temperature, Engineering for Rural Development 21 (2022) 344-350. https://doi.org/10.22616/ERDev.2022.21.TF115

[13] T. Lipiński, J. Pietraszek. Corrosion of the S235JR Carbon Steel after Normalizing and Overheating Annealing in 2.5% Sulphuric Acid at Room Temperature, Mater. Res. Proc. 24 (2022) 102-108.

[14] E. Skrzypczak-Pietraszek, J. Pietraszek. Seasonal changes of flavonoid content in Melittis melissophyllum L. (Lamiaceae), Chem. Biodiv. 11 (2014) 562-570. https://doi.org/10.1002/cbdv.201300148

[15] S. Marković et al. Exploitation characteristics of teeth flanks of gears regenerated by three hard-facing procedures, Materials 14 (20210 art. 4203. https://doi.org/10.3390/ma14154203

[16] M. Krynke et al. Maintenance management of large-size rolling bearings in heavy-duty machinery, Acta Montan. Slovaca 27 (2022) 327-341. https://doi.org/10.46544/AMS.v27i2.04

[17] P. Regulski, K.F. Abramek The application of neural networks for the life-cycle analysis of road and rail rolling stock during the operational phase, Technical Transactions 119 (2022) art. e2022002. https://doi.org/10.37705/TechTrans/e2022002

[18] R. Ulewicz et al. Structure and mechanical properties of fine-grained steels, Period. Polytech. Transp. Eng. 41 (2013) 111-115. https://doi.org/10.3311/PPtr.7110

[19] A. Szczotok et al. Effect of the induction hardening on microstructures of the selected steels. METAL 2018 – 27th Int. Conf. Metall. Mater. (2018), Ostrava, Tanger 1264-1269.

[20] T. Lipinski et al. Influence of oxygen content in medium carbon steel on bending fatigue strength, Engineering for Rural Development 21 (2022) 351-356. https://doi.org/10.22616/ERDev.2022.21.TF116

[21] A. Dudek et al. The effect of alloying method on the structure and properties of sintered stainless steel, Archives of Metallurgy and Materials 62 (2017) 281-287. https://doi.org/10.1515/amm-2017-0042

[22] M. Patek et al. Non-destructive testing of split sleeve welds by the ultrasonic TOFD method, Manuf. Technol. 14 (2014) 403-407. https://doi.org/10.21062/ujep/x.2014/a/1213-2489/MT/14/3/403

Quality Production Improvement and System Safety: QPI 16 - CZOTO 10   Materials Research Forum LLC
Materials Research Proceedings 34 (2023) 53-61   https://doi.org/10.21741/9781644902691-7

[23] N. Radek, J. Pietraszek, A. Goroshko. The impact of laser welding parameters on the mechanical properties of the weld, AIP Conf. Proc. 2017 (2018) art.20025. https://doi.org/10.1063/1.5056288"

[24] N. Radek et al. Properties of Steel Welded with CO2 Laser, Lecture Notes in Mechanical Engineering (2020) 571-580. https://doi.org/10.1007/978-3-030-33146-7_65

[25] N. Radek et al. Technology and application of anti-graffiti coating systems for rolling stock, METAL 2019 28th Int. Conf. Metall. Mater. (2019) 1127-1132. ISBN 978-8087294925

[26] N. Radek, J. Konstanty, J. Pietraszek, Ł.J. Orman, M. Szczepaniak, D. Przestacki. The effect of laser beam processing on the properties of WC-Co coatings deposited on steel. Materials 14 (2021) art. 538. https://doi.org/10.3390/ma14030538

[27] N. Radek et al. Formation of coatings with technologies using concentrated energy stream, Prod. Eng. Arch. 28 (2022) 117-122. https://doi.org/10.30657/pea.2022.28.13

[28] N. Radek et al. The influence of plasma cutting parameters on the geometric structure of cut surfaces, Mater. Res. Proc. 17 (2020) 132-137. https://doi.org/10.21741/9781644901038-20

[29] N. Radek et al. Microstructure and tribological properties of DLC coatings, Mater. Res. Proc. 17 (2020) 171-176. https://doi.org/10.21741/9781644901038-26

[30] N. Radek et al. Influence of laser texturing on tribological properties of DLC coatings, Prod. Eng. Arch. 27 (2021) 119-123. https://doi.org/10.30657/pea.2021.27.15

[31] S. Maleczek et al. Tests of Acid Batteries for Hybrid Energy Storage and Buffering System – A Technical Approach, Energies 15 (2022) art.3514. https://doi.org/10.3390/en15103514

[32] M. Zenkiewicz et al. Electrostatic separation of binary mixtures of some biodegradable polymers and poly(vinyl chloride) or poly(ethylene terephthalate), Polimery/Polymers 61 (2016) 835-843. https://doi.org/10.14314/polimery.2016.835,

[33] M. Zenkiewicz et al. Modeling electrostatic separation of mixtures of poly(ε-caprolactone) with polyfvinyl chloride) or polyfethylene terephthalate), Przemysl Chemiczny 95 (2016) 1687-1692. https://doi.org/10.15199/62.2016.9.6

[34] T. Zuk et al. Modeling of electrostatic separation process for some polymer mixtures, Polymers 61 (2016) 519-527. https://doi.org/10.14314/polimery.2016.519

[35] M. Kekez et al. Modelling of pressure in the injection pipe of a diesel engine by computational intelligence, P. I. Mech. Eng. D.-J. Aut. 225 (2011) art.8766012. https://doi.org/10.1177/0954407011411388

[36] R. Dwornicka et al. The laser textured surfaces of the silicon carbide analyzed with the bootstrapped tribology model, METAL 2017 26th Int. Conf. Metall. Mater. (2017) 1252-1257. ISBN 978-8087294796

[37] L. Cedro Model parameter on-line identification with nonlinear parametrization – manipulator model, Technical Transactions 119 (2022) art. e2022007. https://doi.org/10.37705/TechTrans/e2022007

[38] R. Dwornicka, J. Pietraszek. The outline of the expert system for the design of experiment, Prod. Eng. Arch. 20 (2018) 43-48. https://doi.org/10.30657/pea.2018.20.09

Quality Production Improvement and System Safety: QPI 16 - CZOTO 10   Materials Research Forum LLC
Materials Research Proceedings 34 (2023) 53-61   https://doi.org/10.21741/9781644902691-7

[39] J. Pietraszek et al. Challenges for the DOE methodology related to the introduction of Industry 4.0. Prod. Eng. Arch. 26 (2020) 190-194. https://doi.org/10.30657/pea.2020.26.33

[40] B. Jasiewicz et al. Inter-observer and intra-observer reliability in the radiographic measurements of paediatric forefoot alignment, Foot Ankle Surg. 27 (2021) 371-376. https://doi.org/10.1016/j.fas.2020.04.015

[41] J. Pietraszek. The modified sequential-binary approach for fuzzy operations on correlated assessments, LNAI 7894 (2013) 353-364. https://doi.org/10.1007/978-3-642-38658-9_32

[42] J. Pietraszek et al. Non-parametric assessment of the uncertainty in the analysis of the airfoil blade traces, METAL 2017 26th Int. Conf. Metall. Mater. (2017) 1412-1418. ISBN 978-8087294796

[43] J. Pietraszek et al. The non-parametric approach to the quantification of the uncertainty in the design of experiments modelling, UNCECOMP 2017 Proc. 2nd Int. Conf. Uncert. Quant. Comput. Sci. Eng. (2017) 598-604. https://doi.org/10.7712/120217.5395.17225

[44] J. Pietraszek, L. Wojnar. The bootstrap approach to the statistical significance of parameters in RSM model, ECCOMAS Congress 2016 Proc. 7th Europ. Cong. Comput. Methods in Appl. Sci. Eng. 1 (2016) 2003-2009. https://doi.org/10.7712/100016.1937.9138

[45] J. Pietraszek et al. The bootstrap approach to the statistical significance of parameters in the fixed effects model. ECCOMAS Congress 2016 – Proc. 7th Europ. Congr. Comput. Methods Appl. Sci. Eng. 3, 6061-6068. https://doi.org/10.7712/100016.2240.9206

[46] L. Maslov. Understanding personalities of online students: importance for successful teaching. In: ATINER's Conf. Paper Proc. Series EDU2020-0203, 1-9. Athens Inst. Edu. Res. (2020). [online] Viewed: 31-01-2023. Available from: https://www.atiner.gr/presentations/EDU2020-0203.pdf

Quality Production Improvement and System Safety: QPI 16 - CZOTO 10   Materials Research Forum LLC
Materials Research Proceedings 34 (2023) 62-67   https://doi.org/10.21741/9781644902691-8

The Influence of Hydrogen Embrittlement on Mechanical Properties of Advanced High-Strength Structural Steel S960MC

DRÍMALOVÁ Petra [1,a*], NOVÝ František [1,b], MEDVECKÁ Denisa [1,c] and VÁŇOVÁ Petra [2,d]

[1]University of Žilina, Faculty of Mechanical Engineering, Department of Materials Engineering Univerzitná 8215/1, 010 26 Žilina, Slovak Republic

[2]VŠB Technical University Ostrava, Faculty of Materials and Technologies, Department of Materials Engineering and Recycling, 17. listopadu 2172/15, 708 00 Ostrava – Poruba, Czech Republic

[a]petra.drimalova@fstroj.uniza.sk, [b]frantisek.novy@fstroj.uniza.sk, [c]denisa.medvecka@fstroj.uniza.sk, [d]petra.vanova@vsb.cz

Keywords: Hydrogen Embrittlement, High-Strength Steels, Safety, S960MC, Material Degradation

Abstract. A lot of emphasis is currently being paid to research into how hydrogen affects the mechanical properties of advanced high-strength (AHS) steels. The use of AHS steels in the chemical industry and nuclear technology is heavily influenced by the mechanical properties that result from the impact of hydrogen. Hydrogen, an interstitial element, alters fracture behavior and causes the material to fail earlier than it should. The greatest danger occurs after hydrogen has been absorbed when it diffuses and accumulates into defects such grain boundaries, dislocations, and phase boundaries. The crystal structure of the steel enlarges due to tensile stresses, which increases hydrogen diffusion. Cold forming creates a multitude of components, and even a small amount of hydrogen absorption could result in significant residual stresses in the material. The steel can experience a significant reduction in mechanical properties, strength, and ductility up to fracture at a critical hydrogen concentration level. It is impossible to preventively remove components that have been harmed by hydrogen before they are utilized, which is the primary issue with hydrogen embrittlement from a safety perspective.

Introduction

Hydrogen embrittlement is a phenomenon that occurs in high strength structural steels, when atomic hydrogen is introduced into the material. Numerous procedures, such as electroplating, welding, and hydrogen gas exposure, might cause this. The hydrogen atoms can diffuse into the steel and become trapped at defects in the microstructure, such as dislocations and grain boundaries. After becoming trapped, the hydrogen atoms can make the steel brittle and susceptible to cracking. This can lead to catastrophic failure of the structure and is a major concern in the design of construction of high strength steels [1,2]. The susceptibility to hydrogen embrittlement can be influenced by various factors such as the chemical composition, microstructure, and mechanical properties of the steel, as well as the environmental conditions to which the steel is exposed. Therefore, the material selection, manufacturing processes, and post-treatment procedures such as hydrogen bake-out and stress relief are critical in reducing the risk of hydrogen embrittlement [3,4].

Despite the efforts to reduce the effects of hydrogen embrittlement, it remains a complex and not fully understood phenomenon. There are several theories that attempt to explain the mechanisms of hydrogen embrittlement, such as the decohesion theory, the hydrogen-induced

Quality Production Improvement and System Safety: QPI 16 - CZOTO 10   Materials Research Forum LLC
Materials Research Proceedings 34 (2023) 62-67   https://doi.org/10.21741/9781644902691-8

dislocation emission theory and the hydrogen-induced cracking theory, but none of them provide a complete explanation [5-7].

Furthermore, the susceptibility to hydrogen embrittlement can vary depending on the type of steel and the specific application. For example, high strength steels like S960MC are more susceptible to hydrogen embrittlement than lower strength steels due to their high strength-to-weight ratio and high hardness.

Experimental Material

The material used for this research was thermomechanically rolled advanced high-strength steel S960MC in the form of sheet with a thickness of 3 mm. The chemical composition and mechanical properties are presented in tables Table 1 and 2.

The microstructure of steel was studied in polished and etched condition using Olympus IX70 inversion metallographic light microscope (LM). The investigated material was microalloyed steel with a low carbon content and an increased chromium contend. Grain refinement was also caused by thermomechanical treatment, which resulted in tensile strength values up to 1150 MPa. As a result, a fine martensitic microstructure was created (Fig. 1 and 2).

Table 1. Chemical composition of advanced high-strength steel S960MC [wt. %]

C	Si	Mn	P	S	Al	Nb	V	Ti	Mo	B
0.085	0.205	1.09	0.011	0.009	0.028	<0.004	0.014	0.018	0.122	0.002
Cu	Cr	Ni	N							
0.019	1.17	0.055	0.014							

Table 2. Mechanical properties of advanced high-strength steel S960MC

Yield Strength [MPa]	Tensile Strength [MPa]	Ductility [%]
1014	1162	4

Fig. 1. *Microstructure of S960MC steel (LM, 100x, Nital etch.).*

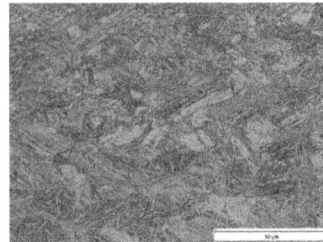

Fig. 2. *Microstructure of S960MC steel (LM, 1000x, Nital etch.).*

Hydrogen Charging

The samples from experimental material were electrolytically hydrogen charged in their as received state (AR) with the oxide layer, after grinding the oxide layer (G) and with a blasted surface (B). The sample was connected as a cathode, platinum-plated wolfram mesh as an anode. The electrolyte was a solution of 0.05 M sulfuric acid (H_2SO_4) with the addition of 1 g of potassium thiocyanate (KSCN) per one liter. The hydrogenation process was carried out for 4 hours at a current density of 1mA·cm^{-2} at the temperature 20±2°C. After hydrogenation, the samples were dried and quickly transferred to the tensile testing machine. The tensile tests were performed using a multifunctional LFV 100 kN servo-hydraulic test machine. The mechanical properties of the samples in the initial state and in the state after hydrogen charging are shown in Table 3 and Figs. 3, 4 and 5.

Results and Discussion

Due to the diffusion and recombination of hydrogen atoms into experimental steel, there was a significant decrease of ductility and tensile strength (Table 3). The effect of hydrogen on the yield strength was different for the different state of surface condition. The highest decrease of yield strength was measured for samples G, while the lowest decrease was measured for samples B. However, changes in mechanical properties are very well evident from the tensile diagrams of each state of samples surface (Figs. 3, 4 and 5).

Table 3.   *Results of mechanical properties in the initial state and after hydrogen charging of samples with every surface condition*

Sample	Test conditions	Yield Strength [MPa]	Tensile Strength [MPa]	Ductility [%]
AR	Initial state	988.05	1157.32	4.11
	After hydrogen charging	936.11	928.97	0.19
G	Initial state	1398.90	1570.73	5.07
	After hydrogen charging	1038.71	1056.39	0.92
B	Initial state	976.72	1056.84	9.31
	After hydrogen charging	975.85	985.18	1.89

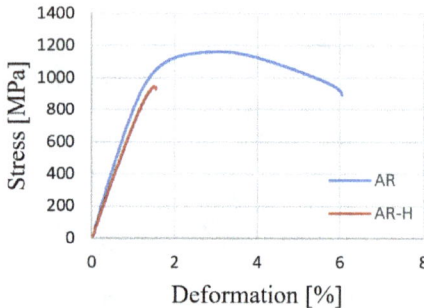

Fig. 3. *Tensile diagrams of samples in their as received state (AR) in initial state and after hydrogen charging (AR-H).*

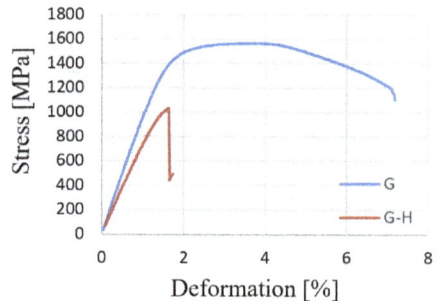

Fig.4. *Tensile diagrams of samples after grinding the oxide layer (G) in initial state and after hydrogen charging (G-H).*

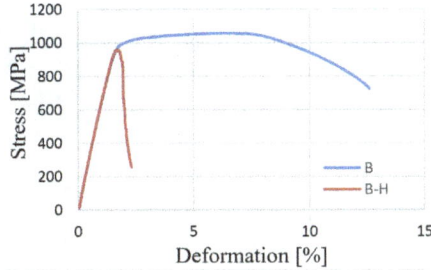

Fig. 5. *Tensile diagrams of samples with a blasted surface (B)*
in initial state and after hydrogen charging (B-H).

Similar results were obtained by Váňová and her coworkers [8], for TRIP 800 and TRIP 780 steels, which were also exposed to hydrogen in the same way and in the same environment as in our case.

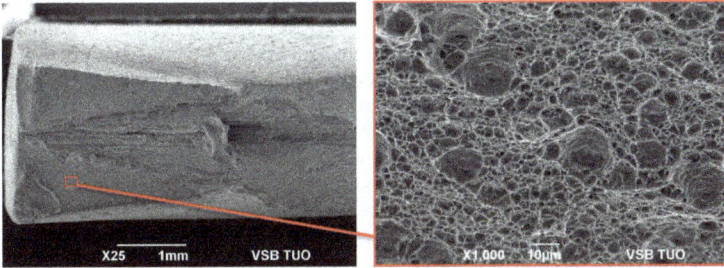

Fig. 6. *The sample fracture surface in the initial state without hydrogen charging,*
transgranular ductile fracture with dimple morphology.

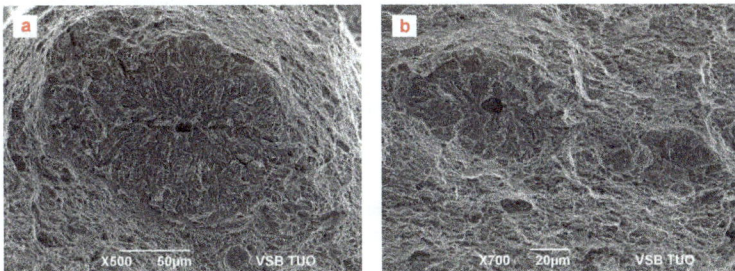

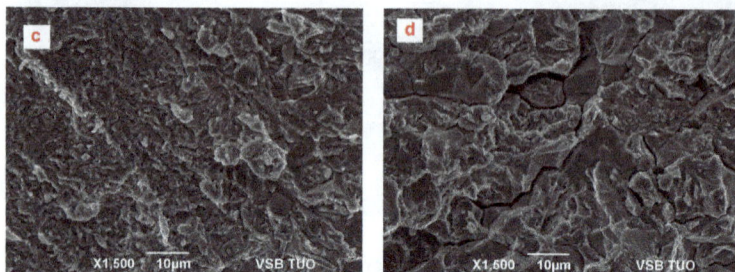

Fig. 7. *The sample fracture surface in the state after hydrogen charging; a) and b) fish-eyes accompanied by quasi-cleavage fracture around inclusions, c) quasi-cleavage fracture, d) intergranular ductile fracture.*

The fracture surfaces of the samples after tensile test were subjected to fractographic analysis. The fractographic analysis was carried out using the JEOL 6490LV scanning electron microscope (SEM) in secondary electron mode. Samples that were not subjected to hydrogen charging shows signs of transgranular ductile fracture with a dimple morphology.

From Fig. 6 and 7 it is possible to see how the presence of hydrogen in microstructure influenced the fracture mechanism during the tensile test. Hydrogen embrittlement typically manifest itself by the occurrence of the quasi-cleavage fracture and by large secondary cracks at the fracture surface for materials with a bcc crystal lattice [9-11]. In our case, hydrogen embrittlement was manifested by the formation of fish-eyes (Fig. 7a and b). Each of these fish-eyes were represented by the area of quasi-cleavage failure, which initiates on a non-metallic inclusion [12]. The inclusions observed in the center of the fish-eyes were identified as elongated sulphide inclusions or as complex globular oxysulphide inclusions. Quasi-cleavage facets are visible on the edge of the samples and around elongated non-metallic inclusions (Fig. 7c). Locally, especially on the edge of the samples, is also possible to see the intergranular ductile fracture (Fig. 7d).

Summary

The study of hydrogen embrittlement of advanced high-strength steel S960MC using tensile tests after hydrogenation confirmed the significant tendency of this steel to hydrogen embrittlement. Even though the electrolytic method of hydrogenation took place at a low current density, compared to structural steels of usual quality, embrittlement due to hydrogen in the S960MC steel during the tensile test was manifested by a more significant reduction in ductility [13]. Therefore, in production processes in which S960MC steel is processed (especially galvanic surface treatments, etc.), where there is a potentially high risk of interaction of atomic hydrogen with the material, it is necessary to pay extra attention to the ingress of hydrogen into the material in order to avoid hydrogen embrittlement.

In conclusion, hydrogen embrittlement is a significant concern in the design of the construction produced from high-strength structural steels. The proper material selection, manufacturing processes, and post-treatment procedures are critical in reducing the risk of hydrogen embrittlement. Despite of the efforts to understand and mitigate the effects of hydrogen embrittlement, it remains a complex and not fully understood phenomenon. Further research is needed to fully understand and reduce the effects of hydrogen embrittlement on high strength structural steels.

Acknowledgement
The research was funded by Slovak Research and Development Agency under contract No. APVV-20-0427 and the Slovak Ministry of Education, Science, Research, and Sport's Scientific Grant Agency under contract VEGA No. 1/0741/21.

References
[1]M.F. McGuire, R. J. Matway. A Mechanism for Hydrogen Embrittlement in Martensitic Steel Based on Hydrogen Dilation, Preprints.org (2022) art. 2022020275. https://doi.org/10.20944/preprints202202.0275.v1

[2]T. Schaupp et al. Hydrogen-assisted cracking of GMA welded 960 MPa grade high-strength steels. Int. J. Hydrog. Energy 45 (2020) 20080-20093. https://doi.org/10.1016/j.ijhydene.2020.05.077

[3]B.L. Enis et al. The role of aluminium in chemical and phase segregation in a TRIP-assisted dual phase steel. Acta Mater. 115 (2016) 132-142. https://doi.org/10.1016/j.actamat.2016.05.046

[4]J. Sojka. Resistance of steel to hydrogen embrittlement. Ostrava: VŠB – Technical University of Ostrava, 2007 (In Czech).

[5]L. Hyspecká, K. Mazanec. Hydrogen embrittlement of structural steels with higher parameters. Praha: Academia – publishing house of the Czechoslovak Academy of Sciences. 1978 (In Czech).

[6]X. Li et al. Review of Hydrogen Embrittlement in Metals: Hydrogen Diffusion, Hydrogen Characterization, Hydrogen Embrittlement Mechanism and Prevention. Acta Metall. Sin.-Engl. 33 (2020) 759-773. https://doi.org/10.1007/s40195-020-01039-7

[7]S. P. Lynch. Hydrogen embrittlement (HE) phenomena and mechanisms. Corr. Rev. 30 (2011) 90-130. https://doi.org/10.1515/corrrev-2012-0502

[8]P.Váňová et al. Comparison of Behaviour of Different Variants of Hydrogenated TRIP Steels at Slow Strain Rate Tests. Key Eng. Mater. 810 (2019) 70-75. https://doi.org/10.4028/www.scientific.net/KEM.810.70

[9]D. Hardie, E.A. Charles, A.H. Lopez. Hydrogen embrittlement of high strength pipeline steels. Corr. Sci. 48 (2006) 4378-4385. https://doi.org/10.1016/j.corsci.2006.02.011

[10] T. Hojo, et al. Effects of Alloying Elements Addition on Delayed Fracture Properties of Ultra High-Strength TRIP-Aided Martensitic Steels. Metals 10 (2020) art.6. https://doi.org/10.3390/met10010006

[11] Q. Liu, et al. The role of the microstructure on the influence of hydrogen of some advanced high-strength steels. Mater. Sci. Eng. A 715 (2018) 370-378. https://doi.org/10.1016/j.msea.2017.12.079

[12] P. Betakova, J. Sojka, L. Hyspecka. Quantitative assessment of hydrogen embrittlement in structural steels, METAL 2002 11th Int. Conf. on Metall. and Mater. (2002), Hradec nad Moravici, 55-57.

[13] P. Váňová. Hydrogen embrittlement and hydrogen diffusion characteristics in TRIP steels. Habilitation thesis. Ostrava: VŠB – Technical University of Ostrava, 2019 (In Czech).

Quality Production Improvement and System Safety: QPI 16 - CZOTO 10   Materials Research Forum LLC
Materials Research Proceedings 34 (2023) 68-76   https://doi.org/10.21741/9781644902691-9

Application of Zinc-Silver Impregnated Activated Carbons in Removal of Lead(II) and Mercury(II) Compounds from Groundwater

KWAK Anna[1,a], POSZWALD Bartosz[1,b], DYSZ Karolina[1,c]
and DYLONG Agnieszka[1,d*]

[1] Military Institute of Engineer Technology, 136 Obornicka Str., 50-961 Wroclaw, Poland

[a] kwak@witi.wroc.pl, [b] poszwald@witi.wroc.pl, [c] dysz@witi.wroc.pl, [d] dylong@witi.wroc.pl

Keywords: Activated Carbon, Impregnation, Water Purification, Adsorbents, Metal-Impregnated Carbonr

Abstract. Nowadays activated carbon is a material generating great interest, as it is characterized by a vast surface area due to a high number of pores in its structure. Therefore, the main purpose behind its use is the filtration of impurities from air and water that can be adsorbed with high efficiency. Activated carbon can be easily modified as well. The paper describes activated carbon modification with copper-, manganese-, silver- and zinc salts. The effects of the selected impregnates and their concentrations were examined. The products included 5 adsorbent samples: four universal adsorbents, impregnated with all the above-mentioned salts, and one specific adsorbent sample, designed to adsorb lead(II) and mercury(II) ions and impregnated with zinc- and silver salts only. The premise was to obtain pure drinking water. Properties, such as bulk density, methylene blue number or iodine number were determined for the modified activated carbons. To test the efficiency of an improved adsorbent, an experiment with water highly contaminated with Pb(II) and Hg(II) was carried out, and its results revealed that absorption efficiency for these heavy metals exceeded 99.9%. The adsorber samples were also observed under a digital microscope to compare their appearance.

Introduction

Activated Carbon (AC), thanks to its vast surface area, is a perfect catalyst carrier (support). Even though ACs impregnated with different compounds change their pore structure, such alteration also allows scientists to obtain a material which can serve as a chemisorbent for specific toxic substances [1]. Many compounds can be used for the impregnation process, in particular copper-, chromium-, silver-, potassium-, sodium-, zinc-, cobalt-, manganese-, vanadium-, molybdenum- and iron salts or some organic compounds e.g. pyridine and aromatic amines [2].

Impregnation process parameters are of key importance; even slight changes to the technological regime may cause a reduction in adsorption capacity by 20-30% and could result in completely different adsorbent properties [3]. Increased efficiency and selectivity in the removal of toxic substances is the effect of physical adsorption and chemical adsorption of substances at the AC's surface or catalytic reactions with the impregnation compounds [4]. One of the most important carbon carriers include Metal-Impregnated Activated Carbons (MIAC). MIACs impregnated with copper, chromium and silver are called copper-chromium-silver (Cu/Cr/Ag) impregnated carbons. They deserve special attention due to their purification performance tested on air samples containing compounds, such as cyanogen chloride, hydrogen cyanide and arsines [5].

Current legal regulations are very restrictive in terms of heavy metal contamination, including mercury contamination of ground- and drinking water, as mercury and its compounds are highly-toxic substances and even in low concentration have a marked detrimental impact on the health and life of living organisms. The contaminants most commonly found in environmental samples include metallic mercury, methylmercury and phenyl mercury derivatives. It is believed that the paper industry, chemical industry, batteries production and agriculture are the main culprits and main sources of Hg contamination of the natural environment [6-8].

The efficiency of adsorptive removal of Hg(II) ions is largely dependent on the type of AC and activation process applied. Steam-activated ACs obtained from wood, coconut shells and coal show great capacity to adsorb mercury(II) ions from pH<5 solutions. It has been also discovered that the pH of the solution affects the amount of Hg(II) adsorbed by AC, where a general rule is the lower the pH, the greater the absorption rate observed [9]. The necessity of protecting waters from these contaminants was recognized a long time ago; the first World Health Organization's Standard for drinking water was published in 1958 [10]. It describes the exact maximum allowable concentration of lead in water, which was 0.1 mg/L, but it only briefly mentions mercury as a health threat. Back then, the issue was no sufficient detection methods. The original WHO Guidelines for Drinking Water Quality 3rd edition from 2008 [11] states mercury's acceptable concentration for 0.006 mg/L. Currently, according to WHO guidelines, the allowable concentration of mercury has been maintained, however, the allowable lead concentration has been reduced 10 times – to 0.001 mg/L [12]. Whilst considering the needs in the field of groundwater purification (the goal of which is to obtain drinking water of proper parameters in line with the current regulations), as well as efficiency in removing lead and mercury ions and innovative character of zinc-impregnated adsorbents, new production technology has been developed by the authors. The final products include two adsorbents: a universal (Cr-Cu-Ag-Zn) one and a silver-zinc-impregnated one.

Experimental Procedure

Materials and sample preparation. *DT0* commercially-available activated carbon (produced by Gryfskand Sp. z o.o. company in Poland) was used in this experiment. It was first being oxidized in 20% nitric acid solution under periodical stirring for 24 hours [13]. Then AC was drained off and washed with deionized water until all the acid was removed. Afterwards, a washed activated carbon was kept a room temperature for 24h. At the end of the preparation process, the product was dried in a fluid bed dryer at 120°C for 20 minutes. The obtained adsorbents were divided into two groups:
– universal adsorbents (UA), containing zinc, copper, manganese, silver, and
– adsorbents containing zinc and silver (S-Hg/Pb) only.
The following metallic catalysts have been selected to produce UA:
– basic copper carbonate (Cu2CO3(OH)2) in aqua ammonia;
– potassium permanganate (KMnO4) in aqueous solution;
– silver nitrate (AgNO3) in aqueous solution;
– zinc nitrate (ZnNO3) in an aqueous solution.
After all the solutions were poured into one, the resultant mixture was poured into a beaker containing dried AC and stirred for a few minutes. The mixture was left over for 2 allowing the salts to precipitate. The next step was crucial for obtaining the desired properties of the MIACs: impregnated carbon was dried for 6 hours in total – 4 hours at the temperature of 120°C and 2 hours at 180°C. Copper was deposited on the carbon surface as a copper monoxide (CuO). CuO was obtained by heating a mixture of basic copper carbonate and ammonium carbonate diluted in an ammonia solution (25 wt%). First, a complex ion of hexaaminecopper(II) ion is formed [14]. Then, by its thermal decomposition, copper monoxide, carbon dioxide and gaseous ammonia are produced. CuO is adsorbed in ACs pores and the AC is thusly becoming a chemisorbent or a catalyst for the decomposition of toxic substances. A similar process can be described for potassium permanganate. However, it does not include any complex compounds, and permanganate ions are reduced to manganese dioxide particles [15]. This compound is a

Quality Production Improvement and System Safety: QPI 16 - CZOTO 10   Materials Research Forum LLC
Materials Research Proceedings 34 (2023) 68-76   https://doi.org/10.21741/9781644902691-9

chemisorbent and catalyst for the decomposition of toxic substances as well. Silver and copper can also be deposited on activated carbon's surface. First, a diamine silver(I) carbonate complex is formed [16], and afterwards, as a result of thermal decomposition, metallic silver particles, silver(I) oxide, carbon dioxide and ammonia are obtained. The concentrations of each metal ion are shown in the table (Table 1).

The last MIAC (S-Hg/Pb) was specially designated to adsorb lead and mercury only. The key factor behind this process is the presence of zinc ions at the carbon surface, as they are known for their substantial heavy metal absorption properties [17]. Silver is added to obtain antibacterial traits [18] because this adsorbent is to be used to obtain drinking water from the surface and ground waters. Oxidized and dried activated carbon was impregnated with an aqueous solution of:
– silver nitrate ($AgNO_3$),
– zinc nitrate ($ZnNO_3$).
Those metals were deposited on the surface of the AC, as was previously mentioned in this chapter. The adsorbent obtained, S-Hg/Pb, was heated for 20 minutes in a fluid bed dryer, at the temperature of 120°C and then for another 2 hours at 350°C in a laboratory oven, with no air permitted. Finally, the sample was allowed to cool down. S-Hg/Pb contained 2.0% zinc (w/w) and 1.0% silver (w/w).

Table 1. Concentrations of metal ions in MIACs

No.	Adsorbent	Copper [% w/w]	Manganese [% w/w]	Silver [% w/w]	Zinc [% w/w]
1.	SU-1	3.0	1.2	0.5	0.5
2.	SU-2	3.0	1.2	0.5	1.0
3.	SU-3	3.0	1.2	-	0.5
4.	SU-4	3.0	1.2	0.5	2.0

Sample characterization. The parameters of obtained adsorbents were characterized by spectrophotometric-, titration- and potentiometric methods. Both the universal adsorbents produced and S-Hg/Pb adsorbed metal ions on their surface, which was corroborated by the methylene blue (MBN) and iodine (IN) numbers study. The two are strictly related to the AC's micro and macropore volume – the greater the volume, the higher the MBN and IN. When these two number decreases, it means that the impregnation salts, or their derivatives, are bound on the surface of carbon grains. The authors experimented according to the Polish Defense Standard [19] which included titration being used to determine IN for all the MIACs samples, as well spectrophotometric measurement of MBN. To find out if there some visual changes on the surface of the impregnated carbons can be seen, the Keyence series VHX-7000 microscope with VH-ZST dual objective zoom lens was used. The authors presumed that the addition of zinc should increase the efficiency of mercury- and lead adsorption capacity, and to corroborate this hypothesis they conducted the required tests. Amount of 1 gram of each SU and S-Hg/Pb were added to two aqueous solutions, one containing lead and one containing mercury, and then they were stirred for 30 minutes. Final concentrations of the tested heavy metals were tested with inductively coupled plasma optical emission spectrometry (ICP OES) using the Thermo Scientific iCAP 7000 series spectrometer. Investigations were carried out with the following spectrometry parameters: lead wavelength: 283.306 nm; mercury wavelength: 253.652 nm or 184.950 nm using the cold vapor method with hydride generation system in case of the latter element.

Results and Discussion

In the beginning, the appearance of *DT0* and S-Hg/Pb was analyzed and the authors found that at the modified adsorbent grain there appeared clearly visible metallic silver aggregates (Fig. 1) that were too large to be deposited inside the pores. Selected parameters of every adsorbent were characterized and compiled in the table (Table 2.). Universal adsorbents' methylene blue number is 20-30% smaller than the MBN of unimpregnated *DT0* signifying that salts used for impregnation, or their derivatives, were deposited in mesopores and greater micropores of AC [20]. SU's iodine number is about 8-13% smaller than IN of plain activated carbon. As iodine is mainly kept in micropores, the lower number points out to the fact that zinc, manganese, copper and silver oxides were deposited in micropores as well [21].

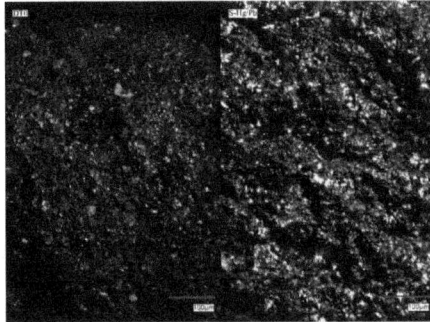

Fig. 1. The appearance of DT0 and S-Hg/Pb MIACs under a microscope

Table 2. *Adsorbent parameters*

No.	Parameter	Units	Adsorbent					
			DT0	SU-1	SU-2	SU-3	SU-4	S-Hg/Pb
1	Water content	%	5	4	4	4	4	3
2	Bulk density	g/cm^3	0.39	0.42	0.42	0.44	0.44	0.39
3	Methylene blue number (MBN)	cm^3	50	35	35	35	40	46
4	Iodine number (IN)	mg/g	1006	883	923	872	895	1045
5	Grain distribution: ≥0.750mm	%	0.40	1.52	1.99	1.16	1.68	2.36
	0.385-0,750mm		96.23	96.57	95.05	95.55	96.32	95.74
	≤0.385mm		3.37	1.92	2.96	2.32	2.00	1.89
6	Concentration in adsorbent: - manganese	%	-	1.2	1.2	1.2	1.2	-
	- copper		-	3.0	3.0	3.0	3.0	-
	- silver		-	0.5	0.5	-	0.5	1.0
	- zinc		-	0.5	1.0	0.5	2.0	2.0

7	Concentration in water (with 10m/h flow): - manganese	mg/dm³	-	0.02	0.02	0.03	0.02	-
	- copper		-	0.02	0.02	0.02	0.02	-
	- ammonia		-	0.9	1.1	1.0	0.9	-
	pH	-	8.5	7.86	9.2	9.4	9.4	7.8

The main goal of this study was to obtain improved properties of activated carbon in terms of lead and mercury adsorption capacity. The test results are shown in Table 3. Even though universal adsorbents are effective adsorbents of heavy metals but not enough to qualify filtered water as drinking water. The most efficient AC proved to be SU-4; it removed almost 80% of lead(II) ions and nearly 90% of mercury(II) ions from water, while other adsorbents adsorbed 58-82% of these elements. Among all the SUs, SU-4 was impregnated with the highest amount of zinc salt, which had been proven to be to be a good heavy metal binder. Thus, a specially designed MIAC – S-Hg/Pb, containing zinc and silver only, removes lead with more than 99.9% efficiency, while in the case of mercury, it is 99.999% efficient.

Table 3. Obtained test results.

Adsorbent	Concentration [mg/l]					
	Lead			Mercury		
	Before adsorption	After adsorption	Percentage removed	Before adsorption	After adsorption	Percentage removed
SU-1		0.32	68%		0.21	79%
SU-2		0.42	58%		0.18	82%
SU-3	1	0.38	62%	1	0.22	78%
SU-4		0.22	78%		0.11	89%
S-Hg/Pb		<0.1	99.9%		0.001	99.999%

Conclusions

In the present study activated carbon was modified with copper-, manganese-, silver- and zinc salts. The effects of selected impregnates and their concentrations were examined. Methylene blue number and iodine number were used to determine if the impregnation process was efficacious or not. The results showed a decrease of MBN and IN by 20-30% and 8-13%, respectively, which means that the process was successfully accomplished and the salts or their derivatives were deposited in AC's pores. There are two more proofs that the deposition occurred; the first of which included the microscopic image o S-Hg/Pb with visible silver aggregates, while the second is the change of bulk density that was caused by the embedding of elements heavier than carbon. The testing of the efficiency of lead(II) and mercury(II) adsorption was done by the comparison of the measurements of SUs and S-Hg/Pb filtrate samples concentrations with ICP OES. While universal adsorbent showed good absorption performance, when heavy metals were concerned, the results for the proprietary adsorbent were excellent. The final concentration of mercury was lower than the reference for drinking water, however, the lead concentration may be higher [11]. Because ICP OES is not accurate enough, another method, e.g. GC-MS, should be used [12]. It can be said that the product designed by the authors performs well, however, further development and testing could be beneficial. Issues related to the removal of heavy metal pollutants from water are strongly linked

Quality Production Improvement and System Safety: QPI 16 - CZOTO 10   Materials Research Forum LLC
Materials Research Proceedings 34 (2023) 68-76   https://doi.org/10.21741/9781644902691-9

to production quality [22-24] and management level in businesses [25]. Such pollutants are a common problem in industries that utilize metals [26-28], including special alloys [29], as well as technologies associated with surface layer modification [30] and the application of special coatings [31-33]. The creation of DLC [34, 35] and ESD [36, 37] coatings involve technologies that generate both liquid and gas pollutants, similar to welding processes [38, 39]. Similar pollutants are also formed during the operation of machinery [40], railway rolling stock [41], chemical installations [42], and the use of chemicals in civil engineering [43-45]. The purification processes need to be highly efficient, requiring optimization and careful planning [46]. These processes are multifactorial, and it is beneficial to perform dimensionality reduction [47] prior to optimizing them using statistical methods in industrial settings [48-50] to avoid detrimental correlations. It is also advantageous to utilize non-parametric methods [51-53] that employ a *data-driven* approach, as they are not limited to predefined model assumptions.

References

[1] S. Choi, J.H. Drese, C.W. Jones. Adsorbent Materials for Carbon Dioxide Capture from Large Anthropogenic Point Sources, ChemSusChem 2 (2009) 796-854. https://doi.org/10.1002/cssc.200900036

[2] Y. Zhi, J. Liu. Surface modification of activated carbon for enhanced adsorption of perfluoroalkyl acids from aqueous solutions, Chemosphere 144 (2016) 1224-1232. https://doi.org/10.1016/j.chemosphere.2015.09.097

[3] M.C.A. Ferraz. Preparation of activated carbon for air pollution control, Fuel 2 (1988) 1237-1241. https://doi.org/10.1016/0016-3261(88)90045-2

[4] H. Lin et al. Simultaneous reductions in antibiotics and heavy metal pollution during manure composting, Sci. Total Environ. 788 (2021) art. 147830. https://doi.org/10.1016/j.scitotenv.2021.147830

[5] J. Choma, M. Kloske. Otrzymywanie i właściwości impregnowanych węgli aktywnych, Ochrona Środowiska 2 (1999) 3-17.

[6] K.V.S. Prasad. Electronic waste – An emerging threat to the environment and health, Pollution Research 35 (2016) 587-593.

[7] F.A. Armah et al. Anthropogenic sources and environmentally relevant concentrations of heavy metals in surface water of a mining district in Ghana: A multivariate statistical approach, J. Environ. Sci. Health A 45 (2010) 1804-1813. https://doi.org/10.1080/10934529.2010.513296

[8] T. Apriantiet al. Heavy metal ions adsorption from pulp and paper industry wastewater using zeolite/activated carbon-ceramic composite adsorbent, AIP Conf. Proc. 2014 (2018) art. 020127. https://doi.org/10.1063/1.5054531

[9] K. Kadirvelu et al. Mercury(II) adsorption by activated carbon made sago waste, Carbon 42 (2004) 745-752. https://doi.org/10.1016/j.carbon.2003.12.089

[10] World Health Organization. (1958). International standards for drinking-water. World Health Organization.

[11] World Health Organization. Guidelines for Drinking-water Quality 3rd edition incorporating the first and second addenda vol.1 Recommendations. Geneva, Switzerland, World Health Organization, 2008

[12] World Health Organization. Guidelines for Drinking-water Quality 4th edition. Geneva, Switzerland, World Health Organization, 2011.

Quality Production Improvement and System Safety: QPI 16 - CZOTO 10   Materials Research Forum LLC
Materials Research Proceedings 34 (2023) 68-76   https://doi.org/10.21741/9781644902691-9

[13] A. Gąskiewicz-Puchalska et al. Improvement od CO_2 uptake of activated carbons by treatment with mineral acids, Chem. Eng. J. 309 (2017) 159-171. https://doi.org/10.1016/j.cej.2016.10.005

[14] L. Velasquez-Yevenes, R. Ram. The aqueous chemistry of the copper-ammonia system and its implications for sustainable recovery of copper, Cleaner Eng. Tech. 9 (2022) art. 100515. https://doi.org/10.1016/j.clet.2022.100515

[15] K. Okitasu et al. Sonochemical reduction of permanganate to manganese dioxide: the effects of H2O2 formed in this sonolysis of water on the rates of reduction, Ultrason. Sonochem. 16 (2009) 387-391. https://doi.org/10.1016/j.ultsonch.2008.10.009

[16] N. Shabanov et al. A Water-Soluble Ink Based on Diamine Silver(I) Carbonate Ammonium Formate, and Polyols for Inkjet Printing of Conductive Patterns, Eur. J. Inorg. Chem. 2019 (2019) 178-182. https://doi.org/10.1002/ejic.201801045

[17] M. Gu et al. The selective heavy metal ions adsorption of zinc oxide nanoparticles from dental wastewater, Chem. Phys. 534 (2020) art. 110750. https://doi.org/10.1016/j.chemphys.2020.110750

[18] N.K. Nasab, Z. Sabouri, S. Ghazal, M. Darroudi. Green-based synthesis of mixed-phase silver nanoparticles as an effective photocatalyst and investigation of their antibacterial properties, J. Mol. Struc. 1203 (2020) art. 127411. https://doi.org/10.1016/j.molstruc.2019.127411

[19] NO-46-A200:2022. Wojskowe urządzenia uzdatniania wody – Materiały eksploatacyjne – wymagania i metody badań. Warszawa, Poland, Ministerstwo Obrony Narodowej, 2012.

[20] R.M. Shrestha. Effect of Preparation Parameters on Methylene blue number of Activated Carbons Prepared from a Locally Available Material, J. Inst. Eng. 12 (2017) 169-174. https://doi.org/10.3126/jie.v12i1.16900

[21] C. Saka. BET, TG-DTG, FT-IR, SEM, iodine number analysis and preparation of activated carbon from acorn shell by chemical activation with ZnCl2, J. Anal. App. Pyrolysis 95 (2012) 21-24. https://doi.org/10.1016/j.jaap.2011.12.020

[22] M. Nowicka-Skowron, R. Ulewicz. Quality management in logistics processes in metal branch, METAL 2015 – 24th Int. Conf. Metall. Mater. (2015) 1707-1712. ISBN 978-8087294628

[23] R. Ulewicz, F. Nový. Quality management systems in special processes, Transp. Res. Procedia 40 (2019) 113-118. https://doi.org/10.1016/j.trpro.2019.07.019

[24] R. Ulewicz et al. Logistic controlling processes and quality issues in a cast iron foundry, Mater. Res. Proc. 17 (2020) 65-71. https://doi.org/10.21741/9781644901038-10

[25] A. Deja et al. The Concept of Location of Filling Stations and Services of Vehicles Carrying and Running on LNG. In: P. Ball, L. Huaccho Huatuco, R. Howlett, R. Setchi (Eds.) Sustainable Design and Manufacturing 2019. KES-SDM 2019. Smart Innovation, Systems and Technologies, 155. Springer, Singapore, 507-520. https://doi.org/10.1007/978-981-13-9271-9_42

[26] R. Ulewicz et al. Structure and mechanical properties of fine-grained steels, Period. Polytech. Transp. Eng. 41 (2013) 111-115. https://doi.org/10.3311/PPtr.7110

[27] D. Klimecka-Tatar, M. Ingaldi. Assessment of the technological position of a selected enterprise in the metallurgical industry, Mater. Res. Proc. 17 (2020) 72-78. https://doi.org/10.21741/9781644901038-11

[28] P. Jonšta et al. The effect of rare earth metals alloying on the internal quality of industrially produced heavy steel forgings, Materials 14 (2021) art. 5160. https://doi.org/10.3390/ma14185160

[29] A. Dudek et al. The effect of alloying method on the structure and properties of sintered stainless steel, Archives of Metallurgy and Materials 62 (2017) 281-287. https://doi.org/10.1515/amm-2017-0042

[30] N. Radek et al. The influence of plasma cutting parameters on the geometric structure of cut surfaces, Mater. Res. Proc. 17 (2020) 132-137. https://doi.org/10.21741/9781644901038-20

[31] N. Radek et al. Technology and application of anti-graffiti coating systems for rolling stock, METAL 2019 28th Int. Conf. Metall. Mater. (2019) 1127-1132. ISBN 978-8087294925

[32] N. Radek et al. The effect of laser beam processing on the properties of WC-Co coatings deposited on steel. Materials 14 (2021) art. 538. https://doi.org/10.3390/ma14030538

[33] N. Radek et al. Formation of coatings with technologies using concentrated energy stream, Prod. Eng. Arch. 28 (2022) 117-122. https://doi.org/10.30657/pea.2022.28.13

[34] N. Radek et al. Microstructure and tribological properties of DLC coatings, Mater. Res. Proc. 17 (2020) 171-176. https://doi.org/10.21741/9781644901038-26

[35] N. Radek et al. Influence of laser texturing on tribological properties of DLC coatings, Prod. Eng. Arch. 27 (2021) 119-123. https://doi.org/10.30657/pea.2021.27.15

[36] N. Radek, J. Konstanty. Cermet ESD coatings modified by laser treatment, Arch. Metall. Mater. 57 (2012) 665-670. https://doi.org/10.2478/v10172-012-0071-y

[37] N. Radek et al. The morphology and mechanical properties of ESD coatings before and after laser beam machining, Materials 13 (2020) art. 2331. https://doi.org/10.3390/ma13102331

[38] N. Radek et al. The impact of laser welding parameters on the mechanical properties of the weld, AIP Conf. Proc. 2017 (2018) art.20025. https://doi.org/10.1063/1.5056288

[39] N. Radek et al. Properties of Steel Welded with CO2 Laser, Lecture Notes in Mechanical Engineering (2020) 571-580. https://doi.org/10.1007/978-3-030-33146-7_65

[40] S. Blasiak et al. Rapid prototyping of pneumatic directional control valves, Polymers 13 (2021) art. 1458. https://doi.org/10.3390/polym13091458

[41] N. Radek, R. Dwornicka. Fire properties of intumescent coating systems for the rolling stock, Commun. – Sci. Lett. Univ. Zilina 22 (2020) 90-96. https://doi.org/10.26552/com.C.2020.4.90-96

[42] E. Skrzypczak-Pietraszek et al. Enhanced accumulation of harpagide and 8-O-acetyl-harpagide in *Melittis melissophyllum* L. agitated shoot cultures analyzed by UPLC-MS/MS. PLoS ONE 13 (2018) art. e0202556. https://doi.org/10.1371/journal.pone.0202556

[43] A. Bakowski et al. Frequency analysis of urban traffic noise, ICCC 2019 20th Int. Carpathian Contr. Conf. (2019) 1660-1670. https://doi.org/10.1109/CarpathianCC.2019.8766012

[44] J.M. Djoković et al. Selection of the Optimal Window Type and Orientation for the Two Cities in Serbia and One in Slovakia, Energies 15 (2022) art.323. https://doi.org/10.3390/en15010323

[45] Ł.J. Orman et al. Analysis of Thermal Comfort in Intelligent and Traditional Buildings, Energies 15 (2022) art.6522. https://doi.org/10.3390/en15186522

[46] L. Cedro. Model parameter on-line identification with nonlinear parametrization – manipulator model, Technical Transactions 119 (2022) art. e2022007. https://doi.org/10.37705/TechTrans/e2022007

[47] J. Pietraszek, E. Skrzypczak-Pietraszek. The uncertainty and robustness of the principal component analysis as a tool for the dimensionality reduction. Solid State Phenom. 235 (2015) 1-8. https://doi.org/10.4028/www.scientific.net/SSP.235.1

[48] R. Dwornicka, J. Pietraszek. The outline of the expert system for the design of experiment, Prod. Eng. Arch. 20 (2018) 43-48. https://doi.org/10.30657/pea.2018.20.09

[49] J. Pietraszek et al. Challenges for the DOE methodology related to the introduction of Industry 4.0. Prod. Eng. Arch. 26 (2020) 190-194. https://doi.org/10.30657/pea.2020.26.33

[50] B. Jasiewicz et al. Inter-observer and intra-observer reliability in the radiographic measurements of paediatric forefoot alignment, Foot Ankle Surg. 27 (2021) 371-376. https://doi.org/10.1016/j.fas.2020.04.015

[51] J. Pietraszek. The modified sequential-binary approach for fuzzy operations on correlated assessments, LNAI 7894 (2013) 353-364. https://doi.org/10.1007/978-3-642-38658-9_32

[52] J. Pietraszek et al. Non-parametric assessment of the uncertainty in the analysis of the airfoil blade traces, METAL 2017 26th Int. Conf. Metall. Mater. (2017) 1412-1418. ISBN 978-8087294796

[53] J. Pietraszek et al. The non-parametric approach to the quantification of the uncertainty in the design of experiments modelling, UNCECOMP 2017 Proc. 2nd Int. Conf. Uncert. Quant. Comput. Sci. Eng. (2017) 598-604. https://doi.org/10.7712/120217.5395.17225

Quality Production Improvement and System Safety: QPI 16 - CZOTO 10   Materials Research Forum LLC
Materials Research Proceedings 34 (2023) 77-86   https://doi.org/10.21741/9781644902691-10

Degradation of R35 Steel in 5% NaCl environment at 10°C

LIPIŃSKI Tomasz [1, a *] and ULEWICZ Robert [2,b]

[1]University of Warmia and Mazury in Olsztyn, Faculty of Technical Sciences Department of Materials and Machines Technology, Olsztyn, Poland

[2]Czestochowa University of Technology, Faculty of Management, Department of Production Engineering and Safety, Częstochowa Poland,

[a]tomekl@uwm.edu.pl, [b]robert.ulewicz@pcz.pl

Keywords: Steel, Carbon Steel, Corrosion, Corrosion Rate, Profile Roughness

Abstract. Carbon steels are willingly used due to the favorable price-performance ratio. They usually do not operate in a non-corrosive state. However, there are products that are hard to protect against corrosion in their entirety. These products include pipes. It is easy to protect the outer layer of the pipe against aggressive environment. The problem with protection is caused by their inner surface. Typically, the working medium is the protective factor, usually filling the internal volume of the pipes completely. It happens, however, that corrosion occurs as a result of long storage of pipes in warehouses, usually in the open air. They are also used as working elements with partial liquid filling. They are then exposed to the environment. One of the corrosive agents is NaCl. The paper presents the results of corrosion rate tests of samples taken from R35 pipes in the environment of a 5% aqueous solution of NaCl at 10°C. The analyzes were carried out based on the determination of mass losses. On the basis of the tests carried out, the relationship between the rate of corrosion and the soaking time of the samples was determined. It was confirmed that on the basis of roughness parameters it is possible to draw conclusions about the suitability of steel for further operation.

Introduction

Non-alloy structural steels are characterized by sufficient mechanical parameters for common applications at a low price. For this reason, they are widely used for structures carrying low loads. One of the examples of the use of unalloyed steels are pipelines for the transmission of media, mainly water. The external surface of the pipeline is protected against the influence of the environment. Paint coatings are usually the main protection. The inner surface is usually not covered with a varnish layer. The reason is the flowing medium, which, while moving in relation to the pipeline, can wash out the applied coating. complete filling of the pipeline with liquid, and thus separating its surface from air access, reduces the corrosion process. It also happens that the medium flows in the pipes and does not completely fill their surface. This is one of the reasons for the corrosion of the inner surface of pipelines. Corrosion-resistant steels created have resistance only in specific conditions. Their corrosion resistance depends i.a. on chemical ingredients, microstructure, surface condition [1-7].

In practice, the process of material degradation attributed to corrosion is more complex. By the term corrosion, researchers describe the degradation of a material resulting from reactions occurring from the metal surface, usually of a chemical or electrochemical nature [8-11]. In exploited pipelines with a flowing medium, the process of wear of the wall surface in contact with the working medium occurs. The literature describes this process as erosion. Thus, the natural wear of the pipeline is a combination of corrosion and mechanical wear, i.e. erosion [12-14].

The corrosion rate first of all is depends of different environment [6-10]. One of the corrosive environmental factors are chlorides. They are mainly found on the coast as an aerosol of sea water, and in large quantities in large industrial areas. Because of this structural steel has been often tested

Quality Production Improvement and System Safety: QPI 16 - CZOTO 10   Materials Research Forum LLC
Materials Research Proceedings 34 (2023) 77-86   https://doi.org/10.21741/9781644902691-10

with the NaCl on corrosion resistance [15-24]. Corrosion mechanism of such steel is not as complex as steel with increased corrosion resistance. Its chemical composition and microstructure indicate susceptibility to surface corrosion [25-27]. Corrosion, though the least dangerous of known corrosion types, causes systematic material destruction by oxidation. Material losesits volume and, therefore strength and stiffness. Although it does not need to carry special loads, it ultimately leads to corrosive wear of the component [18,28].

In the literature, there are a number of test results for corrosion of general-purpose steels in various aggressive environments [29-38]. The obtained results are developed on the basis of various algorithms of their presentation. As a result, there are a number of equations describing the rate of corrosion. In the paper, it was decided to present the research results in a study based on the criteria describing the corrosion of corrosion-resistant steels. The aim of the paper is to develop the relationship between the corrosion rate of R35 steel commonly used for pipelines and the soaking time in 5% NaCl at 10°C.

The phenomenon of corrosion [39-41], influenced by both organic and inorganic factors, is extremely detrimental. It results in significant maintenance and repair costs in the energy sector [42-44], biotechnological industry [45-47], and poses risks in hydraulic power systems [48-50]. Corrosion prevention is crucial to mitigate these negative effects and ensure the efficient operation of various systems. One approach to combat corrosion is the application of special coatings [51-53]. These coatings provide a protective barrier that inhibits the interaction between the corrosive environment and the underlying material.

Another method to prevent corrosion is through the design and control of surface morphology [54-56]. By manipulating the surface features and topography, the corrosive attack can be minimized, thereby extending the lifespan of the materials. Laser technology offers precise control over the surface morphology, allowing for tailored surface properties that resist corrosion.

To ensure the effectiveness of corrosion prevention strategies, the use of appropriate experimental design methodology is essential [57-59]. Design of Experiments (DOE) provides a systematic approach to investigate the effects of various factors on corrosion resistance, even with non-classic approaches [60-61]. By carefully designing experiments and analyzing the results, reliable data can be obtained with reduced uncertainty.

The implementation of effective corrosion prevention measures not only has positive implications for the environment [62] but also influences the anticipated scenarios of potential failures [63-65]. By reducing the risk of corrosion-related failures, the overall reliability and safety of systems are improved. This, in turn, leads to modifications in management processes to incorporate the new corrosion prevention strategies [66,67].

Material and Methods

The experiment was performed on popular carbon steel intended for the production of pipes in the R35 grade designation according to PN-89/H 84023/07 [68] and manufactured in accordance with PN-80/H-74219 [69]. Samples were collected from a pipe measuring 114.3 mm in diameter and 8.0 mm in wall thickness. The samples were mechanically cut to maintain consistent width measurements on both the outer and inner diameters of the pipe. The parallelism of the longitudinal axis of the samples with the axis of symmetry of the pipe was maintained. After cutting, the samples were ground at the cut to Ra less than 1.25 μm. The outer and inner surfaces of the pipe, prior to sampling, were ground in the direction transverse to the axis of the pipe with a rotary disc equipped with sandpaper strips to Ra below 1.25 μm. The real chemical composition tested steel is presented in Table 1. Average mechanical properties at ambient temperature are presented in Table 2.

Before experiments, the specimens after mechanically cut off with an area of 15.2 cm² (40 x 10 x 7.2 mm) were successively cleaned with water and 95% alcohol.

Quality Production Improvement and System Safety: QPI 16 - CZOTO 10   Materials Research Forum LLC
Materials Research Proceedings 34 (2023) 77-86   https://doi.org/10.21741/9781644902691-10

Table 1. Chemical composition of the R35 steel

Mean chemical compositions [wt. %]							
C	Si	Mn	P	S	Cr	Cu	Ni
0.09	0.53	0.21	0.030	0.035	0.01	0.12	0.01

Table 2. Mechanical properties at ambient temperature of the R35 steel

Mechanical properties			
R_{eH}	R_m	A	HB
MPa	MPa	%	HB
224	368	26	168

The samples with ferritic-perlitic microstructure were tested accordance with standard dedicated for stainless steel PN EN ISO 3651-1 [70] corrosive media were represented by 5% NaCl.

The corrosion rare of the S235JR steel measured in mm/year was calculated with the use of the below formula (1), measured in g/m^2 were calculated with the use the below formula (2):

$$r_{corm} = \frac{8760 \cdot m}{S \cdot t \cdot \rho} \tag{1}$$

$$r_{corg} = \frac{10000 \cdot m}{S \cdot t} \tag{2}$$

where:

   t – time of soaking in a corrosive solution of 5% NaCl water solution [hours],
   S - surface area of the sample (the starting value was assumed) [cm^2],
   m – average mass loss in solution (measured as the difference initial mass and mass after corrosion time) [g],
   ρ - sample density [g/cm^3].

The corrosion resistance the R35 steel in 5% water solution NaCl was tested using weight loss. The mass of samples was measured by Kern ALT 3104AM digital laboratory precision scales with accuracy of measurement 0.0001 g.

Profile roughness parameters were analyzed according to the PN-EN 10049:2014-03 standard (Measurement of roughness average Ra and peak count RPc on metallic flat products) by the Diavite DH5 profilometer.

Results

Surface roughness of R35 steel after soaking in 5% NaCl at 10°C with: Ra - arithmetical mean roughness value [μm], Rq - mean peak width [μm] and Rp - maximum roughness depth [μm], Rt - total height of the roughness profile [μm] for time range: 48, 96, 144, 192, 240, 288, 336, 384 and 432 hours of soaking are respectively presented in Fig. 1 and Fig. 2.

Quality Production Improvement and System Safety: QPI 16 - CZOTO 10   Materials Research Forum LLC
Materials Research Proceedings 34 (2023) 77-86   https://doi.org/10.21741/9781644902691-10

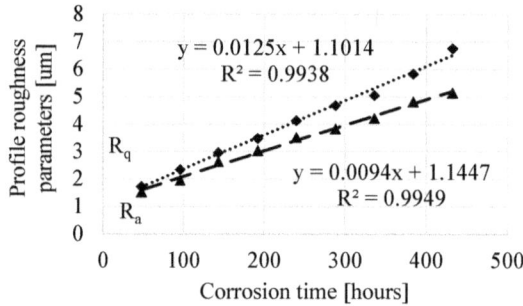

Fig.1. *Profile roughness of R35 steel after corrosion tests in 5% NaCl water solution at 10°C for different corrosion time: Ra - arithmetical mean roughness value [μm], Rq - mean peak width [μm].*

Influence time of soaking the R35 structural steel in 5% NaCl water solution at 10°C on the relative mass loss (RML) with regression equation is presented in Fig. 3.

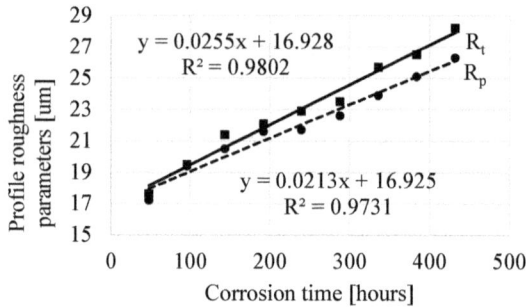

Fig.2. *Profile roughness of R35 steel after corrosion tests in 5% NaCl water solution at 10°C for different corrosion time: Rp - maximum roughness depth [μm], Rt - total height of the roughness profile [μm].*

Influence time of soaking the R35 structural steel in 5% NaCl water solution at 10°C on the corrosion rate measured in mm per year with regression equation is presented in Fig. 4. Influence time of soaking the R35 structural steel in 5% NaCl water solution at 10°C on the corrosion rate measured in gram per m² with regression equation is presented in Fig. 5.

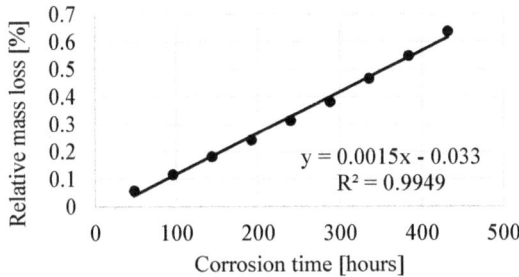

Fig.3. Influence time of soaking the R35 structural steel in 5% NaCl water solution at 10°C on the relative mass loss (RML).

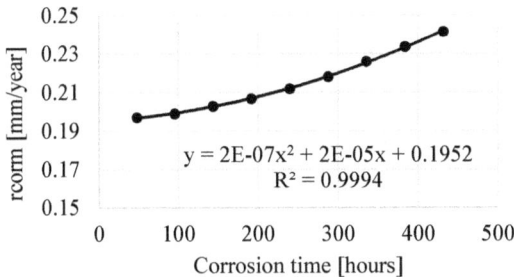

Fig.4. Influence time of soaking the R35 structural steel in 5% NaCl water solution at 10°C on the corrosion rate measured in mm per year.

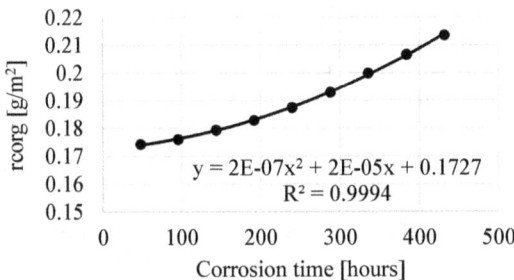

Fig.5. Influence time of soaking the R35 structural steel in 5% NaCl water solution at 10°C on the corrosion rate measured in gram per m².

Summary

Based on the test results and their statistical analysis, it was found that all the analyzed parameters of the surface condition, which are represented by roughness and relative mass loss due to soaking R35 steel in 5% NaCl at 10°C, can be presented with sufficient statistical accuracy using the first order function. This statement confirms the directly proportional relationship between surface roughness and sample soaking time.

Based on the analysis of the corrosion rate in Figs. 4 and 5 and the regression equations describing these relationships, it was found that the corrosion rate increases with the increase in the time of keeping the samples in the NaCl solution. Combining this fact with the increase in roughness, it can be assumed that the increase in the corrosion rate with the passage of soaking time is due to the increasing development of the surface of the tested samples. This development increases with the passage of soaking time, which entails an increase in the rate of corrosion.

References

[1] I. Hren et al. Comprehensive analysis of the coated component from a FORD engine. Manuf. Technol. (2021) 464-470. https://doi.org/10.21062/mft.2021.058

[2] R.W. Pechacek. NDT inspection of insulated vessels and piping for interior corrosion and corrosion under insulation. Mater. Perform. 43 (2004) 28-32.

[3] P. Kováčiková et al. The microstructural study of a damaged motorcycle gear wheel. Manuf. Technol. 21 (2021) 83-90. https://doi.org/10.21062/mft.2021.011

[4] Y. Blikharskyy et al. Corrosion fatigue damages of rebars under loading in time. Materials 14 (2021) art.3416. https://doi.org/10.3390/ma14123416

[5] J. Selejdak et al. The evaluation of the use of a device for producing metal elements applied in civil engineering, In: METAL 2014 23rd Int. Conf. Metall. Mater., Ostrava, Tanger, 2014, 1882-1888. ISBN 978-8087294543

[6] M. Askari et al. A comprehensive review on internal corrosion and cracking of oil and gas pipelines. J. Natur. Gas Sci. Eng. 71 (2019) art.102971. https://doi.org/10.1016/j.jngse.2019.102971

[7] N.G. Thompson et al. Cost of corrosion and corrosion maintenance strategies, Corros. Rev. 25 (2007) 247-262. https://doi.org/10.1515/CORRREV.2007.25.3-4.247

[8] G.A. Zhang, Y.F. Cheng. Electrochemical corrosion of X65 pipe steel in oil–water emulsion Corros. Sci. 51 (2009) 901-907. https://doi.org/10.1016/j.corsci.2009.01.020

[9] Y. Xu et al. Exploring the corrosion performances of carbon steel in flowing natural sea water and synthetic sea waters. Corros. Eng. Sci. Technol. 55 (2020) 579-588. https://doi.org/10.1080/1478422X.2020.1765476

[10] J. Owen et al. An experimental and numerical investigation of CO_2 corrosion in a rapid expansion pipe geometry. Corros. Sci. 165 (2020) art.108362. https://doi.org/10.1016/j.corsci.2019.108362

[11] L. Markovičová et al. Verification of the probability of elastomers degradation in natural environments. Prod. Eng. Arch. 28 (2022) 279-282. https://doi.org/10.30657/pea.2022.28.34

[12] C.G. Telfer et al. Particle concentration and size effects on the erosion-corrosion of pure metals in aqueous slurries. Tribol. Int. 53 (2012) 35-44. https://doi.org/10.1016/j.triboint.2012.04.010

[13] Y. Xu et al. Understanding the influences of pre-corrosion on the erosion-corrosion performance of pipeline steel. Wear 442 (2020 art.203151. https://doi.org/10.1016/j.wear.2019.203151

[14] Y. Blikharskyy et al. Non-uniform corrosion of steel rebar and its influence on reinforced concrete elements` reliability. Prod. Eng. Arch. 26 (2020) 67-72. https://doi.org/10.30657/pea.2020.26.14

[15] A. Dudek et al. Surface remelting of 316 L+434 L sintered steel: microstructure and corrosion resistance. J. Solid State Electr. 18 (2014) 2973-2981. https://do.org/10.1007/s10008-014-2483-2

[16] M. Scendo et al. Influence of laser treatment on the corrosive resistance of Wc-Cu coating produced by electrospark deposition. Int. J. Electrochem. Sci. 8 (2013) 9264-9277.

[17] T. Lipiński. Corrosion Resistance of 1.4362 Steel in Boiling 65% Nitric Acid. Manuf. Technol. 16 (2016) 1004-1009. https://doi.org/10.21062/ujep/x.2016/a/1213-2489/MT/16/5/1004

[18] H.H. Uhlig, R.W. Revie. Corrosion and corrosion control, 3rd Edition. Wiley, Hoboken, 1985.

[19] D. Kocańda et al. Fatigue Behaviour of S235JR Steel after Surface Frictional-Mechanical Treatment in Corrosive Environment. Key Eng. Mater. 598 (2014) 105-112. https://doi.org/10.4028/www.scientific.net/KEM.598.105

[20] T. Lipiński. Corrosion Effect of 20% NaCl Solution on Basic Carbon Structural S235JR Steel. 16th International Scientific Conference Engineering for Rural Development, In: 16th Int. Sci. Conf. Eng. Rural Develop., Jelgava, 24-26 May 2017, 1069-1074. https://doi.org/10.22616/ERDev2017.16.N225

[21] M. Ladan et al. Corrosion protection of AISI 1018 steel using Co-doped TiO2/polypyrrole nanocomposites in 3.5% NaCl solution. Mater. Chem. Phys. 192 (2017) 361-373. https://doi.org/10.1016/j.matchemphys.2017.01.085

[22] M. Alizadeh, S. Bordbar. The influence of microstructure on the protective properties of the corrosion product layer generated on the welded API X70 steel in chloride solution, Corros. Sci. 70 (2013) 170-179. https://doi.org/10.1016/j.corsci.2013.01.026

[23] T. Lipiński. Roughness of 1.0721 steel after corrosion tests in 20% NaCl. Prod. Eng. Arch. 15 (2017) 27-30. https://doi.org/10.30657/pea.2017.15.07

[24] J.J. Santana Rodriquez, J.E. Gonzalez Gonzalez. Identification and formation of green rust 2 as an atmospheric corrosion product of carbon steel in marine atmospheres. Mater. Corros. 57 (2006) 411-417. https://doi.org/10.1002/maco.200503942

[25] E. Naveen et al. Influence of organic corrosion inhibitors on pickling corrosion behaviour of sinter-forged C45 steel and 2% Cu alloyed C45 steel. J. Alloys Compd. 695 (2017) 3299-3309. https://doi.org/10.1016/j.jallcom.2016.11.133

[26] H. Bohni. Corrosion in reinforced concrete structures. Woodhead, Sawston, 2005. ISBN 978-1845690434

[27] P. Szabracki, T. Lipiński. Effect of aging on the microstructure and the intergranular corrosion resistance of X2CrNiMoN25-7-4 duplex stainless steel. Solid State Phenom. 203-204 (2013) 59-62. https://doi.org/10.4028/www.scientific.net/SSP.203-204.59

[28] P. Szabracki, T. Lipiński. Influence of sigma phase precipitation on the intergranular corrosion resistance of X2CrNiMoN25-7-4 super duplex stainless steel. In: METAL 2014 23rd Int. Conf. Metall. Mater. Ostrava, Tanger, 2014, 476-481. ISBN 978-808729454-3

[29] S.A. Al-Duheisat, A.S. El-Amoush. Effect of deformation conditions on the corrosion behavior of the low alloy structural steel girders. Mater. Des. 89 (2016) 342–347. https://doi.org/10.1016/j.matdes.2015.09.160

[30] R. Chandramouli et al. Deformation, densification and corrosion studies on sintered P/M plain carbon steel preforms. Mater. Des. 28 (2007) 2260-2264. http://doi.org/10.1016/j.matdes.2006.05.018

[31] V. Zatkalíková et al. Corrosion resistance of Cr-Ni-Mo Stainless Steel in Chloride and Fluoride Containing Environment, Manuf. Technol. 16 (2016) 1193-1198. https://doi.org/10.21062/ujep/x.2016/a/1213-2489/MT/16/5/119

[32] R. Chandramouli et al. Deformation, densification and corrosion studies on sintered P/M plain carbon steel preforms, Mater. Des. 28 (2007) 2260-2264. http://doi.org/10.1016/j.matdes.2006.05.018

[33] J. Zhang et al. The inhibition mechanism of imidazoline phosphate inhibitor for Q235 steel in hydrochloric acid medium, Corros. Sci. 53 (2011) 3324-3330. https://doi.org/10.1016/j.corsci.2011.06.008

[34] A. Pradityana et al. Effectiveness of myrmecodia pendans extract as eco-friendly corrosion inhibitor for material API 5L grade B in 3,5% NaCl solution, Adv. Mater. Res. 789 (2013) 484-491. https://doi.org/10.4028/www.scientific.net/AMR.789.484

[353] T. Lipinski, J. Pietraszek. Influence of animal slurry on carbon C35 steel with different microstructure at room temperature. In: 21st Int. Sci. Conf. Eng. Rural Develop., 25-27 May 2022, (2022) 344-350. https://doi.org/10.22616/ERDev.2022.21.TF115

[36] A. Y. El-Etre, M. Abdallah. Natural honey as corrosion inhibitor for metals and alloys. II. C-steel in high saline water. Corros. Sci. 42 (2000) 731-738. https://doi.org/10.1016/S0010-938X(99)00106-7

[37] T. Lipiński. Corrosion rate of the X2CrNiMoN22-5-3 duplex stainless steel annealed at 500°C. Acta Phys. Pol. A 130 (2016) 993-995. https://doi.org/10.12693/APhysPolA.130.993

[38] P. Dillmann et al. Advances in understanding atmospheric corrosion of iron. I. Rust characterisation of 767 ancient ferrous artefacts exposed to indoor atmospheric corrosion. Corros. Sci. 46 (2004) 1401-1429. http://doi.org/10.1016/j.corsci.2003.09.027

[39] D. Klimecka-Tatar et al. The effect of consolidation method on elctrochemical corrosion of polymer bonded Nd-Fe-B type magnetic material, Arch. Metall. Mater. 54 (2009) 247-256.

[40] M. Scendo et al. Influence of laser treatment on the corrosive resistance of WC-Cu coating produced by electrospark deposition, Int. J. Electrochem. Sci. 8 (2013) 9264-9277.

[41] T. Lipinski, J. Pietraszek. Influence of animal slurry on carbon C35 steel with different microstructure at room temperature, Engineering for Rural Development 21 (2022) 344-350. https://doi.org/10.22616/ERDev.2022.21.TF115

[42] Ł.J. Orman. Enhancementof pool boiling heat transfer with pin-fin microstructures, J. Enhanc. Heat Transf. 23 (2016) 137-153. https://doi.org/10.1615/JEnhHeatTransf.2017019452

[43] M. Szczepaniak et al. T. Use of the maximum power point tracking method in a portable lithium-ion solar battery charger, Energies 15 (2022) art.26. https://doi.org/10.3390/en15010026

Quality Production Improvement and System Safety: QPI 16 - CZOTO 10   Materials Research Forum LLC
Materials Research Proceedings 34 (2023) 77-86   https://doi.org/10.21741/9781644902691-10

[44] S. Maleczek et al. Tests of Acid Batteries for Hybrid Energy Storage and Buffering System—A Technical Approach, Energies 15 (2022) art.3514. https://doi.org/10.3390/en15103514

[45] E. Skrzypczak-Pietraszek, J. Pietraszek. Seasonal changes of flavonoid content in Melittis melissophyllum L. (Lamiaceae), Chem. Biodiv. 11 (2014) 562-570. https://doi.org/10.1002/cbdv.201300148

[46] E. Skrzypczak-Pietraszek et al. HPLC-DAD analysis of arbutin produced from hydroquinone in a biotransformation process in Origanum majorana L. shoot culture, Phytochem. Lett. 20 (2017) 443-448. https://doi.org/10.1016/j.phytol.2017.01.009

[47] E. Skrzypczak-Pietraszek et al. Enhanced accumulation of harpagide and 8-O-acetyl-harpagide in Melittis melissophyllum L. agitated shoot cultures analyzed by UPLC-MS/MS. PLoS ONE 13 (2018) art. e0202556. https://doi.org/10.1371/journal.pone.0202556

[48] G. Filo, E. Lisowski, M. Domagała, J. Fabiś-Domagała, H. Momeni. Modelling of pressure pulse generator with the use of a flow control valve and a fuzzy logic controller, AIP Conf. Proc. 2029 (2018) art.20015. https://doi.org/10.1063/1.5066477

[49] M. Domagala et al. CFD Estimation of a Resistance Coefficient for an Egg-Shaped Geometric Dome, Appl. Sci. 12 (2022) art.10780. https://doi.org/10.3390/app122110780

[50] M. Domagala et al. The Influence of Oil Contamination on Flow Control Valve Operation, Mater. Res. Proc. 24 (2022) 1-8. https://doi.org/10.21741/9781644902059-1

[51] N. Radek et al. Technology and application of anti-graffiti coating systems for rolling stock, METAL 2019 28th Int. Conf. Metall. Mater. (2019) 1127-1132. ISBN 978-8087294925

[52] N. Radek et al. Influence of laser texturing on tribological properties of DLC coatings, Prod. Eng. Arch. 27 (2021) 119-123. https://doi.org/10.30657/pea.2021.27.15

[53] N. Radek et al. Formation of coatings with technologies using concentrated energy stream, Prod. Eng. Arch. 28 (2022) 117-122. https://doi.org/10.30657/pea.2022.28.13

[54] N. Radek et al. The WC-Co electrospark alloying coatings modified by laser treatment, Powder Metall. Met. Ceram. 47 (2008) 197-201. https://doi.org/10.1007/s11106-008-9005-7

[55] N. Radek et al. Laser Processing of WC-Co Coatings, Mater. Res. Proc. 24 (2022) 34-38. https:10.21741/9781644902059-6

[56] P. Kurp, H. Danielewski Metal expansion joints manufacturing by a mechanically assisted laser forming hybrid method – concept, Technical Transactions 119 (2022) art.e2022008. https://doi.org/10.37705/TechTrans/e2022008

[57] R. Dwornicka, J. Pietraszek. The outline of the expert system for the design of experiment, Prod. Eng. Arch. 20 (2018) 43-48. https://doi.org/10.30657/pea.2018.20.09

[58] J. Pietraszek, N. Radek, A.V. Goroshko. Challenges for the DOE methodology related to the introduction of Industry 4.0. Production Engineering Archives 26 (2020) 190-194. https://doi.org/10.30657/pea.2020.26.33

[59] B. Jasiewicz et al. Inter-observer and intra-observer reliability in the radiographic measurements of paediatric forefoot alignment, Foot Ankle Surg. 27 (2021) 371-376. https://doi.org/10.1016/j.fas.2020.04.015

[60] J. Pietraszek. The modified sequential-binary approach for fuzzy operations on correlated assessments, LNAI 7894 (2013) 353-364. https://doi.org/10.1007/978-3-642-38658-9_32

[61] J. Pietraszek et al. Non-parametric assessment of the uncertainty in the analysis of the airfoil blade traces, METAL 2017 26th Int. Conf. Metall. Mater. (2017) 1412-1418. ISBN 978-8087294796

[62] A. Deja et al. Analysis and assessment of environmental threats in maritime transport, Transp. Res. Procedia 55 (2021) 1073-1080. https://doi.org/10.1016/j.trpro.2021.07.078

[63] J. Fabiś-Domagała, G. Filo, H. Momeni, M. Domagała. Instruments of identification of hydraulic components potential failures, MATEC Web of Conf. 183 (2018) art.03008. https://doi.org/10.1051/matecconf/201818303008

[64] J. Fabis-Domagala et al. A concept of risk prioritization in FMEA analysis for fluid power systems, Energies 14 (2021) art. 6482. https://doi.org/10.3390/en14206482

[65] J. Fabis-Domagala, M. Domagala. A Concept of Risk Prioritization in FMEA of Fluid Power Components, Energies 15 (2022) art.6180. https://doi.org/10.3390/en15176180

[66] D. Klimecka-Tatar. Context of production engineering in management model of Value Stream Flow according to manufacturing industry, Prod. Eng. Arch. 21 (2018) 32-35. https://doi.org/10.30657/pea.2018.21.07

[67] N. Baryshnikova et al. Management approach on food export expansion in the conditions of limited internal demand, Pol. J. Manag. Stud. 21 (2020) 101-114. https://doi.org/10.17512/pjms.2020.21.2.08

[68] PN-89/H 84023/07 Specific application steels. Steels for tubing. Grades.

[69] PN-80/H-74219 Hot rolled seamless steel tubes for general use.

[70] PN EN ISO 3651-1, Determination of resistance to intergranular corrosion of stainless steels. Part 1: Austenitic and ferritic-austenitic (duplex) stainless steels. Corrosion test in nitric acid medium by measurement of loss in mass (Huey test).

Quality Production Improvement and System Safety: QPI 16 - CZOTO 10   Materials Research Forum LLC
Materials Research Proceedings 34 (2023) 87-94   https://doi.org/10.21741/9781644902691-11

Effect of Impregnation of Activated Carbon with Selected Transition Metal Ions on Its Adsorption Properties and Pore Size

POSZWALD Bartosz[1,a]*, KWAK Anna[1,b], DYSZ Karolina[1,c]
and DYLONG Agnieszka[1,d]

[1] Military Institute of Engineer Technology, 136 Obornicka Str., 50-961 Wroclaw, Poland

[a] poszwald@witi.wroc.pl, [b] kwak@witi.wroc.pl, [c] dysz@witi.wroc.pl, [d] dylong@witi.wroc.pl

Keywords: Activated Carbon, Impregnation, Adsorption Isotherm, BET Method, BJH Method, Transition Metals

Abstract. Thanks to their vast surface area, activated carbons are materials of great interest. They are used as adsorbents for both liquid- and gaseous impurities. To improve their properties, impregnation with different substances is conducted. In this paper, such an impregnation was done using copper(II), manganese(II), silver(I), and zinc(II) salts. The aim was to compare the inner structure of the impregnated activated carbon with a non-impregnated one. It was done by Brunauer-Emmett-Teller (BET) adsorption-desorption isotherms that were plotted and described. Lastly, the pore size and volume were determined by Barrett-Joyner-Halenda (BJH) method. The results show that the presence of zinc(II) salts helped to develop the mesoporous structure of activated carbon and resulted in an increase in the surface area of the sorbents, while silver(I) decreased it.

Introduction

Adsorption is a process taking place at the interphase phases [1]. It can be physical and involve covalent bonds, ionic bonds, van der Waals forces, or chemical in character when adsorbate is reacting with adsorbent [2]. The efficiency of the process is highly dependent on how large the surface area of the sorbent is. Many factors may impact this parameter but one of them, more than any other is meaningful – the porosity. Pores are cavities on the surface of a sorbent which multiply its surface by manyfold. Zeolite can serve as an example here, with its surface of almost 130 m^2/g [3], or activated carbon with 2636 m^2/g [4].

Continues process of absorption can be described using graphs called isotherms, showing the relationship between the amount of adsorbed substance and its concentration in an equilibrium state. Many isotherms describe different theories and states. For example, Henry's isotherm is adequate for the process of adsorption of low-concentration fluids and gases, and the Freundlich model is describing a situation in which the surface of the adsorbent is not smooth but porous. One of the most used, thanks to its great adequacy to reality, is Brunauer-Emmett-Teller (BET) theory. Its main assumption is based on Langmuir isotherm [5] but modified, so it is describing a multilayer adsorbate on a porous sorbent's surface. The important thing here is the fact, that the greater the vapour pressure of adsorbate over adsorbent, the higher the number of adsorbed particles. BET is given by a linear equation (Eq. 1) but following further mathematical operations BET isotherms are curves.

$$\frac{1}{X(\frac{p_0}{p} - 1)} = \frac{1}{X_m C} + \frac{C-1}{X_m C}(\frac{p}{p_0})$$

Eq. 1

X – the total adsorbed volume of gas under pressure p
p_0 – saturated vapor pressure of an adsorbate

X_m – monolayer capacity (volume of gas adsorbed in a single layer)
C – equilibrium adsorption state
By knowing the monolayer capacity from the equation (Eq. 1) total surface of the adsorbent can be calculated (Eq. 2):

$$S = \frac{X_m L_{Av} A_m}{M_V}$$
Eq. 2

S – total surface area
$L_{av.}$ – Avogadro's number
A – cross-sectional area of the adsorbate (for nitrogen 0.162 nm^2)
M_v – molar volume

Six types of isotherms can be described, according to the International Union of Pure and Applied Chemistry (IUPAC) [6]. They are depicting (Fig. 1) the adsorption process on different adsorbents with variable pore structures, divided into both pore size and pore shape. Designations (names) for pores by size are determined by IUPAC, a micropore is less than 2nm cavity, a mesopore ranges from 2 to 50 nm, and a macropore is larger than 50 nm. Pore size also affects the condensation mechanism. While there is three-dimensional condensation in micropores and capillary condensation in mesopores, there is no condensation in macropores [7].

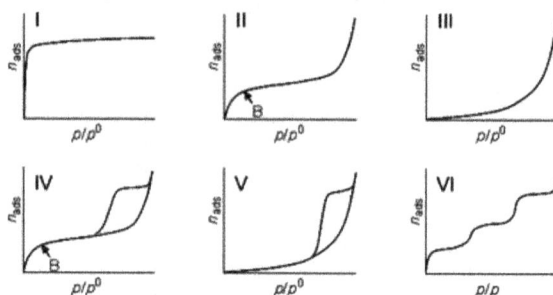

Fig.1. *Types of isotherms by adsorbent outer structure. Type I: microporous; type II: non-porous or macroporous; type III: non-porous or macroporous but with weak interactions; type IV: mesoporous; type V: mesoporous but with weak interactions; type VI: step-by-step multilayer adsorption. (source: [8]).*

Activated carbons (ACs) are being used in many fields. They are obtained from organic matter, through a process called activation. Among the most popular substrates are wood (35% of total substrates used), coal (28%), lignite (14%), peat (10%), and local organic residues such, as coconut shells or fruit stones (10%). Activation can be divided into two types: chemical one and physical one. The former involves two steps, the first step includes mixing the substrate with some reagents, e.g. zinc(II) chloride, phosphoric acid or sodium hydroxide, and the second one is heating. Activation temperature depends on the type of substance that was used. Carbon mixed with $ZnCl_2$ is heated to around 600°C, at which temperature total dehydration takes place. On the other hand, substrates containing H_3PO_4 are heated to 800°C, so that the acid is oxidizes carbon and the structure is changing [9]. Physical activation is heating conducted in two steps. In the beginning, the substrate is carbonized at 400-600°C, as in this temperature rigid carbon structure is obtained.

Quality Production Improvement and System Safety: QPI 16 - CZOTO 10   Materials Research Forum LLC
Materials Research Proceedings 34 (2023) 87-94   https://doi.org/10.21741/9781644902691-11

To get the actual, useful product oxidation has to take place. It is done by increasing the temperature to 800-950°C with the addition of water vapour or carbon dioxide to the reaction environment, which penetrates and expands pores [10]. Such products can be further modified by impregnating them with different compounds so the ACs' properties are altered or upgraded. Metal-impregnated activated carbons (MIACs) are sorbents of great interest due to their relatively simple manufacturing.

Experimental Procedure

DT0 commercially-available activated carbon (produced by Gryfskand Sp. z o.o. company in Poland) was used in this experiment. It was first oxidized in a 20% nitric acid solution and then washed until all the acid was removed. Afterwards, a washed activated carbon was kept at room temperature for 24h. At the end of the process, the product was dried in a fluid bed dryer at 120°C. The following transition metal salts were chosen for impregnation

– universal adsorbents (UA), containing zinc, copper, manganese, silver, and
– adsorbents containing zinc and silver (S-Hg/Pb) only.

The following metallic catalysts have been selected to produce UA:

– basic copper carbonate ($Cu_2CO_3(OH)_2$) in aqua ammonia;
– potassium permanganate ($KMnO_4$) in aqueous solution;
– silver nitrate ($AgNO_3$) in aqueous solution;
– zinc nitrate ($ZnNO_3$) in aqueous solution.

The final step leading to MIACs possessing the desired properties was drying them for 6 hours at the temperature of 120°C and 2 hours at 180°C.

Adsorption-desorption isotherms and inner structure tests were carried out on Autosorb-1C chemisorption - physisorption analyzer by Quantachrome Instruments.

Table 1. Concentrations of metal ions in MIACs

No.	Sorbent	Copper(II) [% w/w]	Manganese(II) [% w/w]	Silver(I) [% w/w]	Zinc(II) [% w/w]
1.	DT0	-	-	-	-
2.	IS-2	3.0	1.2	0.5	1.0
3.	IS-3	3.0	1.2	-	0.5
4.	IS-4	3.0	1.2	0.5	2.0
5.	IS-5	3.5	1.8	1.0	-
6.	IS-6	3.0	1.8	1.0	-
7.	IS-7	3.0	1.8	0.5	-
8.	IS-8	3.0	1.2	0.5	-
9.	IS-9	3.0	1.2	1.0	-
10.	IS-10	2.5	1.2	0.5	-

Results and Discussion

Some of the MIACs impregnated with zinc(II) and other salts contained aggregated copper-blue particles. Those carbons were observed under Keyence series VHX-7000 microscope with VH-ZST dual objective zoom lens (Fig. 2). It seems that some regions adsorbed more copper(II) salt which, during the heating process, was not reduced to copper(II) oxide. This irregularity must be linked to the presence of zinc(II) salt in the samples, but chemistry of this process is not known yet, due to insufficient amount of data.

By comparing the surface areas of DT0 and ISs we see that zinc(II) impregnated ACs have a significantly larger surface than their nonimpregnated precursor. Other adsorbents, IS-5 – IS-10, had their area decreased by 5-10%. Additionally, the highest increase, around 44%, was noted for

IS-3 – the only sorbent which does not contain silver(I). Thus, a conclusion can be made that presence of zinc(II) salt in the impregnation mixture is promoting better performance parameters, while silver(I) presence is accomplishing the opposite.

To specify, which pores were affected the most, the BJH curve was used. It appears that zinc(II) helped to develop the mesoporous structure of activated carbon. The biggest increase is observed for IS-3 and IS-4 sorbents. This implies that the absence of silver(I) and a higher concentration of zinc(II) positively impact the efficiency of adsorption properties. As can be seen in graphs (Fig. 4 b) and c)) micropore capacity has also increased. One unusual situation was noted for IS-8 in which the mesopores disappearance almost totally.

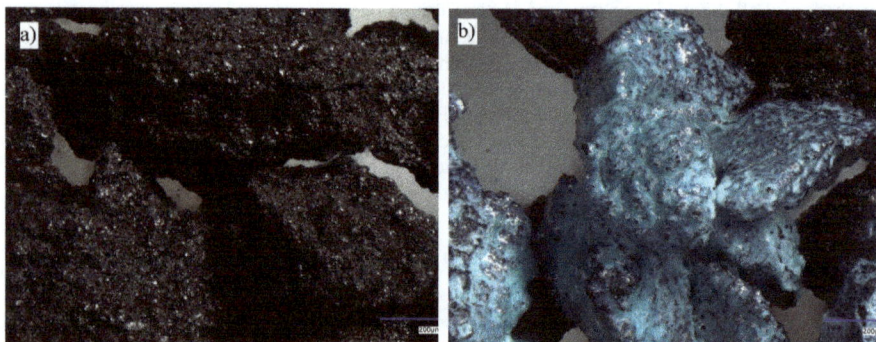

Fig.2. *a) Noncoated IS-5 vs. b) copper(II) salt coated aggregated AC IS-4*

While considering the changes to the structure of all the adsorbents, they are not as large as one could assume. This is proven by BET isotherms (Fig. 4), as all of them are type IV(a) with H4 hysteresis [11], which means that the shape of the pores remains unchanged and only the number of individual pores differentiates the MIACs.

Table 2. *Sorbents parameters*

Sample	DT0	IS-2	IS-3	IS-4	IS-5	IS-6	IS-7	IS-8	IS-9	IS-10
Surface area* [m^2/g]	555.4	796.6	804.3	592.1	523.0	507.8	491.7	498.2	482.7	519.2
Average pore size [nm]	2.100	2.139	2.346	2.296	2.117	2.106	2.076	2.114	2.115	2.112

* BJH method cumulative desorption surface area

Quality Production Improvement and System Safety: QPI 16 - CZOTO 10   Materials Research Forum LLC
Materials Research Proceedings 34 (2023) 87-94   https://doi.org/10.21741/9781644902691-11

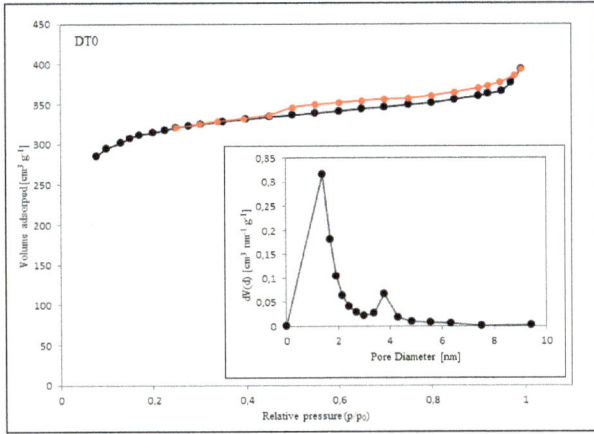

Fig.3. *BET nitrogen adsorption-desorption isotherms and BJH pore size distribution curve of DT0*

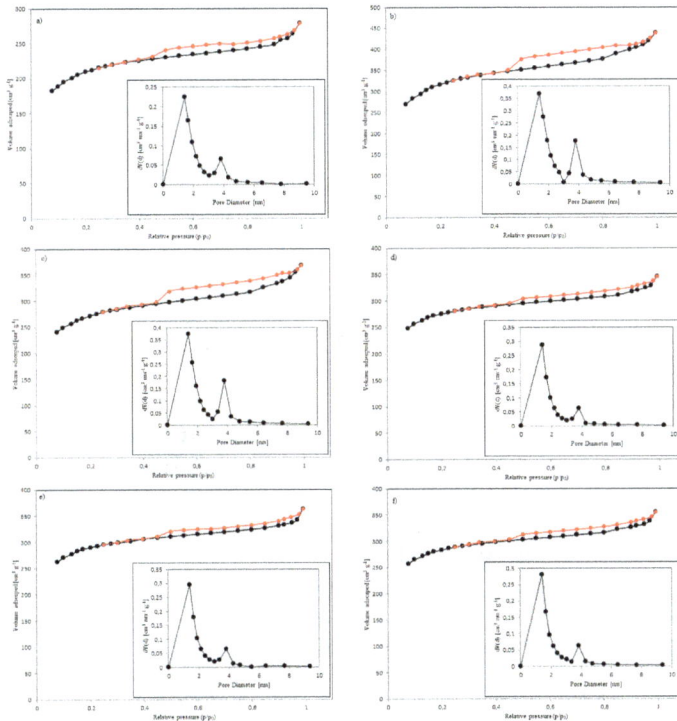

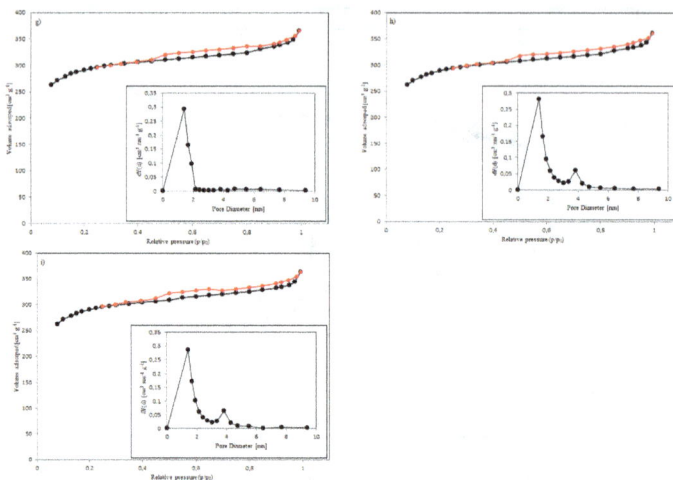

Fig.4. *BET nitrogen adsorption-desorption isotherms and BJH pore size distribution curve of:*
a) IS-2, b) IS-3, c) IS-4, d) IS-5, e) IS-6, f) IS-7, g) IS-8, h) IS-9, i) IS-10

Conclusions

This paper proves that the introduction of transition metals salts is changing the inner structure of activated carbons. For example, silver(I) causes a decrease in the number of pores yielding so smaller surface area and resulting in worse adsorption properties. However, silver is an antiseptic so it would be helpful in the treatment of biologically-contaminated samples. On the opposite side, there is zinc(II), which increases the porosity of carbonaceous adsorbents by as high as 44%. Such modified carbons have more developed mesopores and micropores. More transition metal ions may have a positive impact on the adsorbent structure, however further experiments are required.

Any further work will require in-depth statistical analysis [12], especially regarding possible interactions [13-15]. In cases where the number of available measurements is too small to adequately narrow down the uncertainty range, nonparametric [16-18] and resampling [19, 20] approaches will be useful. The obtained results and the methodology of these approaches can be applicable in metal processing [21], welding [22], and coating applications [23], including special coatings [24, 25].

References

[1] S.J. Chak. Absorption, in: A.D. McNaught, A. Wilkinson (Eds.) IUPAC. Compendium of Chemical Terminology 2nd ed., Blackwell Scientific Publications, Oxford, 1997. https://doi.org/10.1351/goldbook.A00036

[2] Procesy sorpcyjne. Skrypt do ćwiczeń. [online]. 2010. [viewed: 2023-01-25]. Available from: http://www2.chemia.uj.edu.pl/dydaktyka/Procesy_Sorpcyjne.pdf

[3] G. Jozefaciuk, A. Szatanik-Kloc, A. Ambrozewicz-Nita. The Surface area of zeolite-amended soils exceed the sum of the inherent surface areas of soil and zeolites, Eur. J. of Soil Sc. 69(5) (2018) art. 12691. https://doi.org/10.1111/ejss.12691

[4] A. Kumar, H. M. Jena. Adsorption of Cr(VI) from aqueous solution by prepared high surface area activated carbon from Fox nutshell by chemical activation with H_3PO_4, J. of Env. Chem. Eng. 5(2) (2017) 2032-2041. https://doi.org/10.1016/j.jece.2017.03.035

[5] S. Tripathi, N. Arora, P. Gupta, P.A. Pruthi, K.M. Poluri, V. Pruthi. Microalgae: An emerging source for mitigation of heavy metals and their potential implications for biodiesel production, in: A.K. Azad, M. Rasul, Advanced Biofuels Applications, Technologies and Sustainable Development, Woodhead Publishing, Sawstone, 2019, pp. 97-128. https://doi.org/10.1016/B978-0-08-102791-2.00004-0

[6] Standardized reporting of gas adsorption isotherms [online]. 2021. [viewed: 2023-01-27]. Available from: https://iupac.org/project/2021-016-1-024

[7] R.J. White, V. Budarin, J.H. Clark, R.Luque. Tuneable porous carbonaceous materials from renewable resources, Chem. Soc. Rev. 38(12) (2009) 3401-3418. https://doi.org/10.1039/b822668g

[8] M. Kajama. Hydrogen permeation using nanostructured silica membranes, Sustainable Development and Planning 2015 (2015), Istanbul, Turkey, 447-456. https://doi.org/10.2495/SDP150381

[9] Carbonology Chemically activated carbon. [online]. 2019. [viewd: 2023-01-29]. Available from: https://www.desotec.com/en/carbonology/carbonology-academy/chemically-activated-carbon

[10] T. Borowiecki, J. Kijeński, J. Machnikowski, M. Ściążko. Czysta energia, produkty chemiczne i paliwa z węgla- ocena potencjału rozwojowego, Wydawnictwo Instytutu Chemicznej Przeróbki Węgla, Zabrze, 2008.

[11] M. Thommes et al. Physisorption of gases, with special reference to the evaluation of surface area and pore size distribution (IUPAC Technical Report), Pure Appl. Chem. 87(9-10) (2015) 1051-1069. https://doi.org/10.1515/pac-2014-1117

[12] B. Jasiewicz et al. Inter-observer and intra-observer reliability in the radiographic measurements of paediatric forefoot alignment, Foot Ankle Surg. 27 (2021) 371-376. https://doi.org/10.1016/j.fas.2020.04.015

[13] J. Pietraszek et al. The parametric RSM model with higher order terms for the meat tumbler machine process, Solid State Phenom. 235 (2015) 37-44. https://doi.org/10.4028/www.scientific.net/SSP.235.37

[14] J. Korzekwa et al. Tribological behaviour of Al2O3/inorganic fullerene-like WS2 composite layer sliding against plastic, Int. J. Surf. Sci. Eng. 10 (2016) 570-584. https://doi.org/10.1504/IJSURFSE.2016.081035

[15] J. Pietraszek, A. Szczotok, N. Radek. The fixed-effects analysis of the relation between SDAS and carbides for the airfoil blade traces. Arch. Metall. Mater. 62 (2017) 235-239. https://doi.org/10.1515/amm-2017-0035

[16] J. Pietraszek. Fuzzy regression compared to classical experimental design in the case of flywheel assembly, LNAI 7267 LNAI (2012) 310-317. https://doi.org/10.1007/978-3-642-29347-4_36

Quality Production Improvement and System Safety: QPI 16 - CZOTO 10   Materials Research Forum LLC
Materials Research Proceedings 34 (2023) 87-94   https://doi.org/10.21741/9781644902691-11

[17] J. Pietraszek. The modified sequential-binary approach for fuzzy operations on correlated assessments, LNAI 7894 (2013) 353-364. https://doi.org/10.1007/978-3-642-38658-9_32

[18] J. Pietraszek et al. Non-parametric assessment of the uncertainty in the analysis of the airfoil blade traces, METAL 2017 – 26th Int. Conf. Metall. Mater. (2017) 1412-1418. ISBN 978-8087294796

[19] J. Pietraszek, L. Wojnar. The bootstrap approach to the statistical significance of parameters in RSM model, ECCOMAS Congress 2016 Proc. 7th Europ. Congr. Comput. Methods in Appl. Sci. Eng. 1 (2016) 2003-2009. https://doi.org/10.7712/100016.1937.9138

[20] J. Pietraszek et al. Challenges for the DOE methodology related to the introduction of Industry 4.0. Prod. Eng. Arch. 26 (2020) 190-194. https://doi.org/10.30657/pea.2020.26.33

[21] P. Jonšta et al. The effect of rare earth metals alloying on the internal quality of industrially produced heavy steel forgings, Materials 14 (2021) art.5160. https://doi.org/10.3390/ma14185160

[22] N. Radek et al. The impact of laser welding parameters on the mechanical properties of the weld, AIP Conf. Proc. 2017 (2018) art.20025. https://doi.org/10.1063/1.5056288

[23] N. Radek et al. Technology and application of anti-graffiti coating systems for rolling stock, METAL 2019 – 28th Int. Conf. Metall. Mater. (2019) 1127-1132. ISBN 978-8087294925

[24] N. Radek et al. Microstructure and tribological properties of DLC coatings, Mater. Res. Proc. 17 (2020) 171-176. https://doi.org/10.21741/9781644901038-26

[25] N. Radek et al. Influence of laser texturing on tribological properties of DLC coatings, Prod. Eng. Arch. 27 (2021) 119-123. https://doi.org/10.30657/pea.2021.27.15

Quality Production Improvement and System Safety: QPI 16 - CZOTO 10   Materials Research Forum LLC
Materials Research Proceedings 34 (2023) 95-101   https://doi.org/10.21741/9781644902691-12

Effect of KOH Content in the Electrolyte on Corrosion Properties of PEO-Coated EV31 Magnesium Alloy

KNAP Vidžaja [1, a *], OBERTOVÁ Veronika [1, b] and HADZIMA Branislav [2, c]

[1]University of Žilina, Faculty of Mechanical Engineering, Univerzitná 8215/1, 010 26 Žilina, Slovak Republic

[2]University of Žilina, Research centre, Univerzitná 8215/1, 010 26 Žilina, Slovak Republic

[a]vidzaja.knap@fstroj.uniza.sk, [b]veronika.obertova@fstroj.uniza.sk, [c]branislav.hadzima@uniza.sk

Keywords: Plasma Electrolytic Oxidation, PEO, Corrosion, EV31 Magnesium Alloy, Electrochemical Impedance Spectroscopy

Abstract. This study analyzed the influence of electrolyte composition on the corrosion resistance of plasma electrolytic oxidation (PEO) coatings on EV31 magnesium alloy. An electrolyte solution was prepared by mixing 12 g/l Na_3PO_4 and various levels of KOH (1, 2, and 4 g/l). The PEO coating was produced with a direct current (DC) regime of 0.05 mA.cm^{-2} current density and a maximum voltage of 630 V. Electrochemical impedance spectroscopy was performed to assess the corrosion resistance after one hour exposure in 0.1 M NaCl. The results indicated that KOH content has a substantial effect on the morphology and corrosion resistance of the PEO coating and align with previous similar studies. The lowest porosity and highest polarization resistance were observed in the PEO coating prepared with 2 g/l of KOH in the electrolyte, resulting in the best corrosion resistance among the evaluated samples. The presence of only one capacitive loop in the Nyquist diagram and low polarization resistance R_p values of the samples with 4 g/l of KOH in the electrolyte indicated insufficient compactness due to high porosity of the coating.

Introduction

Magnesium and its alloys have gained increasing attention in recent years as lightweight materials with excellent mechanical properties and high strength-to-weight ratios. This has led to a growing interest in their use for a wide range of applications, including aerospace, biomedical, and automotive industries [1, 2]. Despite their attractive properties, the widespread use of magnesium alloys is hindered by their high reactivity and low corrosion resistance, particularly in harsh environments. The low corrosion resistance of magnesium alloys can result in a reduction in the lifetime and performance of magnesium-based components, making it a major concern for their practical applications [3, 4].

To overcome this challenge, surface modification techniques have been developed to improve the corrosion resistance of magnesium alloys. One of the most promising techniques is plasma electrolytic oxidation (PEO), which has gained significant attention due to its environmentally friendly nature and ability to form dense, corrosion-resistant coatings on magnesium alloys. The PEO process involves the formation of a protective oxide layer on the surface of the magnesium alloy through the application of a high voltage electric field in an electrolyte solution. The resulting oxide layer provides excellent protection against corrosion and has been shown to be effective in improving the corrosion resistance of magnesium alloys [5-9].

The PEO process is a complex one, and a wide range of factors can influence the properties of the resulting oxide layer. These factors include the composition of the electrolyte solution, the applied voltage, the duration of the treatment, and the type of magnesium alloy being treated. Understanding the relationships between these factors and the properties of the oxide layer is

Quality Production Improvement and System Safety: QPI 16 - CZOTO 10   Materials Research Forum LLC
Materials Research Proceedings 34 (2023) 95-101   https://doi.org/10.21741/9781644902691-12

crucial for the optimization of the PEO process and the development of high-quality coatings for practical applications [10, 11].

One of the major advantages of the PEO process is its environmentally friendly nature. Unlike traditional surface modification techniques, such as thermal spray, the PEO process does not generate harmful pollutants or require the use of hazardous materials. This makes it an attractive option for industries that are focused on reducing their environmental impact, and it also has the potential to reduce the cost and complexity of the surface modification process [12-14].

Another advantage of the PEO process is its ability to form dense, corrosion-resistant coatings with excellent mechanical properties. The high-quality oxide layer formed by the PEO process provides excellent protection against corrosion and has been shown to significantly improve the overall performance of magnesium alloys in harsh environments. Additionally, the PEO process can be used to create coatings with unique surface textures, such as micro- and nano-structures, which can further improve the corrosion resistance of magnesium alloys [15, 16].

Despite its advantages, the PEO process still faces a number of challenges that need to be addressed in future research and development. For example, the resulting oxide layer can be porous and prone to cracking, which can limit its effectiveness as a corrosion barrier. Additionally, the optimization of the PEO process is an ongoing area of research, as it is important to ensure that the desired properties of the oxide layer are achieved while minimizing the formation of unwanted features, such as porosity and cracks [17].

Therefore, further research and optimization of the PEO process is needed to address the challenges associated with the formation of porous and cracked oxide layers and to develop high-quality coatings for practical applications.

Experimental Methods

Experiments were conducted on samples made from the extruded EV31 alloy. The samples were prepared for PEO treatment through grinding with emery paper p1200. The solution used for the PEO coating was made up of 12g/l of $Na_3PO_4.12H_2O$ and different amounts of KOH (1, 2, and 4 g/l) dissolved in deionized water. The experiments utilized a two-electrode circuit with the EV31 alloy sample as the anode and a stainless-steel plate as the cathode. The electrolyte was kept agitated using a laboratory stirrer operating at 500 rpm to ensure a consistent supply of reactants. The 1.5 L container with the electrolyte was immersed in a 15 L container with a cooling medium to regulate the temperature of the PEO electrolyte, which was maintained at below 50 °C during the entire coating preparation process. The PEO coating preparation utilized DC mode with a constant current density of 50 mA.cm^{-2}.

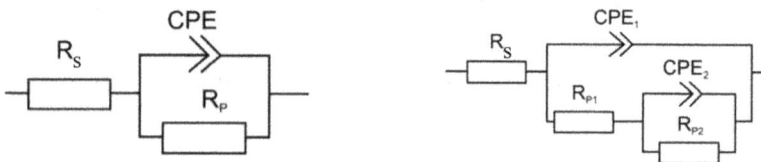

Fig.1. *The equivalent circuits used for analyzing Nyquist diagrams a) circuit with a single capacitive loop and b) circuit with two capacitive loops.*

The electrochemical impedance spectroscopy (EIS) method was used to examine the corrosion properties of PEO coated samples in a corrosive environment after 1 hour of exposure. The EIS measurements were conducted in a frequency range between 100 kHz and 10 mHz, with the frequency changing every decade. The amplitude of the alternating current voltage signal applied was set at 15 mV and the mean value of the voltage component was the same as the open potential.

Quality Production Improvement and System Safety: QPI 16 - CZOTO 10   Materials Research Forum LLC
Materials Research Proceedings 34 (2023) 95-101   https://doi.org/10.21741/9781644902691-12

The results were presented as Nyquist diagrams and analyzed using equivalent circuit method with the EC Lab 10.42 software from Biologic SAS. The equivalent circuits used in the analysis are shown in Fig. 1, with a simple Randles circuit used to describe Nyquist diagrams with one capacitive loop and more complex circuits used to describe diagrams with two capacitive loops. These circuits suggest the presence of different electrochemical behaviors in the regions, which may indicate the presence of a surface layer, a layer of corrosion products, and a subsurface or double layer on the surface. The R_s element in the circuits represents the electrolyte resistance, while the CPE element, which replaces a capacitor, represents the inhomogeneity of the electrode surface. The polarization resistance (R_p) is the most crucial element in determining corrosion resistance, as it directly relates to the corrosion resistance of the measured surface. When two capacitive loops are present, the total polarization resistance is calculated by adding the R_{p1} and R_{p2} partial resistances.

Results and Discussion
The surface micrographs of PEO coated samples with prepared different KOH content in electrolyte are shown in Fig. 2. When comparing the micrographs of individual coatings, common characteristics such as porosity and the presence of microcracks across the layers can be observed. These phenomena are typical for this type of coatings and arise from the very nature of the PEO process [18]. Fig. 2a presents a coating prepared with 1 g/l of KOH, which is commonly [19] used KOH content for PEO coating preparation. When comparing the images of individual coatings, differences between the coatings become apparent. With the increase of KOH content from 1 g/l to 2 g/l, there was a noticeable reduction in both pore size and count. However, further increasing the concentration to 4 g/l did not produce the same result, as the coating with 4 g/l of KOH (Fig. 2c) had the highest porosity and pore size, around 20 μm, of all three samples.

The Nyquist plots of the EV31 samples, surface-treated with PEO coating in electrolyte with different KOH contents, exposed in 0.1M NaCl environment for 1 hour are shown in Fig 2. The values of the electrochemical characteristics obtained from the analysis of the measured data are listed in Table 1.

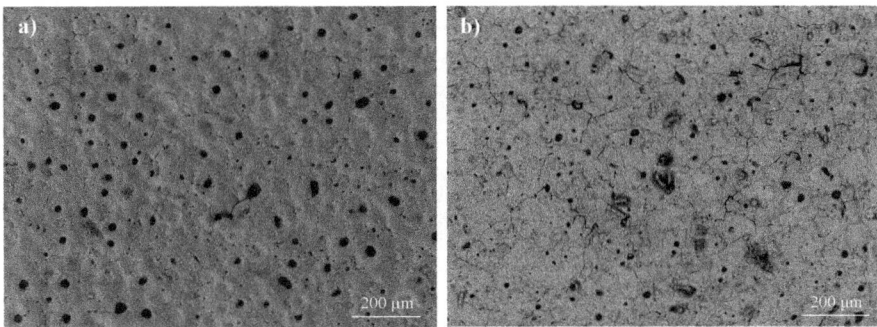

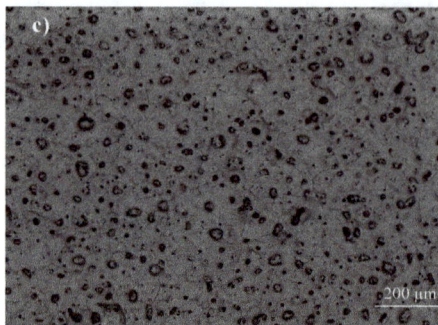

Fig.2. *Surface micrographs for samples with different KOH additions to electrolyte*
a) 1 g/l of KOH b) 2 g/l of KOH c) 4 g/l of KOH.

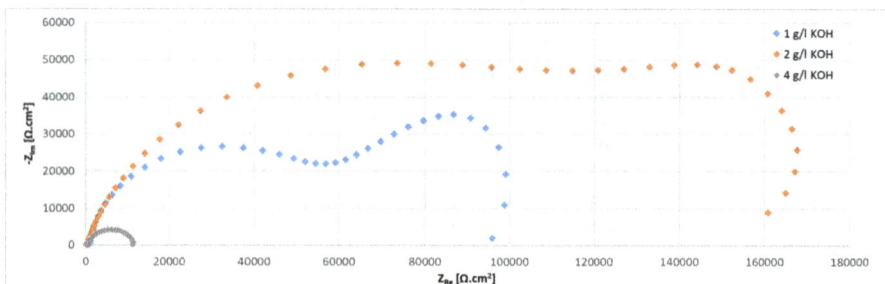

Fig.3. *Nyquist diagrams resulting from EIS measurement after 1h of exposure in 0.1M NaCl for*
1, 2, and 4 g/l of KOH.

Table 1. *Electrochemical characteristics resulting from EIS measurement after 1h of*
exposure in 0.1M NaCl for 1, 2, and 4 g/l of KOH

KOH content	1 g/l KOH	2 g/l KOH	4 g/l KOH
R_s ($\Omega.cm^2$)	107 ± 9	91 ± 14	115 ± 7
R_{p1} ($\Omega.cm^2$)	43908 ± 1353	$114\,371 \pm 5723$	$12\,233 \pm 827$
R_{p2} ($\Omega.cm^2$)	57582 ± 2489	$42\,612 \pm 1472$	-
R_p ($\Omega.cm^2$)	101490 ± 3862	$156\,983 \pm 7195$	$12\,233 \pm 827$
CPE_1 ($F.sn^{-1}.10^{-6}$)	1.7 ± 0.1	0.3 ± 0.02	0.9 ± 0.05
CPE_2 ($F.sn^{-1}.10^{-6}$)	14 ± 0.3	18 ± 0.9	-
n_1	0.70	0.65	0.7
n_2	0.85	0.85	-

The comparison of calculated values of polarization resistance R_p for the samples with different KOH addition to electrolyte (Table 1) shows that samples with addition of 2 g/l of KOH exhibited over a 50 % increase in polarization resistance R_p compared to the values obtained for 1 g/l of KOH samples and over a 10-fold increase compared to the samples with 4 g/l of KOH. Higher values of polarization resistance represent higher corrosion resistance; hence it can be concluded that from the evaluated samples the PEO coating prepared with the 2 g/l of KOH provides the most effective barrier that slows down the degradation of the EV31 magnesium alloy.

As can be seen from the shapes of the respective plots, both 1 g/l and 2 g/l KOH in electrolyte samples' diagrams consist of two capacitance loops. This means that two values of polarization resistance R_p were obtained. The first loop of the Nyquist diagram represents the resistance of the porous layer (resistance R_{p1}), and the second capacitance loop represents the resistance to charge transfer at the sample/electrolyte interface (resistance R_{p2}). In the case of the coating created with 4 g/l of KOH in the electrolyte, due to its insufficient compactness resulting from very high porosity, it shows significant decrease in R_p compared to the other two samples. The presence of only one capacitive loop in the Nyquist diagram of this sample is likely caused by penetration of the electrolyte to the surface of the bare substrate through the defects and pores. Therefore, the PEO coating was not detected significantly during the EIS measurement [20].

These results confirm that KOH content has a significant influence on the morphology and corrosion resistance of PEO coating. In a similar study [21] they observed the influence of KOH content on the PEO coating formed on AZ91 Mg alloy in a silicate electrolyte. They compared the use of 0.09 M and 0.27 M of KOH and the results showed that the use of 0.27 M KOH electrolyte improved the corrosion protection of the alloy due to a significant reduction of the pore size. Although the KOH concentrations in this study were different, their findings are in agreement with the results of the current study and further strengthen the conclusion that the KOH concentration in the electrolyte affects the corrosion resistance and morphology of PEO coatings.

Conclusions

The following conclusions are drawn from the performed experiments, results, and analysis:

- The results showed that KOH content has a significant influence on the morphology and corrosion resistance of PEO coating and the findings are in agreement with similar studies.
- The low values of polarization resistance R_p and the presence of only one capacitive loop in the Nyquist diagram indicate that the PEO coating created with 4 g/l of KOH in the electrolyte shows insufficient compactness due to high porosity of the coating.
- The PEO coating prepared with 2 g/l of KOH in the electrolyte provides the lowest porosity and highest polarization resistance, leading to the highest corrosion resistance from the evaluated samples.

Acknowledgement

This paper was supported under the project of Operational Programme Integrated Infrastructure: Independent research and development of technological kits based on wearable electronics products, as tools for raising hygienic standards in a society exposed to the virus causing the COVID-19 disease, ITMS2014+ code 313011ASK8. The project is co-funding by European Regional Development Fund. The research was also financially supported by the Science Grant Agency of the Slovak Republic through project No. 1/0117/21 and No. 1/0153/21.

References

[1] J. Song et al. Research advances of magnesium and magnesium alloys worldwide in 2021. J. Magnes. Alloys 10 (2022) 863-898. https://doi.org/10.1016/j.jma.2022.04.001

[2] A. Du Plessis Properties and applications of additively manufactured metallic cellular materials: A review. Prog. Mater. Sci. 125 (2021) art.100918. https://doi.org/10.1016/j.pmatsci.2021.100918

[3] S.S. Prasad et al. The role and significance of Magnesium in modern day research – A review. J. Magnes. Alloys 10 (2022) 1-61. https://doi.org/10.1016/j.jma.2021.05.012

[4] D. S. Tsai, Y.C. Tsai, C.C Chou. Corrosion passivation of magnesium alloy with the duplex coatings of plasma electrolytic oxidation and tetrafluoroethylene-based polymers. Surf. Coat. Technol. 366 (2019) 15-23. https://doi.org/10.1016/j.surfcoat.2019.03.015

[5] D. Kajánek et al. Electrochemical impedance spectroscopy characterization of ZW3 magnesium alloy coated by DCPD using LASV deposition technique. Acta Metall. Slovaca 23 (2017) 147-154. https://doi.org/10.12776/ams.v23i2.900

[6] D. Kajánek et al. Corrosion resistance of AZ31 magnesium alloy treated by plasma electrolytic oxidation. KOM – Corros. Mater. Prot. J. 63 (2019) 65-71. https://doi.org/10.2478/kom-2019-0008

[7] M. Štrbák et al. Effect of plasma electrolytic oxidation on the short-term corrosion behaviour of AZ91 magnesium alloy in aggressive chloride environment. Coat. 12 (2022) art.566. https://doi.org/10.3390/coatings12050566

[8] F. Pastorek et al. Corrosion Behaviour of Preserved Peo Coating on Az31 Magnesium Alloy. Communications-Scientific letters of the University of Zilina 23 (2021) B76-B88. https://doi.org/10.26552/COM.C.2021.2.B76-B88

[9] F. Simchen et al. Introduction to plasma electrolytic oxidation—An overview of the process and applications. Coat. 10 (2020) art.628. https://doi.org/10.3390/coatings10070628

[10] J. Dziková et al. Characterization and corrosion properties of fluoride conversion coating prepared on AZ31 magnesium alloy. Coat. 11 (2021) art.675. https://doi.org/10.3390/coatings11060675

[11] M. Aliofkhazraei et al. Review of plasma electrolytic oxidation of titanium substrates: Mechanism, properties, applications and limitations. App. Surf. Sci. Adv. 5 (2021) art.100121. https://doi.org/10.1016/j.apsadv.2021.100121

[12] S. Sikdar et al. Plasma electrolytic oxidation (PEO) process—processing, properties, and applications. Nanomater. 11 (2021) art.1375. https://doi.org/10.3390/nano11061375

[13] T. W. Clyne, S.C. Troughton. A review of recent work on discharge characteristics during plasma electrolytic oxidation of various metals. Int. Mater. Rev. 64 (2019) 127-162. https://doi.org/10.1080/09506608.2018.1466492

[14] T. Zehra, A. Fattah-alhosseini, M. Kaseem. Surface properties of plasma electrolytic oxidation coating modified by polymeric materials: A review. Prog. Org. Coat. 171 (2022) art.107053. https://doi.org/10.1016/j.porgcoat.2022.107053

[15] R.O., Hussein, X. Nie, D.O. Northwood. The application of plasma electrolytic oxidation (PEO) to the production of corrosion resistant coatings on magnesium alloys: a review. Corrosion and Materials 38(1) (2013) 55-65.

[16] H. Sampatirao et al. Developments in plasma electrolytic oxidation (PEO) coatings for biodegradable magnesium alloys. Mater. Today Proc. 46 (2021) 1407-1415. https://doi.org/10.1016/j.matpr.2021.02.650

[17] G.B. Darband et al. Plasma electrolytic oxidation of magnesium and its alloys: Mechanism, properties and applications. J. Magnes. Alloys 5 (2017) 74-132. https://doi.org/10.1016/j.jma.2017.02.004

[18] D. S. Tsai, C.C. Chou. Review of the soft sparking issues in plasma electrolytic oxidation. Metals 8 (2018) art.105. https://doi.org/10.3390/met8020105

[19] B. Hadzima et al. Peo of az31 mg alloy: Effect of electrolyte phosphate content and current density. Metals 10 (2020) art.1521. https://doi.org/10.3390/met10111521

[20] A. Amirudin, D. Thieny. Application of electrochemical impedance spectroscopy to study the degradation of polymer-coated metals. Prog. Org. Coat. 26 (1995) 1-28. https://doi.org/10.1016/0300-9440(95)00581-1

[21] Y.G. Ko, S. Namgung, D.H. Shin, Correlation between KOH concentration and surface properties of AZ91 magnesium alloy coated by plasma electrolytic oxidation. Surf. Coat. Technol. 205 (2010) 2525-2531. https://doi.org/10.1016/j.surfcoat.2010.09.055

Quality Production Improvement and System Safety: QPI 16 - CZOTO 10   Materials Research Forum LLC
Materials Research Proceedings 34 (2023) 102–108   https://doi.org/10.21741/9781644902691-13

Effect of PEO Surface Treatment on Biodegradable Magnesium Alloy ZK60

OBERTOVÁ Veronika [1, a *], KNAP Vidžaja [2,a], KAJÁNEK Daniel [3,b]
and HADZIMA Branislav [4,b]

[1]University of Žilina, Faculty of Mechanical Engineering, Department of Materials Engineering, Univerzitná 8215/1, 010 26 Žilina, Slovakia

[2]University of Žilina, Faculty of Mechanical Engineering, Department of Materials Engineering, Univerzitná 8215/1, 010 26 Žilina, Slovakia

[3] Research Centre of the University of Žilina, University of Žilina, Univerzitná 8215/1, Žilina, Slovakia

[4]Research Centre of the University of Žilina, University of Žilina, Univerzitná 8215/1, Žilina, Slovakia

[a]veronika.obertova@fstroj.uniza.sk*, [b]vidzaja.knap@fstroj.uniza.sk,

Keywords: Extruded Magnesium Alloy ZK60, Biodegradable, Surface Treatment, Plasma Electrolytic Oxidation (PEO), Potentiodynamic Polarization Test (PD)

Abstract. Magnesium alloy ZK60 is a biocompatible and biodegradable material that has gained attention for its potential use in biomedical applications due to its high mechanical strength, lightweight properties, and biocompatibility. However, the high chemical reactivity and poor resistance to corrosion of magnesium alloys limit their use. This study evaluates the results from a potentiodynamic polarization test (PD) on ZK60 with surface treatments of plasma electrolytic oxidation (PEO) and PEO+PVA (Polyvinyl Alcohol with Glycerin). The aim is to understand the impact of these treatments on the corrosion behavior of ZK60 and to improve its properties for medical applications. The results showed that the PEO surface treatment significantly improved the corrosion resistance of ZK60, and the combination of PEO and PVA resulted in further improved corrosion behavior. This study highlights the potential of PEO and PEO+PVA treatments for improving the corrosion resistance of ZK60 and its suitability for use in biomedical applications.

Introduction

Magnesium alloys have gained recognition for their biodegradability and biocompatibility, making them a potential material for use in the biomedical field. Among the different magnesium alloys available, ZK60 stands out due to its relatively high mechanical strength and lightweight properties. The high strength-to-weight ratio of ZK60 allows it to be used in applications where weight reduction is critical without compromising on strength. This makes it an ideal material for applications such as sutures, screws, intramedullary nails, and plates, which are used as temporary support for damaged biological tissue or bone fractures [1].

ZK60 is an attractive option for biomedical applications due to its excellent biocompatibility, which means that it is well tolerated by the human body and does not cause adverse reactions. The biodegradability of ZK60 ensures that it does not pose long-term health risks and it can be gradually absorbed by the body, reducing the need for additional surgeries to remove it [2].

Despite the many desirable properties of ZK60, its high chemical reactivity and poor resistance to corrosion have limited its use in many applications. The high chemical reactivity of ZK60 means that it reacts readily with other substances, and it has poor resistance to corrosion in aqueous solutions, making it susceptible to galvanic corrosion. To overcome these limitations, researchers

Quality Production Improvement and System Safety: QPI 16 - CZOTO 10   Materials Research Forum LLC
Materials Research Proceedings 34 (2023) 102--108   https://doi.org/10.21741/9781644902691-13

have been exploring various surface modification techniques to improve the corrosion resistance of ZK60 [3].

One promising approach is the use of Plasma Electrolytic Oxidation (PEO) process for electrochemical surface treatment. PEO is a process that creates a rough, hard, and dense ceramic coating on lightweight metal construction materials, such as magnesium. In addition, the PEO process provides increased resistance to wear and corrosion, thermal stability, dielectric properties, improved bioactivity, and biocompatibility suitable for medical applications [4]. The PEO process involves anodic electrochemical dissolution, the combination of metal ions with anions to form ceramic compounds, and spark-induced condensation on the substrate [5]. The PEO technology is used to apply ceramic coatings to ZK60 to improve its tribological properties and protect it from corrosion [6].

The performance of the coatings produced by the PEO process strongly depends on process parameters such as the chemical composition of the material, the current density and voltage of the PEO process, the conductivity of the cathode metal, the electrolyte used, and the pH of the electrolyte.

In conclusion, ZK60 is a promising material for use in the biomedical field due to its high mechanical strength, lightweight properties, and excellent biocompatibility. While its high chemical reactivity and poor resistance to corrosion limit its use in many applications, the use of PEO surface treatment and the combination of PVA polymer layer with PEO layering provide improved corrosion properties, making ZK60 a more suitable material for use in biomedical applications [4].

Material and Methods

The extruded magnesium alloy ZK60 was selected as the experimental material. The ZK60 experimental material was treated with plasma electrolytic oxidation (PEO) in an environmentally friendly electrolyte on the surface of the extruded magnesium alloy. Furthermore, the PEO-processed surface was treated with a water-soluble polymer coating PVA (polyvinyl alcohol + glycerin) to be applied on the resulting porous PEO layer, in order to achieve higher corrosion resistance of the biodegradable experimental material ZK60.

The elemental composition of the alloy is listed in Table 1, with the composition data based on the analysis using a hand-held X-ray analyzer type VANTA VCR with an SDD detector with GRAPHENE window.

Table 1. Chemical composition of extruded alloy ZK60

Elements	Zn	Al	Fe	Ni	Cu	Zr	Mg
[wt. %]	5.54	<0.01	<0.01	<0.01	<0.01	0.55	Bal.

The microstructure of the extruded alloy ZK60 was studied using a ZEISS AXIO Imager.A1m optical microscope. The preparation of metallographic samples for optical analysis was carried out on grinding papers with grits of P800 and P1200. In the next step, the samples were polished on a polishing plate with the simultaneous application of a diamond paste with a particle size of 1 and 3 microns. Then, the samples were rinsed with demineralized water, ethanol, and dried with airflow and etched with a solution of 2.1 g of picric acid, 2.5 ml of acetic acid, 35 ml of ethanol, and 5 ml of distilled water. The exposure time was 15 seconds (1).

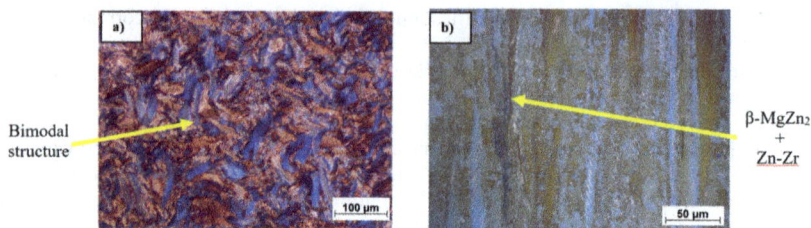

Fig.1. *Microstructure of Mg alloy ZK60: a) transverse section b) longitudinal section, etchant: picral, polarized light.*

The microstructure of extruded ZK60 alloy is shown in Fig.1, with a transverse section in (a) and a longitudinal section in (b). The polarized light method was used to observe the grain morphology. The structure is characterized by a bimodal grain morphology of deformed and fine grains, which is a typical result of the extrusion process [8]. The extrusion process usually takes place at a temperature of 400°C, near 0.56 times the melting temperature of the alloy (923 K). The grain size is not uniform and small grains can be seen between the larger grains. This is due to dynamic recrystallization that occurs during extrusion and results in the formation of fine-grained structures [8], [9]. The recrystallized grain size is estimated to be between 2-8 μm.

In the longitudinal section of the extruded ZK60 alloy, the grain orientation is along the extrusion direction (Fig.1b). The intermetallic compound β-$MgZn_2$ is visible as chains, as well as a small amount of Zn-Zr compound. These intermetallic particles may act as nucleation sites for dynamic recrystallization and prevent grain growth, leading to fine recrystallized grains near the intermetallic compounds [10].

The samples of extruded ZK60 alloy for potentiodynamic (PD) tests were ground on gridding papers with P1200 grain size. Similarly, samples of ZK60 alloy for the surface treatment PEO process were prepared. Keysight N8762A was used as the power supply source. The ZK60 sample was connected as the anode in a two-electrode system, with the cathode secured by a stainless steel plate. Both electrodes were placed in the PEO electrolyte, which was a solution of 12 g/l $Na_3PO_4.12H_2O$ and 1g/l KOH at room temperature (22 ± 1°C) and pH = 12.4. The electrolyte was continuously cooled with water and stirred up during the PEO procedure, both to improve the distribution of active ions and to maintain the temperature below 50°C. The current density was set to 0.05 A/cm^2 for 14 minutes [11]. After the PEO process, the samples were immersed into a polymer bath of PVA with glycerine solution. The corrosion resistance of ground samples of extruded ZK60 alloy and samples with PEO and PEO+(PVA+glycerin) layers were evaluated using potentiodynamic (PD) tests, which were performed on the SP-300 potentiostat with the samples placed in a corrosion cell at a temperature of 37 ± 2°C (Fig.30). The stable temperature was maintained using the Heating Bath Circulator.

The corrosive environment selected was a 0.9% NaCl solution, simulating the presence of chloride ions in a body environment with pH 6.8. Potentiodynamic polarization corrosion tests started after 1 hour of potential stabilization between the experimental sample and the 0.9% NaCl electrolyte being tested. The range of applied potential was from -200 mV to +500 mV, and the potential range was set with consideration for the open-circuit potential (OCP), and the scanning speed was 1.0 mV/s. The data obtained in the form of potentiodynamic curves were analyzed using the Tafel extrapolation method with the EC Lab V10.40 software. For statistical processing, three grounded samples ZK60, samples with PEO conversion layer, and samples after the PEO process with a created PVA+glycerin polymer coating were measured.

Quality Production Improvement and System Safety: QPI 16 - CZOTO 10   Materials Research Forum LLC
Materials Research Proceedings 34 (2023) 102--108   https://doi.org/10.21741/9781644902691-13

Results and Discussion

The PEO process involves the application of high voltage and current between the electrode and the magnesium alloy, leading to the creation of a ceramic-like oxide layer on the surface of the metal. The SEM image of the PEO coating on the ZK60 biodegradable magnesium alloy (as demonstrated in Fig.2) reveals its visibly porous structure, with pore sizes ranging from 2-10 micrometers. This is a typical characteristic of the PEO process, as the formation of micropores is associated with the presence of molten oxides and gas bubbles produced during the discharges that occur on the surface of the samples.

This porosity is a result of the formation of oxide pores during the PEO process. The PEO process involves the application of high voltage and current, which releases oxygen from the electrolyte solution. This oxygen reacts with the magnesium surface to form oxide pores, resulting in a porous structure on the surface of the metal. The size of the pores depends on the process parameters, such as voltage and current, as well as the composition of the electrolyte solution [13].

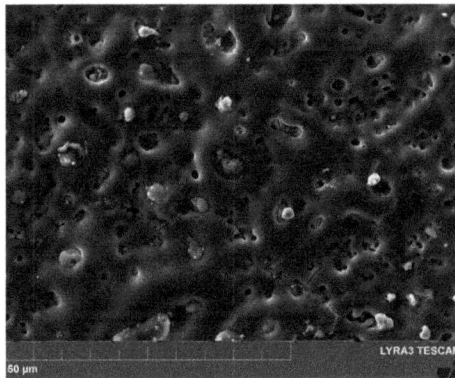

Fig.2. Surface morphology of PEO coatings
formed in phosphate-based electrolyte.

It is important to note that the porous structure of the PEO-treated layer has both advantages and disadvantages. On one hand, the pores provide a large surface area for the adsorption of corrosion inhibitors, which can improve the corrosion resistance of the magnesium alloy. On the other hand, the pores can also act as corrosion initiation sites if not properly filled or sealed. In general, this porous structure can provide benefits such as improved corrosion resistance, enhanced surface hardness and wear resistance, and improved biocompatibility.

The potentiodynamic curves obtained for the grounded sample, the PEO treated sample, and the PEO+(PVA+glycerin) treated sample are shown in Fig.3. The electrochemical characteristics obtained from the Tafel analysis are listed in Table 2. We can observe from the potentiodynamic curves and the electrochemical characteristics that the corrosion potential of the ground sample, E_{corr}, was -1458 mV. The PEO and PEO+(PVA+glycerin) treated samples had slightly higher E_{corr} values compared to the ground sample, which indicating lower thermodynamic stability. However, the key factor for corrosion resistance is the kinetic aspect of the process, specifically the corrosion current density i_{corr}, which is directly proportional to the corrosion rate r_{corr}. The PEO coating achieved over 5 times lower corrosion rate (0.447 mm/year) compared to the ground surface (2.547 mm/year). An even more significant improvement in the corrosion rate was achieved with the PEO coating in combination with the PVA+glycerin polymer coating, with the corrosion rate being 7 times lower (0.356 mm/year) compared to the grounded sample surface.

The results of electrochemical tests in 0.9% NaCl at a temperature of $37 \pm 2°C$ showed that the sample treated with PEO showed significant improvement in corrosion resistance compared to the ground sample. However, when compared to a study [15] with the same electrical parameters (constant current density of 0.05 A/cm^2 and DC mode) and temperature conditions for PD tests ($37 \pm 2°C$), the PEO layer or the PEO+PVA+glycerol composite layer on the ZK60 magnesium alloy showed slightly lower corrosion resistance. One possible explanation for this could be the extruded state of the sample, which had finer grains resulting in a higher density of grain boundaries, which can cause localized corrosion in Mg alloys. Despite this, the results of the PEO+PVA sample are promising as the goal is to slow down the degradation of the alloy, not to completely prevent it, so that the implants can be safely absorbed within an acceptable timeframe once they have served their purpose.

On the other hand, PD analysis showed that the porous structure of the PEO surface combined with the PVA+glycerol polymer coating had higher corrosion resistance compared to the pure PEO layer. Adding PVA to the PEO coating can fill its porous structure, potentially leading to improved corrosion resistance compared to the grounded P1200 sample and the PEO-only sample. This is based on the idea that filling the pores could prevent active surface exposure during corrosion and maintain the integrity of the PEO layer

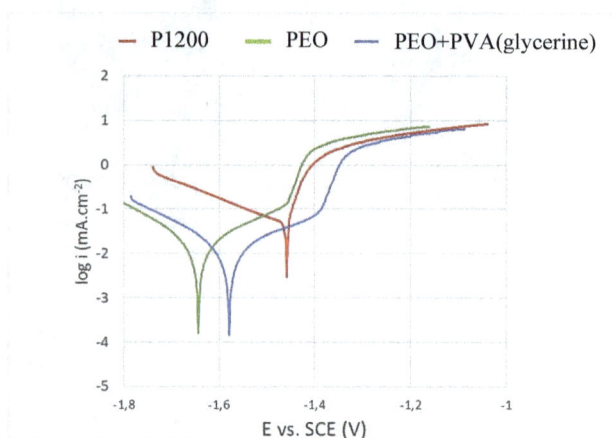

Fig.3. Polarization curves for different surface treated samples measured in a 0.9% NaCl solution at a temperature of 37 ± 2 °C.

Table 2. Determined electrochemical characteristics of the untreated and coated extruded ZK60 magnesium alloy.

	Ground P1200	**PEO**	**PEO+(PVA+glycerine)**
E_{corr} **[mV]**	-1458 ± 29	-1644 ± 35	-1578 ± 25
i_{corr} **[μA/cm^{-2}]**	111.7 ± 5.2	20.6 ± 1.8	15.9 ± 1.6
r_{corr} **[mm.yr^{-1}]**	2.547 ± 0.118	0.447 ± 0.039	0.356 ± 0.008

Quality Production Improvement and System Safety: QPI 16 - CZOTO 10   Materials Research Forum LLC
Materials Research Proceedings 34 (2023) 102--108   https://doi.org/10.21741/9781644902691-13

Conclusion

The results of potentiodynamic polarization confirmed the improved corrosion resistance of the PEO layer with the PVA+glycerine polymer coating compared to the grounded sample, with a significant reduction in corrosion current density (i_{corr} = 111.7 μA.cm^{-2} > 15.99 μA.cm^{-2}) in the case of the PEO-coated sample compared to the grounded sample (i_{corr} = 20.6 μA.cm^{-2}), leading to a similar reduction in corrosion rate for the sample treated with only the PEO process.

The created PEO layer achieved more than 5 times lower corrosion rate (0.447 mm/year) compared to the not treated grounded surface (2.547 mm/year), The PEO layer combined with the PVA+glycerine polymer coating achieved a significantly better corrosion rate, 7 times lower (0.356 mm/year) compared to the grounded surface of the sample.

Acknowledgments

This paper was supported under the project of Operational Programme Integrated Infrastructure: Independent research and development of technological kits based on wearable electronics products, as tools for raising hygienic standards in a society exposed to the virus causing the COVID-19 disease, ITMS2014+ code 313011ASK8. The project is co-funding by European Regional Development Fund. The research was also financially supported by the Science Grant Agency of the Slovak Republic through projects No. 1/0117/21 and No. 1/0153/21.

References

[1] Z. Sheikh et al. Biodegradable Materials for Bone Repair and Tissue Engineering Applications, Materials 8 (2015) 5744-5794. https://doi.org/10.3390/ma8095273

[2] N.T. Kirkland et al. Assessing the corrosion of biodegradable magnesium implants: A critical review of current methodologies and their limitations, Acta Biomater. 8 (2012) 925-936. https://doi.org/10.1016/j.actbio.2011.11.014

[3] J.Telegdi et al. Biocorrosion—Steel, In: Encyclopedia of Interfacial Chemistry, Surface Science and Electrochemistry, 2018, 28-42. https://doi.org/10.1016/B978-0-12-409547-2.13591-7

[4] T. Schilling et al. Cardiovascular Applications of Magnesium Alloys, In: M. Aliofkhazraei (Ed.) Magnesium Alloys, IntechOpen, 2017. http://doi.org/10.5772/66182

[5] S. Sikdar et al. Plasma electrolytic oxidation (PEO) process – processing, properties, and applications, Nanomaterials 11 (2021) art.1375. https://doi.org/10.3390/nano11061375

[6] H. Hu et al. Corrosion and Surface Treatment of magnesium alloys, In: M. Aliofkhazraei (Ed.) Magnesium Alloys, IntechOpen, 2017. https://doi.org/10.5772/58929

[7] N.T. Kirkland et al. In-vitro dissolution of magnesium-calcium binary alloys: Clarifying the unique role of calcium additions in bioresorbable magnesium implant alloys, J. Biomed. Mater. Res. Part B Appl. Biomater. 95 (2010) 91-100. https://doi.org/10.1002/jbm.b.31687

[8] B. Chen, J. Zhang. Microstructure and mechanical properties of ZK60-Er magnesium alloys, Mater. Sci. Eng. A 633 (2015) 154-160. https://doi.org/10.1016/j.msea.2015.03.009

[9] J. Chen et al. Effect of heat treatment on mechanical and biodegradable properties of an extruded ZK60 alloy, Bioact. Mater. 2 (2017) 19-26. https://doi.org/10.1016/j.bioactmat.2016.12.002

[10] Y. Xue et al. Corrosion Protection of ZK60 Wrought Magnesium Alloys by Micro-Arc Oxidation, Metals 12 (2022) art.449. https://doi.org/10.3390/met12030449

[11] D. Kajanek et al. Effect of applied current density of plasma electrolytic oxidation process on corrosion resistance of AZ31 magnesium alloy, Commun. – Sci. Lett. Univ. Zilina 21 (2019) 32-36. https://doi.org/10.26552/COM.C.2019.2.32-36

[12] F. Pastorek et al. Corrosion behaviour of preserved PEO coating on AZ31 magnesium alloy, Commun. – Sci. Lett. Univ. Zilina 23 (2021) B76-B88. https://doi.org/10.26552/COM.C.2021.2.B76-B88

[13] B. Hadzima. Korózia zliatin Mg-Al-Zn., Ph.D. thesis, Žilinská univerzita v Žiline, Žilina, 2003. [Online]. Available from: http://katalog.utc.sk/ukzu/epubl/ddz-Hadzima-KoroziaZliatinMg-Al-Zn-low.pdf

[14] Z. Shi et al. Measurement of the corrosion rate of magnesium alloys using Tafel extrapolation, Corros. Sci. 52 (2010) 579-588. https://doi.org/10.1016/j.corsci.2009.10.016

Quality Production Improvement and System Safety: QPI 16 - CZOTO 10   Materials Research Forum LLC
Materials Research Proceedings 34 (2023) 109-119   https://doi.org/10.21741/9781644902691-14

The Biodegradation and the Rheological Properties of Polypropylene/Hyperbranched Polyester Blends for Industrial Applications

AL-MUTAIRI Nabeel Hasan [1, a], AL-ZUBIEDY Ali [2, b], AL-ZUHAIRI Ali J. [3,c] and IDZIKOWSKI Adam [4,d]

[1,2]Polymer and Petrochemical Industries Department, University of Babylon, Hilla, Iraq

[3]Department of Energy, College of Engineering, University of Babylon, Al-Musayyab, Iraq

[4]Department of Production Engineering and Safety, Faculty of Management, Czestochowa University of Technology

[a] nabeeleng90@gmail.com, [b] mat.ali.alzubiedy@uobabylon.edu.iq, [c] alijassim33@yahoo.com, [d] adam.idzikowski@pcz.pl

Keywords: Polypropylene PP, Hyperbranched Polyester HBP, Contact Angle CA, Biodegradation, MFR

Abstract. In this research, four novel types of hyperbranched polyester polymers (HBPs) were used and blended with polypropylene polymer PP, virgin VPP and recycled rPP, using a twin-screw extrusion machine. Hyperbranched polyester was added in different weight ratios (5%, 10%, and 20%). The Fourier transform spectroscopy FTIR, the water contact angle CA, the biodegradation in soil, and the rheological properties (melt flow rate MFR) of the prepared blends were investigated. The results showed that the contact angle of VPP and rPP blends has improved with the addition of HBPs, and in addition, the biodegradation results in soil showed that weight loss increased as the amount of HBPs increased in VPP and rPP blends. The rheological properties, melt flow rate MFR, showed that the addition of HBPs increased the MFR for both PP blends, VPP and rPP. In addition, it was found that the blends' viscosity was decreased and their shear rate was increased. This is an indication that the HBPs work as a processing aid additive by increasing the shear thinning behavior.

Introduction

New materials have made significant contributions to economic and technical advancement throughout the past century. There are numerous metals, alloys, composites, ceramics, and polymers in the list of materials. Out of these, polymers have significantly influenced this process of development and have emerged as a significant class of engineering materials [1]. High molecular mass compounds and numerous repeating units make up the structure of polymers. Today, polymer materials are so crucial to human life that every aspect of it depends on them [2].

Polymers have a wide range of qualities that make them the perfect material for a wide range of uses, from polyethylene in sandwich and garbage bags to poly (p-phenylene terephthal amide) in bulletproof vests [3]. Since they combine flexibility, toughness, excellent barrier properties, ease of manufacturing, and good chemical resistance, polyolefins have drawn a lot of attention recently. This makes them excellent materials for various packaging applications, particularly in food packaging. Examples of this include polypropylene (PP) and polyethylene (PE), which are widely used in commodities and packaging applications [4],[5]. Due to their low cost, lightweight, high mechanical strength, water resistance, and strong barrier properties, PP is increasingly used in the plastic packaging industry [6].

It has been estimated that the volume of plastic packaging will expand by two and three times, respectively, in 2030 and 2050, compared to the current level of global plastic manufacturing [7].

Because of their superior physical and mechanical qualities, non-biodegradable polymers such as polyethylene (PE), polypropylene (PP), ethylene vinyl alcohol, poly (ethylene terephthalate), polystyrene, expanded polystyrene, polyamides, polyurethane, and poly (vinyl chloride) have previously dominated the packaging industry. These polymers typically remain stable in the environment for a very long time and are a major waste management challenge since they are inert and resistant to microbial attacks [8].

Eco-friendly polymeric materials should be used for potential short-lifespan applications to prevent the environmental and ecological damage caused by post-consumer plastic trash. For applications requiring disposable materials, such as packaging, consumer goods, and hygiene items, biodegradable polymeric materials make good choices. However, due to their comparatively expensive cost and inadequate mechanical and thermomechanical properties in comparison to some non-biodegradable commodity polymers, the use of biodegradable polymers is restricted for several applications [9].

Therefore, these restrictions can be circumvented by creating biodegradable polymer blends with suitable characteristics. Improving adhesion between blended components, lowering interfacial tension, and producing limiting inclusion phase size are the three key issues with polymer melt blending [10].

Additionally, when processing polymers, manufacturers want to enhance flow rate, lower energy costs, and limit the likelihood of extrudate abnormalities. The upper limiting production rate is greatly increased by using polymer processing aids (PPAs) to effectively remove processing abnormalities that lead to issues with products [11]. A number of chemicals, including polymers, have been used as processing aids in an attempt to minimize surface roughness in polyolefin processing. Hyperbranched polymers (HBP) have received significant interest as PPAs, owing to their unique dendrimer like branched structure with lack of chain entanglements and the resulting dramatically reduced melt viscosity [12].

In this study, four novel hyperbranched polyester polymers HBPs were blended in different weight percentages (5%, 10%, and 20%) with both VPP and rPP. The blends were prepared using a twin-screw extruder. The effect of HBPs on the biodegradation and rheological properties were investigated.

Materials and Methods
Materials. VPP and rPP were purchased from sabic (KSA), and used as the major phase for the preparation of polymeric blends. A novel Hyperbranched polyester polymers HBPs (HBP-TA, HBP-AD, HBP-MA, and HBP-PA) were synthesized by polycondensation polymerization in our laboratory (polymer department/Babylon University) according our previous works [13] and [14]. Table 1 shows the properties of VPP and rPP.

Table 1. Properties of the used PP

Properties	Value		Testing Standard
	VPP	rPP	
Tensile strength (MPa)	20-35	20-30	ASTM D-638
Modulus of Elasticity (GPa)	0.13	0.12	
MFR (g/10min)	5-10	15-25	ASTM D-1238

The preparation of PP/HBPs blends. The HBPs and PP pellets (VPP and rPP) were mixed at various weight ratios, as shown in Table 2.

Quality Production Improvement and System Safety: QPI 16 - CZOTO 10   Materials Research Forum LLC
Materials Research Proceedings 34 (2023) 110-120   https://doi.org/10.21741/9781644902691-14

Table 2. *The percentages of polymeric blends*

VPP/HBPs (%)	rPP/HBPs (%)
80/20	80/20
90/10	90/10
95/5	95/5
100/0	100/0

The two polymers, PP and HBPs, were first mechanically mixed in the dry state for 10 minutes before being processed into blends using a twin-screw extruder. With a screw speed of 25 rpm and temperatures of 165–175 °C for VPP and 155–165 °C for rPP, the blends were extruded. A plastic sheet is created by passing the molten material that exits the extruder between two moving rollers. These sheets were cut in accordance with ASTM standards to provide various samples for various tests.

Characterization. FTIR were used identify the changes in chemical structure by the change in band values. Instrument, type IR Affinity-1 range (400-4000 cm-1) was used for this purpose.

MFR was used to measure the effect of HBPs on the melt flow rate of PP (VPP and rPP) using Shi Jia Zhuang MFI tester based on ASTM-D1238. The samples tested at 230 °C and 2.16 Kg load with 8mm die. The melt flow rate, the viscosity (η) and shear rate ($\gamma^{\cdot}{}_w$) were evaluated according to the equations below [15].

$$MFR = t_{ref} * w/t \dots\dots\dots\dots\dots Eq\,(1)$$

$$\eta = (4.98 * 10^4 * \rho * L)/MFR \dots\dots\dots\dots Eq\,(2)$$

$$\gamma^{\cdot}{}_w = (1840/\rho) * MFR \dots\dots\dots\dots Eq\,(3)$$

Where: - t_{ref} =600 sec, t= time (sec), w= average weight of of sample (g), $\gamma^{\cdot}{}_w$= Shear rate at the wall (s^{-1}), ρ= Density (kg /m³).

The wettability of PP (VPP, rPP) and its blends with HBPs at different wt. % percentages have been investigated using the device, SL 200C - Optical Dynamic I Static Interfacial Tensiometer & Contact Angle Meter, which using circle fitting method of water on samples surface. The tested data have been collected at two intervals after 3 sec and 60 sec. The aim of this test is to study the effect of the HBP on the wettability of polypropylene PP.

The biodegradation of the blends in soil were measured according to ASTM-D 5988, by burying in soil with pH 7.56 at a distance 10 cm from the surface for 90 days. The weight loss of samples was calculated 10 days according to the equation below [16].

$$\% \text{ weight loss} = [W_b – W_a)/W_b] * 100 \dots\dots\dots\dots Eq\,(4)$$

Where: - W_b and W_a are initial mass before and after degradation in the soil respectively.

Results and Discussion

Characterization of PP and PP/HBP Blends – FTIR analysis result. FTIR spectroscopy was used to analyze the effect of the prepared HBP addition on the chemical structure of VPP and rPP. Figure 1 and 2 shows the bands values and the changes in these values with the addition of HBP. Table 3 and 4 lists the most important bands of PP and its blends with HBP as derived from Figure 1 and 2 and compare them with the bands mentioned in [17, 18]. The following bands were identified for VPP and rPP: 2954.95 cm^{-1} and 2885.51 cm^{-1} for –CH$_3$ asymmetric stretching vibration and –CH$_3$ symmetrical stretching vibration, bands at 995.27 cm^{-1}, 972.12 cm^{-1} and 1165 cm^{-1} related to –CH$_3$ rocking vibration. The peaks at 1458.18 cm^{-1}, 2839.22 cm^{-1} and 2924.09 cm^{-1} are related to –CH$_2$ symmetric bending, –CH$_2$ symmetric stretching and –CH$_2$ asymmetric stretching. The peak at 1373.32 cm^{-1} is attributed to –CH$_3$ symmetric bending vibration mode. The peak at 840.96 cm^{-1} for C–CH$_3$ stretching vibration and peak at 810.10 cm^{-1} is for CH$_2$ rock, C–C stretch, C–CH stretch. The addition of the prepared HBP to vPP and rPP had some shifting on the bands of PP; also, the bands of HBP appears in the IR spectrum. From these results, it can be concluded that the interaction between the two polymers, PP and HBP, is physical and not chemical.

MFR Results. The rheological properties (MFR, viscosity, and shear rate) of PP and its blends with HBP have been characterized using MFR. The MFR of PP and PP/HBP blends shown by Figure 3. From figure 3A for VPP and VPP/HBP blends, it's found that the MFR of all blends has been increased with the increase of HBP content. It's shown that the VPP has MFR of 10 (g/10 min), as the HBP blended with VPP the MFR was increased by 2.6, 6.23, and 4.14 g/10 min for 5%, 10%, and 20% of HBP-TA samples and increased by 2.8, 28.5, and 15.7 g/10 min for 5%, 10%, and 20% of HBP-AD samples respectively. While, it increased by 5.7, 12.1, and 6.1 g/10 min for 5%, 10%, and 20% of HBP-MA samples and increased by 7.8, 11.24, and 9.6 g/10 min for 5%, 10%, and 20% of HBP-PA samples respectively.

However, for rPP and rPP/HBP blends shown in Figure 3B, it was found that the MFR improved by the addition of HBP. The MFR of rPP is 26.7 g/10min, as the HBP added the MFR improved by 6.2, 20.8, and 10.3 g/10min for 5%, 10%, and 20% of HBP-TA samples and improved by 37, 65.45, and 56.8 g/10min for 5%, 10%, and 20% of HBP-AD samples respectively. While, it improved by 15.4, 48.7 and 34.9 g/10min for 5%, 10%, and 20% of HBP-MA samples and improved by 21.7, 31.2 and 10.1 g/10min for 5%, 10%, and 20% of HBP-PA samples respectively.

From the above results, it's found that HBP addition improve the MFR for rPP better that VPP due to the less entanglements and shorter chains of rPP. In addition, it is found that 10% of HBP has the highest MFR value for VPP and rPP blends due to the better compatibility and distribution of HBP within PP matrix as indicated by SEM. These results attributed to the fact that hyperbranched polymers HBP have relatively lower viscosity than linear polymers PP, which mean that these additives act as viscosity reducers (processing aids additives). So that the presence of HBP in the molten PP generate a rolling effect which facilitates sliding of melt on the cylinder wall of MFI tester, serve as ball bearings, and reducing the interlayer interaction.

According to the results above, the viscosity, shear rate and the power law index (n) of VPP, rPP, and their blends were calculated easily from equations 2 and 3 to give the data in table 5.

Table 3.IR transmission bands for VPP and its blends

Bond type	PP[17], [18]standard	VPP	PP/HBP-TA	PP/HBP-AD	PP/HBP-MA	PP/HBP-PA
C-C stretch	808	810.1	802.39	802.39	802.39	802.39
C-H rocking	840	840.96	848.68	840.96	843.55	840.96
CH_3 rocking C-C stretch	973	972.12	972.12	972.12	972.12	972.12
CH_3 rocking C-C stretch	996	995.27	996.07	995.27	=	995.27
CH wagging CH_2 rocking	1166	1165	1165	1165	1141.86	1165
CH_3 sym.bend	1376 1456	1373.32 1458.18	1381.03 1458.18	1381.03 1458.18	1404.18 1458.18	1373.32 1458.18
CH_3 stretch	2870	2885.51	2877.19	2885.51	2877.19	2877.79
CH_3 asym.stretch	2920	2924.09	2924.09	2931.8	2931.8	2924.09
CH_3 asym.stretch	2950	2954.95	2962.66	2954.95	2962.66	2954.95

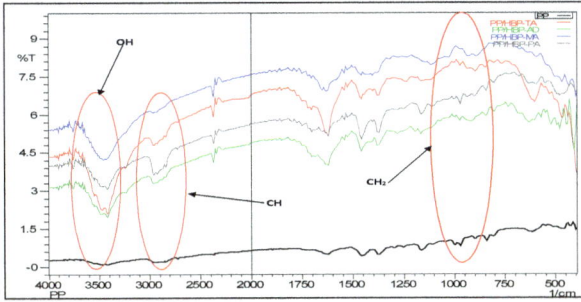

Fig. 1. *IR spectrum for VPP and PP/HBP.*

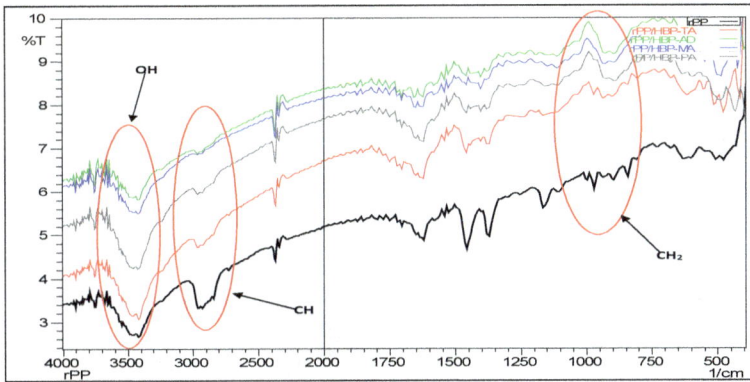

Fig. 2. *IR spectrum of rPP and rPP/HBP.*

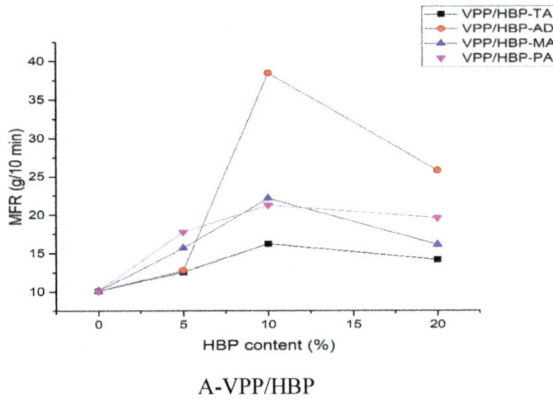

A-VPP/HBP

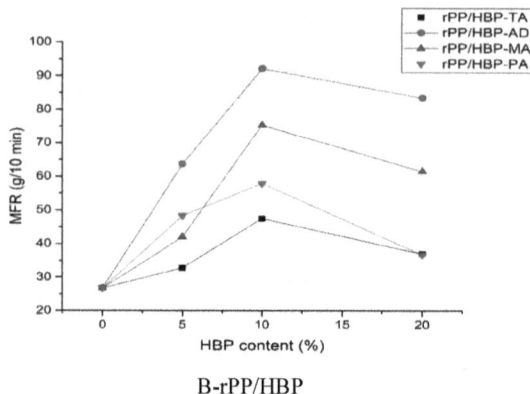

B-rPP/HBP

Fig. 3. *MFR of VPP, rPP, and their blends with HBP.*

Table 4. *IR transmission bands for rPP and its blends*

Bond type	PP[17], [18]standard	rPP	PP/HBP-TA	PP/HBP-AD	PP/HBP-MA	PP/HBP-PA
C-C stretch	808	810.1	810.1	802.39	802.39	802.39
C-H rocking	840	840.96	840.96			
CH_3 rocking C-C stretch	973	972.12	972.12	972.12	972.12	972.12
CH_3 rocking C-C stretch	996	995.27	=	=	=	=
CH wagging CH_2 rocking	1166	1165	1165	1149.57	1188.15	1149.57
CH_3 sym.bend	1376 1456	1373.32 1458.18	1381.03 1458.18	1381.03 1458.18	1381.03 1458.18	1381.03 1458.18
CH_3 stretch	2870	2877.79	2877.79	2877.79	2883.11	2889.37
CH_3 asym.stretch	2920	2924.09	2931.8	2924.09	2931.8	2924.09
CH_3 asym.stretch	2950	2954.95	2962.66	2962.66	2962.66	2962.66

Table 5. *The calculated data from the MFR for VPP, rPP, and their blends*

Samples	Viscosity (Pa.s)	Shear rate (S^{-1})	Power law index (n)
VPP	34903.5	21002.4	0.43
VPP/5%HBP-TA	28836.2	25421.4	0.392
VPP/10%HBP-TA	22382.1	32752	0.343
VPP/20%HBP-TA	26326.2	27845.1	0.374
VPP/5%HBP-AD	28205.97	21002.4	0.388
VPP/10%HBP-AD	9564.3	25989.4	0.195
VPP/20%HBP-AD	14281.42	76645.1	0.262
VPP/5%HBP-MA	22888.92	32026.7	0.348

VPP/10%HBP-MA	16896.91	43384.03	0.292
VPP/20%HBP-MA	23388.7	31342.4	0.352
VPP/5%HBP-PA	20471.84	35808.02	0.327
VPP/10%HBP-PA	17215.23	42581.83	0.295
VPP/20%HBP-PA	18097.1	40506.92	0.304
rPP	13141	55783.9	0.39
rPP/5%HBP-TA	10850.8	67557.8	0.366
rPP/10%HBP-TA	7713.2	95039.3	0.311
rPP/20%HBP-TA	10205.7	71828.2	0.355
rPP/5%HBP-AD	5619.28	130453.73	0.263
rPP/10%HBP-AD	3891.9	188356.12	0.211
rPP/20%HBP-AD	4432.6	165379.9	0.229
rPP/5%HBP-MA	8537.14	85866.7	0.327
rPP/10%HBP-MA	4808.84	152439.21	0.24
rPP/20%HBP-MA	6073.82	120691.13	0.275
rPP/5%HBP-PA	7364.21	99543.01	0.303
rPP/10%HBP-PA	6288.02	116579.7	0.28
rPP/20%HBP-PA	10242.1	71572.9	0.356

From Table 5, it is shown that the viscosity of all blends PP/HBP were decreased by the increase of HBP while the shear rate increased with HBP increasing, which proves that HBP addition increases the flow behavior for PP. The shear thinning increased with the increase of HBP content as indicated by table 5, the values of power law index decreased for all blends.

From these results, it can be concluded that the addition of HBP has improved the processability of PP and the energy consumption during the preparation of blends, these results are in agreement with Guzmán [19] and Mesias [20].

Contact Angle Results. The water contact angle gives an indication of the nature of the surface, whether it is hydrophilic or hydrophobic. The higher the contact angle the lower the wettability and the surface is hydrophobic and vice versa. In table 6 and 7, the water contact angle data for PP and PP/HBP blends are shown at different time intervals. For VPP and its blends with HBP, its found that the contact angle decreased with the increase of HBP content from 0% to 20%, the contact angle after 3 sec for VPP decreased from 78.38° to 60.17° for HBP-TA and decreased to 62.1° for HBP-AD and decreased to 61.61° and 67.87° for HBP-MA and HBP-PA respectively after 60 sec the contact angle for VPP decreased from 78.38° to 69.4° and decreased further for the other blends, as shown in Table 6.

Table 6. Contact angle data for VPP, and their blends with HBP

Samples	Contact angle	
	After 3 sec	After 60 sec
VPP	78.38°	69.4°
VPP/5%HBP-TA	65.66°	54.8°
VPP/10%HBP-TA	63.18°	54.18°
VPP/20%HBP-TA	60.17°	49.95°
VPP/5%HBP-AD	62.24°	60.32°
VPP/10%HBP-AD	61.68°	57.07°
VPP/20%HBP-AD	62.1°	58.12°
VPP/5%HBP-MA	65.51°	59.06°

VPP/10%HBP-MA	63.36°	58.85°
VPP/20%HBP-MA	61.61°	57.32°
VPP/5%HBP-PA	65.95°	61.24°
VPP/10%HBP-PA	63.89°	61.43°
VPP/20%HBP-PA	67.87°	64.05°

Table 7. Contact angle data for rPP, and their blends with HBP

Samples	Contact angle	
	After 3 sec	After 60 sec
rPP	70.77°	66.5°
rPP/5%HBP-TA	65.02°	48.82°
rPP/10%HBP-TA	63.92°	42.57°
rPP/20%HBP-TA	58.23°	47.95°
rPP/5%HBP-AD	68.2°	59.25°
rPP/10%HBP-AD	65.89°	53.33°
rPP/20%HBP-AD	65.37°	58.53°
rPP/5%HBP-MA	65.41°	58.49°
rPP/10%HBP-MA	63.38°	55.57°
rPP/20%HBP-MA	64.1°	55.21°
rPP/5%HBP-PA	68.23°	59.61°
rPP/10%HBP-PA	67.3°	55.45°
rPP/20%HBP-PA	68.2°	59.85°

For rPP and rPP/HBP blends, it's found that the contact angle also decreased with HBP increasing. The contact angle after 3 sec for rPP decreased from 70.77° to 58.23° and 65.37° for HBP-TA and HBP-AD respectively and decreased to 64.1° and 68.2° for HBP-MA and HBP-PA respectively after 60 sec the contact angle for rPP decreased from 70.77° to 66.5° and decreased further for the other blends, as shown in Table 7.

From the results, it's found that the addition of HBP to PP has improve its hydrophilicity (the surface wettability increased) by reducing the contact angle, this improvement is due to the presence of large number of hydroxyl group and carboxyl group in the prepared HBP (the presence of polar group can improve the wettability and hydrophilicity of the surface) these results are in agreement with Zhang [21] and Caicedo [22]. Also, it's found that the rPP/HBP blends have better wettability and lower contact angle than VPP/HBP, this due to the shorter the chains and the less entanglement in rPP than VPP. In addition, the blends with HBP-TA have the lowest contact angle, the best surface wettability and the best hydrophilicity, this is due to the higher content of hydroxyl and carboxyl group in HBP-TA when compared with the other types of the prepared HBPs.

Biodegradation Results (Soil Burial). Most types of polymers when they meet the required need and after the expiry of their use, they will accumulate in large quantities, which will have a dangerous impact on the environment as well as on the general appearance. In order to reduce the accumulation of these materials, these polymers are usually mixed with other types of materials that have the ability to decompose in order to improve the decomposition of these polymers and to reduce their danger to the environment and human life.

The biodegradation through the weight loss of PP and its blends with HBP in soil are shown in Figure 4 and 5. From the results, its note that both VPP and rPP did not show any weight loss during the 90-day examination period. These results indicate that this type of polymer does not have the ability to degrade in short periods, but needs very long periods, because the chemical structure of this type of polymer does not contain any groups that aid in degradation. For VPP/HBP

Quality Production Improvement and System Safety: QPI 16 - CZOTO 10   Materials Research Forum LLC
Materials Research Proceedings 34 (2023) 110-120   https://doi.org/10.21741/9781644902691-14

blends it is found that there is an improvement in the degradation of PP through the test period. It is shown from Figure 4 that the weight loss increases as both test period and the amount of HBP increase. It is found that the blend with HBP-TA increases the weight loss to 6.5%, 7.7%, and 8.5% as the HBP increase from 5% to 20%, and the blend with HBP-MA increases the weight loss to 3.5%, 4.2%, and 4% as the HBP increase from 5% to 20%. While for the blends with HBP-AD and HBP-PA it's found that the weight loss are in the range of 2.5% to 3.25%.

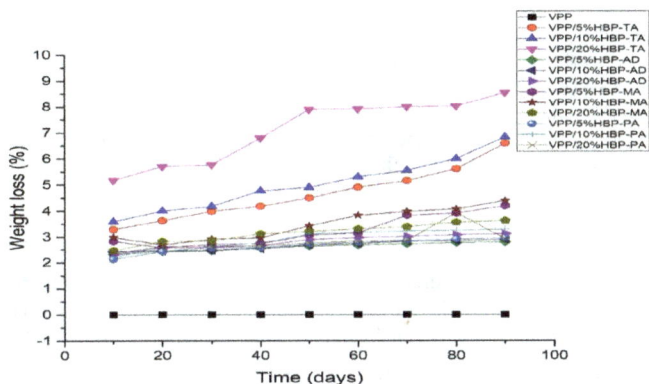

Fig. 4. *The biodegradability of VPP and VPP/HBP in soil environment.*

While in Figure 5, for rPP/HBP blends it is found that the weight loss for blend with HBP-TA is 9%, 12.2%, and 13.1% as the HBP increase from 5% to 20%, and the blend with HBP-MA increases the weight loss to 4.5%, 5.2%, and 5.8% as the HBP increase from 5% to 20%. While for the blends with HBP-AD and HBP-PA it's found that the weight loss is in the range of 3.9% to 4.25% for blend with HBP-AD and the weight loss are in the range of 4.8% to 5.4% for blend with HBP-PA.

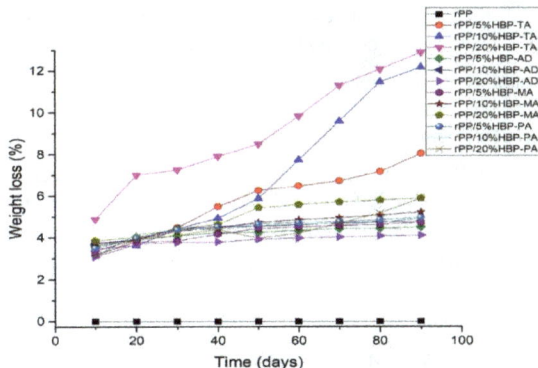

Fig. 5. *The biodegradability of rPP and rPP/HBP in soil environment.*

From the above results, it's found that the blends with HBP-TA has the best results when compared with other blends this is due to the higher content of OH groups in HBP-TA. Also, it is found that the result of rPP is better than VPP this is because rPP has shorter chains and lower entanglements which permit in better absorption of water and make the degradation easier. The presence of HBP molecule can reduce the polymer crystallinity, the reduction in the crystallinity means the polymer ability to withstand the degradation reduced, and the hydrolysis within amorphous regions are higher than crystalline regions. The presence of esters in HBP can result in faster degradation, these results are in agreement with Bakhshi and Agarwal [23], and Gu et al. [24].

Conclusion

From the results above, it can be concluded that the FTIR results show the absence of chemical interaction between PP and the four HBPs and only physical interaction with a slight shifting in band values. In addition, it is found that the biodegradation in soil increased as the amount of HBPs increased and that the rate of weight loss in rPP blends was higher than that in VPP blends, and the same results were obtained from the water contact angle. Furthermore, the MFR increased as the HBPs increased while the viscosity decreased, indicating that the HBPs addition improved the processability of PP (VPP and rPP).

References

[1] P. R. Deviprasad Varma. Studies on the fracture behaviour of polymer blends with special reference to PP/HDPE and PS/HIPS blends, Cochin University of Science & Technology, 2010. [online]. 2010. [viewed: 2023-01-31]. Available from: http://purl.org/purl/2625

[2] A.M.P.B. Samarasekara, E.A.P.C.D. Jayasuriya. Synthesis of biodegradable polyolefins based polymer composites using degradable natural materials, in: Proc. Int. Forestry and Environment Symp. 18 (2013) 61-72. https://doi.org/10.31357/fesympo.v18i0.1913

[3] E. Eastwood et al. Methods to improve the properties of polymer mixtures: optimizing intermolecular interactions and compatibilization, Polymer 46 2005) 3957-3970. https://doi.org/10.1016/j.polymer.2005.02.073

[4] G. Scott, D.M. Wiles. Programmed-life plastics from polyolefins: a new look at sustainability, Biomacromolecules 2001 (2) 615-622. https://doi.org/10.1021/bm010099h

[5] A. Ammala et al. An overview of degradable and biodegradable polyolefins, Prog. Polym. Sci. 2011 (36) 1015-1049. https://doi.org/10.1016/j.progpolymsci.2010.12.002

[6] D.K. Mandal et al. Radiation-induced grafting of acrylic acid onto polypropylene film and its biodegradability, Radiat. Phys. Chem. 2016 (123) 37-45. https://doi.org/10.1016/j.radphyschem.2016.02.011

[7] D.E. MacArthur, D. Waughray, M.R. Stuchtey. The new plastics economy, rethinking the future of plastics, World Economic Forum 2016. [online]. 2016. [viewed: 2023-01-31]. Available from: https://www3.weforum.org/docs/WEF_The_New_Plastics_Economy.pdf

[8] D.K. Mandal et al. Optimization of acrylic acid grafting onto polypropylene using response surface methodology and its biodegradability, Radiat. Phys. Chem. 132 (2017) 71-81. https://doi.org/10.1016/j.radphyschem.2016.12.003

[9] R. Muthuraj, M. Misra, A.K. Mohanty. Biodegradable compatibilized polymer blends for packaging applications: A literature review, J. App. Polymer Sci. 135 (2018) art. 45726. https://doi.org/10.1002/app.45726.

[10] A.I. Buthaina, K.M. Kadum. Influence of polymer blending on mechanical and thermal properties, Mod. Appl. Sci. 4 (2010) 157-161. https://doi.org/10.5539/mas.v4n9p157

Quality Production Improvement and System Safety: QPI 16 - CZOTO 10   Materials Research Forum LLC
Materials Research Proceedings 34 (2023) 110-120   https://doi.org/10.21741/9781644902691-14

[11] Ye. Hong et al. A novel processing aid for polymer extrusion: Rheology and processing of polyethylene and hyperbranched polymer blends, J. Rheol. 43 (1999) 781-793. https://doi.org/10.1122/1.550999

[12] J. Wang et al. Chain-topology-controlled hyperbranched polyethylene as effective polymer processing aid (PPA) for extrusion of a metallocene linear-low-density polyethylene (mLLDPE), J. Rheol. 52 (2008) 243-260. https://doi.org/10.1122/1.2807445

[13] N.H. Al-Mutairi, A. Al-Zubiedy, A.J. Al-Zuhairi. Hyperbranched Polyester Polymer Preparation and Study Its Effect on Some Properties of Polypropylene, Egypt. J. Chem. 65 (2022) 35-43. https://doi.org/10.21608/ejchem.2022.98747.4594

[14] N.H. Al-Mutairi, A. Al-Zubiedy. Preparation and Characterization of a Novel Hyperbranched Polyester Polymers Using A2+B3 Monomers, Prod. Eng. Arch. 29 (2023) 28-36. https://doi.org/10.30657/pea.2023.29.5

[15] A. Shenoy. Thermoplastic melt rheology and processing, 1st Ed. CRC Press, Boca Raton, 1996.

[16] H.C. Obasi, I.O. Igwe. Cassava starch-mixed polypropylene biodegradable polymer: Preparation, characterization, and effects of biodegradation products on growth of plants, Int. J. Sci. Res. 3 (2014) 802-807.

[17] M.R. Jung et al. Validation of ATR FT-IR to identify polymers of plastic marine debris, including those ingested by marine organisms, Mar. Pollut. Bull. 127 (2018) 704-716. https://doi.org/10.1016/j.marpolbul.2017.12.061

[18] M. Gopanna et al. Fourier transform infrared spectroscopy (FTIR), Raman spectroscopy and wide-angle X-ray scattering (WAXS) of polypropylene (PP)/cyclic olefin copolymer (COC) blends for qualitative and quantitative analysis, Polym. Bull. 76 (2019) 4259-4274, 2019. https://doi.org/10.1007/s00289-018-2599-0

[19] M. Guzmán, D. Giraldo, E. Murillo. Hyperbranched polyester polyol plasticized tapioca starch/low density polyethylene blends, Polimeros 27 (2017) 1-7. https://doi.org/10.1590/0104-1428.04816

[20] R. Mesias, E. Murillo. Hyperbranched polyester polyol modified with polylactic acid as a compatibilizer for plasticized tapioca starch/polylactic acid blends, Polimeros 28 (2018) 44-52. https://doi.org/10.1590/0104-1428.09516

[21] J. Zhang et al. Novel waterborne UV-curable hyperbranched polyurethane acrylate/silica with good printability and rheological properties applicable to flexographic ink, ACS Omega 2 (2017) 7546-7558. https://doi.org/10.1021/acsomega.7b00939

[22] C. Caicedo, E.A. Murillo. Structural, thermal, rheological, morphological and mechanical properties of polypropylene functionalized in molten state with maleinized hyperbranched polyol polyester, Eur. Polym. J. 118 (2019) 254-264. https://doi.org/10.1016/j.eurpolymj.2019.06.005

[23] H. Bakhshi, S. Agarwal. Hyperbranched polyesters as biodegradable and antibacterial additives, J. Mater. Chem. B 5 (2017) 6827-6834. https://doi.org/10.1039/c7tb01301a

[24] W. Gu et al. Tough, strong, and biodegradable composite film with excellent UV barrier performance comprising soy protein isolate, hyperbranched polyester, and cardanol derivative, Green Chem. 21 (2019) 3651-3665. https://doi.org/10.1039/c9gc01081e

Quality Production Improvement and System Safety: QPI 16 - CZOTO 10   Materials Research Forum LLC
Materials Research Proceedings 34 (2023) 120-126   https://doi.org/10.21741/9781644902691-15

The Influence of Mn Addition on Corrosion Resistance in Secondary Aluminium Alloy AlSi7Mg0.3 after the Heat Treatment

ŠURDOVÁ Zuzana[1,a]*, KUCHARIKOVÁ Lenka[1,b], TILLOVÁ Eva[1,c] and CHALUPOVÁ Mária[1,d]

[1]University of Žilina, Faculty of Mechanical Engineering, Department of Materials Engineering, Univerzitná 8215/1, 010 26 Žilina, Slovak Republic

[a] zuzana.surdova@fstroj.uniza.sk *, [b] lenka.kucharikova@fstroj.uniza.sk,
[c] eva.tillova@fstroj.uniza.sk, [d] maria.chalupova@fstroj.uniza.sk

Keywords: AlSi7Mg0.3; Al5FeSi; Heat Treatment, 3.5% NaCl, Corrosion Resistance

Abstract. This paper aims to study the secondary aluminium alloy AlSi7Mg0.3 with higher Fe content after heat treatment. The addition of Mn is also investigated. The quantitative analysis was performed to evaluate the effect of higher Fe content on the amount and shape of platelet-like Fe-rich intermetallic phases. As a method of corrosion resistance testing an immersion test in 3.5% NaCl was used. This test was done in order to evaluate the influence of higher Fe content and Mn addition in heat-treated AlSi7Mg0.3 alloys. The results show that the Mn addition increases the corrosion resistance of AlSi7Mg0.3 alloys, but not significantly. AlSi7Mg0.3 alloys have numerous uses within the automobile and other sectors.

Introduction

Aluminium alloys have gained widespread use in various industrial applications due to their exceptional properties, such as low density, high thermal conductivity, and good corrosion resistance. Among aluminium alloys, secondary aluminium alloys are particularly attractive due to their low production cost and high recyclability. These alloys are produced by casting processes using recycled aluminium scrap as the raw material [1,2].

AlSi7Mg0.3 alloy is a popular secondary aluminium alloy, which is widely used in various applications such as automotive, aerospace, and construction. It is an Al-Si-Mg-based alloy with a nominal composition of 7.0 wt. % Si and 0.3 wt. % Mg, which gives it good casting properties, high strength, and good corrosion resistance [3,4]. The heat-treatable nature of AlSi7Mg0.3 also enables further improvement of its mechanical properties through solution heat treatment and ageing. Despite its advantages, the presence of Fe and other impurities in the recycled aluminium scrap can negatively impact the microstructure and properties of AlSi7Mg0.3 alloys, thus, requiring careful selection of raw materials and precise control of casting conditions [5-8].

Intermetallic phases play a critical role in determining the mechanical and corrosion properties of the secondary Al-Si-Mg-based AlSi7Mg0.3 alloy. The microstructure of AlSi7Mg0.3 consists of a complex network of intermetallic phases and eutectic structures, which result from the reaction between the solid aluminium matrix and the dispersed silicon particles. The presence and distribution of intermetallic phases significantly affect the properties of the alloy, and therefore, understanding their formation and growth is important for optimizing the properties of the AlSi7Mg0.3 alloy. The intermetallic phases in AlSi7Mg0.3 can be classified into several types, including Fe-rich Al_5FeSi, $Al_{15}(FeMn)_3Si_2$, and Mg_2Si, each with unique chemical and physical properties that contribute to the overall behaviour of the alloy [8-10].

In AlSi7Mg0.3 alloys, an increase in Fe content can have a negative impact on the microstructure and properties of the material. The presence of Fe in these alloys can also result in the formation of undesired phases, such as Fe-rich Al_5FeSi, which can negatively affect the mechanical properties and corrosion resistance of the material. Therefore, controlling the Fe

Quality Production Improvement and System Safety: QPI 16 - CZOTO 10   Materials Research Forum LLC
Materials Research Proceedings 34 (2023) 120-126   https://doi.org/10.21741/9781644902691-15

content and mitigating its negative effects are important considerations in the production and application of said alloys [10-12].

Manganese is an important element added to AlSi7Mg0.3 alloy for the purpose of improving its properties. The addition of Mn has been found to have a significant effect on the microstructure and mechanical properties of the alloy, including enhanced corrosion resistance and improved formability. Mn addition can also alter the precipitation kinetics of intermetallic phases and change the type of phases formed in the alloy [6,13,14].

Secondary AlSi7Mg0.3 aluminium alloy is widely used in various industries such as aerospace, transportation, and construction. In the automotive industry, it is commonly used in wheel production, engine components and other structural parts. The alloy's good mechanical properties and corrosion resistance make it an attractive choice for various engineering applications, including high-temperature high-stress, and high-corrosion environments. The widespread use of this alloy highlights its versatility and adaptability [3,15].

Materials and Methods

The secondary AlSi7Mg0.3 aluminium cast alloy samples being studied were produced by UNEKO spol. s.r.o, Zátor, Czech Republic. They were supplied in the form of circular rods, 300 mm long and 20 mm in diameter (Fig. 1), made by gravity casting into sand moulds treated to prevent liquid metal penetration. A protective spray was applied to improve surface quality.

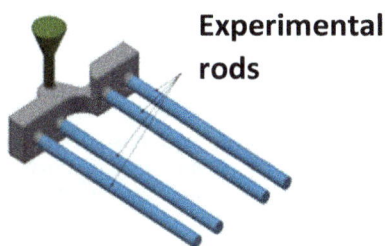

Fig.1. The casting model of experimental rods.

To assess the impact on Fe-rich intermetallic phases and corrosion resistance caused by increased Fe content and Mn addition in heat-treated samples, 4 alloys with different Fe contents were cast: 0.123 % Fe (alloy A), 0.454 % Fe (alloy B), 0.679 % Fe (alloy C), and 1.209 % Fe (alloy D). Table 1 displays the chemical composition of the alloys and the calculated critical iron value. Each melt was cast at 750°C and refined at 740-745°C. As a refining salt, ECOSAL AL 113S was used. All experimental bars were subjected to heat treatment T6 – solution annealing at a temperature of 530°C ± 5 °C with a holding time of 7 hours, rapid cooling to a temperature of 50°C and artificial ageing at a temperature of 160°C for 6 hours. These samples were compared to determine whether the heat treatment negates the negative effect of higher Fe content.

Microstructure analysis was performed to understand the connection between microstructure and corrosion resistance, and the impact of heat treatment on the AlSi7Mg0.3 alloys. The samples were prepared using a metallographic process involving grinding with 500 and 1 200-grit Strues SiC paper, followed by polishing with 3 µm diamond paste and Strues Op-S. Chemical etching was done using 0.5% HF and H_2SO_4 to highlight Fe-rich phases. The samples were then observed using a NEOPHOT 32 optical microscope. An optical microscope and NIS Elements 5.2 software were utilized for the quantitative analysis of Fe-rich phases to assess the impact of Fe content Mn addition and heat treatment (T6) on their length, particularly those in the plate-like (needle) shape. The length of needle-shaped Fe-rich intermetallic phases was measured 50 times on all experimental samples with and without Mn addition.

To study the corrosion resistance of the secondary AlSi7Mg0.3 aluminium alloy, an immersion test using a 3.5% NaCl solution was employed. The specimens for the test were 18 mm in diameter and 10 mm long, with 3 specimens per melt. They were cleaned in distilled water and ethanol, dried with hot air, then weighed with analytical scales, accurately to 4 decimal places, and immersed in the 3.5% NaCl solution for three weeks.

Table 1. *The chemical composition of experimental alloy AlSi7Mg0.3 [wt. %]*

Alloy	Si	Fe	Cu	Mn	Mg	Cr	Zn	Ti	Ga	V	Al	$Fe_{crit.}$
A	7.340	**0.454**	0.021	**0.01**	0.3	0.002	0.02	0.12	0.01	0.01	ball.	0.501
A^{Mn}	7.051	**0.450**	0.021	**0.122**	0.26	0.002	0.02	0.11	0.01	0.01	ball.	0.479
B	7.346	**0.679**	0.0096	**0.01**	0.4	0.002	0.03	0.11	0.01	0.01	ball.	0.501
B^{Mn}	7.039	**0.681**	0.008	**0.337**	0.31	0.002	0.03	0.12	0.01	0.01	ball.	0.478
C	7.340	**1.209**	0.01	**0.01**	0.31	0.002	0.01	0.12	0.01	0.01	ball.	0.501
C^{Mn}	7.212	**1.200**	0.0098	**0.585**	0.28	0.002	0.01	0.12	0.01	0.01	ball.	0.491

To prevent interference with the corrosion process, the samples were positioned in the glass container so that they did not touch each other or the walls. In order to avoid a change in the concentration of 3.5% of NaCl solution and that the electrolyte level remains throughout the duration of the test (3 weeks) constant, distilled water was carefully poured down the sides of the vessel. In the room where the immersion test was carried out, a constant temperature of 20 ± 2 °C was maintained. If the measures described above are not followed, the kinetics may be altered of the corrosion processes and therefore erroneous results.

After the test time had elapsed, the samples were rinsed in distilled water and ethanol, dried with hot air, reweighed, and evaluated gravimetrically. The extent of the corrosion attack was observed using an Olympus Stereo microscope SZX16 and the impact of increased Fe content and Mn addition was evaluated.

Results and Discussion
The microstructure of the alloys studied in this research is displayed in Fig. 2. The standard microstructure of the alloys includes α-phase eutectic (seen as dark Si crystals in α-phase), and various intermetallic phases. Eutectic silicon is seen after heat treatment as small, round particles. The identified intermetallic phases in the alloys include plate-like Al_5FeSi skeleton-like $Al_{15}(FeMn)_3Si_2$ and Mg-rich (Mg_2Si) very fine particles. The needle-like Al_5FeSi phase is fragmented into smaller needles after heat treatment.

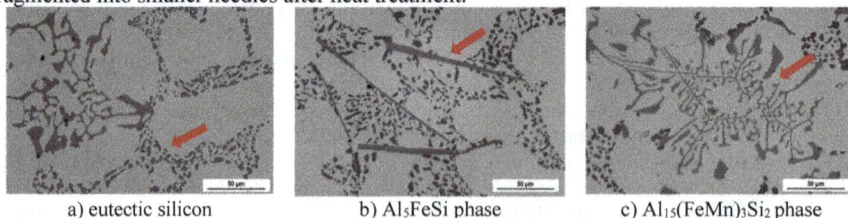

a) eutectic silicon   b) Al_5FeSi phase   c) $Al_{15}(FeMn)_3Si_2$ phase

Fig.2. *Microstructure of experimental alloys.*

Quality Production Improvement and System Safety: QPI 16 - CZOTO 10   Materials Research Forum LLC
Materials Research Proceedings 34 (2023) 120-126   https://doi.org/10.21741/9781644902691-15

An increase in Fe content usually results in a higher number of Fe-rich needle-like phases in the microstructure and in the growth in their length, but this trend was not seen in the alloys used in this study. The growth of Al_5FeSi phases was not exponential as was expected (Table 2). The highest average length of the Fe-rich phase Al_5FeSi was observed in alloy A. In alloys B and C with increasing iron content, larger acicular phases were not observed. Probably after exceeding 0.5 % iron content, iron no longer has such a negative effect, the phases are rather thicker and shorter. The smallest average length was measured in alloys B^{Mn} and C^{Mn}. This is because these experimental alloys met the Mn/Fe = 1:2 condition (a condition that if met, results in a change in intermetallic morphology from needle-like phases to skeleton-like phases).

Table 2. *The results of the quantitative analysis of experimental alloys*

Alloy	Fe [wt. %]	Mn [wt. %]	Mn/Fe	Minimum length [μm]	Maximum length [μm]	Average length [μm]
Alloy A	0.454	0.01	0.022	7.97	269.16	55.02
Alloy A^{Mn}	0.450	0.122	0.271	15.26	117.41	37.56
Alloy B	0.679	0.01	0.015	6,57	155.32	36.25
Alloy B^{Mn}	0.681	0.337	**0.495**	11.63	75.87	34.52
Alloy C	1.209	0.01	0.008	5.85	141.8	40.57
Alloy C^{Mn}	1.200	0.585	**0.488**	6.01	59.36	17.41

Table 3. *The weight changes of experimental alloys after immersion test [g]*

Sample	A	A^{Mn}	B	B^{Mn}	C	C^{Mn}
Weight loss	-0.0039	-0.0044	-0.0035	-0.0014	-0.0023	-0.0013

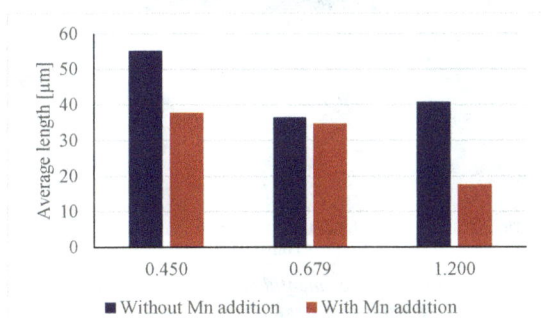

Fig.3. *The effect of Mn addition on the length of Al_5FeSi needles.*

Thus, quantitative analysis proved that in melts, which had Mn addition, there were on average shorter Fe phases, in the form of needles, than in the melts which were without Mn addition, only with increased Fe content. A graph was created to visually compare the average length of Fe phases in the experimental alloys with and without Mn addition (Fig. 3).

The immersion test using 3.5% NaCl was conducted to study the corrosion behaviour of the AlSi7Mg0.3 alloys. After three weeks the samples were removed from the solution and washed, dried, weighed, and evaluated gravimetrically by calculating average mass changes. The gravimetric results are shown in Table 3.

Quality Production Improvement and System Safety: QPI 16 - CZOTO 10   Materials Research Forum LLC
Materials Research Proceedings 34 (2023) 120-126   https://doi.org/10.21741/9781644902691-15

All experimental samples showed weight loss. The immersion test using 3.5% NaCl showed that the addition of Mn enhances corrosion resistance, particularly in alloy B. Nevertheless, it should be noted that the Mn addition does not result in a substantial improvement in corrosion resistance in the alloys studied in this particular corrosion test. Subsequently, a macroscopic examination of the corrosion damage was conducted. The macroscopic evaluation of the surface of the samples after the immersion corrosion test showed the presence of pitting corrosion in all samples examined (Fig.4).

In the experimental alloys after the immersion test, a heterogeneous corrosion attack of the material was preferentially observed. Chloride anions, by passing through the passive layer of the material, cause the formation of a corrosion pit. The pits are formed on the surface of the material, especially where there are various inhomogeneities, and impurities and their formation also occur at grain boundaries, i.e., in the eutectic. The intensity of the corrosion attack of the experimental alloys also depends on whether the alloys are heat treated or not. The change in the morphology of Si particles after heat treatment (the so-called spheroidization of Si) positively affects the dissolution rate of α-phase in the eutectic; it slows it down, compared to alloys in the as-cast state. Therefore, a lower density of corrosion pits can generally be observed in heat-treated alloys.

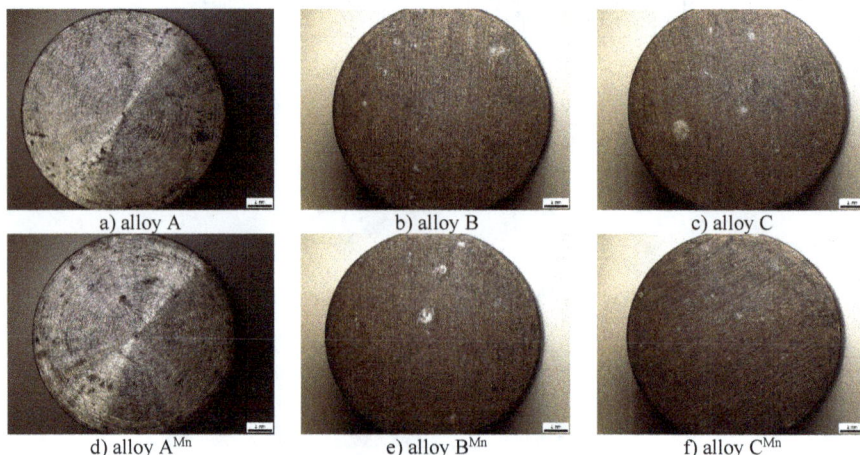

| a) alloy A | b) alloy B | c) alloy C |
| d) alloy A^{Mn} | e) alloy B^{Mn} | f) alloy C^{Mn} |

Fig.4. *Macrographs of experimental samples after immersion corrosion test*
a), b), c) – alloys without Mn addition; d), e), f) – with Mn addition.

However, it cannot be claimed that the level of corrosion attack increased with increasing Fe content in each melt. In the melts with Mn addition, the observed surface was less attacked compared to the samples without Mn addition. Alloy A showed the most severe corrosion damage, with the largest and deepest pits, regardless of the Mn addition (Fig. 4a, d).

It can be observed in the cross-section of the test specimens that the pitting corrosion propagates to the depth of the material through the α-phase in the eutectic.

Quality Production Improvement and System Safety: QPI 16 - CZOTO 10
Materials Research Proceedings 34 (2023) 120-126

Materials Research Forum LLC
https://doi.org/10.21741/9781644902691-15

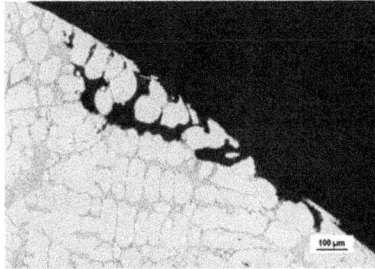

Fig.5. *Presence of pitting corrosion in experimental alloy C.*

Conclusion

The study aimed to determine the impact of increased Fe content and Mn addition on the microstructure and corrosion resistance of AlSi7Mg0.3 cast aluminium alloys after heat treatment T6. The results led to the following conclusions:

- The AlSi7Mg0.3 experimental alloys microstructure is composed of α-phase eutectic and various intermetallic phases such as Fe-rich needle-like Al_5FeSi skeleton-like $Al_{15}(FeMn)_3Si_2$ and Mg_2Si.
- Quantitative analysis revealed that in experimental alloys with Mn addition, there were on average shorter Fe phases. The smallest average length was measured in alloys B^{Mn} and C^{Mn}. Presumably, because the Mn/Fe ratio is close to 0.5.
- The immersion test using 3.5% NaCl was performed in order to evaluate the influence of Mn addition on alloys after heat treatment. The test revealed that in the samples that had Mn addition, the observed surface was less attacked compared to the samples without Mn addition.
- The macrographs of the experimental samples showed pitting corrosion as the type of corrosion that attacked the samples.
- It can be said that Mn addition overall increases corrosion resistance and reduces the size and amount of Fe-rich intermetallic phases even in the alloys after the heat treatment.

Acknowledgements

The research was supported by Grand Agency of Ministry of Education of Slovak Republic and Slovak Academy of Sciences, project KEGA 004ŽU-4/2023, project to support young researchers at UNIZA, ID project 12715 (Kucharíková) and project 313011ASY4 "Strategic implementation of additive technologies to strengthen the intervention capacities of emergencies caused by the COVID-19 pandemic".

References

[1] D. Varshney, K. Kumar. Application and use of different aluminium alloys with respect to workability, strength and welding parameter optimization, Ain Shams Eng. J. 12 (2021) 1143-1152. https://doi.org/10.1016/j.asej.2020.05.013

[2] F. Czerwinski. Thermal Stability of Aluminum Alloys, Materials 13 (2020) art.3441. https://doi.org/10.3390/ma13153441

[3] L. Lattanzi et al. Room Temperature Mechanical Properties of A356 Alloy with Ni Additions from 0.5 Wt to 2 Wt%, Metals 8 (2018) art.224. https://doi.org/10.3390/met8040224

[4] E. Erzi et al. Determination of Acceptable Quality Limit for Casting of A356 Aluminium Alloy: Supplier's Quality Index (SQI), Metals 9 (2019) art.957. https://doi.org/10.3390/met9090957

[5] D. Manickam et al. Effect of Solution Heat Treatment and Artificial Aging on Compression Behaviour of A356 Alloy, Medziagoryra 25 (2019) 281-285. https://doi.org/10.5755/j01.ms.25.3.20442

[6] H. Nunes et al. Adding Value to Secondary Aluminum Casting Alloys: A Review on Trends and Achievements, Materials 16 (2023) art.895. https://doi.org/10.3390/ma16030895

[7] L. Stanček et al. Structure and properties of silumin castings solidified under pressure after heat treatment. Met. Sci. Heat Treat. 56 (2014) 197-202. https://doi.org/10.1007/s11041-014-9730-0

[8] B. Vanko, L. Stanček. Utilization of heat treatment aimed to spheroidization of eutectic silicon for silumin castings produced by squeeze casting. Arch. Foundry Eng. 12 (2012) 111-114. https://doi.org/10.2478/v10266-012-0021-1

[9] T. Xu et al. Microstructure and mechanical properties of in-situ nano γ-Al_2O_{3p}/A356 aluminum matrix composite, J. Alloys Compd. 787 (2019) 72-85. https://doi.org/10.1016/j.jallcom.2019.02.045

[10] W.S. Ebhota, J. Tien-Chien. Effects of Modification Techniques on Mechanical Properties of Al-Si Cast Alloys, In: S. Sivasankaran (Ed.), Aluminium Alloys – Recent Trends in Processing, Characterization, Mechanical Behavior and Applications, IntechOpen, 2017. https://doi.org/10.5772/intechopen.70391

[11] M.S. Kaiser et al. Study of Mechanical and Wear Behaviour of Hyper-Eutectic Al-Si Automotive Alloy Through Fe, Ni and Cr Addition, Mater. Res. 21 (2018) art.e20171096. https://doi.org/10.1590/1980-5373-MR-2017-1096

[12] M.A. Moustafa. Effect of iron content on the formation of β-Al5FeSi and porosity in Al-Si eutectic alloys, J. Mater. Proces. Technol. 209 (2009) 605-610. https://doi.org/10.1016/j.jmatprotec.2008.02.073

[13] J. Kozana et al. The Effect of Tin on Microstructure and Properties of the Al-10 wt.% Si Alloy, Materials 15 (2022) art.6350. https://doi.org/10.3390/ma15186350

[14] A.Y. Algendy et al. Formation of intermetallic phases during solidification in Al-Mg-Mn 5xxx alloys with various Mg levels, MATEC Web of Conf. 326 (2020) art.02002. https://doi.org/10.1051/matecconf/202032602002

[15] C. Berlanga-Labari et al. Corrosion of Cast Aluminum Alloys: A Review, Metals 10 (2020) art.1384. https://doi.org/10.3390/met10101384

Quality Production Improvement and System Safety: QPI 16 - CZOTO 10   Materials Research Forum LLC
Materials Research Proceedings 34 (2023) 127-138   https://doi.org/10.21741/9781644902691-16

Thermal Stability of Ammonium Nitrate in Two-Component Mixtures with Powdered and Fine-Grained Materials

DYSZ Karolina[1,a]*, POSZWALD Bartosz[1,b], KWAK Anna[1,c]
and DYLONG Agnieszka[1,d]

[1] Military Institute of Engineer Technology, 136 Obornicka Str., 50-961 Wroclaw, Poland

[a] dysz@witi.wroc.pl, [b] poszwald@witi.wroc.pl, [c] kwak@witi.wroc.pl, [d] dylong@witi.wroc.pl

Keywords: Ammonium Nitrate, Thermal Analysis, TG, DSC, Fertilizers, Powdered Materials

Abstract. Ammonium nitrate(V) (AN, NH_4NO_3) is widely and widely used in the chemical industry, in agriculture as a fertilizer, explosive for military and civil purposes (e.g. in mining) or as a solid propellant [1, 2]. Storage of ammonium nitrate poses many problems, as it may be hazardous. This was proven, for example, by the explosion in 2020 in Beirut. Ammonium nitrate was stored in a warehouse at the port among other wares and an unfortunate turn of events caused a huge explosion. The explosion contributed to the formation of a 140-meter crater and an earthquake with a magnitude of 3.3 on the Richter scale. This explosion was classified as the third most destructive urban explosion of all time, after the atomic bombs in Hiroshima and Nagasaki at the end of World War II [3, 4], as the mixtures of oils (fuel or gas) and a concentrated form of nitrogen fertilizer – ammonium nitrate form explosives [5]. The dangerous properties of AN have been extensively studied. It is known that pure AN is stable at room temperature but may explode when mixed with impurities in a confined space or under fire-hazard conditions [1]. The research aimed to analyze the changes occurring in two-component mixtures with ammonium nitrate and powdered or fine-grained materials and to assess the effect of such an admixture on the fertilizer. Thermal analysis was used to carry out the TG-DSC tests.

Introduction

Ammonium nitrate(V) is a colorless, crystalline salt, well soluble in water. AN is hygroscopic and does not form hydrates. In explosives, the nitrate ion from ammonium nitrate is a source of oxygen and is used as an oxidant [6]. The most commonly used AN-based explosive is ANFO (Ammonium Nitrate Fuel Oil) due to its easy and inexpensive manufacturing. ANFO is produced by mixing ammonium nitrate with fuel oil (combustible component), typically in a weight ratio of 94:6. There are also other AN-based explosives, e.g. dynamon K (mixture with easily oxidized combustible substances, e.g.: 90% AN, 10% wood flour, ANNM (Ammonium Nitrate – NitroMethane) and ammonals (explosive mixtures consisting of TNT, AN and aluminum powder) [7-9].

The storage of ammonium nitrate for agricultural purposes is primarily connected with the problem of caking of such material or the hygroscopic nature of this compound. The clumping of AN is the result of adhesive forces and the process itself causes too much pressure on the layers located in the areas located lower during storage. Pressure contributes to changes in the physicochemical properties of AN. The chemical reactivity of AN has been well documented throughout the last century. Ammonium nitrate(V) can explode between 260°C and 300°C [6]. Interestingly, AN fertilizer, being non-flammable by nature, does not have explosive properties, unlike coal, wood, grain-based or other organic powders [10]. At sufficiently high temperatures, ammonium nitrate may decompose rapidly on its own. As a result, gases are formed, including nitrogen oxides and water **vapor**, and the rapid release of these gases causes an explosion [6, 10].

Different mechanisms of AN degradation have been described in the literature [6, 10-12], and the most accepted reactions are summarized below. The reversible reaction can occur at relatively low temperatures (i.e., ca 170°C). It is believed that the evaporation of molten AN leads to the formation of ammonia and nitric acid, which can initiate the decomposition of AN by the following reaction:

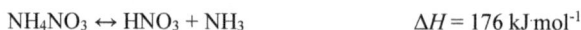

$$NH_4NO_3 \leftrightarrow HNO_3 + NH_3 \qquad \Delta H = 176 \ kJ \cdot mol^{-1}$$

At higher temperatures (between 170°C and 280°C) irreversible exothermic reactions occur:

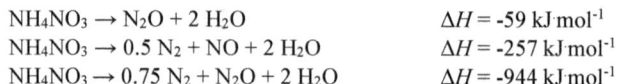

$$NH_4NO_3 \rightarrow N_2O + 2 \ H_2O \qquad \Delta H = -59 \ kJ \cdot mol^{-1}$$
$$NH_4NO_3 \rightarrow 0.5 \ N_2 + NO + 2 \ H_2O \qquad \Delta H = -257 \ kJ \cdot mol^{-1}$$
$$NH_4NO_3 \rightarrow 0.75 \ N_2 + N_2O + 2 \ H_2O \qquad \Delta H = -944 \ kJ \cdot mol^{-1}$$

If the material is suddenly heated, explosive decomposition will occur [6, 10-12].

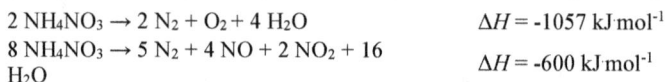

$$2 \ NH_4NO_3 \rightarrow 2 \ N_2 + O_2 + 4 \ H_2O \qquad \Delta H = -1057 \ kJ \cdot mol^{-1}$$
$$8 \ NH_4NO_3 \rightarrow 5 \ N_2 + 4 \ NO + 2 \ NO_2 + 16 \ H_2O \qquad \Delta H = -600 \ kJ \cdot mol^{-1}$$

Ammonium nitrate(V) under standard pressure exists in five stable polymorphic forms (designated as phases I, II, II, IV and V; see Table 1).

Table 1. Crystallographic forms of ammonium nitrate [12]

Crystalline Phase	Temperature Range	Crystal System
I	169.5°C … 125.2°C	*Cubic*
II	125.2°C … 84.2°C	*Tetragonal*
III	84.2°C … 32.3°C	*Orthorhombic*
IV	32.3°C … -18°C	*Orthorhombic*
V	-18°C … -103°C	*Orthorhombic*

Various types of admixtures for mixtures with AN can act in three ways: as an inhibitor, promoter or neutral additive. When the decomposition onset temperature is higher than that of pure AN, the additive is an inhibitor. When the decomposition onset temperature is lower than that for pure AN, the additive is considered a promoter. There is still a third option, the addition does not change the "initial" temperature of decomposition. The additives behave as inert materials that only thin the AN [12]. Additives in the form of inhibitors can mitigate the risk of AN explosion. As the literature data show, such properties showed:
- sodium, potassium, ammonium and calcium salts;
- sulphates, phosphates, carbonates, organic substances (urea, oxalate, methanoate, guanidine salts);
- sodium salts of weak acids (carbonic acid, acetic acid, formic acid, oxalic acid).

On the other hand, chemical additives called promoters had completely different properties, they accelerate the potential explosion of AN. These include:
- nitrocellulose, aromatic nitro compounds;
- non-explosive combustible substances: sulphur, charcoal, flour, sugar or oil;
- pyrite, aluminum, zinc, cadmium and copper filings;
- chloride salts: NH_4Cl, KCl, $NaCl$, $BaCl_2$, $CaCl_2$;
- chromium, iron cations;

- inorganic acid, e.g. hydrochloric acid;
- organic impurities: animal fats, cotton, waste paper, bleaching powder, jute bags, caustic soda;
- coke, charcoal, coal, cork;
- camphor, fibers of all kinds;
- fish oil, fish meal, lubricating oil, linseed oil or drying oils, vegetable oil, naphthalene;
- hay, sawdust, and wood shavings [12].

Experimental Procedure

Materials. Ammonium nitrate used for making mixtures with various types of admixtures was N34 ammonium nitrate WE fertilizer (Polish name: *nawóz WE azotan amonu N34*) produced by Anwil Grupa ORLEN. In ammonium nitrate nitrogen occurs in two forms: nitrate and ammonia. This fertilizer was enriched with magnesium. In the initial form, it has the form of white granules (Fig. 1, 2). The total nitrogen content is 34.0 +/-0.6% (m/m).

Fig. 1. *Granular form of ammonium nitrate*

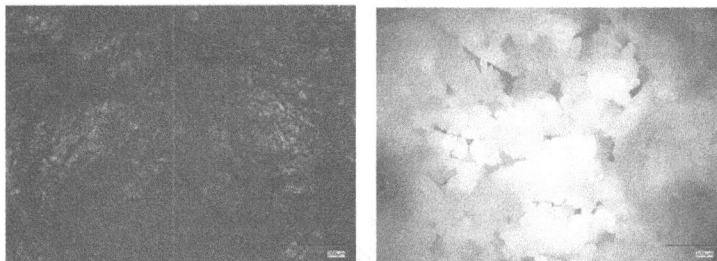

Fig. 2. *Ammonium nitrate granule surface (left side) and after grinding in a mortar (right side)*

Quality Production Improvement and System Safety: QPI 16 - CZOTO 10   Materials Research Forum LLC
Materials Research Proceedings 34 (2023) 127-138   https://doi.org/10.21741/9781644902691-16

Table 2. Details of sample identification

No.	Additive	Additive identification	AN%/A%	Identification of samples with AN (mixtures)
1	Pure ammonium nitrate	-	AN100	AN
2	DT0 carbon	A_1	AN80A20	Mix_1
3	Kaolin (powdered)	A_2	AN80A20	Mix_2
4	Instant coffee	A_3	AN80A20	Mix_3
5	Sand (very fine)	A_4	AN80A20	Mix_4
6	Sand (grainy)	A_5	AN80A20	Mix_5
7	Cosmetic talc	A_6	AN80A20	Mix_6
8	Face powder (loose)	A_7	AN80A20	Mix_7
9	Activated charcoal (medical)	A_8	AN80A20	Mix_8
10	Cocoa	A_9	AN80A20	Mix_9
11	Powdered sugar	A_10	AN80A20	Mix_10
12	Chalk powder	A_11	AN80A20	Mix_11

The admixture material consisted of selected materials available on the market, such as DT0 commercially available carbon (A_1), kaolin in the form of powder (A_2), instant coffee (A_3), very fine sand (dust-like) (A_4), powdered chalk, kaolin powder, sand grains, talc grains (A_5), talc (A_6), cosmetic powder (A_7), activated carbon (A_8), cocoa powder (A_9), powdered sugar (A_10) and powdered chalk (A_11). Details of the sample determination are presented in Table 2. Photographs of dopant materials are shown in Fig. 3. All the photographs with very high zoom (for impurities and AN) were taken using the KEYENCE VHX-7000N digital microscope.

A_1 – DT0 carbon	**A_2 - Kaolin**	**A_3 – Instant coffee**
A_4 – Very fine sand	**A_5 - Sand**	**A_6 - Talc**
A_7 – Cosmetic powder (loose)	**A_8 – Activated carbon**	**A_9 - Cocoa**
A_10 – Powdered sugar	**A_11– Chalk powder**	

Fig. 3. Optical microscope photographs of powdered and loose materials serving as admixtures in the created two-component mixtures with AN

Sample preparations and methods. Simultaneous thermal gravimetric analysis and differential scanning calorimetry (TG-DSC) were performed. The samples were analyzed using the Setline STA+ thermogravimeter by SETARAM company. Thermal analysis was performed using the Calisto software dedicated to this device. All the samples prepared were heated at the temperature range of 30-400°C in crucibles without a lid. Samples with an initial weight of 5.0-8.5 mg were heated at 5°C.min^{-1} under a nitrogen atmosphere.

Samples (mixtures) were prepared based on 80 wt. ammonium nitrate (AN) and 20 wt. dopant material (A). Before the preparation of two-component mixtures, the materials used were dried for min. 24 hours at 45-50°C. Before taking the samples, ammonium nitrate in the form of spheres (Fig. 1) was ground to powder (AN100) (Fig. 2). The mixtures were prepared by grinding both components in a mortar.

Results and Discussion
The purpose of the thermogravimetric measurements was to assess the thermal stability of ammonium nitrate (AN) in two-component mixtures with selected powdered or fine-grained materials. The samples were mixtures of 80 wt. AN and 20 wt. selected admixture (A). The results of DSC thermal analysis for mixtures marked Mix_1 – Mix_11 are shown in Fig.4. The values of characteristic physicochemical changes for each sample are presented in Table 3.

Table 3. The result of characteristic physicochemical transformations for each sample

Sample	Phase transition temperature [°C] (*onset*)				Endotherms [°C] (*onset*)	
	IV→III	III→II	II→I	Melting point		
AN	47.3	87.5	125.2	162.5	256.5	
Mix_1	48.3	88.0	123.2	-	146.9	
Mix_2	48.7	86.8	125.2	164.8	244.3	
Mix_3	46.1	88.4	125.4	152.8	181.6	191.2 225.5
Mix_4	45.0	87.5	125.1	162.6	249.8	
Mix_5	46.4	83.9	122.2	160.7	256.0	
Mix_6	48.8	87.2	125.2	162.6	238.8	
Mix_7	49.2	86.0	124.6	160.7	240.2	
Mix_8	51.0	88.6	124.5	-	151.7	
Mix_9	46.2	88.1	125.2	158.2	190.8	230.6
Mix_10	46.7	87.9	100.5	-	138.2	
Mix_11	46.9	87.3	125.2	155.5	269.7	

The tests carried out using the TG-DSC method for a 100% AN sample are shown in Fig. 5. The obtained results indicate that the thermal properties of AN samples do not differ significantly from the data presented in the literature [14]. The average values of the signals obtained on the DSC curve (from 3 measurements) indicate endothermic phase transitions at temperatures of 47.3°C, 87.5°C, 125.2°C, and 162.5°C.

Similarly, as did Popławski et al. [15], in studies using differential scanning calorimetry, we obtained a signal of AN decomposition as endothermic decomposition. The authors measured AN samples in uncovered and covered crucibles (with a lid with a small hole in the middle). Both DSC curves obtained by the authors differ. When an open crucible is used, the weight loss of the sample is accompanied by the endothermic effect only. However, the use of a crucible with a lid made it impossible to remove the decomposition products of ammonium nitrate from the measurement system. This effect is clearly visible in the form of a strongly exothermic effect at temperatures above 200°C. However, in an open crucible, the TG curve confirms that the mass loss begins at a much lower temperature [15].

In the case of the tested fertilizer, the beginning of the AN decomposition exotherm starts at 181.5°C and its maximum signal is at 299.6°C (Fig. 4).

Quality Production Improvement and System Safety: QPI 16 - CZOTO 10   Materials Research Forum LLC
Materials Research Proceedings 34 (2023) 127-138   https://doi.org/10.21741/9781644902691-16

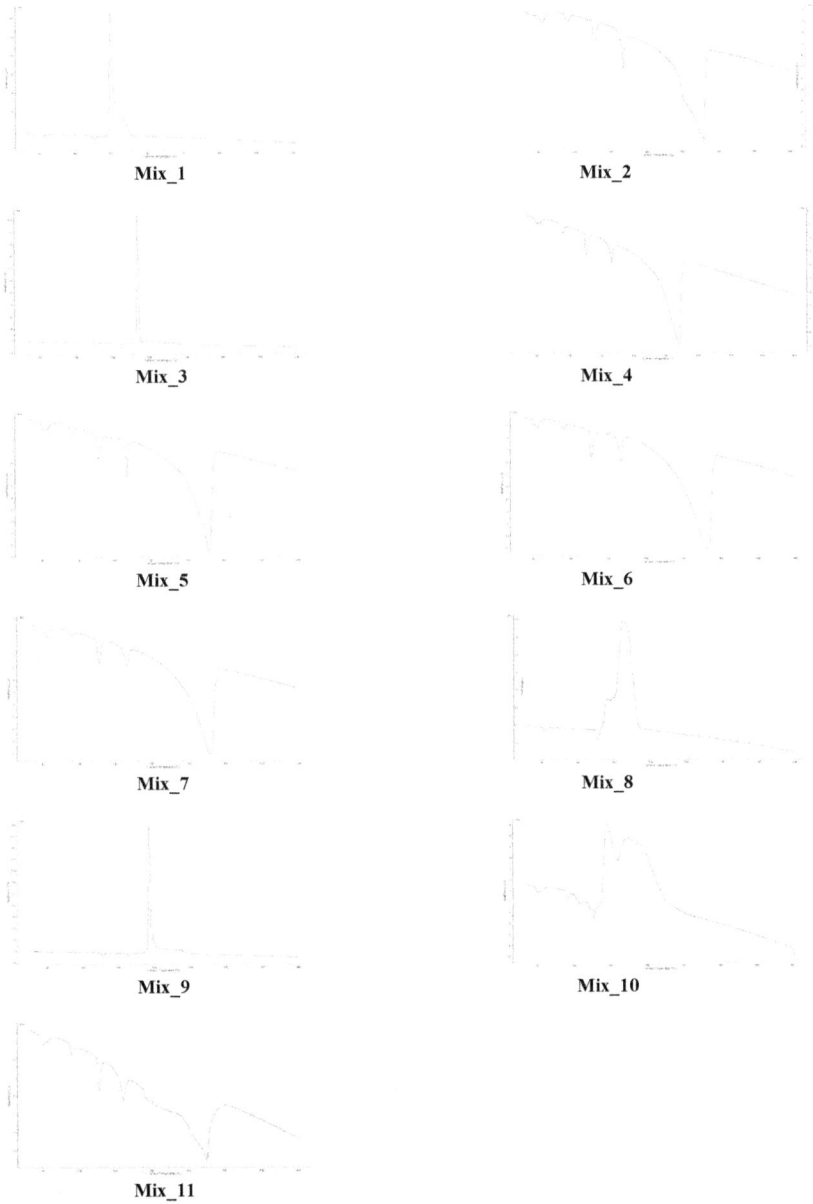

Fig.4. *DSC measurement results of samples Mix_1 − Mix_11 in two-component mixtures 80 wt.% AN at 20 wt. admixtures (AN80A20)*

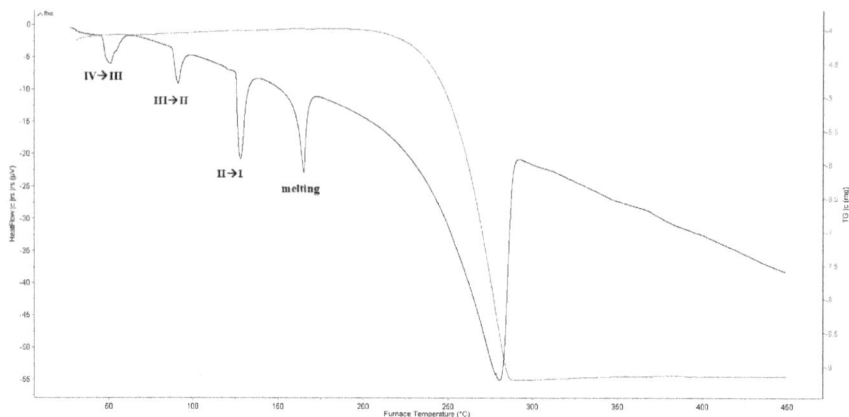

Fig.5. *The results of TG-DSC measurements for non-doped AN*

Based on the results of the results listed in Table 3, it should be noted that only some of the tested compounds showed the ability to reduce the thermal stability of ammonium nitrate. This effect is observed for samples labelled as Mix_1, Mix_3, Mix_8, Mix_9 and Mix_10.

In the case of samples Mix_1 (DT0), Mix_3 (instant coffee) and Mix_9 (cocoa powder) we can see a very strong decrease in the thermal stability of such mixtures. The exothermic signal observed on the DSC curves appears at 146.9°C, 181.6°C, and 190.8°C for Mix_1, Mix_3 and Mix_9, respectively. Such additives in two-component mixtures with ammonium nitrate(V) undoubtedly play the role of promoters of the analysis and reduce the thermal stability of the tested fertilizer. In the case of the Mix_3 sample, two low-intensity endothermic signals were also observed on the DSC curve, which is probably related to the successive distributions of the components contained in the sample.

In the case of mixtures of AN with powdered sugar and activated carbon (medicinal), the decomposition of samples begins at a lower temperature than the AN decomposition temperature. For Mix_8 and Mix_10, the exothermic signal splits into two vertices. For Mix_10 the reading from DSC (onset) is 138.2°C and for Mix_8 it is 151.7°C. However, this distribution is less energetic than in the case of samples Mix_1, Mix_3, and Mix_9. Therefore, powdered sugar and medicinal activated carbon decrease the stability of AN the most out of the tested mixtures. This is consistent with the data from the literature, which indicates that coal dust and organic dust are combustible components [10]. However, sugar itself is known for its effectiveness in improvised explosives [16].

Summary

The tests carried out with the use of TG-DSC made it possible to assess the effect of selected compounds on the thermal stability of ammonium nitrate. Addition of 20% (by weight) kaolin, sand regardless of its granularity, talc and cosmetic powder did not significantly affect the thermal stability of the tested samples. For the other admixtures, such as DT0 carbon, instant coffee, medicinal activated carbon, cocoa and powdered sugar (20 wt.% admixture), a shift at the beginning of the exothermic reaction accompanied by a thermal effect or a strong thermal effect was recorded. These admixtures were considered to promote decomposition, which accelerates the potential explosion of ammonium nitrate. Based on the obtained results, it can be concluded that the interaction of AN and DT0 carbon or instant coffee, medicinal activated carbon, cocoa and

Quality Production Improvement and System Safety: QPI 16 - CZOTO 10   Materials Research Forum LLC
Materials Research Proceedings 34 (2023) 127-138   https://doi.org/10.21741/9781644902691-16

powdered sugar reduces the thermal stability of the nitrate salt, as the exothermic decomposition begins at a lower temperature and is much more violent in the presence of these admixtures.

Storage and handling of explosive materials require maintaining the highest possible quality standards [17-19]. This is particularly important for ammonium nitrate, which, in its commercial form as a fertilizer [20], is not perceived as particularly hazardous, and the safety standards in agriculture are much lower than those in ammunition production [21, 22], which can pose a threat to buildings [23, 24]. The stability of explosive materials is crucial as they are also used in the metalworking industry [25-27] and alloys [28]. One can mention explosive welding as an alternative to conventional welding [29] for joining non-weldable metals, such as steel with aluminum, lead, or aluminum with copper. The resulting connections have specific surface characteristics [30], which can be a basis for their modification through the application of special coatings [31, 32], such as ESD [33, 34] or DLC [35, 36].

Further detailed examination of the mixtures mentioned in the article will require the use of special data analysis methods [37], particularly dimensionality reduction techniques [38], to avoid harmful correlations, and the implementation of statistical techniques [39-41] for identifying potential higher-order interactions, characteristic of chemical and thermomechanical problems. Non-parametric methods [42-44] and resampling techniques [45] are likely to be useful.

References

[1] X. Wang et al. Experiment-based cause analysis of secondary explosion of ammonium nitrate in fire conditions, J. Loss Prev. Process Ind. 77 (2022) art. 104780, https://doi.org/10.1016/j.jlp.2022.104780

[2] T. Sałaciński, A. Maranda. Thermal sensitivity of mixtures of ammonium nitrate(V) with some components of explosives applied in the mining industry, Przegląd Górniczy 73(3) (2017) 71-76.

[3] Y. Yue, W. Gai, G. Boustras. Exploration of the causes of ammonium nitrate explosions: Statistics and analysis of accidents over the past 100 years, Safety Science 158 (2023) art. 105954. https://doi.org/10.1016/j.ssci.2022.105954

[4] S. Al-Hajj et al. Beirut Ammonium Nitrate Explosion, blast Analysis, Review and Recommendations., J. Front. Public Health. 9 (2021) art. 657996. https://doi.org/10.3389/fpubh.2021.657996

[5] C. Oommen, S.R. Jain. Ammonium nitrate: a promising rocket propellant oxidizer, J. Hazard. Mater. 67 (1999) 253-281. https://doi.org/10.1016/S0304-3894(99)00039-4

[6] S. Pragnesh, P.N. Dave, Review on Thermal Decomposition of Ammonium Nitrate, Journal of Energetic Materials, 31, 1 (2013) 1-26, https://doi.org/10.1080/07370652.2011.573523

[7] Ł. Kuterasiński et al. Variously Prepared Zeolite Y as a Modifier of ANFO, Materials 15 (2022) art. 5855. https://doi.org/10.3390/ma15175855

[8] A. Maranda, R. Szymański. Tests on critical diameter and detonation velocity of mixtures of ammonium nitrate (V) and selected organic substances, Chemik 67(1) (2013), 13-18.

[9] M. Fabin, T. Jarosz. Improving ANFO: Effect of Additives and Ammonium Nitrate Morphology on Detonation Parameters, Materials 14 (2021) art. 5745. https://doi.org/10.3390/ma14195745

[10] W. Wiśniewski. Wybuchowość azotanu amonu, Przegląd pożarniczy (access date: 0112.2022 https://www.ppoz.pl/czytelnia/rozpoznawanie-zagrozen/Wybuchowosc-azotanu-amonu-/idn:2151)

[11] D. Buczkowski. Ammonium nitrate – a treat of accidental explosion and terrorist attack, Chemik 66(3) (2012) 227-234.

[12] Z. Han et al. Ammonium nitrate thermal decomposition with additives, J. Loss Prev. Process Ind. 35 (2015) 307-315. https://doi.org/10.1016/j.jlp.2014.10.011

[13] M. Kaniewski et al. Crystalline Phase Transitions and Reactivity of Ammonium Nitrate in Systems Containing Selected Carbonate Salts, Crystals 11 (2021) art. 1250. https://doi.org/10.3390/cryst11101250.

[14] M. Kaniewski et al. Study of thermal stability of ammonium nitrate and potassium chloride mixtures, Proc. of ECOpole 11(1) (2017) 149-155. https://doi.org/10.2429/Proc.2017.11(1)016

[15] D. Poplawski, M. Kaniewski, J. Hoffmann, K. Hoffmann, Application of thermal analysis and calorimetry for assessment of safety and quality of nitrogen fertilizers, Scientific Journal of the Military University of Land Forces 50(3) (2018) 206-217. https://doi.org/10.5604/01.3001.0012.6238

[16] S. Doyle. Explosions, in: J. Siegel, P. Saukko (Eds.) Encyclopedia of Forensic Sciences (Second Edition), Academic Press, 2013, 443-448.

[17] M. Nowicka-Skowron, R. Ulewicz. Quality management in logistics processes in metal branch, METAL 2015 – 24th Int. Conf. Metall. Mater. (2015) 1707-1712. ISBN 978-8087294628

[18] D. Siwiec et al. Improving the non-destructive test by initiating the quality management techniques on an example of the turbine nozzle outlet, Mater. Res. Proc. 17 (2020) 16-22. https://doi.org/10.21741/9781644901038-3

[19] R. Ulewicz et al. Logistic controlling processes and quality issues in a cast iron foundry, Mater. Res. Proc. 17 (2020) 65-71. https://doi.org/10.21741/9781644901038-10

[20] E. Skrzypczak-Pietraszek et al. Enhanced accumulation of harpagide and 8-*O*-acetyl-harpagide in *Melittis melissophyllum* L. agitated shoot cultures analyzed by UPLC-MS/MS. PLoS ONE 13 (2018) art. e0202556. https://doi.org/10.1371/journal.pone.0202556

[21] W. Przybył et al. Virtual Methods of Testing Automatically Generated Camouflage Patterns Created Using Cellular Automata, Mater. Res. Proc. 24 (2022) 66-74. https://doi.org/10.21741/9781644902059-11

[22] W. Przybył et al. Microwave absorption properties of carbonyl iron-based paint coatings for military applications, Def. Technol. 22 (2023) 1-9. https://doi.org/10.1016/j.dt.2022.06.013

[23] A. Bakowski et al. Frequency analysis of urban traffic noise, ICCC 2019 20th Int. Carpathian Contr. Conf. (2019) 1660-1670. https://doi.org/10.1109/CarpathianCC.2019.8766012

[24] J.M. Djoković et al. Selection of the Optimal Window Type and Orientation for the Two Cities in Serbia and One in Slovakia, Energies 15 (2022) art.323. https://doi.org/10.3390/en15010323

[25] M. Mazur et al. The structure and mechanical properties of domex 700 MC steel, Comm. - Sci. Lett. Univ. Žilina 15 (2013) 54-57. https://doi.org/10.26552/com.C.2013.4.54-57

[26] P. Szataniak et al. HSLA steels – Comparison of cutting techniques, METAL 2014 – 23rd Int. Conf. Metallurgy and Materials (2014), Ostrava, Tanger, 778-783.

[27] P. Jonšta et al. The effect of rare earth metals alloying on the internal quality of industrially produced heavy steel forgings, Materials 14 (2021) art.5160. https://doi.org/10.3390/ma14185160

[28] A. Dudek et al. The effect of alloying method on the structure and properties of sintered stainless steel, Arch. Metall. Mater. 62 (2017) 281-287. https://doi.org/10.1515/amm-2017-0042

[29] N. Radek et al. The impact of laser welding parameters on the mechanical properties of the weld, AIP Conf. Proc. 2017 (2018) art.20025. https://doi.org/10.1063/1.5056288

[30] N. Radek et al. The influence of plasma cutting parameters on the geometric structure of cut surfaces, Mater. Res. Proc. 17 (2020) 132-137. https://doi.org/10.21741/9781644901038-20

[31] N. Radek et al. Technology and application of anti-graffiti coating systems for rolling stock, METAL 2019 – 28th Int. Conf. Metall. Mater. (2019) 1127-1132. ISBN 978-8087294925

[32] N. Radek et al. Formation of coatings with technologies using concentrated energy stream, Prod. Eng. Arch. 28 (2022) 117-122. https://doi.org/10.30657/pea.2022.28.13

[33] N. Radek et al. The effect of laser treatment on operational properties of ESD coatings, METAL 2021 – 30th Ann. Int. Conf. Metall. Mater. (2021) 876-882. https://doi.org/10.37904/metal.2021.4212

[34] N. Radek et al. The impact of laser processing on the performance properties of electro-spark coatings, 14th World Congr. Comput. Mech. and ECCOMAS Congr. 1000 (2021) 1-10. https://doi.org/10.23967/wccm-eccomas.2020.336

[35] N. Radek et al. Microstructure and tribological properties of DLC coatings, Mater. Res. Proc. 17 (2020) 171-176. https://doi.org/10.21741/9781644901038-26

[36] N. Radek et al. Influence of laser texturing on tribological properties of DLC coatings, Prod. Eng. Arch. 27 (2021) 119-123. https://doi.org/10.30657/pea.2021.27.15

[37] L. Cedro. Model parameter on-line identification with nonlinear parametrization – manipulator model, Technical Transactions 119 (2022) art. e2022007. https://doi.org/10.37705/TechTrans/e2022007

[38] J. Pietraszek, E. Skrzypczak-Pietraszek. The uncertainty and robustness of the principal component analysis as a tool for the dimensionality reduction. Solid State Phenom. 235 (2015) 1-8. https://doi.org/10.4028/www.scientific.net/SSP.235.1

[39] J. Pietraszek, A. Szczotok, N. Radek. The fixed-effects analysis of the relation between SDAS and carbides for the airfoil blade traces. Arch. Metall. Mater. 62 (2017) 235-239. https://doi.org/10.1515/amm-2017-0035

[40] R. Dwornicka, J. Pietraszek. The outline of the expert system for the design of experiment, Prod. Eng. Arch. 20 (2018) 43-48. https://doi.org/10.30657/pea.2018.20.09

[41] J. Pietraszek, N. Radek, A.V. Goroshko. Challenges for the DOE methodology related to the introduction of Industry 4.0. Prod. Eng. Arch. 26 (2020) 190-194. https://doi.org/10.30657/pea.2020.26.33

[42] J. Pietraszek. The modified sequential-binary approach for fuzzy operations on correlated assessments, LNAI 7894 (2013) 353-364. https://doi.org/10.1007/978-3-642-38658-9_32

[43] J. Pietraszek et al. Non-parametric assessment of the uncertainty in the analysis of the airfoil blade traces, METAL 2017 – 26th Int. Conf. Metall. Mater. (2017) 1412-1418. ISBN 978-8087294796

[44] J. Pietraszek et al. The non-parametric approach to the quantification of the uncertainty in the design of experiments modelling, UNCECOMP 2017 Proc. 2nd Int. Conf. Uncert. Quant. Comput. Sci. Eng. (2017) 598-604. https://doi.org/10.7712/120217.5395.17225

[45] J. Pietraszek, L. Wojnar. The bootstrap approach to the statistical significance of parameters in RSM model, ECCOMAS Congr. 2016 Proc. 7th Europ. Congr. Comput. Methods in Appl. Sci. Eng. 1 (2016) 2003-2009. https://doi.org/10.7712/100016.1937.9138

Quality Production Improvement and System Safety: QPI 16 - CZOTO 10   Materials Research Forum LLC
Materials Research Proceedings 34 (2023) 139-144   https://doi.org/10.21741/9781644902691-17

Assessment of the Application of Lumpy Steel Slag as an Aggregate Replacement in Concrete

HARWAT Artur[1] and RESPONDEK Zbigniew[2, a] *

[1] Czestochowa University of Technology, Faculty of Civil Engineering, Poland (Graudate)

[2] Czestochowa University of Technology, Faculty of Civil Engineering, Poland

[a] zbigniew.respondek@pcz.pl

Keywords: Waste Management, Recycling in Construction, Cement Concrete, Aggregate Replacement, Steel Slag

Abstract. The need to protect the Earth's resources necessitates pro-ecological activities. One of the aspects of these activities is the rational management of post-production and post-consumer waste. One of the materials whose production significantly pollutes the natural environment is concrete. On the other hand, the analysis of the literature carried out in the work showed that it is a material that can absorb a significant amount of processed waste used as a substitute for cement or aggregate. These are polymer materials of various origins, residues from biomass combustion and, above all, steel slag, which has long been used in construction in various forms. One of the types of this material is lumpy steel slag, which has been used for years mainly in road construction as a replacement for crushed stone in the construction of the road base. The paper presents instrumental tests of the influence of the lump slag content used as a replacement for basalt aggregate on selected parameters of concrete (compressive strength, water absorption, volume density). The research showed that such a modification of the mixture composition does not significantly change the tested parameters. The lump slag can therefore be successfully used in the production of normal and heavy concrete.

Introduction

Waste is inseparable from human industrial activity (post-production waste) and everyday life (post-consumer waste). In the past decades, its recovery, partial or total re-use has not been sufficiently taken into account. The reason for this - apart from the lack of ecological awareness and modernist admiration for industrialization - was the wide availability of fossil resources from shallow deposits or opencast sources. Along with the growing demand for raw materials, the need to exploit deeper and deeper seams and the shrinking sources of their extraction, the use of materials previously considered waste and deposited in landfills occupying more and more areas began to be seriously considered. Waste, especially construction and post-demolition waste, was also often thrown in random places, burned or irresponsibly managed [1]. Analyses of the composition and properties of the waste were conducted, and it was noticed that what was previously considered useless could be useful. It has been noticed that apart from the recovery of raw materials that can be easily reused in the production process (primary or otherwise), especially after processing, they can be used in building construction and road construction. Over the years, the environmental awareness of both governments and societies has grown, leading to the introduction of legislation in many countries, especially in Europe and North America, forcing a continuous increase in activities aimed at protecting the natural environment. Note that the recovery and recycling of materials does not always bring direct economic benefits. Therefore, the ecological context is important here - the reduction of the use of natural resources reduces the burden on the surrounding environment [2]. Also, not all waste material can be reused. For

Quality Production Improvement and System Safety: QPI 16 - CZOTO 10   Materials Research Forum LLC
Materials Research Proceedings 34 (2023) 139-144   https://doi.org/10.21741/9781644902691-17

example, asbestos-cement panels or reclaimed road pavement containing tar can be harmful to your health. In this case, the only alternative is wise and safe disposal.

In modern construction, concrete is a material whose production significantly burden the environment. Therefore, many research centers are conducting research on the possibility of using various materials, including waste materials, in the production of concrete [3]. There are two main directions of this search: reducing cement consumption and replacing it with another binding material [4] and the use of waste products as a replacement for aggregate. Recycling materials used as an additive can also advantageously change some parameters of the concrete.

Among the various waste materials, waste polymer materials are of great interest. One of the best-studied waste economies in this area is polyethylene terephthalate, PET [5]. It was used for concrete as a replacement for aggregate or in the form of fibers as dispersed reinforcement [6]. Research was also conducted on the use of recycled polyethylene and polypropylene [7], polystyrene [8], rubber [9] or a thermoplastic elastomer from the production process of car mats [10].

The use of combustion materials is also very popular. These materials can be divided into those coming from production processes and from biomass combustion. The first of these types, which has been used for many years, will be discussed in the next section. As for the residues from biomass combustion, there has been a recent increase in interest in the use of these products. Plant waste, which so far has not been considered for use in construction, can be used as a replacement for cement – here one can mention ashes from rice hulls, oil palm, sugar cane pomace or corn cobs [11] – or even as a replacement for aggregate [12]. In Poland, research was conducted on biomass ash – a mixture of wood and sunflower waste [13].

It should be mentioned that concrete may not only be a material absorbing processed waste, but also a post-consumer waste used, after processing, as a substitute for aggregate in the composition of the concrete mix or in the construction of the road base (Fig. 1).

Fig. 1. The use of crushed concrete in the construction of the pavement base (photo: Z. Respondek)

The aim of the article is to assess the legitimacy of the use of waste material on the example of lumpy steel slag for the production of concrete. Therefore, instrumental studies of the impact of the content of processed waste, used as a replacement for basalt aggregate, on selected concrete parameters were performed.

Management of Waste from Combustion in Production Processes

As already mentioned, processed industrial waste, which is a by-product of the combustion process, has been used in construction for many years. Here we can list, among others: silica fume, fly ash and steel slag.

Silica fume (microsilica) is obtained during the production of metallic silicon and ferrosilicon alloys in electric arc furnaces. The use of microsilica has a very positive effect on the increase in

Quality Production Improvement and System Safety: QPI 16 - CZOTO 10   Materials Research Forum LLC
Materials Research Proceedings 34 (2023) 139-144   https://doi.org/10.21741/9781644902691-17

concrete strength in the initial period, viscosity, increases resistance to alkali and sulphates and the absorption of $Ca(OH)_2$ in the chemical process. This is due to the large specific surface area of microsilica particles [14].

Fly ash is a fine-grained fraction of ashes produced during the combustion of solid fuels. Currently, they are the most commonly used waste materials used as a partial replacement for Portland cement because its properties are similar to those of Portland clinker (pozzolanic activity). However, due to the insufficient content of calcium hydroxide in the chemical composition, they cannot function as a hydraulic binder on their own. Fly ash as a partial replacement for cement initially reduces the compressive strength of concrete, but after about 90 days, an increase in this strength is observed in relation to the base concrete [15].

Steel slag (blast furnace slag) can be used in several forms.

Granulated steel slag is obtained by rapidly cooling molten slag. A granular material with a glassy amorphous structure and hydraulic properties is obtained. This material is used as an additive to Portland cement (slag cement). Also, cooling, but with a smaller amount of water, results in the production of pumice slag – this material has the form of larger lumps with a porous-spongy amorphous structure. After mechanical processing in crushers and division into fractions, light aggregate is obtained, used primarily for the production of bricks and blocks [16].

Lumpy steel slag is a slag slowly cooled with air, which, after solidification, is crushed and separated into fractions. For many years, this material has been used instead of natural stone, mainly in road construction, because its physical properties are equivalent to those of aggregates obtained from natural rocks. In the context of the use of blast furnace slag for heavy concretes and reinforced concrete, this material initially raised doubts – there was a concern that the reinforcement would corrode as a result of the sulfur contained in it. Currently, it is assumed that there is no concern about the negative impact of this aggregate when the sulfur content, converted into SO_3, does not exceed 2 wt.%. It should be added that all metallurgical waste may have a different chemical composition, therefore each time there is a need to research the possible negative impact of these materials on human health and the environment [17].

Methodology of research

Instrumental tests of concrete were carried out at the Faculty of Civil Engineering of the Czestochowa University of Technology. The tests were preceded by the design of the base concrete for the assumed grade C30 / 37:

- Portland cement CEM I 42,5 R (Górażdże) – 296 kg/m^3,
- water - 160 kg/m^3,
- sand - 836 kg/m^3,
- basalt aggregate 2-8 mm - 836 kg/m^3,
- basalt aggregate 8-16 mm - 836 kg/m^3.

Two sets of five series of 15×15×15 cm cubic samples were made: base concrete and four series in which 8-16 mm basalt aggregate was replaced with lumpy steel slag in the amount of: 25, 50, 75 and 100 wt.%. Each series consisted of 3 samples.

Fig.2. *Test preparation: a) 8-16 mm lumpy steel slag used, b) samples after demolding (photo: A Harwat)*

The slag used for the test was imported from one of the Ukrainian smelters. The 8-16 mm fraction was used (Fig. 2a). The research program included:
- compressive strength after 28 days,
- water absorption after 28 days,
- volume density.

One of the prepared sets of samples was used to determine the compressive strength, the other - for water absorption and volume density. Preparation of samples (Fig. 2b) and execution of the test were carried out according to the procedures included in the standards of the series "EN 12390 Testing hardened concrete". In order to finally select the quantitative composition of the materials used, a trial preparation was made for each series and the amount of water was adjusted due to the improvement of the workability of the mixture. Stachement 2050 superplasticizer was also used. ToniTechnik 2030 test machine (Berlin, Germany) was used for the compressive strength test.

Results

The test results are shown in Figures 3-5. In all graphs, the solid line represents the average value of three samples, while the dashed line shows the minimum and maximum value of a given parameter obtained for individual samples from each series. Table 1 illustrates the percentage change in the average values of the tested parameters in relation to the base concrete.

The compressive strength test shows that the content of 25 wt.% of slag as a replacement for basalt aggregate results in a slight increase in compressive strength. Higher slag content causes a slight deterioration of this parameter. However, these fluctuations do not affect the change of concrete grade. Determined according to EN 206 "Concrete. Specification, performance, production and conformity" is in any case C30/37. With regard to the water absorption, fluctuations of several percent were also observed. The worst parameters were obtained for the series of 50 wt.%, best for 100 wt.%. According to the literature recommendations [18], the water absorption of concrete should not exceed 5% for concretes exposed to the external environment and 9% for concrete protected against this effect. Score 4.98 for 100 wt.% is therefore near the limit value.

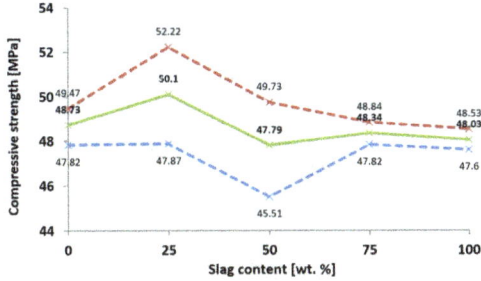

Fig.3. *The influence of slag content on the compressive strength of concrete [own research].*

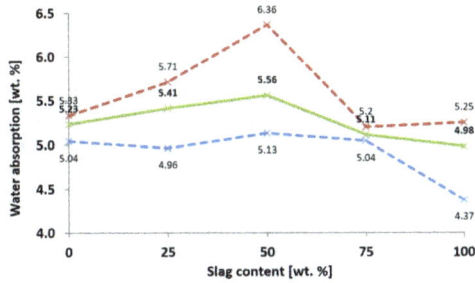

Fig.4. *The influence of slag content on the water absorption of concrete [own research].*

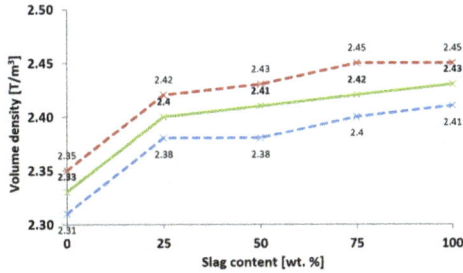

Fig.5. *The influence of slag content on the volume density of concrete [own research].*

Table 1.   *Percentage change of the tested parameters (average values) in relation to the base concrete [own research]*

Parameter	Slag content			
	25 wt. %	50 wt. %	75 wt. %	100 wt. %
Compressive strength	2.8%	-1.9%	-0.8%	-1.4%
Water absorption	3.4%	6.3%	-2.3%	-4.8%
Volume density	3.0%	3.4%	3.9%	4.3%

Quality Production Improvement and System Safety: QPI 16 - CZOTO 10   Materials Research Forum LLC
Materials Research Proceedings 34 (2023) 139-144   https://doi.org/10.21741/9781644902691-17

The tests also showed an increase in the slag content, resulting in an increase to 4.3 wt.%, the concrete volume density in relation to the base concrete with basalt aggregate.

Conclusions

The literature analysis carried out in the article showed a wide spectrum of waste materials that can be used as a component of concrete. This use is limited, first of all, by the harmful effects on human health. The condition for the proper use of waste as a replacement for cement or aggregate in concrete is the prior processing of this waste and carrying out tests confirming or disproving the legitimacy of using such modified materials in construction. The instrumental tests of concrete with the content of lump slag carried out in the work are an example of such an action.

In common understanding, slag concrete is associated with a cheaper and worse type of concrete. For example, the commonly used slag concrete blocks are characterized by high water absorption, reduced strength and often insufficient durability. The conducted research has shown that lumpy steel slag, commonly used as a replacement for aggregate in the road-base layers, can be successfully used as a substitute for basalt aggregate in classic concrete.

The compressive strength of the modified concrete is not more than 2% worse than that of the base concrete. The aggregate replacement does not change the concrete grade.

The water absorption of concrete when replacing basalt aggregate with processed waste of the same fraction does not differ significantly from the parameters of the base concrete. All series show water absorption at the limit of applicability for concrete exposed to moisture.

The aggregate replacement increases the volume density of the concrete, although all series should be classified as normal concrete. It seems possible, however, that with the appropriate design of the mixture composition, heavy concrete (specific gravity above 2.6 T/m^3) can be achieved with the content of lumpy slag. Such a composite can be used, for example, as ballast concrete.

References

[1] N. Brycht. Construction Waste Management in Rural Areas of the Czestochowa District in the Aspect of Environmental Safety, Quality Production Improvement, R. Ulewicz, B. Hadzima (Eds.) 2(1) (2020) 60-68. https://doi.org/10.2478/cqpi-2020-0008

[2] M. Tomov, C. Velkoska. Contribution of the quality costs to sustainable development, Prod. Eng. Arch. 28 (2022) 164-171. https://doi.org/10.30657/pea.2022.28.19

[3] K. Kishore, N. Gupta. Application of domestic & industrial waste materials in concrete: A review, Mater. Today: Proc. 26 (2020) 2926-2931. https://doi.org/10.1016/j.matpr.2020.02.604

[4] B. Dębska, J. Krasoń, L. Lichołaj. The evaluation of the possible utilization of waste glass in sustainable mortars, Construction of Optimized Energy Potential 9(2) (2021) 7-15. https://doi.org/10.17512/bozpe.2020.2.01

[5] Y.W. Choi et al. Effects of waste PET bottlers aggregate on the properties of concrete, Cem. Concr. Res. 35 (2005) 776–781. https://doi.org/10.1016/j.cemconres.2004.05.014

[6] P. Helbrych. Effect of Dosing with Propylene Fibers on the Mechanical Properties of Concretes, Construction of Optimized Energy Potential 10(2) (2021) 39-44. https://doi.org/10.17512/bozpe.2021.2.05

[7] M. Chaudhary, V. Srivastava, V. Agarwal. Effect of waste low density polyethylene on mechanical properties of concrete, J. Acad. Ind. Res. 3(3) (2014) 123-126.

[8] A. Herki, J. Khatib, E. Negim. Lightweight concrete made from waste polystyrene and fly ash, World Appl. Sci. J. 21 (2013) 1356-1360. https://doi.org/10.5829/idosi.wasj.2013.21.9.20213

[9] N. Holmes, K. Dunne, J. O'Donnell. Longitudinal shear resistance of composite slabs containing crumb rubber in concrete toppings, Constr. Build. Mater. 55 (2014) 365-378. https://doi.org/10.1016/j.conbuildmat.2014.01.046

[10] A. Pietrzak, M. Ulewicz. Properties and Structure of Concretes Doped with Production Waste of Thermoplastic Elastomers from the Production of Car Floor Mats, Materials 14 (2021) art.872. https://doi.org/10.3390/ma14040872

[11] E. Aprianti. A huge number of artificial waste material can be supplementary cementitious material (SCM) for concrete production – a review part II, J. Clean. Prod. 142 (2017) 4178-4194. https://doi.org/10.1016/j.jclepro.2015.12.115

[12] I.M. Aslam et al. Benefits of using blended waste coarse lightweight aggregates in structural lightweight aggregate concrete, J. Clean. Prod. 119 (2016) 108-117. https://doi.org/10.1016/j.jclepro.2016.01.071

[13] J. Jura, M. Ulewicz. Assessment of the Possibility of Using Fly Ash from Biomass Combustion for Concrete, Materials 14 (2021) art.6708. https://doi.org/10.3390/ma14216708

[14] J.M. Paris et al. A review of waste products utilized as supplements to Portland cement in concrete, J. Clean. Prod. 121 (2016) 1-18. https://doi.org/10.1016/j.jclepro.2016.02.013

[15] Z. Giergiczny. Właściwości popiołu lotnego a trwałość betonu, Budownictwo, Technologie, Architektura 3 (2007) 44-48.

[16] E. Ganjian, G. Jallul, H. Sadeghi-Pouya. Using waste materials and by-products to produce concrete paving blocks, Constr. Build. Mater. 77 (2015) 270-275. https://doi.org/10.1016/j.conbuildmat.2014.12.048

[17] T. Lis, K. Nowacki. Pro-ecological possibilities of using metallurgical waste in the production of aggregates, Prod. Eng. Arch. 28 (2022) 252-256. https://doi.org/10.30657/pea.2022.28.31

[18] A.M. Neville. Properties of Concret, Pearson, Harlow, 2011. ISBN 978-0273755807

Quality Production Improvement and System Safety: QPI 16 - CZOTO 10   Materials Research Forum LLC
Materials Research Proceedings 34 (2023) 146-153   https://doi.org/10.21741/9781644902691-18

The Use of Dispersed Plastic Reinforcement in Concrete

HELBRYCH Paweł[1, a] *

[1] Czestochowa University of Technology, Faculty of Civil Engineering, Poland

[a] pawel.helbrych@pcz.pl

Keywords: Polypropylene Fibers, Steel Fibres, Fiber-Reinforced Concrete, Concrete, Mechanical Properties

Abstract. The article deals with the issue of the usefulness of plastic reinforcement dispersed in concrete, including polypropylene fibers, and the impact of the amount of dosing of such fibers was assessed. The amount of addition was analyzed in the range of 0.5 to 2.0 kg/m^3. Fibrillated fibers arranged in bundles and crimped fibers were used in the tests. Fibers with a length of 50 mm were analyzed. The results were compared with test series with the addition of steel barbed fibers 50 mm long and 1.0 mm in diameter. The tests showed that steel fibers do not adversely affect the designed consistency of the concrete mix, unlike polypropylene fibers. The tests showed that, for strength and technological reasons, in the case of concrete reinforced with the addition of up to 20 kg/m^3 of steel fibers, polypropylene fibers in the amount of 1.5 kg/m^3 can be successfully used, with a decrease in bending and compressive strength not exceeding 10%.

Introduction

Continuous development of technology makes construction products meet the highest quality requirements. Taking care of the quality of the processes that are necessary to produce a construction product easily translates into the quality of the final product [1]. New technologies are introduced in every industry, including construction. In order for the investor to be able to evaluate new technological solutions, including new building materials, an analytical model should be developed that will enable the investor to assess the profitability and the degree of risk of introducing the solution or building material on the construction market. In this assessment, the key role is played by the final quality of the building material or the quality of building technology processes [2].

Continuous technological development also causes a systematic increase in the amount of waste in industry. This generates a problem with their disposal. Therefore, waste management is carried out in urban and rural areas to ensure environmental safety [3]. Some of the waste is used in the production processes of building materials. Recyclates currently used in the construction industry are fly ashes and slags from the combustion of fossil fuels added to concrete or cement mortars [4] and, for example, metallurgical waste used for the production of aggregates [5]. Concrete and mortar as composite materials with a cement matrix are commonly used in construction. Due to their composite characteristics, they are perfect for adding additives and admixtures to them, which can be recyclants. Research was carried out on such additives as: waste glass [6], PET waste [7], thermoplastic elastomers from the production of car mats [8] or fly ashes from biomass combustion [9], in all the mentioned studies, these additives in the form of recyclates did not deteriorate the properties of concretes and mortars.

The use of waste materials in the production of concrete does not guarantee an increase in the basic parameters of concrete, such as compressive strength, tensile strength in a bending test, which are crucial in terms of obtaining the appropriate quality of concrete in the context of its intended use, or consistency class and workability, which are crucial for from the point of view of the quality of the concrete placement technology at the destination. Additives in the form of fibers

Quality Production Improvement and System Safety: QPI 16 - CZOTO 10   Materials Research Forum LLC
Materials Research Proceedings 34 (2023) 146-153   https://doi.org/10.21741/9781644902691-18

are used to improve the strength characteristics of concrete and to reduce its cracking. Most often, steel, polymer, glass and natural fibers are added [10]. According [11] steel fibers are straight or deformed fragments of cold-drawn steel wire, straight or deformed fragments of cut steel sheet, melt fibers, cut and rolled fibers from steel blocks, while polymer fibers according to [12] are straight or deformed fragments of extruded, oriented and cut polymeric material, used to make a homogeneous concrete mix. Polymer fibers are added to concrete in the form of short thin threads (up to 30 mm) of small diameter (0.02-0.05 +/- 0.05 mm) or rigid rods with a diameter of (0.2-0.5 +/- 0.5 mm) and lengths up to 60 mm. Polypropylene, polyethylene, polyester, nylon, polyacrylic or aramid are most often used to make polymer fibers [13]. In the case of steel fibers, the standard length is 10-60 mm. Fibers of this type usually have a round, flat or oval cross-section. The diameter of typical steel fibers is usually 0.2-1.5 mm. The most common shapes used for concrete are: straight, smooth, hooked, flattened at the ends and with split ends [14].

Methodology of Research
The aim of the work was to determine the effect of dosing fibers made of plastics (polypropylene) of different structure, the same length and the same material on the mechanical properties of the concrete mix and the finished concrete product. In addition, during the research work, the feature of the fresh concrete mix in the form of consistency class was determined. The results were compared with test series with the addition of hooked steel fibers and without the addition of fibers (control series).

Materials. The research used ingredients of natural origin, commonly used in construction, easily available. The concrete mix was designed using the successive approximation method (Kuczyński's method). Due to the burning of fiber additives, after reading the guidelines contained in [14–18] and based on preliminary tests, a limited W/C ratio of 0.4 was chosen. In order to obtain the required consistency class, the superplasticizer BASF MasterEase 5051 was used in the production of concrete mix in the amount of 2.5% of the cement mass. In addition, the following ingredients were used to produce the mix: Portland cement CEM V/A 32.5R, a mixture of gravel aggregate of the 2-8 mm and 8-16 mm fractions, sand of the 0-2 mm fraction and tap water from the intake in Częstochowa. The mix was designed in the S3 consistency class, and the aggregate mix was selected so that the graining curve was in the upper range of the field of graining boundary curves [14,15]. The design class of concrete is C30/37. Table 1 shows the concrete mix recipe.

Table 1. Concrete mix recipe

Type of ingredients	Amount of ingredients per 1 m³ of the mix [kg]
Cement CEM V/A 32.5R	444
Water	178
Sand 0-2mm	934
Gravel 2 - 8 mm	536
Gravel 8 - 16 mm	646
Superplasticizer	11
W/C	0,40
Consistency class	S3

Quality Production Improvement and System Safety: QPI 16 - CZOTO 10   Materials Research Forum LLC
Materials Research Proceedings 34 (2023) 146-153   https://doi.org/10.21741/9781644902691-18

Table 2. *Properties of the polypropylene fibers used*

Properties	Unit	P1 fibers	P2 fibers
Type of polymer	-	polypropylene	polypropylene
Density	g/dm^3	0.91	0.91
Diameter	mm	1.5	1.0
Length	mm	50	50
Modulus of elasticity	kN/mm^2	2	2
Tensile strength	N/mm^2	600	600
Flash-point	°C	160	160
Melting temperature	°C	160	160

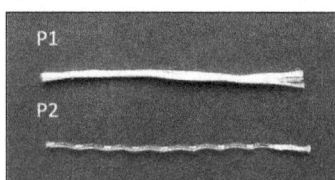

Fig. 1. *Polypropylene fibers used in the tests (P1 fiber at the top, P2 at the bottom)*
[own study].

Two types of polypropylene fibers were used in the research. P1 in the form of a twisted bundle, fibrillated sections had a length of 50 mm (±1.5 mm), and a crimped fiber marked in the tests as P2 with a length of 50 mm (±1.5 mm). Fiber properties are presented in table 2.

In addition, non-galvanized steel reinforcement (fig. 2) in the form of a barbed wire was used for the tests. The fibers used in the tests were made of steel with a tensile strength of 1000-1020 MPa. They had a length of 50 mm and a diameter of 1.0 mm. The barbed fibers are designated S1. Table 3 shows the geometric parameters of the fibers and their properties.

Table 3. *Properties of the steel fibers used*

Properties	Unit	S1 fibers
Material	-	steel
Length	mm	50
Diameter	mm	1.0
Slenderness	L/d	50
Shape	-	hooked
Tensile strength	MPa	1000÷1020
Modulus of elasticity	GPa	210

Fig. 2. *Steel fibers used in the tests (S1 fiber) [own study].*

Methodology. The components of the concrete mix were measured by weight. The dosing of ingredients took place in exactly the same order in each test series. The concrete mix production procedure was consulted with the local concrete plant in order to reflect the industrial method of concrete production. First, 50% of the volume of aggregate with a fraction of 2-8 mm was dosed into the mixer, then pears with a fraction of 8-16 mm, and then 50% of the volume of sand was added. The ingredients were mixed for 1 minute. The superplasticizer was then mixed with water, and this mixture was added to the previously mixed aggregate mix. The ingredients were mixed for a period of 2 minutes. After this time, the remaining aggregate was added (50% of gravel of the 2-8 mm fraction and 50% of the gravel of the 8-16 fraction and 50% of sand), and then mixed for another 2 minutes. For the fiber-added batch, a pre-measured amount of fiber was added at this point. The mixing time of the test series with added fibers was 2 minutes for all series. The SK control lot contained no fibres. In total, 13 research series were performed. For each, the consistency was determined by the falling cone method in accordance with [19]. Each test series contained at least 6 samples (3 samples with dimensions of 150x150x150 mm and 3 samples with dimensions of 150x150x600). The amount of dosed fibers for individual test series is shown in Table 4.

Table 4. List of research series

Series name	Type of added fibers	Number of fibers per 1 m^3 of the mixture [kg]
SK	-	-
P1-S1	Polypropylene P1	0,5
P1-S2	Polypropylene P1	1,0
P1-S3	Polypropylene P1	1,5
P1-S4	Polypropylene P1	2,0
P2-S1	Polypropylene P2	0,5
P2-S2	Polypropylene P2	1,0
P2-S3	Polypropylene P2	1,5
P2-S4	Polypropylene P2	2,0
S1-S1	Steel S1	10
S1-S2	Steel S1	15
S1-S3	Steel S1	20
S1-S4	Steel S1	25

Standards in accordance with the requirements were used in the tests [20]. The mixture was placed in two equal layers and then vibrated for 25 seconds. The procedure for making test samples met the requirements contained in [21]. After 24 hours, the samples were removed from the molds and stored in water at 20±2°C until testing. After 28 days from the preparation of the samples, compressive strength tests were carried out according to [22] and tensile strength of concrete in a bending test - a simply supported beam loaded symmetrically with one force according to [23]. All tests were performed in a Toni Technik type 2030 testing machine in accordance with the requirements [24].

Research Results and Their Discussion
The results of consistency tests are presented in table 5

Table 5. *Summary of consistency classes of tested concrete mixes*

Series name	Cone drop [mm]	Consistency class
SK	145	S3
P1-S1	120	S3
P1-S2	110	S3
P1-S3	70	S2
P1-S4	35	S1
P2-S1	140	S3
P2-S2	130	S3
P2-S3	110	S3
P2-S4	70	S2
S1-S1	140	S3
S1-S2	130	S3
S1-S3	115	S3
S1-S4	100	S3

The largest cone slump was obtained for a series of control concretes (SK). The consistency class here was consistent with the designed consistency. In the series with the addition of polypropylene fibers, a tendency to change the consistency class with the amount of added fibers was observed, while in the series with the addition of steel fibers this phenomenon was not observed.

Compressive strength tests were performed in accordance with the requirements of the standard [22] on a Toni Technik 2030 machine. The load increase was 0.5 MPa/sec. The averaged results along with the standard deviation are shown in Figure 3.

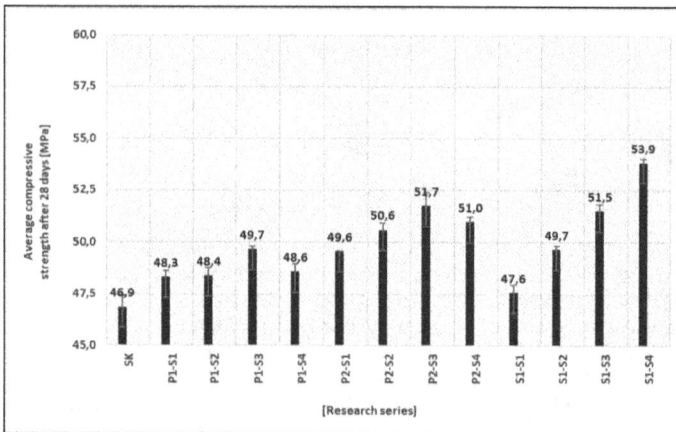

Fig. 3. *Compressive strength of tested concrete series.*

The average compressive strength of concrete was determined for all series after 28 days of curing of the samples. For the control series, fcm=50.80 MPa was obtained. Depending on the amount of polypropylene fibers P1 and P2 dosed, the value of the average compressive strength of concrete ranged from 48.3 MPa (P1-S1 series) to 51.7 MPa (P2-S3 series), while the standard deviation for all series was in the range of ± 0.01MPa for the P2-S1 series to ± 0.67MPa for the

Quality Production Improvement and System Safety: QPI 16 - CZOTO 10   Materials Research Forum LLC
Materials Research Proceedings 34 (2023) 146-153   https://doi.org/10.21741/9781644902691-18

P2-S3 series. For the series with the addition of steel reinforcement, the average compressive strength of concrete ranged from 47.6 MPa (series S1-S1) to 53.9 (series S1-S4), and the standard deviation was in the range of ± 0.18 MPa for S1-S4 series to ± 0.37 MPa for S1-S1 series.

In the case of samples with steel reinforcement, the compressive strength increased with the amount of fibers added, while in the case of the series with the addition of polypropylene reinforcement, no such relationship was observed. The graph shows that the corrugated reinforcement (P2) had a better effect on the compressive strength of the concrete than the fibrillated reinforcement (P1). The addition of 2kg/m3 of this type of reinforcement caused a slight decrease in compressive strength. Tensile strength tests in the bending test were performed in accordance with the requirements of the standard [23] on a Toni Technik 2030 machine. The load increase was 0.5 MPa/sec. The averaged results along with the standard deviation are shown in Figure 4.

The average flexural strength of concrete was determined for all series after 28 days of curing of the samples. For the control series, f_{ct}=3.7 MPa was obtained. Depending on the amount of polypropylene fibers P1 and P2 dosed, the value of the average concrete flexural strength ranged from 3.8 MPa (P1-S1 series) to 4.7 MPa (P2-S3 series), while the standard deviation for all series ranged from ± 0.04 MPa for P1-S4 series to ± 0.14 MPa for P1-S3 series. For the series with the addition of steel reinforcement, the average concrete flexural strength ranged from 4.1 MPa (series S1-S1) to 5.0 (series S1-S4), and the standard deviation was in the range of ± 0.05 MPa for S1-S1 series to ± 0.10 MPa for S1-S4 series.

In the case of samples with steel reinforcement, the bending strength increased with the amount of fibers added, while in the case of the series with the addition of polypropylene reinforcement, no such relationship was observed. The graph shows that the corrugated reinforcement (P2) had a better effect on the bending strength of the concrete than the fibrillated reinforcement (P1). The addition of 2kg/m^3 of this type of reinforcement caused a slight decrease in bending strength.

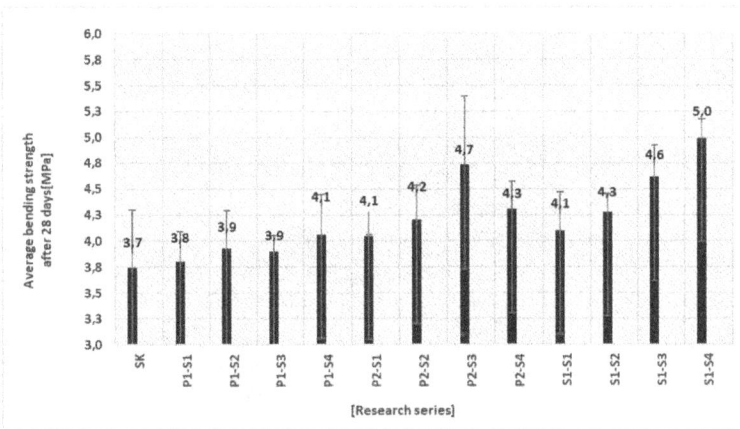

Fig.4. *Tensile strength of concrete in a bending test - simply supported beam symmetrically loaded with one force.*

Conclusions
In the case of compressive strength, the best series of concretes with the addition of polypropylene fibers was 4.5% lower than the best result of the series with the addition of steel fibers.

Quality Production Improvement and System Safety: QPI 16 - CZOTO 10   Materials Research Forum LLC
Materials Research Proceedings 34 (2023) 146-153   https://doi.org/10.21741/9781644902691-18

In the case of flexural strength, the best series of concretes with the addition of polypropylene fibers was 7% lower than the best result of the series with the addition of steel fibers.

Micromeshed polypropylene fibers do not significantly improve performance compared to crimped polypropylene fibers.

The addition of too much polypropylene fibers causes deterioration of the concrete mix structure and strength results. Tests have shown that the permissible value of the addition of polypropylene fibers is 1.5 kg/m^3.

On the basis of the tests carried out, it can be assumed that, for strength and technological reasons, in the case of concrete reinforced with the addition of up to 20 kg/m^3 of steel fibers, polypropylene fibers in the amount of 1.5 kg/m^3 can be successfully used, with a decrease in bending and compressive strength not exceeding 10 %.

The polypropylene reinforcement protruding above the surface is not dangerous for users and workers due to its rigidity, unlike the protruding steel reinforcement.

References

[1] J. Prasalska-Nikoniuk et al. ISO 9001:2015 vs. Factory Production Control (FPC) to ensure the quality of construction products used in road and bridge engineering, BoZPE. 11 (2022) 103–111. https://doi.org/10.17512/bozpe.2022.11.12

[2] L. Osnovina, I. Maltsevich. Assessment of innovative solutions and risks of development in the construction industry, BoZPE. 10 (2021) 7-14. https://doi.org/10.17512/bozpe.2021.2.01

[3] N. Brycht. Construction Waste Management in Rural Areas of the Czestochowa District in the Aspect of Environmental Safety, QPI 2 (2020) 60-68. https://doi.org/10.2478/cqpi-2020-0008

[4] J. Popławski. Influence of biomass fly ash blended with bituminous coal fly ash on the properties of concrete, BoZPE 9 (2020) 89-96. https://doi.org/10.17512/bozpe.2020.1.10

[5] T. Lis, K. Nowacki. Pro-ecological possibilities of using metallurgical waste in the production of aggregates, Prod. Eng. Arch. 28 (2022) 252-256. https://doi.org/10.30657/pea.2022.28.31

[6] B. Dębska, J. Krasoń. The evaluation of the possible utilization of waste glass in sustainable mortars, BoZPE 10 (2020) 7-15. https://doi.org/10.17512/bozpe.2020.2.01

[7] A. Pietrzak. Ocena wpływu recyklatów z butelek pet na wybrane właściwości betonu, BoZPE 7 (2018) 51-56. https://doi.org/10.17512/bozpe.2018.1.07

[8] M. Ulewicz, A. Pietrzak. Properties and Structure of Concretes Doped with Production Waste of Thermoplastic Elastomers from the Production of Car Floor Mats, Materials 14 (2021) art. 872. https://doi.org/10.3390/ma14040872

[9] J. Jura, M. Ulewicz. Assessment of the Possibility of Using Fly Ash from Biomass Combustion for Concrete, Materials. 14 (2021) 6708. https://doi.org/10.3390/ma14216708

[10] N. Makul. Principles of Fiber-Reinforced Concrete, in: N. Makul (Ed.), Principles of Cement and Concrete Composites, Springer, 2021, 79–98. https://doi.org/10.1007/978-3-030-69602-3_4

[11] PN-EN 14889-1:2007 Włókna do betonu - Część 1: Włókna stalowe - Definicje, wymagania i zgodność, (n.d.).

[12] PN-EN 14889-2:2007 Włókna do betonu - Część 2: Włókna polimerowe - Definicje, wymagania i zgodność, (n.d.).

[13] M.R. Latifi et al. Effect of the addition of polypropylene fiber on concrete properties, J. Adhes. Sci. Technol. 36 (2022) 345-369. https://doi.org/10.1080/01694243.2021.1922221

[14] M. Glinicki. Beton ze zbrojeniem strukturalnym, in: Proc. XXV WPPK Szczyrk, Poland, 2010, 279-308. [online; viewed: 2023-01-31] Available from: https://www.ippt.pan.pl/repository/open/o70.pdf

[15] Z. Jamroży. Beton i jego technologie, PWN, Warszawa, 2020. ISBN 978-8301182106

[16] A. Bentur, S. Mindess. Fibre reinforced cementitious composites, CRC Press, London, 2019. ISBN 978-0429177293

[17] G.A. Pikus. Steel Fiber Concrete Mixture Workability, Procedia Eng. 150 (2016) 2119-2123. https://doi.org/10.1016/j.proeng.2016.07.250

[18] T.M.D. Do, T.Q.K. Lam. Design parameters of steel fiber concrete beams, Mag. Civ. Eng. 102 (2021) 10207-10207. https://doi.org/10.34910/MCE.102.7

[19] PN-EN 12350-2:2019-07 – Badania mieszanki betonowej - Część 2: Badanie konsystencji metodą opadu stożka, (n.d.).

[20] PN-EN 12390-1:2021-12 – Badania betonu - Część 1: Kształt, wymiary i inne wymagania dotyczące próbek do badań i form, (n.d.).

[21] PN-EN 12390-2:2019-07 – B dania betonu - Część 2: Wykonywanie i pielęgnacja próbek do badań wytrzymałościowych, (n.d.).

[22] PN-EN 12390-3:2019-07 – Badania betonu - Część 3: Wytrzymałość na ściskanie próbek do badań, (n.d.).

[23] PN-EN 12390-5 – Badania betonu - Część 5: Wytrzymałość na zginanie próbek do badań, (n.d.).

[24] PN-EN 12390-4:2020-03 – Badania betonu - Część 4: Wytrzymałość na ściskanie - Wymagania dla maszyn wytrzymałościowych, (n.d.).

Quality Production Improvement and System Safety: QPI 16 - CZOTO 10   Materials Research Forum LLC
Materials Research Proceedings 34 (2023) 154-160   https://doi.org/10.21741/9781644902691-19

Laser Metal Deposition of Inconel 625 Alloy – Comparison of Powder and Filler Wire Methods

DANIELEWSKI Hubert[1, a *], RADEK Norbert[1, b *], ORMAN Łukasz[1,c],
PIETRASZEK Jacek [2,d] and BRONČEK Jozef [3,e]

[1]Kielce University of Technology, Faculty of Mechatronics and Mechanical Engineering,
Al. 1000-lecia Państwa Polskiego 7, 25-314 Kielce, Poland

[2]Cracow University of Technology, Faculty of Mechanical Engineering, Al. Jana Pawła II 37,
31-864 Cracow, Poland

[3]University of Zilina, Univerzitna 1, 01026 Zilina, Slovakia

[a]hdanielewski@tu.kielce.pl, [b]norrad@tu.kielce.pl, [c]orman@tu.kielce.pl,
[d]jacek.pietraszek@pk.edu.pl, [e]jozef.broncek@fstroj.utc.sk

Keywords: Laser Metal Deposition, Metal Powder and Filler Wire, Additive Manufacturing, Inconel 625, Process Analysis

Abstract. This article presents comparison of laser metal deposition methods. The main methods presented in this work assume use additional materials in a form of metal powder and filler wire. An important differences and problem with deposition of two different additional materials was presented. A comparison of abovementioned metal deposition methods of Inconel 625 alloy was shown. Microstructure analysis performed on manufactured specimens was presented. Differences in deposition mechanism for both methods were shown.

Introduction

Due to rising of material properties requirements, a rapid development of Additive Manufacturing (AM) was seen in a recent years. The high application potential encouraged researchers to study these processes nowadays. One of this new advanced material is nickel based super alloy Inconel 625, with high content of nickel (greater than 50%) and other elements improving its characteristics. The AM process, particularly using laser as a heat source combine unique laser beam properties with potential of building metal elements [1]. The deposition process was perform using methods such as: Laser Engineering Net Shaping (LENS), Direct Laser Deposition (DLD), Laser Metal Deposition (LMD), Selective Laser Sintering (SLS), Laser Powder-Bed Fusion (LPBF) [2,3]. In prototyping of high alloy materials, with high temperature and corrosion resistance, that are used in nuclear, energetic, and aerospace applications very important aspects is to achieve uniform structure, good properties and avoid presence of defects [4]. One of the most important aspects affected on those properties is a fusion mechanism, used during prototyping of each process. Intensity of beam interaction with additional and based material affects on diffusion of alloying elements from substrate to deposited layers or vice versa [5]. When we consider prototyping of small series production use same material for substrate as an additional high alloy material cost can be very high, however using different material for substrate can affect properties of printed element.

Depending on the deposited material's form (powder or filler wire), the density of obtained material are slightly different. Less power delivered to material occur during deposition of metal powders; however, density of the developed material is limited. More problematic is prototyping with a filler wire is considered. Shape of wire affect on laser reflectivity which is especially difficult when CO_2 laser is used. In this case bonding of filler wire with the substrate is obtained by using deep material penetration via the keyhole effect [6]. Another important problem is a

Quality Production Improvement and System Safety: QPI 16 - CZOTO 10   Materials Research Forum LLC
Materials Research Proceedings 34 (2023) 154-160   https://doi.org/10.21741/9781644902691-19

mixing ratio of alloying elements between the substrate and the deposited material. Nevertheless, density of printed metal elements using this process are higher, and potential application of this process can't be omitted [7]. Another difference between those two methods is surface condition, after laser metal powder deposition surface smoothness in better, however roughness is lower only in one direction – parallel to deposition direction.

The properties of the deposited material are related to the thermal cycles and chemical composition of the used materials. In this article two main methods of LMD were presented, the deposition of additional material in the form of filler wire and metal powders. In both presented processes additional material is deposited directly on the substrate. Mixing of the alloying elements between substrate and deposited material for analyzed methods is different; therefore, metallographic analysis was performed. Micro and macrostructure study was carried out using Scanning Electron Microscopy (SEM) and elements diffusion was tested with Energy Dispersive Spectroscopy (EDS) analysis. A visual microscopic test was carried out in order to detect potential defects in printed materials. The metallographic structure of characteristic areas was tested, including precipitations [8,9].

The issue discussed in the article holds significant technological and practical importance. The applied method can have a significant impact on improving the quality and durability of machinery in the metal industry [10,11], automotive parts manufacturing [12,13], quality management in the automotive industry [14,15], as well as in the wood industry [16], where machine wear and tear significantly affect worker safety [17,18]. The results obtained during the work can be useful in similar processes utilizing laser processing techniques [19,20], including those in particularly demanding high-quality surgical implants [21-23] and surfaces prone to the deposition of contaminants [24,25]. The LMD technique is clearly competitive compared to the powder-based technique [26,27], although the desired characteristics of the obtained surface layer remain an open question [28,29], especially for specific material coatings [30,31] or alloy substrates [32,33]. An undeniable outcome will be a significant reduction in machine part wear [34-36] and the provision of new material possibilities for research on more efficient batteries [37,38].

Experimental Procedure
The experimental procedure was divided into two separate steps. For every part, Trumpf CO2 laser integrated with TruLaserCell 1005 machine was used. The first experiment concerned deposition of Inconel 625 alloy in the form of filler wire on the S355J2 substrate. The second experiment shows powder deposition of Inconel 625 alloy on the same substrate material. In order to reduce the plasma ionization effect, as a shielding gas helium (5.0) with a flow rate equal to 20 l/min was used. For metal powder deposition additional argon (5.0) was used as a powder conveying gas.

Fig. 1. LMD prototyping using additional material in form of: I - metal powder, II - filler wire.

Quality Production Improvement and System Safety: QPI 16 - CZOTO 10   Materials Research Forum LLC
Materials Research Proceedings 34 (2023) 154-160   https://doi.org/10.21741/9781644902691-19

Prototyping processes were performed using two process heads: I - a welding head (focal length equal to 270 mm), with side wire feeder and II – a cladding head (focal length equal to 250 mm) with coaxial metal powder delivery system GTV M-PF 2/2. Deposition of metal powder on the substrate surface was performed with output power equal to 2.3 kW, speed of process head moving equal to 0.8 m/min, and single layer interval – 2.5 mm. The powder feed rate approximately 15 g/min (Fig.1). For the filler wire deposition, output power was set as 3 kW, with process speed equal to 1.5 m/min (Fig.1). Speed of wire feed rate was similar, while wire was oriented at an angle of 45 degrees [39].

Results and Discussion
The obtained specimens have similar dimensions; however, the structures of bead geometries for the both mentioned methods are significantly different. Global observation showed higher substrate material distortion in sample prototyped with filler wire. Moreover, the fusion mechanisms in the deposition process are different for these methods [40]. The microstructure of the obtained specimens showed significant differences; therefore, further metallographic study is presented (Fig.2).

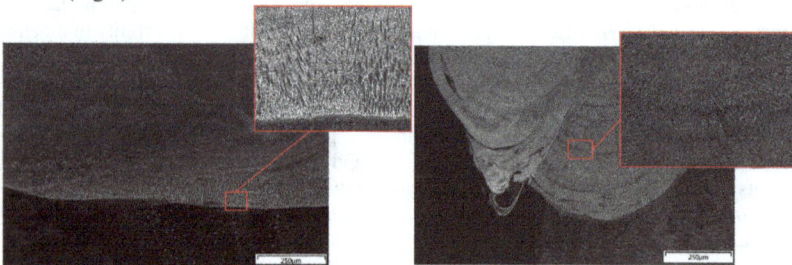

Fig. 2. *Microstructure of manufactured specimens with metal powders (left), filler wire (right)*

The investigated fusion mechanisms are different; in wire deposition there are keyhole effect and high-power density process with depth substrate material penetration, meanwhile in powder deposition the mechanism is based on conductivity phenomenon of melting a thin layer of substrate with molten powder. According to global observation, certain separate regions for all the mentioned methods can be identified. Microstructure in the material manufactured using the wire deposition method is complex, changing according to specimen height; differences in dendritic growth can be observed in all methods [41]. However, in the powder deposition material structure are more uniform, with differences visible only in dendritic growth direction. For both analyzed methods, the upper layers reveal differences only in dendritic growth direction related to each padding bead and dendrite size [42].

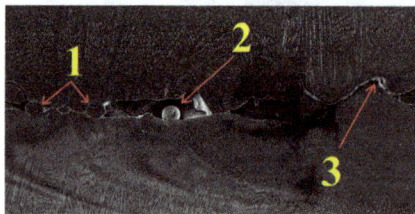

Fig. 3. *Discontinuities in material manufactured using the powder deposition method.*

Quality Production Improvement and System Safety: QPI 16 - CZOTO 10   Materials Research Forum LLC
Materials Research Proceedings 34 (2023) 154-160   https://doi.org/10.21741/9781644902691-19

During metallographic analysis mentioned in introduction section, discontinuities was revealed (Fig.3), where (1) partial molten powder (2), not molten powder particles and (3) gap between printed materials were identified. This type of defects occurs between layers when some sort of disruption appear during melting of metal powder, usually when cooling rate is too high or laser power too low.

Structure of sample prototyped using laser beam and filler wire have complex microstructure (Fig.4), due to occur of keyhole effect the severe fluctuations in the flow field of molten material appears. Structure with typical pillar dendritic (D) alongside with cellular dendritic structure (C) was identified.

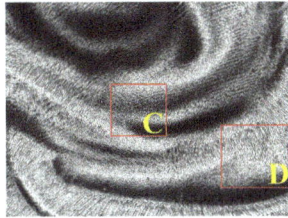

Fig. 4. Complex microstructure of filler wire deposition specimen.

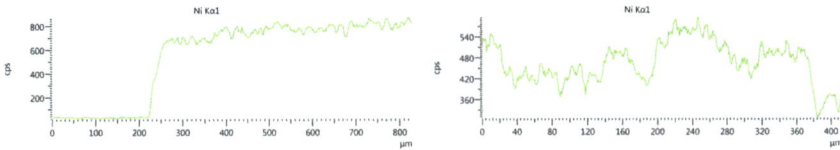

Fig. 5. Nickel distribution in substrate to deposited material section for: metal powders deposition (left) filler wire deposition (right).

The materials used for deposition have important differences in chemical composition and structure compared to the substrate material. The aforementioned differences affect diffusion and solidification processes. Deposition parameters were developed according to a separate preliminary study. Global observation showed significant differences in the fusion zone between the substrate and the deposited materials depending on the deposition process. Moreover, macroscopic analysis showed a potentially non-uniform structure in the fusion zones for the wire deposition method and cavity with partially molten powder for the powder deposition method [43]. Microstructure of the manufactured materials is dendritic; however, dendritic orientation differs depending on the deposition method. In the top part of the layers typical fine dendritic structure with secondary dendrites are observed. The cooling rate of the molten pool, and consequently the solidification velocity is higher at the bottom part of the layer and relatively slower at the top part. Non uniform distribution of nickel in Inconel 625 (Fig.5) deposited layers can reduce its properties, especially high temperature resistance and corrosion resistance. The wire deposition alone method via keyhole deep penetration affects intensive mixing of deposited material with the substrate, when the keyhole penetrates the bottom workpiece.

Quality Production Improvement and System Safety: QPI 16 - CZOTO 10   Materials Research Forum LLC
Materials Research Proceedings 34 (2023) 154-160   https://doi.org/10.21741/9781644902691-19

Summary

Analyzed deposition methods are related to direct laser deposition, where filler wire and metal powder were used as an additional material. Deposition of high alloy material is associated with a complex structure formation and numerous inclusions related to the solidification process. In the wire deposition method, a greater thermal gradient appears alongside with more rapid phenomena. Homogenous structure of manufactured materials using metal powder deposition method was confirmed, however, some deposition defects in the form of gas pore and cavity can be observed.

The presented results show significant differences in bonding mechanism between the substrate and deposited materials for both methods. Advantages and limitation of aforementioned methods was presented, where developing metal parts using LMD methods those aspects should be considered in order to avoid part failure or poor properties of prototyped elements.

References

[1] B.E. Carroll. et. all. Functionally graded material of 304L stainless steel and Inconel 625 fabricated
by directed energy deposition: Characterization and thermodynamic modeling. Acta Mater. 108 (2016) 46–54. https://doi.org/10.1016/j.actamat.2016.02.019

[2] D. Verdi, M. Garrido, C. Munez, P. Poza. Mechanical properties of Inconel 625 laser cladded coatings: Depth sensing indentation analysis. Mater. Sci. Eng. A 598 (2014) 15–21. https://doi.org/10.1016/j.msea.2014.01.026

[3] C. Li, Y. Guo, J. Zhao. Interfacial phenomena and characteristics between the deposited material and substrate in selective laser melting Inconel 625. J. Mater. Process. Technol. 243 (2017) 269–281. https://doi.org/10.1016/j.jmatprotec.2016.12.033

[4] G. Marchese, et. all. Characterization and Comparison of Inconel 625 Processed by Selective Laser Melting and Laser Metal Deposition. Adv. Eng. Mater. 19 (2017) art. 1600635. https://doi.org/10.1002/adem.201600635

[5] H. Lee, et. all. Lasers in additive manufacturing: A review. Int. J. Precis. Eng. Manuf. Technol. 4 (2017) 307–322. https://doi.org/10.1007/s40684-017-0037-7

[6] M. Teli, et. all. Study for Combined Wire + Powder Laser Metal Deposition of H11 and Niobium. Procedia Manuf. 25 (20180 426–434. https://doi.org/10.1016/j.promfg.2018.06.113

[7] N. Radek, et. all. The impact of laser processing parameters on the properties of electro-spark deposited coatings, Archives of Metallurgy and Materials 2 (2018) 809-816. https://doi.org/10.24425/122407

[8] K. Amato, et. all. Microstructures and mechanical behavior of Inconel 718 fabricated by selective laser melting. Acta Mater. 60 (2012) 2229–2239. https://doi.org/10.1016/j.actamat.2011.12.032

[9] C.P. Alvarães, et. all. Microstructural Characterization of Inconel 625 Nickel-Based Alloy Weld Cladding Obtained by Electroslag Welding Process. J. Mater. Eng. Perform. 29 (2020) 3004-3015. https://doi.org/10.1007/s11665-020-04861-3

[10] M. Nowicka-Skowron, R. Ulewicz. Quality management in logistics processes in metal branch, METAL 2015 24th Int. Conf. Metall. Mater. (2015) 1707-1712. ISBN 978-8087294628

[11] R. Ulewicz et al. Logistic controlling processes and quality issues in a cast iron foundry, Mater. Res. Proc. 17 (2020) 65-71. https://doi.org/10.21741/9781644901038-10

[12] D. Siwiec, R. Dwornicka, A. Pacana. Improving the non-destructive test by initiating the quality management techniques on an example of the turbine nozzle outlet, Materials Research Proceedings 17 (2020) 16-22. https://doi.org/10.21741/9781644901038-3

[13] D. Nowakowski et al. Application of machine learning in the analysis of surface quality – the detection the surface layer damage of the vehicle body, METAL 2021 – 30th Int. Conf. Metallurgy and Materials (2021), Ostrava, Tanger 864-869. https://doi.org/10.37904/metal.2021.4210

[14] R. Ulewicz. Outsorcing quality control in the automotive industry, MATEC Web of Conf. 183 (2018) art.03001. https://doi.org/10.1051/matecconf/201818303001

[15] A. Pacana et al. Analysis of quality control efficiency in the automotive industry, Transp. Res. Procedia 55 (2021) 691-698. https://doi.org/10.1016/j.trpro.2021.07.037

[16] K. Knop, R. Ulewicz. Assessment of technology, technological resources and quality in the manufacturing of timber products, Proc. Digitalisation and Circular Economy: Forestry and Forestry Based Industry Implications (2019) 251-256. ISBN 978-9543970421

[17] M. Ingaldi. Overview of the main methods of service quality analysis, Production Engineering Archives 18 (2018) 54-59. https://doi.org/10.30657/pea.2018.18.10

[18] R. Ulewicz, F. Nový. Quality management systems in special processes, Transp. Res. Procedia 40 (2019) 113-118. https://doi.org/10.1016/j.trpro.2019.07.019

[19] B. Antoszewski et al. Assessment of technological capabilities for forming al-c-b system coatings on steel surfaces by electrospark alloying method, Materials 14 (2021) art.739. https://doi.org/10.3390/ma14040739

[20] L. Cedro Model parameter on-line identification with nonlinear parametrization – manipulator model, Technical Transactions 119 (2022) art.e2022007. https://doi.org/10.37705/TechTrans/e2022007

[21] A. Dudek. Surface properties in titanium with hydroxyapatite coating, Optica Applicata 39 (2009) 825-831.

[22] A. Dudek. Investigations of microstructure and properties in bioceramic coatings used in medicine, Arch. Metall. Mater. 56 (2011) 135-140. https://doi.org/10.2478/v10172-011-0015-y

[23] D. Klimecka-Tatar. Electrochemical characteristics of titanium for dental implants in case of the electroless surface modification, Arch. Metall. Mater. 61 (2016) 923-926. https://doi.org/10.1515/amm-2016-0156

[24] S. Michna et al. Research the causes of surface stains after eloxal coating for the profile from the AlMgSi alloy using substructural analysis, Manuf. Technol. 15 (2015) 620-624. https://doi.org/10.21062/ujep/x.2015/a/1213-2489/MT/15/4/620

[25] M. Opydo et al. The effect of biomass co-combustion in a CFB boiler on solids accumulation on surfaces of P91 steel tube samples, Biomass and Bioenergy 85 (2016) 61-68. https://doi.org/10.1016/j.biombioe.2015.12.011

[26] D. Klimecka-Tatar. The powdered magnets technology improvement by biencapsulation method and its effect on mechanical properties, Manuf. Technol. 14 (2014) 30-36. https://doi.org/10.21062/ujep/x.2014/a/1213-2489/MT/14/1/30

[27] D. Klimecka-Tatar et al. The effect of powder particle biencapsulation with Ni-P layer on local corrosion of bonded Nd-(Fe, Co)-B magnetic material, Arch. Metall. Mater. 60 (2015) 154-157. https://doi.org/10.1515/amm-2015-0024

[28] D. Klimecka-Tatar et al. The kinetics of Ti-1Al-1Mn alloy thermal oxidation and charcteristic of oxide layer, Arch. Metall. Mater. 60 (2015) 735-738. https://doi.org/10.1515/amm-2015-0199

[29] J. Korzekwa et al. The influence of sample preparation on SEM measurements of anodic oxide layers, Pract. Metallogr. 53 (2016) 36-49. https://doi.org/10.3139/147.110367

[30] K. Milewski et al. The interaction between diamond like carbon (DLC) coatings and ionic liquids under boundary lubrication conditions, Metalurgija 56 (2017) 55-58.

[31] P. Kurp, D. Soboń. The influence of laser padding parameters on the tribological properties of the Al2O3 coatings, METAL 2018 27th Int. Conf. Metall. Mater. (2018) 1157-1162. ISBN 978-808729484-0

[32] T. Lipiński, P. Szabracki. Modification of the hypo-eutectic Al-Si alloys with an exothermic modifier, Arch. Metall. Mater. 58 (2013) 453-458. https://doi.org/10.2478/amm-2013-0017

[33] A. Dudek, B. Lisiecka, R. Ulewicz. The effect of alloying method on the structure and properties of sintered stainless steel, Archives of Metallurgy and Materials 62 (2017) 281-287. https://doi.org/10.1515/amm-2017-0042

[34] K. Trzewiczek et al. Evaluation of the state for the material of the live steam superheater pipe coils of V degree. Adv. Mater. Res. 874 (2014) 35-42. https://doi.org/10.4028/www.scientific.net/AMR.874.35

[35] M. Ingaldi, S.T. Dziuba. Modernity evaluation of the machines used during production process of metal products, METAL 2015 - 24th Int. Conf. Metallurgy and Materials (2015), Ostrava, Tanger 1908-1914.

[36] S. Marković et al. Exploitation characteristics of teeth flanks of gears regenerated by three hard-facing procedures, Materials 14 (2021) art. 4203. https://doi.org/10.3390/ma14154203

[37] U. Antoniw et al. Logarithmic analog-to-digital converter with accumulation of charge and pulse feedback, Przeglad Elektrotechniczny 89 (2013) 277-281.

[38] M. Szczepaniak et al. T. Use of the maximum power point tracking method in a portable lithium-ion solar battery charger, Energies 15 (2022) art.26. https://doi.org/10.3390/en15010026

[39] J. Heigel, M. Gouge P. Michaleris, T. Palmer. Selection of powder or wire feedstock material for the laser cladding of Inconel® 625. J. Mater. Process. Technol. 231 (2016) 357-365. https://doi.org/10.1016/j.jmatprotec.2016.01.004

[40] B. Dubiel, J. Sieniawski. Precipitates in Additively Manufactured Inconel 625 Superalloy. Materials 12 (2019) art. 1144. https://doi.org/10.3390/ma12071144

[41] J. Huebner, et. all. Microstructural and mechanical study of Inconel 625 – tungsten carbide composite coatings obtained by powder laser cladding. Archives of Metallurgy and Materials 62 (2017) 531-538. https://doi.org/10.1515/amm-2017-0078

[42] F. Khodabakhshi, et. all. Effects of laser additive manufacturing on microstructure and crystallographic texture of austenitic and martensitic stainless steels. Addit. Manuf. 31 (2020) art. 100915. https://doi.org/10.1016/j.addma.2019.100915

[43] B. Antoszewski, H. Danielewski. Microstructure and Properties of Laser Additive Deposited of Nickel Base Super Alloy Inconel 625, Archives of Foundry Engineering, 20 (2020) 53-59. https://doi.org/10.24425/afe.2020.133330

Quality Production Improvement and System Safety: QPI 16 - CZOTO 10
Materials Research Proceedings 34 (2023) 161-168

Materials Research Forum LLC
https://doi.org/10.21741/9781644902691-20

Operational Properties of Heterogeneous Surfaces

RADEK Norbert[1, a *], DANIELEWSKI Hubert[1,b], PIETRASZEK Jacek [2,c],
ORMAN Łukasz[3,d], GONTARSKI Dariusz[1,e], RADEK Maria[4,f] and
PARASKA Olga[5,g]

[1]Kielce University of Technology, Faculty of Mechatronics and Mechanical Engineering,
Al. 1000-lecia Państwa Polskiego 7, 25-314 Kielce, Poland

[2]Cracow University of Technology, Faculty of Mechanical Engineering, Al. Jana Pawła II 37, 31-864 Cracow, Poland

[3]Kielce University of Technology, Faculty of Geomatics and Energy Engineering,
Al. 1000-lecia P.P. 7, 25-314 Kielce, Poland

[4]F. H. Barwa, ul. Warkocz 3-5, 25-253 Kielce, Poland

[5]Khmelnitskiy National University, Department of Chemical Technology, st. Instytutska 11,
29016, Khmelnickiy, Ukraine

[a]norrad@tu.kielce.pl, [b]hdanielewski@tu.kielce.pl, [c]jacek.pietraszek@pk.edu.pl,
[e]orman@tu.kielce.pl, [e]gontar@tu.kielce.pl, [f]mariaradek@wp.pl, [g]olgaparaska@gmail.com

Keywords: ESD Coating, Laser Beam Machining, Heterogeneous Surfaces, Properties

Abstract. The paper is concerned with testing Cu-Mo coatings deposited over carbon steel C45, which were then eroded with a laser beam. A combination of electro-spark deposition (ESD) process and laser treatment has been developed, and tested, and improvement of certain surface properties has been demonstrated. The analysis involved measuring the macrogeometry, microhardness roughness and corrosion resistance of selected areas after laser treatment. The coatings were deposited by means of the ELFA-541 and they were laser treated with the Nd:YAG, the laser parameters being variable. The properties of heterogeneous surfaces, based on laser treated ESD, are largely dependent upon material combination systems, manipulating methods, ESD and laser parameters as well as process control.

Introduction

There is an ever-increasing requirement for low-cost coatings with high quality tribological properties of its surfaces for wider applications with combined requirements. Examples are machine elements subjected to sever conditions, such as friction and wear, corrosion, or exposure to high temperature. For example, coatings of shafts of rotating machinery have combined requirements. There is a need to increase the hardness of the surfaces rotating inside the bearings to resist wear, and increase the load capacity of the surface, while the core of the shaft must retain its original plasticity, in order to prevent failure due to brittle cracking under the impact forces in operating machinery. In addition, the coating must have good bonding to the substrate material of the machine element in order to avoid undesired peeling (delamination). It has been already realized that heterogeneous surfaces, are advantageous for such combined requirements. They are designed to have the desired distribution of composition and gradients of various properties, such as microhardness, along the thin width of the coating.

There are many methods for surface coatings such as electroplating or plasma spraying. Very thin layers can be deposited by vapor deposition. Various surface treatment techniques have been developed to improve the desired properties of the deposited layers, based on the substrate material. One important low-cost method is the electro-spark deposition (ESD), which has been

Quality Production Improvement and System Safety: QPI 16 - CZOTO 10   Materials Research Forum LLC
Materials Research Proceedings 34 (2023) 161-168   https://doi.org/10.21741/9781644902691-20

recognized and widely applied as an economically effective surface coating [1-4]. It is already widely used process for its expedient way of achieving the desired properties of surfaces. ESD has been known by several other terms such as *spark hardening, electric spark toughening*, and *electro-spark alloying*. Electro-spark deposited coatings have also some disadvantages but these can be easily eliminated by laser beam machining (LBM), which can be used for polishing the surface, sealing and modifying its topography, and chemical homogenization of coatings [5, 6]. This paper reports on the effects of laser treatment on microstructure, microhardness, roughness and corrosion resistance of electro-spark Cu-Mo coatings.

The technique presented in this study involves the use of lasers to modify the surface morphology, thereby impacting heat transfer. This approach has broader implications beyond the specific case studied, as it can be applied to other laser processing scenarios where surface morphology plays a critical role [7-9]. Altering the characteristics of the surface layer is a commonly employed solution in various fields [10-12], and image analysis serves as a valuable diagnostic tool in this process [13-15]. These modifications are not only relevant to the energy sector [16], but also to hydraulic power systems [17,18] and the broader field of material engineering [19-21] and corrosion protection [22,23]. By modifying the surface layer morphology, significant advancements can be made in reducing wear and tear on machine parts [24-26]. Additionally, this technique has the potential to enhance the quality of products in the automotive industry [27,28] and railway sector [29,30], which, in turn, can inspire the development of new quality control methods [31-33] implemented in both the automotive [34,35] and metal industries [36,37]. Implementing these quality enhancements has a significant impact on minimizing potential failure scenarios and their associated consequences [38-40], allowing for the effective implementation of lean manufacturing principles [41-43]. By employing these techniques, manufacturers can streamline their processes, reduce waste, and improve overall efficiency.

Methodology and Results

The testing process consisted of two stages: first, Cu-Mo coatings were electro-spark deposited on standard steel samples (C45 steel); then, they were modified with a laser beam. The electro-spark deposition of Cu-Mo wires with a diameter of 1 mm was performed by means of an ELFA-541, a modernized device made by a Bulgarian manufacturer. The subsequent laser treatment was performed with the aid of a BLS 720 laser system employing the Nd:YAG type laser operating in the pulse mode.

The parameters of the electro-spark deposition, established during the experiment, include: current intensity I = 16 A (for Cu I = 8A); table shift rate v = 0.5 mm/s; rotational speed of the head with electrode n = 4200 rev/min; number of coating passes L = 2 (for Cu L = 1); capacity of the condenser system C = 0.47 μF; pulse duration T_i = 8 μs; interpulse period T_p = 32 μs; frequency f = 25 kHz.

The produced heterogeneous coatings were eroded by laser beam after the electro-spark deposition. The laser surface treatment was performed by an Nd:YAG laser (impulse mode), model BLS 720, and operating in the pulse mode under the following conditions: laser spot diameter, d = 0.462-0.739 mm; laser power, P = 10-150 W; beam shift rate, v = 1200 mm/min; nozzle-sample distance, h = 1 mm; pulse duration, t_i = 0.8 ms, 1.2 ms, 1.48 ms, 1.8 ms, 5.5 ms, 8 ms; frequency, f = 8 Hz.

A typical methodology used for surface analysis was established by using optical image analyzer, Vickers microhardness tester, and computer controlled surfoanalyzer with computer data acquisition. This equipment was used to measure the surface finishing as well as other mechanical properties of the applied coating on the outer surface and properties distribution inside the coating. All metallographic examinations were carried out on samples polished and etched by Nital.

The Vickers microhardness tests, along the depth cross section of all zones (as shown in Fig. 3) used 40 G load, while for crater cavity cross section, 100 G load was applied (Fig. 4).

Results and Disscussion

Surface topography of produced by ESD and followed by laser spot treatment specimens is shown on Fig.1. Coated substrate by ESD has a matt appearance, with "small craters", due to local roughening by individual sparks. Noticeable features are pores and erosion pits on the surface, in particular for EDS treated surface at 0.5 mm/sec EDS applicator speed, as shown in Fig. 1a. Surface roughness was measured by using a group of parameters: Ra, Rz, Rq, Ry, Rsk and tp. Some parameters, such as Ra and Rz, have been increased by the ESD over 10 times in comparison with the original surface. Surface roughness results for various ESD treatments are summarized in Table1.

Table 1. Surface roughness parameters (Ra = roughness average; Rz = average maximum height of the profile; Rq = root mean square roughness; Ry = maximum height of the profile; Rsk = skewness)

Specimens	Ra, µm	Rq, µm	Rz, µm	Ry, µm	Rsk
Original as machined surface - curve A	0.36	0.45	2.28	1.25	-0.58
ESD treated surface, Fig1a, Sample 1 - curve B	3.01	4.07	23.84	23.99	-0.61
ESD treated surface, Sample 2 - curve C	3.90	4.92	30.35	30.88	-0.54

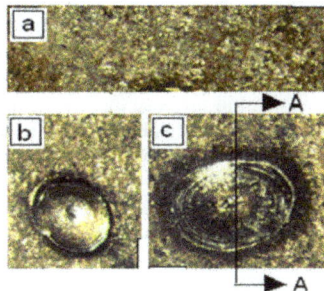

Fig. 1. Surface topography produced by ESD and laser treatment (x50): a-EDS treated surface at 0.5mm/sec applicator speed (Sample 1), b & c-laser treated surface with craters of Sample 1:
(b)- 20W; (c) - 100W.

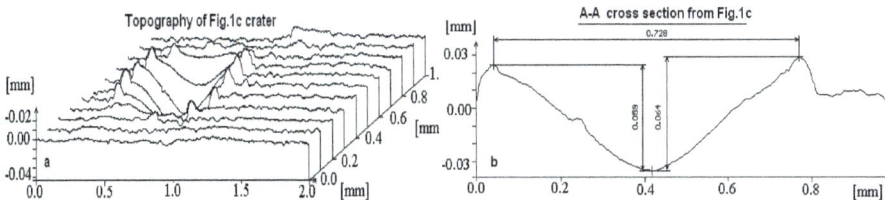

Fig. 2. Macrogeometry and cross section of the crater formed under the influence of laser: a-3D crater topography; b-A-A cross section on Fig.1c.

Quality Production Improvement and System Safety: QPI 16 - CZOTO 10   Materials Research Forum LLC
Materials Research Proceedings 34 (2023) 161-168   https://doi.org/10.21741/9781644902691-20

A 3D macrogeometry of the developed heterogeneous surface, eroded by the laser craters, for the used specimens with build in 2-D crater cross section A-A (Fig. 1c) is shown on Fig. 2a, b.

As can be concluded from these built graphs, crater edges are sharp and are advanced up to 0.03mm above an average height, just treated by ESD surface, what is within a range of tolerances for designed clearance fit. The average size of the crater, shown on Fig.1c, produced by laser power 100 W has diameter about 0.7 mm and the total depth about 0.06 mm. The crater is going below so-called "ground zero level" by down to 0.030mm. For instant, crater displayed on Fig.1b, produced by laser power 20W has diameter about 0.05mm and depth of 0.015mm. Produced crater profile (picks and valleys) and also order of craters location, depending on the required or desired surface performance, could be controlled and adjusted to acceptable level.

The microhardness test results, concerning the Cu-Mo coating before and after laser treatment, is presented in Fig. 3. After the indentation was made on metallographic specimens parallel in three zones: the coating, the remelted coating, and the heat affected zone. The original material was also tested. The electro-spark deposition process caused some changes in the material. Laser treatment had a favorable effect on the changes in the microhardness of the electro-spark deposited coatings.

In Fig. 4, a microhardness-profile is shown along the depth of obtained craters measured across the cross section shown on Fig.1c. The surface of this cross section was treated by ESD and then post finished by laser erosion. Its microhardness profile reveals distribution of properties from the fusion zone into the bulk materials. It is shown that there is a soft region in the bottom valley of the fusion zone while a big increase is achieved across a fusion line. Since the samples were deposited by a two-stage process, the softening effect can be caused by insufficient alloying and reheat affection. This saying is confirmed already by the microstructural examination which shows very limited penetration of alloy components into the bottom fused layer. The thermal cracks presence, due to rapid solidification, could increase their contributions toward the low hardness values. The microhardness profile shows gradient transition from outmost surface to the bulk materials while the variation in the outmost layer reveals the non-uniformity and the existence of micro cracks. The highest hardness value of about 800 HV achieved for the melted zone during the laser process is due to the fast-quenching effect of highly alloyed surface. By comparison, surface softening effect, near the bottom of the fusion line, is achieved due to a milder extent so that better bonding can be expected.

Moreover, the microhardness values above the fusion lines present less variation corresponding to the concentration distribution in the EDXA profiles. Therefore, the formation of compounds and effect of diffusion can be confirmed. Measured average microhardness of the crater's tips of about 800HV, HAZ zone about 650HV and substrate of 300HV shows that the hardness of the working surface could be almost tripled by preparation of the heterogeneous surface.

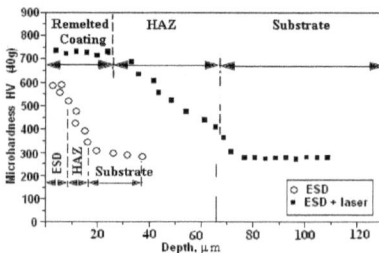

Fig. 3. *Results of microhardness tests.*

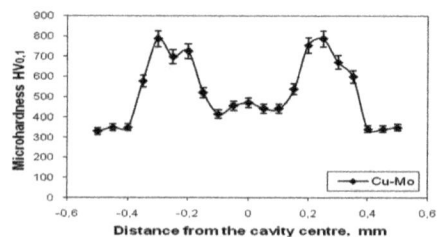

Fig. 4. *Microhardness distribution along crater cross section.*

Corrosion resistance tests were carried out by the computerized Atlas'99 electrochemical analysis system using the potentiodynamic method. The cathodic and anodic polarization curves were acquired by polarizing the tested specimens at 0.2 mV/s (within the area of ±200 mV from the corrosion potential) and 0.4 mV/s (within the area of higher potentials). Specimens with a 10 mm diameter separated area were polarized to 500 mV. In order to establish the corrosion potential, the polarization curves were acquired 24 hours after exposure to the test solution (0.5M NaCl). All tests were carried out at 21±1°C. The corrosion resistance results are shown in Fig. 5.

The Cu-Mo coating was reported to have the highest corrosion resistance. The corrosion current density of the coating was 42.9 $\mu A/cm^2$, while that of the C45 steel substrate was 112 $\mu A/cm^2$. Applying the Cu-Mo coating improved the sample corrosion resistance by approx. 162%. The fusion of the coating and the substrate resulted in a considerable heterogeneity of electrochemical potentials on the coating surface. The microcracks in the surface layer also contributed to the intensification of the corrosion processes.

There was some improvement in the corrosion resistance of the electro-spark deposited coatings after laser treatment. The healing of microcracks resulted in higher density and therefore better sealing properties. The highest corrosion resistance after laser treatment was reported for the Cu-Mo coating (I_k=30.7 $\mu A/cm^2$). For the C45 steel substrate, I_k was 6.4 $\mu A/cm^2$. Thus, the corrosion resistance increased by about 30 % after laser treatment.

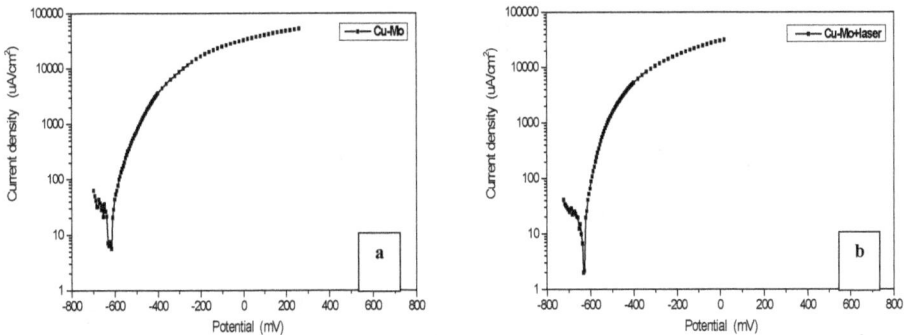

Fig. 5. *Curves of the Cu-Mo coating polarization:*
a) before laser treatment, b) after laser treatment.

Summary
1. The process of creating technological surface layers by the ESD method is associated with the transfer of mass and energy and the phenomenon of the formation of low-temperature plasma.
2. A concentrated laser beam can effectively modify the state of the surface layer, i.e. the functional properties of electro-spark coatings.
3. Laser radiation causes an improvement in the functional properties of the two-layer electro-spark deposited Cu-Mo coatings, i.e. they exhibit higher microhardness and higher resistance to corrosion.
4. Laser treatment of ESD coatings resulting in crater formation made the surface stronger and more resistant to wear.
5. The surface heterogeneity (i.e. the cavities) are desirable in sliding friction pairs. They may be used as reservoirs of lubricants as well as sources of hydrodynamic forces increasing the capacity of a sliding pair.

References

[1] N. Radek, A. Sladek, J. Broncek, I. Bilska, A. Szczotok. Electrospark alloying of carbon steel with WC-Co-Al_2O_3: deposition technique and coating properties, Advanced Materials Research 874 (2014) 101-106. https://doi.org/10.4028/www.scientific.net/AMR.874.101

[2] Z. Chen, Y. Zhou. Surface modification of resistance welding electrode by electro-spark deposited composite coatings: Part I. Coating characterization, Surface and Coatings Technology 201 (2006) 1503-1510. https://doi.org/10.1016/j.surfcoat.2006.02.015

[3] A. V. Ribalko, O. A. Sahin. A modern representation of the behaviour of electrospark alloying of steel by hard alloy, Surface and Coatings Technology 201 (2006) 1724-1730. https://doi.org/10.1016/j.surfcoat.2006.02.044

[4] M. Salmaliyan, F.M. Ghaeni, M. Ebrahimnia. Effect of electro spark deposition process parameters on WC-Co coating on H13 steel, Surface and Coatings Technology 321 (2017) 81-89. https://doi.org/10.1016/j.surfcoat.2017.04.040

[5] N. Radek, A. Szczotok, A. Gądek-Moszczak, R. Dwornicka, J. Bronček, J. Pietraszek. The impact of laser processing parameters on the properties of electro-spark deposited coatings, Archives of Metallurgy and Materials 2 (2018) 809-816. https://doi.org/10.24425/122407

[6] M. Scendo, J. Trela, N. Radek. Influence of laser power on the corrosive resistance of WC-Cu coating, Surface and Coatings Technology 259 (2014) 401-407. https://doi.org/10.1016/j.surfcoat.2014.10.062

[7] P. Kurp et al. The influence of treatment parameters on the microstructure, properties and bend angle of laser formed construction bars, Arch. Metall. Mater. 61 (2016) 1151-1156. https://doi.org/10.1515/amm-2016-0192

[8] P. Kurp, D. Soboń. The influence of laser padding parameters on the tribological properties of the Al2O3 coatings, METAL 2018 27th Int. Conf. Metall. Mater. (2018) 1157-1162. ISBN 978-808729484-0

[9] P. Kurp, H. Danielewski Metal expansion joints manufacturing by a mechanically assisted laser forming hybrid method – concept, Technical Transactions 119 (2022) art.e2022008. https://doi.org/10.37705/TechTrans/e2022008

[10] D. Klimecka-Tatar. Electrochemical characteristics of titanium for dental implants in case of the electroless surface modification, Arch. Metall. Mater. 61 (2016) 923-926. https://doi.org/10.1515/amm-2016-0156

[11] D. Nowakowski et al. Application of machine learning in the analysis of surface quality - the detection the surface layer damage of the vehicle body, METAL 2021 - 30th Int. Conf. Metallurgy and Materials (2021), Ostrava, Tanger 864-869. https://doi.org/10.37904/metal.2021.4210

[12] B. Antoszewski et al. Assessment of technological capabilities for forming al-c-b system coatings on steel surfaces by electrospark alloying method, Materials 14 (2021) art.739. https://doi.org/10.3390/ma14040739

[13] A. Gądek et al. Application of computer-aided analysis of an image for assessment of reinforced polymers structures, Polymers 51 (2006) 206-211. https://doi.org/10.14314/polimery.2006.206

Quality Production Improvement and System Safety: QPI 16 - CZOTO 10   Materials Research Forum LLC
Materials Research Proceedings 34 (2023) 161-168   https://doi.org/10.21741/9781644902691-20

[14] A. Gadek-Moszczak, P. Matusiewicz. Polish stereology - A historical review, Image Analysis and Stereology 36 (2017) 207-221. https://doi.org/10.5566/ias.1808

[15] I. Jastrzębska, A. Piwowarczyk. Traditional vs. Automated Computer Image Analysis—A Comparative Assessment of Use for Analysis of Digital SEM Images of High-Temperature Ceramic Material, Materials 16 (2023) art. 812. https://doi.org/10.3390/ma16020812

[16] K. Trzewiczek et al. Evaluation of the state for the material of the live steam superheater pipe coils of V degree. Adv. Mater. Res. 874 (2014) 35-42. https://doi.org/10.4028/www.scientific.net/AMR.874.35

[17] G. Filo, E. Lisowski, M. Domagała, J. Fabiś-Domagała, H. Momeni. Modelling of pressure pulse generator with the use of a flow control valve and a fuzzy logic controller, AIP Conf. Proc. 2029 (2018) art.20015. https://doi.org/10.1063/1.5066477

[18] M. Domagala et al. The Influence of Oil Contamination on Flow Control Valve Operation, Mater. Res. Proc. 24 (2022) 1-8. https://doi.org/10.21741/9781644902059-1

[19] A. Dudek, B. Lisiecka, R. Ulewicz. The effect of alloying method on the structure and properties of sintered stainless steel, Archives of Metallurgy and Materials 62 (2017) 281-287. https://doi.org/10.1515/amm-2017-0042

[20] R. Ulewicz et al. Structure and mechanical properties of fine-grained steels, Period. Polytech. Transp. Eng. 41 (2013) 111-115. https://doi.org/10.3311/PPtr.7110

[21] D. Klimecka-Tatar, M. Ingaldi. Assessment of the technological position of a selected enterprise in the metallurgical industry, Mater. Res. Proc. 17 (2020) 72-78. https://doi.org/10.21741/9781644901038-11

[22] D. Klimecka-Tatar, R. Dwornicka. The assembly process stability assessment based on the strength parameters statistical control of complex metal products, METAL 2019 28th Int. Conf. Metall. Mater. (2019) 709-714. ISBN 978-808729492-5

[23] D. Klimecka-Tatar, G. Pawlowska, R. Orlicki, G.E. Zaikov. Corrosion characteristics in acid, alkaline and the ringer solution of Fe68-xCoxZr10Mo5W2B 15 metallic glasses, J. Balk. Tribol. Assoc. 20 (2014) 124-130.

[24] S. Marković et al. Exploitation characteristics of teeth flanks of gears regenerated by three hard-facing procedures, Materials 14 (20210 art. 4203. https://doi.org/10.3390/ma14154203

[25] M. Krynke et al. Maintenance management of large-size rolling bearings in heavy-duty machinery, Acta Montan. Slovaca 27 (2022) 327-341. https://doi.org/10.46544/AMS.v27i2.04

[26] P. Regulski, K.F. Abramek The application of neural networks for the life-cycle analysis of road and rail rolling stock during the operational phase, Technical Transactions 119 (2022) art. e2022002. https://doi.org/10.37705/TechTrans/e2022002

[27] D. Siwiec, R. Dwornicka, A. Pacana. Improving the non-destructive test by initiating the quality management techniques on an example of the turbine nozzle outlet, Materials Research Proceedings 17 (2020) 16-22. https://doi.org/10.21741/9781644901038-3

[28] K. Czerwinska et al. Improving quality control of siluminial castings used in the automotive industry, METAL 2020 29th Int. Conf. Metall. Mater. (2020) 1382-1387. https://doi.org/10.37904/metal.2020.3661

[29] T. Lipiński, R. Ulewicz. The effect of the impurities spaces on the quality of structural steel working at variable loads, Open Eng. 11 (2021) 233-238. https://doi.org/10.1515/eng-2021-0024

[30] K. Czerwińska, A. Piwowarczyk. The use of combined quality management instruments to analyze the causes of non-conformities in the castings of the cover of the rail vehicle bearing housing, Prod. Eng. Arch. 28 (2022) 289-294. https://doi.org/10.30657/pea.2022.28.36

[31] M. Ingaldi. Overview of the main methods of service quality analysis, Production Engineering Archives 18 (2018) 54-59. https://doi.org/10.30657/pea.2018.18.10

[32] R. Ulewicz, F. Nový. Quality management systems in special processes, Transp. Res. Procedia 40 (2019) 113-118. https://doi.org/10.1016/j.trpro.2019.07.019

[33] D. Siwiec, A. Pacana. Method of improve the level of product quality, Prod. Eng. Arch. 27 (2021) 1-7. https://doi.org/10.30657/pea.2021.27.1

[34] R. Ulewicz. Outsorcing quality control in the automotive industry, MATEC Web of Conf. 183 (2018) art.03001. https://doi.org/10.1051/matecconf/201818303001

[35] A. Pacana et al. Analysis of quality control efficiency in the automotive industry, Transp. Res. Procedia 55 (2021) 691-698. https://doi.org/10.1016/j.trpro.2021.07.037

[36] M. Nowicka-Skowron, R. Ulewicz. Quality management in logistics processes in metal branch, METAL 2015 24th Int. Conf. Metall. Mater. (2015) 1707-1712. ISBN 978-8087294628

[37] R. Ulewicz et al. Logistic controlling processes and quality issues in a cast iron foundry, Mater. Res. Proc. 17 (2020) 65-71. https://doi.org/10.21741/9781644901038-10

[38] J. Fabiś-Domagała, G. Filo, H. Momeni, M. Domagała. Instruments of identification of hydraulic components potential failures, MATEC Web of Conf. 183 (2018) art.03008. https://doi.org/10.1051/matecconf/201818303008

[39] K. Knop et al. Evaluating and Improving the Effectiveness of Visual Inspection of Products from the Automotive Industry, Lecture Notes in Mechanical Engineering (2019) 231-243. https://doi.org/10.1007/978-3-030-17269-5_17

[40] J. Fabis-Domagala, M. Domagala. A Concept of Risk Prioritization in FMEA of Fluid Power Components, Energies 15 (2022) art.6180. https://doi.org/10.3390/en15176180

[41] A. Maszke, R. Dwornicka, R. Ulewicz. Problems in the implementation of the lean concept at a steel works - Case study, MATEC Web of Conf. 183 (2018) art.01014. https://doi.org/10.1051/matecconf/201818301014

[42] R. Ulewicz, M. Ulewicz. Problems in the Implementation of the Lean Concept in the Construction Industries, LNCE 47 (2020) 495-500. https://doi.org/10.1007/978-3-030-27011-7_63

[43] R. Ulewicz et al. Implementation of Lean Instruments in Ceramics Industries, Manag. Sys. Prod. Eng. 29 (2021) 203-207. https://doi.org/10.2478/mspe-2021-0025

Quality Production Improvement and System Safety: QPI 16 - CZOTO 10   Materials Research Forum LLC
Materials Research Proceedings 34 (2023) 169-177   https://doi.org/10.21741/9781644902691-21

Radar Recognition: Paint Coatings with Absorption Properties in the Microwave Range

PRZYBYŁ Wojciech[1,a*], MICHALSKI Marek[2,b], MAZURCZUK Robert[1,c], BOGDANOWICZ Krzysztof A.[1,d] and RADEK Norbert[3,e]

[1]Military Institute of Engineer Technology, ul. Obornicka 136, 50-961 Wrocław, Poland

[2] F. H. Barwa, ul. Warkocz 3-5, 25-253 Kielce, Poland

[3] Kielce University of Technology, Faculty of Mechatronics and Machine Building, Al. 1000-lecia P. P. 7, 25-314 Kielce, Poland

[a]przybyl@witi.wroc.pl, [b]mmichalski@barwa.kielce.pl, [c] mazurczuk.r@witi.wroc.pl, [d]bogdanowicz@witi.wroc.pl, [e]norrad@tu.kielce.pl

Keywords: Radar, Coatings, Absorption, Microwave

Abstract. The article presents the characteristics of modern reconnaissance systems in the radar range and camouflage methods for this range. Two absorbers of electromagnetic radiation in the 4-18 GHz range were tested, measuring their attenuation properties. Carbonyl iron and thin-walled hollow microspheres based on soda-lime-borosilicate glass were tested, on the basis of which paint coatings with different shares of absorbers and different coating thicknesses were produced. The attenuation properties of both absorbers were determined and attention was paid to the maximum values and frequencies for which they occur. Further directions of research were also proposed in order to obtain varnish coatings that are an effective camouflage agent in radiolocation.

Introduction

On the contemporary battlefield, as shown by recent full-scale conflicts, artillery, both barreled and rocket artillery, as well as armored forces with accompanying mechanized infantry units, still play a decisive role. On the fronts of modern conflicts, the improvement of means of destruction is observed, increasing the share of "intelligent" weapon components. However, what ensures their effective attack, even for the most modern system, is effective reconnaissance. In close range and direct combat, the human eye is still the basic reconnaissance device, but since the invention of the rifle, later cannons, then rockets and airplanes, the area of direct combat has been constantly increasing. For reconnaissance purposes, a number of sensors are used that operate in different ranges of electromagnetic radiation (Fig.1).

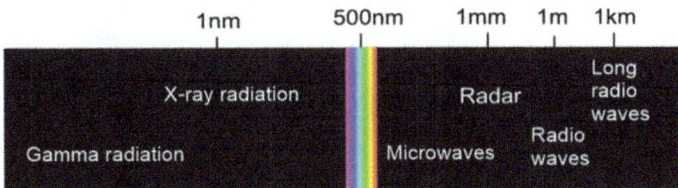

Fig.1. Scheme of the electromagnetic radiation spectrum

Sensors in the optical and thermal bands are commonly used – especially for closer distances. On the other hand, for longer distances, radars operating in the microwave band up to hectometer waves are used. Due to various properties, e.g. reflection or absorption by water vapor, penetration through selected media, etc., they find various applications: telecommunications, meteorological,

SAR or military reconnaissance. Therefore, the frequency was divided into different bands (Table 1),

Table 2. Frequency division into bands (source: [1])

Band name	Frequency range [GHz]	Wavelength range [cm]	Band's symbol
VHF	0.1 - 0.3	300 - 100	A
UHF	0.3 - 0.5 0.5 - 1.0	100 - 60 60 - 30	B C
L	1 - 2	30 - 15	D
S	2 - 3 3 - 4	15 - 10 10 - 7.5	E F
C	4 - 6 6 - 8	7.5 - 5 5 - 3.75	G H
X	8 - 10 10 - 12	3.75 - 3 3 - 2.5	I J
Ku	12 - 18	2.5 - 1.67	J
K	18 – 26.5	1.67 - 1.1	J (do 20 GHz)
Ka	26.5 - 40	1.1 - 0.75	K
Millimeter wave	40 - 100	0.75 - 0.3	L (do 60 GHz) M (>60 GHz)

Several bands are used in reconnaissance applications. Here are some example uses:
* S band – NUR-15 – three-coordinate complementary radar for radar coverage and detection of low-flying targets; the AN/APY-2 radar station of the E-3/AWACS aircraft;
* Band C – LIWIEC artillery reconnaissance radar set;
* Band X – AN/APG-66 radar station of the F-16 aircraft; RPW-10 – battlefield radar (Figure 2);
* Ku-band – AN/PPS-5C – battlefield radar;
* L band – 59N6E Protiwnik-GE – long range radars.

Fig.2. PGSR-3i Beagle battlefield reconnaissance radars (source: [2])

Quality Production Improvement and System Safety: QPI 16 - CZOTO 10   Materials Research Forum LLC
Materials Research Proceedings 34 (2023) 169-177   https://doi.org/10.21741/9781644902691-21

The main task of the radars is to raise the situational awareness in the theatre of war by detecting, locating and identifying ground and air objects, thanks to the registration of their radar echoes by the antennas. Some of them that use the Doppler effect can also contain information about the speed of a moving object. Others using synthetic aperture radar (SAR) are able to produce a three-dimensional image of the terrain, and sometimes also determine the type of material, shape and size of the object. Currently, the most modern radars are AESA – Active Electronically Scanned Array, whose antenna consists of many independently, electronically controlled modules, which makes it possible to track many targets. They are also characterized by greater resolution and accuracy (Fig.3).

Fig.3. *C-band AESA radar used in the ZDPSR BYSTRA radiolocation station (source: [3])*

The vast majority of radars are equipped with receiving antennas as well as emitting electromagnetic pulses, which makes them easy targets to track. Passive radars, which do not send their own signals, do not have this handicap, and they receive the radar echo from other sources, which may be, for example, telecommunication transmitters (telephony, radio, terrestrial or satellite TV).

An important element of the modern battlefield is to hide your own forces from the enemy's radar reconnaissance. One way to achieve this goal is to construct your own objects in such a way that their radar echo is as small as possible. The measure used to determine and compare the size of signals in echolocation is the Radar Cross Section (RCS). It describes the measure of the energy that would be reflected from the object in relation to the total energy incident on the object and referred to a unit sphere with an area of 1 m² and perfectly reflecting waves in all directions. Thus, this quantity is described in m² or in dB. Table 3 shows estimated RCS values for sample objects.

Table 3. *Example RCS values for selected objects (source: [4])*

Object	RCS [m²]	RCS [dB]
Bird	0.01	-20
Human	1	0
Motorboat	10	10
Passenger car	100	20
Truck	200	23
corner	20379	43.1

Quality Production Improvement and System Safety: QPI 16 - CZOTO 10   Materials Research Forum LLC
Materials Research Proceedings 34 (2023) 169-177   https://doi.org/10.21741/9781644902691-21

Thus, the basic features affecting the RCS are: the size of the object, its shape and material properties in terms of absorption and reflection of electromagnetic waves. Already today there are constructions in the stealth technology, which, thanks to their construction (specific shape) and the materials used for the outer shells, are able to effectively reduce the RCS, and thus the radar echo, so as to become almost invisible to radars or significantly reduce the detection distance. For example, B-2 Spirit heavy bomber with a wingspan of 52 m has RCS approx. 0.1 m² [5]. Table 4 shows a few selected RCS for modern military aircraft.

Fig.4. B-2 Spirit heavy bomber, constructed in stealth technology (source: [6])

Table 4. RCS for selected modern models of military aircraft (source: [7])

Aircraft model	RCS	Rmax
	[m²]	[km]
F-15C; Su-27	10–15	450–600
MIG-29	5	370–450
F/A-18C	3	330–395
F-16C	1,2	260–310
Su-47	0.3	185–220
F-18E	0.1	140–170
F-35A	0.0015	50–60

The authors of the article are also working on reducing RCS. They focus on material properties and, in this article, they tested and compared sets of paint coatings that were selected as potential radar echo reducing. Wave absorbers were sought, which, as thin varnish coatings, would attenuate electromagnetic waves in the C, X, Ku ranges, which are used in radiolocation, as previously mentioned.

Radiation energy in the paint coating is decomposed into 3 components: reflection, absorption and multiple reflection [8].

$$SE=SE_R+SE_A+SE_{MR} \tag{1}$$

Due to the study of attenuation, only the absorption part of the SEA was considered in this article. The presence of magnetic and electric dipoles enables the conversion of electromagnetic energy into energy dissipated in the layer (e.g. heat), which contributes to increasing electromagnetic attenuation.

Paint coatings were tested - a lossy, absorbing layer with a thickness of s and the assumed reflection coefficient $\Gamma = 0$ and wave impedance $Zn = 1$, applied to metal plates (reflective layer). Thus, the model system can be described by the equation:

$$\Gamma = \frac{Z_n - 1}{Z_n + 1} \qquad (2)$$

For a plane wave it was assumed:

$$Z_n = \sqrt{\frac{\mu_r}{\varepsilon_r}} \cdot \tanh \left(j \cdot \frac{2 \cdot \pi}{\lambda} \cdot \sqrt{\varepsilon_r \cdot \mu_r} \cdot s \right) \qquad (3)$$

λ – wavelength in free space

$\varepsilon_r(f) = \varepsilon'(f) - j\,\varepsilon''(f)$ – complex permittivity

$\mu_r(f) = \mu'(f) - j\,\mu''(f)$ – complex magnetic permeability

j – imaginary part $\sqrt{-1}$

Two different materials were tested as absorbers:

- BASF's spherical carbonyl iron, designated EB, with a particle size of 3 - 4 μm (Table 5 contains basic properties)

- Thin-walled empty microspheres manufactured by 3M marked as S22 with particle size D50 - 35 μm (Table 6 contains basic properties).

Table 5. Properties of the carbonyl iron used in the study

Type	Form	Size	Bulk density [kg/m³]	Fe atom content [%]	C atom content [%]	N atom content [%]	O atom content [%]
BASF EB	powder	3 ÷ 4 μm with a wider grain size distribution	4 200	> 97,3	< 1	< 1	< 0,4

Table 6. Properties of the glass microspheres used in the study

Type	Form	Particle size D$_{50}$ [μm]	Density [kg/m³]	Scrush test [MPa]	Minimum survivability [%]
3M S22	Powder	35	0.22	2.76	80

On the basis of these absorbers, varnish pastes were made, which were used to paint aluminum plates with dimensions of 300 × 300 × 4 mm.

In the case of carbonyl iron, Epikote 828 epoxy resin and Ancamine 1618 hardener were used as a binder. The composition also included additives produced by Amepox Microelectronic Co Ltd.: AX-R, reducing viscosity, and AX-S - increasing flexibility.

For 3M S22 glass microspheres, XX0606 polyurethane resin was selected as a binder, with XPH80002 hardener and BYK 969 dispersant additive.

The share of carbonyl iron as an absorber in the varnish paste was 75% for the first series of samples and 80% for the second series of samples. The thickness of the applied paste layer is 05, 1, 1.5, 2.0 mm for each series, respectively.

On the other hand, in the samples with the participation of 3M S22 glass microspheres, 3 series of samples with the absorber content of 20.5%, 35.0 and 54.0%, respectively, were made. The thickness of the applied paint layer was for 20.5% 4w - 0.814mm, 6w - 1.041mm, 8w - 1.273 mm.

For 35% 4w - 0.987mm, 6w - 1.416mm, 8w - 1.803mm and for 54% 4w - 1.457mm, 6w - 1.798mm, 8w - 2.278mm.

The tests consisted in measuring the radar signal reflected from the samples with coatings and comparing them to the reference sample - without the varnish coating. The tests were carried out on a measuring stand consisting of a reflectometer operating in the 4-18 GHz range and a control computer (Figure 5).

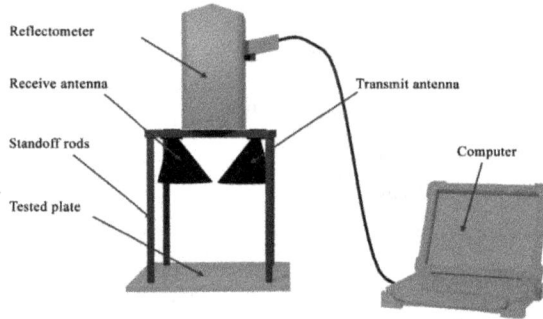

Fig.5. *Scheme of the measuring station (description of individual elements in the diagram).*

The results of attenuation measurements for carbonyl iron as an absorber are presented in Table 7, while the graphs in Fig.6 and Fig.7 show the attenuation tendency depending on the thickness of the paint layer with the absorber.

Table 7. *Results of attenuation measurements for EB carbonyl iron*

Frequency [GHz]	content 75%				content 80%			
	0.5mm	1.0mm	1.5mm	2.0mm	0.5mm	1.0mm	1.5mm	2.0mm
4	-4.5	-5.0	-4.5	-18.0	-1.7	-2.6	-5.0	-6.0
6	-4.5	-7.0	-7.0	-12.5	-1.9	-3.0	-7.5	-8.0
8	-7.0	-12.0	-12.0	-9.0	-2.1	-5.0	-11.5	-14.0
10	-7.0	-13.0	-11.0	-7.0	-2.3	-7.0	-12.5	-12.0
12	-5.0	-10.0	-7.0	-5.0	-2.5	-9.0	-11.0	-6.0
14	-4.5	-5.0	-4.5	-4.0	-2.8	-10.0	-7.0	-5.0
16	-4.0	-4.5	-4.0	-4.0	-5.0	-10.0	-6.0	-5.0
18	-4.0	-4.5	-4.0	-4.0	-6.0	-10.0	-6.0	-5.0

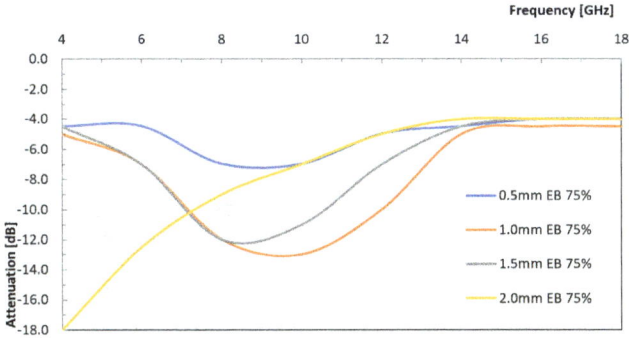

Fig.6. *Influence of thickness on attenuation - EB absorber, 75% share*

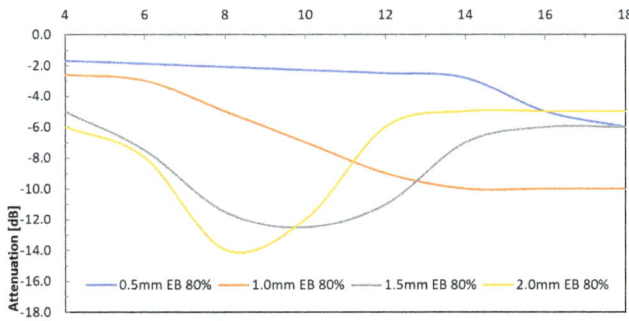

Fig.7. *Influence of thickness on attenuation - EB absorber, 80% share*

For 3M S22 glass microspheres, the results are presented in Table 8, and the damping tendency in Fig.8, Fig.9 and Fig.10.

The thickness of the applied paint layer was for 20.5% 4w - 0.814mm, 6w - 1.041mm, 8w - 1.273 mm. For 35% 4w - 0.987mm, 6w - 1.416mm, 8w - 1.803mm and for 54% 4w - 1.457mm, 6w - 1.798mm, 8w - 2.278mm.

Table 8. *Results of attenuation measurements for S22 glass microspheres*

Frequency [GHz]	Content 3M S22 20.5%			Content 3M S22 35%			Content 3M S22 54%		
	4w 0.814 mm	6w 1.041 mm	8w 1.273 mm	4w 0.987 mm	6w 1.416 mm	8w 1.803 mm	4w 1.457 mm	6w 1.798 mm	8w 2.278 mm
4.0	-0.282	-0.200	-0.380	-0.252	-0.222	-0.371	-0.256	-0.310	-0.481
6.0	-0.706	-0.096	-0.176	-0.097	-0.120	-0.165	-0.112	-0.733	-0.179
8.0	-0.074	-0.100	-0.183	-0.096	-0.141	-0.257	-0.142	-0.228	-0.172
10.0	-0.086	-0.153	-0.313	-0.204	-0.215	-0.335	-0.182	-0.282	-0.289
12.0	-0.102	-0.269	-0.376	-0.177	-0.362	-0.454	-0.267	-0.484	-0.506

14.0	-0.073	-0.200	-0.312	-0.117	-0.310	-0.596	-0.244	-0.479	-0.656
16.0	-0.459	-0.671	-0.891	-0.519	-0.787	-0.657	-0.830	-0.650	-0.403
18.0	-0.538	-0.746	-0.948	-0.554	-0.711	-1.250	-0.849	-0.989	-1.726

From the presented results, it can be seen that the absorbers used show attenuation properties in the range of the tested frequencies (4-18GHz). It should be added that carbonyl iron has 10 times better damping properties than S22 glass microspheres, achieving a reduction of even -18 dB for selected frequencies and coating thicknesses (4GHz, 2mm, 75% EB) compared to a sample without a paint coating with an absorber. The S22 glass microsphere absorber reached a maximum value of -1.73 dB for a share of 54% and a coating thickness of 2.278 mm. At the same time, it can be stated that the maximum attenuation for both absorbers depends on the thickness of the layer - usually for a thicker layer there is greater attenuation and the type, share and thickness also affect the position of the maximum attenuation in the spectrum of the tested frequencies.

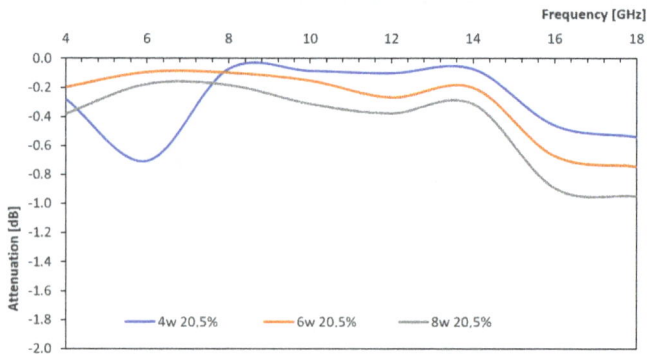

Fig.8. *Influence of thickness on attenuation - 3M S22 absorber, share 20.5%*

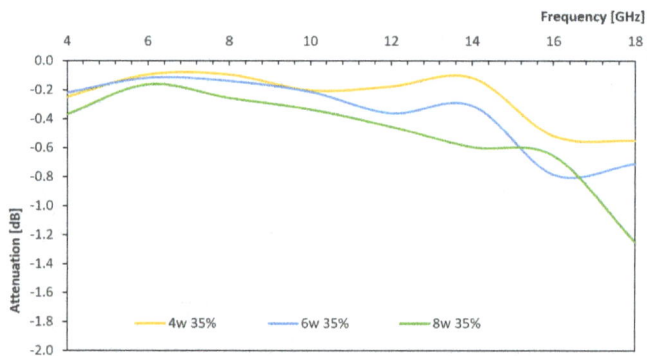

Fig.9. *Influence of thickness on attenuation - 3M S22 absorber, share 35%*

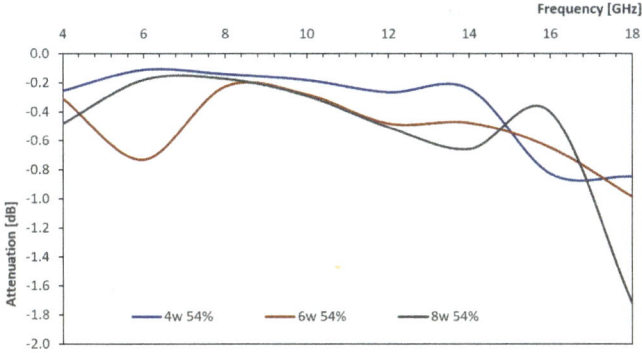

Fig.10. *Influence of thickness on attenuation - 3M S22 absorber, share of 54%*

Conclusions

Carbonyl iron as an absorber provides good attenuation properties (-18dB), which cannot be stated in relation to S22 glass microspheres, for which the highest attenuation was -1.73 dB. At the same time, good attenuation properties for carbonyl iron occur in a fairly narrow frequency range and do not cover the C, X, Ku bands, but occur for different frequency values depending on the thickness of the layer and the share of the absorber. In connection with the above, it can be assumed that attenuation in a wider range could be obtained by using multi-layer systems with different shares of absorbers, as well as different thicknesses of individual layers forming the paint coating. In addition, further research with greater precision and a wider frequency range would be advisable to confirm the trends of "shifting" of the maximum attenuation.

In conclusion, carbonyl iron has great potential to be used as a radar absorber in paint coatings in camouflage applications.

References

[1] Based on IEEE, https://www.radartutorial.eu/07.waves/wa04.po.html, 2023-01-23

[2] https://polska-zbrojna.pl/Mobile/ArticleShow/30589, 2023-01-23

[3] Materiały producenta – PIT-RADWAR, https://www.pitradwar.com/oferta/425,zdolna-do-przerzutu-stacja-radiolokacyjna-bystra#, 2023-01-20

[4] M. Skolnik, « Introduction to radar systems », wydanie drugie, McGraw-Hill Inc, 1980, strona 44.

[5] D. Richardson: Stealth Warplanes, Zenith Press, 2001

[6] https://www.af.mil/About-Us/Fact-Sheets/Display/Article/104482/b-2-spirit/

[7] Prasad N.E., Wanhill R.J.H., Aerospace Materials and Material Technologies Volume 1: Aerospace Materials, Springer Nature, 2016, s. 1-343. https://doi.org/10.1007/978-981-10-2143-5

[8] Sankaran S., Deshmukh K., Ahamed M.B., Khadheer Pasha S.K., Recent advances in electromagnetic interference shielding properties of metal and carbon filler reinforced flexible polymer composites: A review, Composites Part A: Applied Science and Manufacturing, t. 114, 2018, s. 49-71. https://doi.org/10.1016/j.compositesa.2018.08.006

Quality Production Improvement and System Safety: QPI 16 - CZOTO 10   Materials Research Forum LLC
Materials Research Proceedings 34 (2023) 178-185   https://doi.org/10.21741/9781644902691-22

Laser Treatment Technique for Boiling Heat Transfer Application

ORMAN Łukasz J.[1, a *], RADEK Norbert[1, b], PIETRASZEK Jacek[2, c],
HONUS Stanislav[3] and DZIEDZIC Joanna[4, d]

[1]Kielce University of Technology, al. Tysiaclecia P.P.7, 25-314 Kielce, Poland

[2]Cracow University of Technology, Al. Jana Pawła II 37, 31-864 Cracow, Poland

[3]Technical University of Ostrava, 17. listopadu 2172/15, 708 00 Ostrava-Poruba, Czech Republic

[4]Jan Kochanowski University of Kielce, ul. Uniwersytecka 15, 25-406 Kielce, Poland

[a]orman@tu.kielce.pl, [b]norrad@tu.kielce.pl, [c]jacek.pietraszek@mech.pk.edu.pl,
[d]s145704@ujk.edu.pl

Keywords: Boiling, Laser Treatment, Surface Treatment

Abstract. The article discusses the use of laser treatment to produce surfaces that enhance boiling heat transfer. The laser beam has been applied on copper substrates to generate longitudal fins of regular geometry. The altered morphology pattern enables to dissipate higher heat flux values than the smooth, untreated surface – in the case of distilled water and ethyl alcohol as boiling agents. The issue of modeling nucleate pool boiling heat flux on laser processed surfaces is discussed in the paper considering two selected models of boiling.

Introduction

Boiling is a highly complex phenomenon. It allows to exchange considerable heat fluxes at small temperature differences and – as a result – has a significant practical application potential. It is commonly known that the application of additional layers or surface modifications can increase the heat flux values being exchanged. Piasecka et al. [1, 2] considered laser texturing as the method of enhancement for flow boiling conditions, while Kaniowski and Pastuszko [3, 4] focused on the pool boiling mode. In [3] water boiling on samples with microchannels containing long fins of 0.2 - 0.5 mm depth was analyzed. It turned out that the largest heat flux was almost 2.5 times bigger then the in the case of the smooth surface. A different medium (namely, FC-72) was used on the same type of microstructures in [4]. Similarly, enhancement of boiling was observed. The heat transfer coefficient was higher by over 500% - reaching values comparable to those obtained when nanotubes are applied. Pranoto et al. [5] tested sub-cooled boiling of HFE-7100 and the influence of the geometrical shape of pin fins on heat flux. The authors studied surfaces containing 196 and 144 microfins and reported that rectangular fins performed better than the circular ones. Moreover, the flow resistance of both types of pin fins was compared and the rectangular ones show ca. 1.06 times higher resistance, which leads to smaller bubble departure frequency. It was concluded that both kinds of microstructures enhanced boiling due to agitated movement caused by evaporation as well as the rewetting process at bubble detachment. Chinnov et al. [6] experimentally analyzed water boiling on heaters with different patterns of hydrophobic spots made on the copper surface as well as the surface covered with a matrix of micrococoons made of silicon oxide nanowires. It was reported that biphillic heaters showed better performance that non-biphillic ones. Moreover, the morphological texture of the surface, the shape of the hydrophobic areas or their production method did not have a considerable influence of the heat exchange process. However, the heat flux values depend on the hydrophobic spot size, the distance between the spots as well as the number of those spots. Belyaev et al. [7] tested the augmentation of flow boiling caused by modifying the inner surface of the vertical channel with the laser. The laser treatment developed formations of

Quality Production Improvement and System Safety: QPI 16 - CZOTO 10   Materials Research Forum LLC
Materials Research Proceedings 34 (2023) 178-185   https://doi.org/10.21741/9781644902691-22

various heights and diameters. It was stated that the treatment of the surface with laser improved heat transfer conditions. The largest impact was recorded at the reduced pressure of 0.43 with the maximal increase in the critical heat flux of over twenty percent. Considerable rise in heat transfer coefficient was also reported.

Due to the fact that there exist large opportunities offered by the laser treatment the present paper aims to address the issue of surface modification with laser and provide data on the boiling performance of the laser treated heater with the focus on modeling boiling heat flux. The problem is especially important for the proper design of heat exchangers produced with the laser technology.

The technique presented here, which utilizes lasers to modify the surface morphology affecting heat transfer, can also find application in other laser processing cases where surface morphology is a crucial process or technological factor [8-10]. Changing the characteristics of the surface layer is a frequently employed solution [11-13], with image analysis serving as a vital diagnostic tool [14-16]. Such modifications are of interest not only in the energy sector [17] but also in hydraulic power systems [18,19] and the broader field of material engineering [20-22]. Modifying the surface layer morphology significantly contributes to reducing machine part wear [23-25], improving the quality of products in the automotive [26,27] and railway industries [28,29], which, in turn, serves as inspiration for new quality control methods [30-32] applied in the automotive [33,34] and metal industries [35,36]. Such quality enhancements substantially limit the range of potential failure scenarios and their consequences [37-39], enabling the effective implementation of lean manufacturing principles [40-42].

Samples and Experimental Method
The preparation of the samples begins with cutting copper discs – 3 cm in diameter and 3 mm in height. Afterwards, the laser treatment occurs with the use of SPI G3.1 SP20P pulsed fiber laser with the focal spot size of 35 μm and power of 20 W. The pattern of the longitudal fins is made on the specimen within ca. 2 hours according to the design drawn using the appropriate software. Fig. 1 presents the sample with microfins.

Fig. 1. Laser treated sample with longitudal fins.

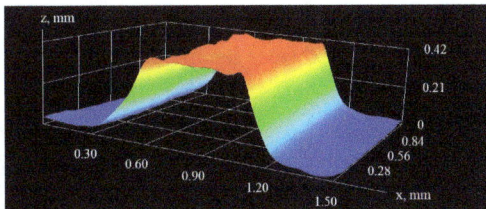

Fig. 2. Optical microscope image of the laser made microfin.

Quality Production Improvement and System Safety: QPI 16 - CZOTO 10   Materials Research Forum LLC
Materials Research Proceedings 34 (2023) 178-185   https://doi.org/10.21741/9781644902691-22

The laser treatment evaporated some of the material and, thus, the fins and grooves are created. However, the shape of the fins is not rectangular, but they are thinner at the top and broader at the base of the sample. Fig. 2 presents the optical microscope image of the fin, which confirms the statement regarding the shape of the produced element. It also needs to be added that the laser beam produces rough surface at the base of the sample (between the fins). It is the additional advantage for boiling heat transfer enhancement due to the fact that this type of surface morphology provides additional nucleation sites (locations where the vapor bubbles can develop). This leads to more heat being exchanged, especially at small temperature differences, due to large heat of vaporization and the motion of the bubbles that improve convective heat transfer.

The shape of the fin that becomes thinner at the tip is favorable because the vapor bubbles generated at the base of the sample can flow into the pool of liquid (and later to the surface) more easily. Consequently, heat transfer is further improved. This advantageous shape of the microfins can be naturally obtained due to the laser beam interaction with the material of the sample and would otherwise be difficult to obtain using mechanical methods. The drawback of the laser technique might be the time-consuming process of samples' preparation.

The tests are done on the experimental set-up, which consists of the electric heater. The sample is soldered to this heater and above it the pool of liquid is located. Two kinds of boiling liquids are considered by the authors: distilled water and ethyl alcohol. The rising electric power supplied to the heater leads to increasing the temperature of the sample and, consequently, boiling curves can be prepared as a dependence of the heat flux (q) vs. temperature difference (θ), calculated at the difference between the sample temperature and the saturation temperature of the boiling fluid.

Results and Discussion

The test result of the laser processed surface will be presented for the sample of microfin height of 0.25 mm, microfin width of 0.5 mm and groove width of 1.5 mm. The process of bubbles' formation on the sample as well as their movement in the pool of water has been shown in Fig. 3. The vapor bubbles form on the surface and typically merge together before a departure into the liquid. As can be seen the process is very intense and heat exchange occurs both due to a change of phase (involving large values of the heat of vaporization) and convection mode of heat transfer.

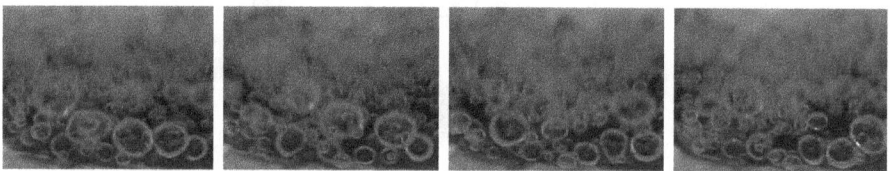

***Fig. 3.** Boiling phenomenon on the laser treated sample – time interval between each picture 0.02 s.*

The heat transfer analysis of the sample has been presented and discussed by the authors in [43]. It was observed that the laser treatment provided significant augmentation of heat flux in relation to the untreated (smooth) reference surface. However, an important element for the design of heat exchangers is the proper calculation of heat flux based on the geometrical parameters of the surface morphology and its material properties. The models and correlations can be based on a number of concepts [44] and can show a different level of determination precision. In order to assess and compare the correctness and applicability of selected models of boiling, two correlations have been selected from the literature – one proposed by Smirnov and Afanasiev [45] and the other by Nishikawa et al. [46]. Fig. 4 and 5 present the comparison of the calculation results according to the above-mentioned models with the experimental data from [43] for water and ethanol, respectively.

Analysis of the above graphs indicates that the model proposed by Nishikawa et al. [46] provides calculation results which are much higher than the actual experimental data. On the other hand, the model by Smirnov and Afanasiev [45] is accurate, but only in the range of low temperature differences. Consequently, in order to obtain more precise calculation results a new model for the laser processed samples should be considered.

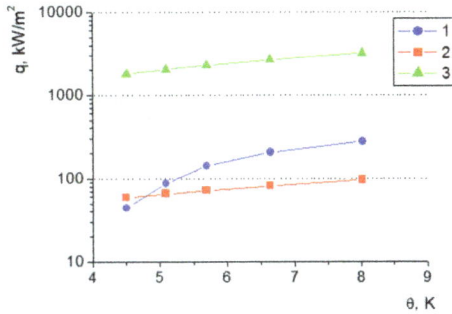

Fig. 4. *Comparison of the experimental data and model calculations for water: 1 – experimental data, 2 – calculation results according to the equation by Smirnov and Afanasiev [45], 3 – calculation results according to the equation by Nishivawa et al. [46].*

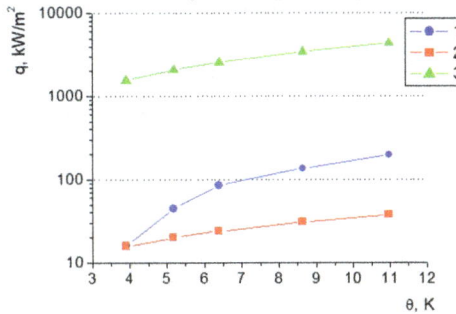

Fig. 5. *Comparison of the experimental data and model calculations for ethanol: 1 – experimental data, 2 – calculation results according to the equation by Smirnov and Afanasiev [45], 3 – calculation results according to the equation by Nishivawa et al. [46].*

It needs to be emphasized that all laser processed specimens tested by the authors in [43] largely outperformed the reference surface (the one without any modifications). Furthermore, the most efficient were the samples of the highest height of the longitudal fins as well as of the roughest surface at the base (two surface roughness values were considered). The authors reported that the biggest augmentation could have been observed for small temperature differences – both for the influence regarding the height of the fin elements and for the surface roughness.

Another vital problem that needs to be addressed in the future studies of the authors is the issue of bubble dynamics during boiling on laser modified heaters. As mentioned in [5] based on visual studies of dielectric liquid boiling, in the case of pin fins there is a bigger probability of occurrence at a cavity site, which increases nucleation site density.

Summary and Conclusions

The laser technology enables to alter the morphology of the heat exchanging surfaces. As a result, improved conditions for boiling might occur, which leads to higher heat fluxes being exchanged. It is an advantageous phenomenon due to the increase in the overall efficiency of heat exchangers. The problem might only be with the proper calculation of the heat flux values based on the geometrical and material properties of the heater surface. The selected models adopted from literature proved not particularly accurate and a new one needs to be developed to address this issue.

References

[1] M. Piasecka, K. Strąk. Influence of the surface enhancement on the heat transfer in a minichannel, Heat Transfer Engineering 40(13-14) (2019) 1162-1175. https://doi.org/10.1080/01457632.2018.1457264

[2] M. Piasecka, K. Strąk, B. Grabas. Vibration-assisted laser surface texturing and electromachining for the intensification of boiling heat transfer in a minichannel, Archives of Metallurgy and Materials 62(4) (2017) 1983-1990. https://doi.org/10.1515/amm-2017-0296

[3] R. Kaniowski, R. Pastuszko. Pool boiling of water on surfaces with open microchannels, Energies 14(11) (2021) art. 3062. https://doi.org/10.3390/en14113062

[4] R. Kaniowski, R. Pastuszko. Boiling of FC-72 on surfaces with open copper microchannel, Energies 14(21) (2021) art. 7283. https://doi.org/10.3390/en14217283

[5] I. Pranoto, M.A. Rahman, P.A.P. Mahardhika. Pool boiling heat transfer performance and bubble dynamics from pin fin-modified surfaces with geometrical shape variation, Energies 15(5) (2022) 1847. https://doi.org/10.3390/en15051847

[6] E.A. Chinnov, S.Y. Khmel, V.Y. Vladimirov, A.I. Safonov, V.V. Semionov, K.A. Emelyanenko, A.M. Emelyanenko, L.B. Boinovich. Boiling beat transfer enhancement on biphilic surfaces, Energies 15(19) (2022) 7296. https://doi.org/10.3390/en15197296

[7] A.V. Belyaev, A.V. Dedov, N.E. Sidel'nikov, P. Jiang, A.N. Varava, R. Xu. Flow boiling heat transfer intensification due to inner surface modification in circular mini-channel, Water 14(24) (2022) 4054. https://doi.org/10.3390/w14244054

[8] P. Kurp et al. The influence of treatment parameters on the microstructure, properties and bend angle of laser formed construction bars, Arch. Metall. Mater. 61 (2016) 1151-1156. https://doi.org/10.1515/amm-2016-0192

[9] P. Kurp, D. Soboń. The influence of laser padding parameters on the tribological properties of the Al2O3 coatings, METAL 2018 27th Int. Conf. Metall. Mater. (2018) 1157-1162. ISBN 978-808729484-0

[10] P. Kurp, H. Danielewski Metal expansion joints manufacturing by a mechanically assisted laser forming hybrid method – concept, Technical Transactions 119 (2022) art.e2022008. https://doi.org/10.37705/TechTrans/e2022008

[11] D. Klimecka-Tatar. Electrochemical characteristics of titanium for dental implants in case of the electroless surface modification, Arch. Metall. Mater. 61 (2016) 923-926. https://doi.org/10.1515/amm-2016-0156

Quality Production Improvement and System Safety: QPI 16 - CZOTO 10   Materials Research Forum LLC
Materials Research Proceedings 34 (2023) 178-185   https://doi.org/10.21741/9781644902691-22

[12] D. Nowakowski et al. Application of machine learning in the analysis of surface quality - the detection the surface layer damage of the vehicle body, METAL 2021 - 30th Int. Conf. Metallurgy and Materials (2021), Ostrava, Tanger 864-869. https://doi.org/10.37904/metal.2021.4210

[13] B. Antoszewski et al. Assessment of technological capabilities for forming al-c-b system coatings on steel surfaces by electrospark alloying method, Materials 14 (2021) art.739. https://doi.org/10.3390/ma14040739

[14] A. Gądek et al. Application of computer-aided analysis of an image for assessment of reinforced polymers structures, Polymers 51 (2006) 206-211. https://doi.org/10.14314/polimery.2006.206

[15] A. Gadek-Moszczak, P. Matusiewicz. Polish stereology - A historical review, Image Analysis and Stereology 36 (2017) 207-221. https://doi.org/10.5566/ias.1808

[16] I. Jastrzębska, A. Piwowarczyk. Traditional vs. Automated Computer Image Analysis—A Comparative Assessment of Use for Analysis of Digital SEM Images of High-Temperature Ceramic Material, Materials 16 (2023) art. 812. https://doi.org/10.3390/ma16020812

[17] K. Trzewiczek et al. Evaluation of the state for the material of the live steam superheater pipe coils of V degree. Adv. Mater. Res. 874 (2014) 35-42. https://doi.org/10.4028/www.scientific.net/AMR.874.35

[18] G. Filo, E. Lisowski, M. Domagała, J. Fabiś-Domagała, H. Momeni. Modelling of pressure pulse generator with the use of a flow control valve and a fuzzy logic controller, AIP Conf. Proc. 2029 (2018) art.20015. https://doi.org/10.1063/1.5066477

[19] M. Domagala et al. The Influence of Oil Contamination on Flow Control Valve Operation, Mater. Res. Proc. 24 (2022) 1-8. https://doi.org/10.21741/9781644902059-1

[20] A. Dudek, B. Lisiecka, R. Ulewicz. The effect of alloying method on the structure and properties of sintered stainless steel, Archives of Metallurgy and Materials 62 (2017) 281-287. https://doi.org/10.1515/amm-2017-0042

[21] R. Ulewicz et al. Structure and mechanical properties of fine-grained steels, Period. Polytech. Transp. Eng. 41 (2013) 111-115. https://doi.org/10.3311/PPtr.7110

[22] D. Klimecka-Tatar, M. Ingaldi. Assessment of the technological position of a selected enterprise in the metallurgical industry, Mater. Res. Proc. 17 (2020) 72-78. https://doi.org/10.21741/9781644901038-11

[23] S. Marković et al. Exploitation characteristics of teeth flanks of gears regenerated by three hard-facing procedures, Materials 14 (20210 art. 4203. https://doi.org/10.3390/ma14154203

[24] M. Krynke et al. Maintenance management of large-size rolling bearings in heavy-duty machinery, Acta Montan. Slovaca 27 (2022) 327-341. https://doi.org/10.46544/AMS.v27i2.04

[25] P. Regulski, K.F. Abramek The application of neural networks for the life-cycle analysis of road and rail rolling stock during the operational phase, Technical Transactions 119 (2022) art. e2022002. https://doi.org/10.37705/TechTrans/e2022002

[26] D. Siwiec, R. Dwornicka, A. Pacana. Improving the non-destructive test by initiating the quality management techniques on an example of the turbine nozzle outlet, Materials Research Proceedings 17 (2020) 16-22. https://doi.org/10.21741/9781644901038-3

Quality Production Improvement and System Safety: QPI 16 - CZOTO 10   Materials Research Forum LLC
Materials Research Proceedings 34 (2023) 178-185   https://doi.org/10.21741/9781644902691-22

[27] K. Czerwinska et al. Improving quality control of siluminial castings used in the automotive industry, METAL 2020 29th Int. Conf. Metall. Mater. (2020) 1382-1387. https://doi.org/10.37904/metal.2020.3661

[28] T. Lipiński, R. Ulewicz. The effect of the impurities spaces on the quality of structural steel working at variable loads, Open Eng. 11 (2021) 233-238. https://doi.org/10.1515/eng-2021-0024

[29] K. Czerwińska, A. Piwowarczyk. The use of combined quality management instruments to analyze the causes of non-conformities in the castings of the cover of the rail vehicle bearing housing, Prod. Eng. Arch. 28 (2022) 289-294. https://doi.org/10.30657/pea.2022.28.36

[30] M. Ingaldi. Overview of the main methods of service quality analysis, Production Engineering Archives 18 (2018) 54-59. https://doi.org/10.30657/pea.2018.18.10

[31] R. Ulewicz, F. Nový. Quality management systems in special processes, Transp. Res. Procedia 40 (2019) 113-118. https://doi.org/10.1016/j.trpro.2019.07.019

[32] D. Siwiec, A. Pacana. Method of improve the level of product quality, Prod. Eng. Arch. 27 (2021) 1-7. https://doi.org/10.30657/pea.2021.27.1

[33] R. Ulewicz. Outsorcing quality control in the automotive industry, MATEC Web of Conf. 183 (2018) art.03001. https://doi.org/10.1051/matecconf/201818303001

[34] A. Pacana et al. Analysis of quality control efficiency in the automotive industry, Transp. Res. Procedia 55 (2021) 691-698. https://doi.org/10.1016/j.trpro.2021.07.037

[35] M. Nowicka-Skowron, R. Ulewicz. Quality management in logistics processes in metal branch, METAL 2015 24th Int. Conf. Metall. Mater. (2015) 1707-1712. ISBN 978-8087294628

[36] R. Ulewicz et al. Logistic controlling processes and quality issues in a cast iron foundry, Mater. Res. Proc. 17 (2020) 65-71. https://doi.org/10.21741/9781644901038-10

[37] J. Fabiś-Domagała, G. Filo, H. Momeni, M. Domagała. Instruments of identification of hydraulic components potential failures, MATEC Web of Conf. 183 (2018) art.03008. https://doi.org/10.1051/matecconf/201818303008

[38] K. Knop et al. Evaluating and Improving the Effectiveness of Visual Inspection of Products from the Automotive Industry, Lecture Notes in Mechanical Engineering (2019) 231-243. https://doi.org/10.1007/978-3-030-17269-5_17

[39] J. Fabis-Domagala, M. Domagala. A Concept of Risk Prioritization in FMEA of Fluid Power Components, Energies 15 (2022) art.6180. https://doi.org/10.3390/en15176180

[40] A. Maszke, R. Dwornicka, R. Ulewicz. Problems in the implementation of the lean concept at a steel works - Case study, MATEC Web of Conf. 183 (2018) art.01014. https://doi.org/10.1051/matecconf/201818301014

[41] R. Ulewicz, M. Ulewicz. Problems in the Implementation of the Lean Concept in the Construction Industries, LNCE 47 (2020) 495-500. https://doi.org/10.1007/978-3-030-27011-7_63

[42] R. Ulewicz et al. Implementation of Lean Instruments in Ceramics Industries, Manag. Sys. Prod. Eng. 29 (2021) 203-207. https://doi.org/10.2478/mspe-2021-0025

Quality Production Improvement and System Safety: QPI 16 - CZOTO 10   Materials Research Forum LLC
Materials Research Proceedings 34 (2023) 178-185   https://doi.org/10.21741/9781644902691-22

[43] Ł.J. Orman, N. Radek, J. Pietraszek, J. Wojtkowiak, M. Szczepaniak. Laser treatment technology of the surface for pool boiling heat transfer enhancement, Materials 16(4) (2023) 1365; https://doi.org/10.3390/ma16041365

[44] M. Krechowicz, A. Krechowicz, A. Risk assessment in energy infrastructure installations by horizontal directional drilling using machine learning, Energies, 14(2) (2021) 289. https://doi.org/10.3390/en14020289

[45] G. Smirnov, B.A. Afanasiev. Investigation of vaporisation in screen wick – capillary structures", in Advances in Heat Pipe Technology, VI Int. Heat Pipe Conference Proceedings (1982) London, 405 - 413.

[46] K. Nishikawa, T. Ito, K. Tanaka. Enhanced heat transfer by nucleate boiling on a sintered metal layer, Heat transfer – Japanese Research 8 (1979), 65 – 81.

Quality Production Improvement and System Safety: QPI 16 - CZOTO 10　　　　Materials Research Forum LLC
Materials Research Proceedings 34 (2023) 186-198　　　　https://doi.org/10.21741/9781644902691-23

Checkpoint Efficiency Analysis Method

CZERWIŃSKA Karolina[1,a], DWORNICKA Renata[2,b] * and PACANA Andrzej[3,c]

[1]Rzeszow University of Technology, Rzeszów, Poland; ORCID: 0000-0003-2150-0963

[2]Cracow University of Technology, Kraków, Poland; ORCID: 0000-0002-2979-1614

[3]Rzeszow University of Technology, Rzeszów, Poland; ORCID: 0000-0003-1121-6352

[a]k.czerwinska@prz.edu.pl, [b]renata.dwornicka@pk.edu.pl, [c]app@prz.edu.pl

Keywords: Quality Management, Mechanical Engineering, Control Point, Efficiency, NDT Tests

Abstract. The use of multifaceted quality analyses contributes to increasing the efficiency of production processes and quality control as part of maintaining competitiveness. The aim of the study was to identify an integrally configured method for analysing the effectiveness of control points in the context of their ranking in terms of given variables. Verification of the model was carried out against the production process of an aluminium casting. The obtained ranking of the detection methods indicated the MT method as the most effective, which was influenced by the significant detection of critical non-conformities. The study observed little difference between the performance parameters of visual inspection and endoscopic testing, which is largely due to the detection of non-conformities of lower significance and the relatively low cost of testing. Further work will relate to the implementation of the model against other processes carried out in the company.

Introduction

One of the key challenges currently facing manufacturing enterprises is the increasing level of market competitiveness. Such conditions force business units to comprehensively meet the changing demands of buyers while ensuring a significant level of flexibility of the processes implemented [1-3]. Both the volatility of demand and the challenges relating to shorter product and technology life cycles, expectations of shorter lead times, satisfactory product quality levels result in the quality of manufacturing processes and systems being the basic parameters for assessing the effectiveness of an organisation's functioning [4-6].

Within the scope of the issue of functioning of production systems, the basic problem of the performed processes is the management and use of all necessary resources to produce the final product from the supplied raw materials [7, 8]. Consequently, it is possible to define the reliability of a production process by the ability of the production system to completely fulfil the production plan by producing fully valuable final products under specified operational conditions and within a specified time frame. Operational conditions include: the correct functioning of machinery, production equipment and maintenance and logistical support infrastructure, information flows (along with measures for their correct execution), decision-making processes, the human factor, and also include the possibility of internal and external risks [9]. A comprehensive reliability analysis should include an assessment of all facets of production systems in the context of their efficiency, including in particular the assessment of quality control activities.

When considering efficiency, reference should be made to the concept of efficiency of action, which is defined on the level of praxeology. In this aspect, the actions undertaken have the attribute of efficiency only when they are characterised by effectiveness, advantageousness and economy [10]. The term construction of efficiency determinants is based on the specification of the relationship between objective, effect and input. An important point is that beneficence indicates

Quality Production Improvement and System Safety: QPI 16 - CZOTO 10   Materials Research Forum LLC
Materials Research Proceedings 34 (2023) 186-198   https://doi.org/10.21741/9781644902691-23

the absolute advantage of a system, while economy indicates the relative advantage. A review of the literature indicates that efficiency is usually equated with effectiveness. However, according to [11, 12], special attention should be paid to the difference in the interpretation of these two concepts, as 'efficiency is concerned with getting things done in the right way, while effectiveness is concerned with getting the right things done. According to the theory of efficient action, in a strict sense efficiency corresponds to economy. Within economic reality, the desired state is a combination of efficiency, economy and expediency. Therefore, it is possible to take actions that are efficient but not economical, or actions that are efficient and yet will be economically damaging or beneficial [13].

An analysis of the literature shows that manufacturing companies tend to focus on introducing quality engineering activities [14-16] or implementing mixed quality and reliability engineering approaches [17-19] to improve efficiency. By limiting improvement activities to quality engineering activities and the combined approach of quality engineering and reliability, the achievable benefits of implementing a multidimensional approach are neglected. Which may be due to:

- the lack of guidance relating to the correlation between the quality intactness of production systems and operational variables,
- lack of data and information necessary to implement a multidimensional assessment of the intactness of production systems,
- lack of awareness among managers of the benefits of a combined approach to assessing the performance of production systems.

The common ground for both approaches to organisational management and quality assurance is an effectively implemented quality control process. For this reason, the main objective of the study is to propose an integrally configured method for the multidimensional analysis of the effectiveness of control points, taking into account the established criteria. The method is based on the integration of selected diagnostic tests from the non-destructive testing (NDT) group with a cycle of analyses relating to the effectiveness and cost, time and reliability of checkpoints. The method allows the total effectiveness of the checkpoints to be indicated together with their ranking.

The proposed method, which falls under the category of organizational methods [20-22], plays a significant role in enhancing product quality across various industries. Its impact extends to sectors such as energy [23-25], machinery manufacturing [26,27], including heavy-duty machinery [28-30], and military equipment [31-33]. These changes have substantial implications for the applied technologies in these domains, encompassing special coatings [34-36], modified functional and technological layers [37], as well as the methods employed for their application [38] and modification [39,40], alongside the selection of suitable substrate materials [41-43]. Implementing such extensive changes across multiple areas necessitates a meticulous approach, utilizing supporting methodologies such as experimental design [44,45], even with non-classical approaches [46,47]. The analytical microscopic techniques [48] as well as image analysis [49] further enhance the efficacy of the implementation process. These methodologies enable researchers and practitioners to gain in-depth insights into the materials, surfaces, and processes involved, facilitating comprehensive improvements. The benefits of the proposed method manifest in numerous ways. Firstly, it contributes to a significant reduction in the wear rate of machine parts [50,51], leading to enhanced durability and longevity. Additionally, it enhances the fatigue resistance of welded joints [52-54] and separation membranes [55-57], thereby improving their performance under demanding operational conditions. The combined impact of these improvements enhances the overall reliability and efficiency of the systems. By integrating the proposed method into the existing production systems, organizations can achieve notable advancements in product quality. Furthermore, by addressing potential failure scenarios and their consequences [58-60], it enables the development of proactive strategies to minimize risks and

Quality Production Improvement and System Safety: QPI 16 - CZOTO 10   Materials Research Forum LLC
Materials Research Proceedings 34 (2023) 186-198   https://doi.org/10.21741/9781644902691-23

optimize operations. The integration of these changes also fosters a more sustainable approach, reducing environmental impact and promoting responsible practices.

Checkpoint Efficiency Analysis Method

The proposed method is divided into three areas: test preparation, analysis of checkpoint indications and analysis of total effectiveness in the context of checkpoint rankings. With the implementation of sequential control (diagnostic-analytical), the method enables research in an area broader than passive control. Fig.1 shows the assumptions of the method.

Stage 1 – selection of the research subject, team of experts and definition of the research objective

Due to the specifics of the detection tests applied in the method, the selection of the test subject should concern the production system within which the production of ferromagnetic alloy castings is realised. The non-destructive tests assumed in the method make it impossible to detect products made of non-ferromagnetic metals and non-metals.

The appointment of an appropriate expert team is a necessary step for the successful implementation of the developed method. The members of the expert team should have a broad knowledge of the object of testing and of the process within which it is produced, as well as experience in carrying out NDT tests included in the method.

The aim of the model's implication should be to improve the selected production process in terms of the quality of manufactured products and optimisation in terms of the selection of inspection methods, their distribution and frequency of application according to the Kaizen concept.

Stage 2 – visual, ultrasonic, endoscopic, magnetic-powder testing

Visual inspection is carried out as part of the initial visual inspection. Preparation includes a thorough familiarisation with the product (e.g. shape, geometry, type of object, material, mass, surface condition). The visual inspection and evaluation of the object informs the identification or not of discontinuities in the object under examination. Identified discontinuities are classified by specifying their number, the severity of the discontinuity, their type, their size and their designation [61,62].

Ultrasonic testing belongs to the volumetric testing group. They consist of introducing ultrasonic waves into the product, which are reflected by the discontinuities, scattered and deflected at the edges of the discontinuities. The test makes it possible to detect cracks, collapses, delaminations, porosity, penetration leaks and other discontinuities within components. The method can also be used to measure the thickness of objects [63,64].

Quality Production Improvement and System Safety: QPI 16 - CZOTO 10   Materials Research Forum LLC
Materials Research Proceedings 34 (2023) 186-198   https://doi.org/10.21741/9781644902691-23

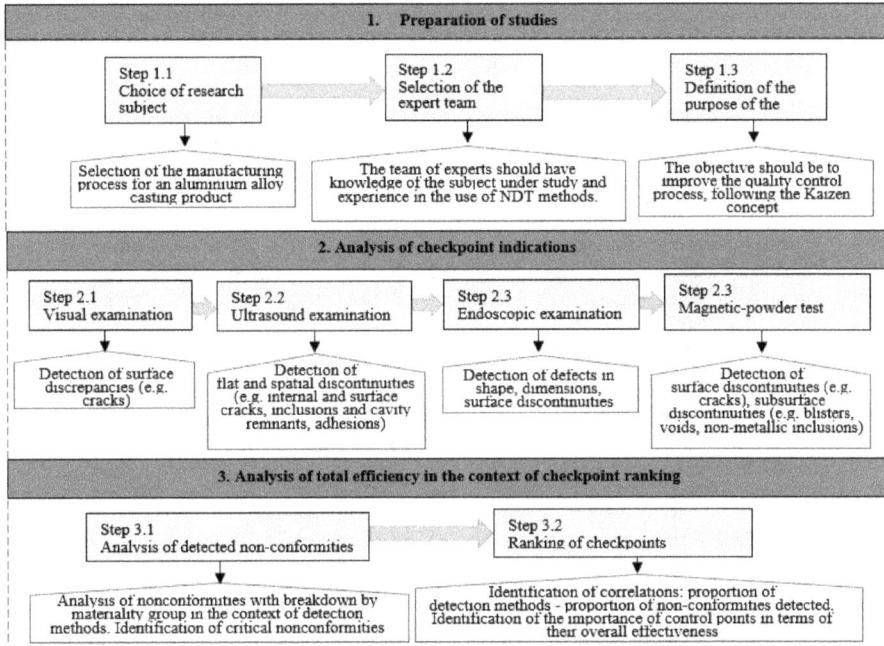

Fig.1. Concept of the checkpoint efficiency analysis method

Endoscopic examinations are based on viewing the internal areas of products using apparatuses that allow the supply of light and optics. A variety of equipment is used for this purpose (e.g. inspection mirrors, magnifying glasses, joint meters, microscopes and video endoscopes). Testing allows the identification of discrepancies caused by dimensional deviations, shape defects, surface discontinuities or operational damage [65,66].

Magnetic-powder testing - when a non-conformity occurs, magnetic flux scattering takes place and the magnetic powder is rearranged in this area. With this method, narrow and shallow subsurface and surface discontinuities of up to about 2 mm can be detected. The magnetic flux source in yoke testing (for hard-to-reach or small object testing) or current generators. The test material used takes the form of oil or water suspensions or coloured or fluorescent dry magnetic powders [67].

Stage 3 – analysis of detected inconsistencies, ranking of control points

After collecting data such as the type of nonconformity detected, the percentage, the cumulative value of the nonconformity, the assignment to one of three groups indicating the frequency of occurrence, and the indication of the detection method under which they were identified. A Pareto-Lorenz diagram taking into account the ABC principle is constructed. This activity allows the correlation between non-conformities and the type of quality control to be analysed in terms of effectiveness, time and cost of the test implementation and the possibility of immediate repair of the detection device. In this step, the type of correlation is identified, which is based on ranks. Table 1 shows the formulae used in the analysis.

Quality Production Improvement and System Safety: QPI 16 - CZOTO 10   Materials Research Forum LLC
Materials Research Proceedings 34 (2023) 186-198   https://doi.org/10.21741/9781644902691-23

The main idea of this stage is to rank the inspection points from the most to the least effective (in terms of the set variables), i.e. to rank the NDT methods. This will enable the optimisation of the quality control process within the production process under study.

Table 1. *Formulas used to identify the correlations analysed*

No.	Correlation studied	Design and markings	
1.	Checkpoint efficiency – relationship between frequency of non-compliance and frequency of inspection methods	$S = CN \cdot (1 - F)$ where: S – checkpoint effectiveness, CN – non-compliance detection rate, F – control method frequency	(1)
2.	Cost effectiveness – relationship between checkpoint efficiency and the cost of a unit detection	$EK = S \cdot (1 - K)$ where: EK – checkpoint cost effectiveness, S – checkpoint efficiency, K – unit detection cost	(2)
3.	Time effectiveness – relationship between checkpoint efficiency and time to complete a unit detection	$EC = S \cdot (1 - Cz)$ where: EC – checkpoint time efficiency, S – checkpoint effectiveness, Cz – unit detection time	(3)
4.	Overall effectiveness – relationship between the efficiency, cost, time per unit detection and reparability of the detection device	$E = S \cdot K \cdot Cz$ where: E – total efficiency, S – checkpoint efficiency, K – unit detection cost, Cz – unit test execution time	(4)

The developed method of analysing the effectiveness of control points makes use of the diversity and complementarity of NDT methods and techniques and quality analysis. The aim of the synergic linking of activities is to optimise the number of inspection points, their distribution (incoming inspection, inter-operational inspection, final inspection) and detection frequency (random inspection, 100% inspection). The method promotes a reduction in the level of diagnostic uncertainty through a step-by-step approach to the detection tests carried out and an increase in the efficiency of quality control throughout the production process.

Verification and Test of the Method
Verification of the universal checkpoint efficiency analysis model was carried out for a production process that had lost the quality stability of the manufactured products. The selected process is implemented in a foundry company located in the south-eastern part of Poland. The test covered production data from 4 months of the year 2021.

Stage 1 – selection of the research subject, team of experts and definition of the research objective

Verification of the method was carried out by means of its implication to the production process of the casting responsible for the water jet inlet used within engine and car technology. The product, with dimensions of 1330 x 600 x 420 and a weight of 66 kg, is cast in AlSi7Mg0.3 alloy. A model of the tested product is shown in Fig. 2.

The company observed a decrease in the quality level of the jet inlet casting after the implementation of reorganisation measures and structural changes to the casting.

Expert team members were selected for their skills and knowledge of the process and detection testing. The following were appointed to the team: a quality control manager, an NDT specialist, a quality control employee and a claims specialist.

The aim of applying the model to the process presented was to propose a course of action for assessing the level of overall efficiency of the individual inspection points, thus optimising the entire quality control process in terms of their efficiency, time and cost effectiveness and the reparability of the machines and detection equipment used.

Quality Production Improvement and System Safety: QPI 16 - CZOTO 10   Materials Research Forum LLC
Materials Research Proceedings 34 (2023) 186-198   https://doi.org/10.21741/9781644902691-23

Stage 2 – visual (VT), ultrasonic (UT), endoscopic (IVT), magnetic-powder (MT) inspection

Quality control of the waterjet inlet production process is carried out in accordance with the control plan. The quality control plan for the production process of the waterjet inlet has been developed taking into account the key parameters of the product, which are specified by the customer and standards and reflected in the technological documentation. The control plan includes information on the location and number of quality gates (preliminary, inter-operational and final control) along with an indication of detection methods (visual, ultrasound, endoscopic, magnetic-powder) after specific technological operations. The plan also contains information on the scope of controlled characteristics: name of product, scope of control, names of measuring and detection instrumentation, measuring/verification method, expected values with parameter tolerance, standards, specified sample sizes, testing frequency, relevant regulations and standards.

The detection results of the indicated methods within the control points were the input data of step 3 of the method.

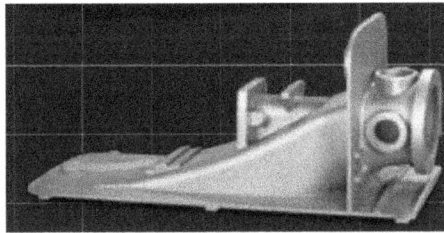

Fig.2. *Water jet inlet model*

Stage 3 – analysis of the detected non-conformities, ranking of control points

In order to identify the correlation between the incidence of quality control during waterjet inlet detection and the proportion of identified non-conformities, a Pareto-Lorenz diagram taking into account the ABC principle was drawn up, showing the critical non-conformities (occurring in the waterjet inlet in terms of their incidence). The resulting diagram is shown in Fig.3.

In Fig.3, the types of incompatibility have the following designations: 1) Presence of shrinkage cavities, 2) Presence of oxides, 3) Presence of rows, 4) Cracks, 5) Edge spalling, 6) Porosity, 7) Blistering, 8) Sintering, 9) Underfilling, 10) Scratching. The diagram also includes the type of detection method that is most likely to detect a particular non-conformity.

The critical non-conformities of waterjet inlet castings, from group "A", are three of the ten listed, i.e. presence of shrinkage cavities, presence of oxides, presence of ripples. These account for 84.7% of the quality problems identified. Further non-conformities from group "A" are detected using ultrasonic and magnetic-powder testing.

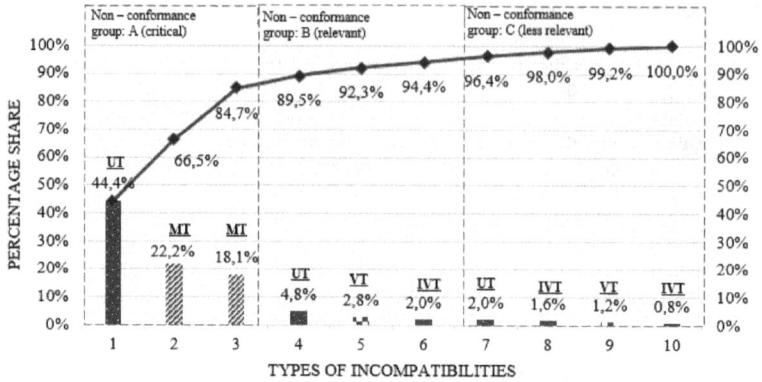

Fig.3. *Pareto-Lorenz diagram taking into account the ABC principle of casting incompatibility*

The results of the analyses of the occurrence of quality control points in the identification of casting non-conformities and the proportion of identified non-conformities by these detection methods are included in Table 2.

Table 2. *Types and number of waterjet inlet casting incompatibilities*

Method used at the checkpoint	Contribution of the control method to the detection of non-compliance	Detection of non-compliance by inspection method
Visual inspection (VT)	20.0%	4.0%
Ultrasound examination (UT)	30.0%	51.2%
Endoscopic examination (IVT)	30.0%	4.4%
Magnetic-powder testing (MT)	20.0%	40.3%

Table 3. *Checkpoint ranking*

Parameter	Value achieved by individual checkpoints		Ranking
Effectiveness of checkpoints	• $S_{VT} = 80.0\% \cdot 4.0\% = 3.2\%$ • $S_{UT} = 70.0\% \cdot 51.2\% = 35.8\%$ • $S_{IVT} = 70.0\% \cdot 4.4\% = 3.1\%$ • $S_{MT} = 80.0\% \cdot 40.3\% = 32.3\%$	(5) (6) (7) (8)	UT > MT > VT >IVT (9)
Cost-effectiveness of checkpoints	• $EK_{VT} = 3.2\% \cdot 93.0\% = 3.0\%$ • $EK_{UT} = 35.8\% \cdot 31.9\% = 11.7\%$ • $EK_{IVT} = 3.1\% \cdot 87.5\% = 2.7\%$ • $EK_{MT} = 32.3\% \cdot 62.8\% = 20.3\%$	(10) (11) (12) (13)	MT >UT> VT >IVT (14)
Time efficiency of checkpoints	• $EC_{VT} = 3.2\% \cdot 84.2\% = 2.7\%$ • $EC_{UT} = 35.8\% \cdot 79.9\% = 28.6\%$ • $EC_{IVT} = 3.1\% \cdot 45.1\% = 1.4\%$ • $EC_{MT} = 32.3\% \cdot 50.6\% = 16.3\%$	(15) (16) (17) (18)	UT >MT> VT >IVT (19)
Total efficiency of checkpoints	• $E_{VT} = 3.2\% \cdot 93.7\% \cdot 84.2\% = 2.53\%$ • $E_{UT} = 35.8\% \cdot 31.9\% \cdot 79.9\% = 9.14\%$ • $E_{IVT} = 3.1\% \cdot 87.5\% \cdot 45.1\% = 1.23\%$ • $E_{MT} = 32.3\% \cdot 62.8\% \cdot 50.6\% = 10.25\%$	(20) (21) (22) (23)	MT>UT> VT >IVT (24)

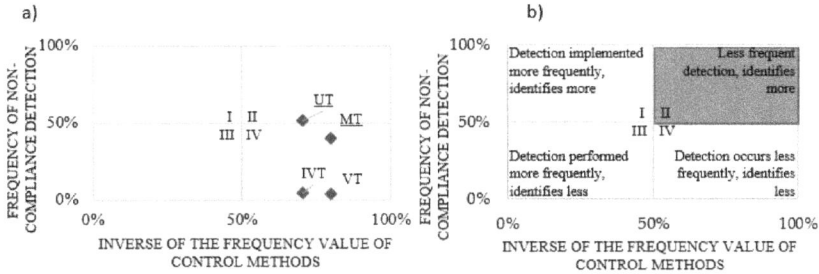

Fig.4. *Matrix diagram indicating (a) effectiveness - relationship between frequency of non-compliance identification and frequency of checkpoint occurrence (b) interpretation of the diagram*

The quality control method most frequently involved in the detection of workpiece nonconformities (3 out of 10 cases) was ultrasound and endoscopic inspection, while the method identifying the most nonconformities is ultrasound (51.2%).

The results of the correlation analysis of the effectiveness of the method (Table 2) are presented in a matrix diagram (Fig. 4). In the diagram, the quadrant in which the individual detection methods should be located is shaded. In order to realise the objective function: the more the better, the function values on the X-axis are expressed as (1-Frequency of control methods).

The location of the control points within the defined planes of the matrix diagram indicates that the control points are located in the 2nd and 4th quadrant. This positioning indicates:

- II quadrant - methods with relatively infrequent detection identifying relatively many nonconformities (ultrasonic method),
- IV quadrant - methods with relatively infrequent detection identifying relatively few nonconformities (magnetic-powder method, visual examination, endoscopic examination).

Parameters such as efficiency, cost-effectiveness, time-effectiveness and total checkpoint efficiency are shown in Table 3.

According to series (9) and (19), the control point with the highest efficiency and at the same time the highest time efficiency is the control point within which ultrasonic detection takes place. The high level of efficiency of the UT method is influenced by the significant level of identification of critical casting non-conformities, while in terms of time efficiency the determinant is the automation of detection by the UT method. The MT method test shows the highest level of cost-effectiveness and the highest level of overall efficiency. The checkpoint within which the endoscopic examination is implemented shows the lowest performance of the analysed indicators. However, this test is necessary due to the complex geometry of the product and thus the lack of possibility for effective detection.

The presented concept of checkpoint efficiency analysis is not only concerned with the identification of a group of critical nonconformities, but also allows the monitoring of the total checkpoint efficiency level. Such an approach makes it possible to identify the detection methods with the highest level of efficiency (i.e. those occurring relatively rarely and detecting a significant number of nonconformities with particular emphasis on critical defects). The proposed method also makes it possible to optimise the placement of checkpoints and the frequency of their use, taking into account the cost and time of detection. Checkpoints with low cost-effectiveness can, for example, be transformed into 100 per cent inspection.

Conclusions

Continuous supervision and monitoring of the processes carried out in a production enterprise, together with quality control, are the key to success for any organisation wishing to maintain a competitive position in an ever-changing market. The aim of the developed method is to propose an integrally configured, multidimensional analysis of the effectiveness of control points, taking into account the assumed criteria.

The checkpoint efficiency analysis method presented in the study allows for the ranking of detection methods in terms of effectiveness, cost efficiency, time efficiency, total efficiency. This makes it possible to manage quality by improving checkpoints through their optimisation, relocation, application of selective control, which influences the maintenance of stability of the production process. As part of the study, a model test was carried out. By verifying the model against the waterjet inlet production process and the quality control points within it, it was shown that the magnetic powder method has the highest level of cost efficiency and the highest level of overall efficiency - thus representing a key control point influencing the level of product quality. The endoscopic test, on the other hand, shows the lowest performance of the indicators analysed. However, this testing is necessary due to the complex geometry of the product and thus the inability of other detection methods to detect it effectively.

Due to the ever-increasing requirements for quality, price, reliability and the relatively short period of time in which components are available on the automitiv market, the issue is important and topical.

Further research directions will concern the implementation of a model for the pro-quality analysis of the company's casting processes, the aim of which is to ensure a high level of product quality while reducing process times and ensuring cost optimisation.

References

[1] M. Plewa. M. Assessment of influence of products' reliability on remanufacturing processes. Int. J. Perform. Eng 5 (2009) 463-470.

[2] A. Tubis. The Potentials in the Integration of Planning Enterprise Activity, Logistics and Transport 2010/1 (2010) 105-113.

[3] R. Ulewicz. Quality control system in production of the castings from spheroid cast iron, Metalurgija 42 (2003) 61-63.

[4] A. Pacana, K. Czerwińska. Analysis of the causes of control panel inconsistencies in the gravitational casting process by means of quality management instruments, Prod. Eng. Arch. 25 (2019) 12-16. https://doi.org/10.30657/pea.2019.25.03

[5] R. Wolniak. Operation manager and its role in the enterprise, Prod. Eng. Arch. 24 (2019) 1-4. https://doi.org/10.30657/pea.2019.24.01

[6] W. Dai, P.G. Maropoulos, Y. Zhao. Reliability modelling and verification of manufacturing processes based on process knowledge management. Int. J. Comput. Integr. Manuf. 28 (2015) 98-111. https://doi.org/10.1080/0951192X.2013.834462

[7] D. Klimecka-Tatar, M. Ingaldi. Assessment of the technological position of a selected enterprise in the metallurgical industry, Mater. Res. Proc. 17 (2020) 72-78. https://doi.org/10.21741/9781644901038-11

[8] M. Fertsch, K. Grzybowska, A. Stachowiak. Models of manufacturing systems – classification framework. Research in Logistics and Production 1 (2011) 45-51.

[9] M. Chlebus, S. Werbińska-Wojciechowska. Issues on Production Process Reliability Assessment – Review, Research in Logistics and Production 6 (2016) 481-497.

[10] A. Gazda et al. Study on improving the quality of stretch film by Taguchi method, Przemysl Chemiczny, 92 6(2013) 980-982.

[11] M.M. Helms. Encyclopedia of Management 5th Edition, Thompson Gale, Detroit, 2006, 211-211. ISBN 978-0787665562

[12] B. Clark. Managerial perceptions of marketing performance: efficiency, adaptability, effectiveness and satisfaction, J. Strategic Mark. 8 (2000) 3-25. https://doi.org/10.1080/096525400346286

[13] A. Pacana et al. Effect of selected factors of the production process of stretch film for its resistance to puncture, Przemysl Chemiczny 93 (2014), 2263-2264. https://doi.org/10.12916/przemchem.2014.2263

[14] W. Zong, X.H. Shen, L.C. Wang. Discussion on Enhancing Engineering Quality Supervision and Management by Information Technology, EBM 2011 Int. Conf. on Engineering and Business Management, 22-24 March 2011, Wuhan, China, 1440-1443. ISBN 978-1629932620

[15] E.A. Elsayed. Perspectives and challenges for research in quality and reliability engineering, Int. J. Prod. Res. 38 (2000) 1953-1976. https://doi.org/10.1080/002075400188438

[16] T. Hosokawa, Z. Miyagi. Quality engineering-based management: a proposal for achieving total optimisation of large systems, TQM Bus. Excel. 30 (2019) 182-194. https://doi.org/10.1080/14783363.2019.1665843

[17] G.J. Savage, S.M. Carr. Interrelating Quality and Reliability in Engineering Systems. Qual. Eng. 14 (2001) 137-152. https://doi.org/10.1081/QEN-100106893

[18] D. Malindžák, A. Pacana, H. Pačaiova. An effective model for the quality of logistics and improvement of environmental protection in a cement plant, Przemysl Chemiczny 96 (2017) 1958-1962.

[19] A. Jodejko-Pietruczuk, M. Plewa. Components' rejuvenation in production with reused elements, Int. J. Perform. Eng. 10 (2014) 567-575.

[20] S. Borkowski, R. Ulewicz, J. Selejdak, M. Konstanciak, D. Klimecka-Tatar. The use of 3x3 matrix to evaluation of ribbed wire manufacturing technology, METAL 2012 - 21st Int. Conf. Metallurgy and Materials (2012), Ostrava, Tanger 1722-1728.

[21] R. Ulewicz. Outsorcing quality control in the automotive industry, MATEC Web of Conf. 183 (2018) art.03001. https://doi.org/10.1051/matecconf/201818303001

[22] R. Ulewicz, F. Nový. Quality management systems in special processes, Transp. Res. Procedia 40 (2019) 113-118. https://doi.org/10.1016/j.trpro.2019.07.019

[23] Ł.J. Orman Ł.J., N. Radek, J. Pietraszek, M. Szczepaniak. Analysis of enhanced pool boiling heat transfer on laser-textured surfaces. Energies 13 (2020) art. 2700. https://doi.org/10.3390/en13112700

[24] M. Szczepaniak et al. T. Use of the maximum power point tracking method in a portable lithium-ion solar battery charger, Energies 15 (2022) art.26. https://doi.org/10.3390/en15010026

[25] S. Maleczek et al. Tests of Acid Batteries for Hybrid Energy Storage and Buffering System—A Technical Approach, Energies 15 (2022) art.3514. https://doi.org/10.3390/en15103514

[26] A. Goroshko et al. Construction and practical application of hybrid statistically-determined models of multistage mechanical systems, Mechanika 20 (2014) 489-493. https://doi.org/10.5755/j01.mech.20.5.8221

[27] R. Ulewicz, M. Mazur. Economic aspects of robotization of production processes by example of a car semi-trailers manufacturer, Manuf. Technol. 19 (2019) 1054-1059. https://doi.org/10.21062/ujep/408.2019/a/1213-2489/MT/19/6/1054

[28] G. Filo, E. Lisowski, M. Domagała, J. Fabiś-Domagała, H. Momeni. Modelling of pressure pulse generator with the use of a flow control valve and a fuzzy logic controller, AIP Conf. Proc. 2029 (2018) art.20015. https://doi.org/10.1063/1.5066477

[29] M. Domagala et al. CFD Estimation of a Resistance Coefficient for an Egg-Shaped Geometric Dome, Appl. Sci. 12 (2022) art.10780. https://doi.org/10.3390/app122110780

[30] M. Domagala et al. The Influence of Oil Contamination on Flow Control Valve Operation, Mater. Res. Proc. 24 (2022) 1-8. https://doi.org/10.21741/9781644902059-1

[31] W. Przybył et al. Virtual Methods of Testing Automatically Generated Camouflage Patterns Created Using Cellular Automata, Mater. Res. Proc. 24 (2022) 66-74. https://doi.org/10.21741/9781644902059-11

[32] N. Radek et al. Operational tests of coating systems in military technology applications, Eksploat. i Niezawodn. 25 (2023) art.12. https://doi.org/10.17531/ein.2023.1.12

[33] W. Przybył et al. Microwave absorption properties of carbonyl iron-based paint coatings for military applications, Def. Technol. 22 (2023) 1-9. https://doi.org/10.1016/j.dt.2022.06.013

[34] N. Radek et al. Technology and application of anti-graffiti coating systems for rolling stock, METAL 2019 28th Int. Conf. Metall. Mater. (2019) 1127-1132. ISBN 978-8087294925

[35] N. Radek et al. Influence of laser texturing on tribological properties of DLC coatings, Prod. Eng. Arch. 27 (2021) 119-123. https://doi.org/10.30657/pea.2021.27.15

[36] N. Radek et al. Formation of coatings with technologies using concentrated energy stream, Prod. Eng. Arch. 28 (2022) 117-122. https://doi.org/10.30657/pea.2022.28.13

[37] N. Radek et al. The influence of plasma cutting parameters on the geometric structure of cut surfaces, Mater. Res. Proc. 17 (2020) 132-137. https://doi.org/10.21741/9781644901038-20

[38] N. Radek et al. The effect of laser treatment on operational properties of ESD coatings, METAL 2021 30th Ann. Int. Conf. Metall. Mater. (2021) 876-882. https://doi.org/10.37904/metal.2021.4212

[39] N. Radek et al. The WC-Co electrospark alloying coatings modified by laser treatment, Powder Metall. Met. Ceram. 47 (2008) 197-201. https://doi.org/10.1007/s11106-008-9005-7

[40] P. Kurp, H. Danielewski Metal expansion joints manufacturing by a mechanically assisted laser forming hybrid method – concept, Technical Transactions 119 (2022) art. e2022008. https://doi.org/10.37705/TechTrans/e2022008

[41] R. Ulewicz et al. Structure and mechanical properties of fine-grained steels, Period. Polytech. Transp. Eng. 41 (2013) 111-115. https://doi.org/10.3311/PPtr.7110

[42] D. Klimecka-Tatar, M. Ingaldi. Assessment of the technological position of a selected enterprise in the metallurgical industry, Mater. Res. Proc. 17 (2020) 72-78. https://doi.org/10.21741/9781644901038-11

[43] P. Jonšta et al. The effect of rare earth metals alloying on the internal quality of industrially produced heavy steel forgings, Materials 14 (2021) art.5160. https://doi.org/10.3390/ma14185160

[44] J. Pietraszek, N. Radek, A.V. Goroshko. Challenges for the DOE methodology related to the introduction of Industry 4.0. Production Engineering Archives 26 (2020) 190-194. https://doi.org/10.30657/pea.2020.26.33

[45] B. Jasiewicz et al. Inter-observer and intra-observer reliability in the radiographic measurements of paediatric forefoot alignment, Foot Ankle Surg. 27 (2021) 371-376. https://doi.org/10.1016/j.fas.2020.04.015

[46] J. Pietraszek. The modified sequential-binary approach for fuzzy operations on correlated assessments, LNAI 7894 (2013) 353-364. https://doi.org/10.1007/978-3-642-38658-9_32

[47] J. Pietraszek et al. Non-parametric assessment of the uncertainty in the analysis of the airfoil blade traces, METAL 2017 26th Int. Conf. Metall. Mater. (2017) 1412-1418. ISBN 978-8087294796

[48] J. Korzekwa et al. The influence of sample preparation on SEM measurements of anodic oxide layers, Pract. Metallogr. 53 (2016) 36-49. https://doi.org/10.3139/147.110367

[49] A. Gadek-Moszczak, P. Matusiewicz. Polish stereology – A historical review, Image Analysis and Stereology 36 (2017) 207-221. https://doi.org/10.5566/ias.1808

[50] M. Krynke et al. Maintenance management of large-size rolling bearings in heavy-duty machinery, Acta Montan. Slovaca 27 (2022) 327-341. https://doi.org/10.46544/AMS.v27i2.04

[51] P. Regulski, K.F. Abramek The application of neural networks for the life-cycle analysis of road and rail rolling stock during the operational phase, Technical Transactions 119 (2022) art. e2022002. https://doi.org/10.37705/TechTrans/e2022002

[52] M. Patek et al. Non-destructive testing of split sleeve welds by the ultrasonic TOFD method, Manuf. Technol. 14 (2014) 403-407. https://doi.org/10.21062/ujep/x.2014/a/1213-2489/MT/14/3/403

[53] I. Miletić, A. Ilić, R.R. Nikolić, R. Ulewicz, L. Ivanović, N. Sczygiol. Analysis of selected properties of welded joints of the HSLA Steels, Materials 13 (2020) art.1301. https://doi.org/10.3390/ma13061301

[54] N. Radek et al. Properties of Steel Welded with CO2 Laser, Lecture Notes in Mechanical Engineering (2020) 571-580. https://doi.org/10.1007/978-3-030-33146-7_65

[55] M. Ulewicz et al. Transport of lead across polymer inclusion membrane with p-tert - butylcalix[4]arene derivative , Physicochem. Probl. Miner. Process. 44 (2010) 245-256.

[56] M. Zenkiewicz et al. Modeling electrostatic separation of mixtures of poly(ϵ-caprolactone) with polyfvinyl chloride) or polyfethylene terephthalate), Przemysl Chemiczny 95 (2016) 1687-1692. https://doi.org/10.15199/62.2016.9.6

[57] M. Zenkiewicz, T. Zuk, J. Pietraszek, P. Rytlewski, K. Moraczewski, M. Stepczyńska. Electrostatic separation of binary mixtures of some biodegradable polymers and poly(vinyl chloride) or poly(ethylene terephthalate), Polimery/Polymers 61 (2016) 835-843. https://doi.org/10.14314/polimery.2016.835

[58] J. Fabiś-Domagała, G. Filo, H. Momeni, M. Domagała. Instruments of identification of hydraulic components potential failures, MATEC Web of Conf. 183 (2018) art.03008. https://doi.org/10.1051/matecconf/201818303008

[59] K. Knop et al. Evaluating and Improving the Effectiveness of Visual Inspection of Products from the Automotive Industry, Lecture Notes in Mechanical Engineering (2019) 231-243. https://doi.org/10.1007/978-3-030-17269-5_17

[60] P. Lempa, G. Filo. Analysis of Neural Network Training Algorithms for Implementation of the Prescriptive Maintenance Strategy, Mater. Res. Proc. 24 (2022) 281-287. https://doi.org/10.21741/9781644902059-41

[61] M. Gupta et al. Advances in applications of Non-Destructive Testing (NDT): A review, Adv. Mater. Process. Technol. 8 (2022) 2286-2307. https://doi.org/10.1080/2374068X.2021.1909332

[62] A. Sophian, G.Y. Tian, D. Taylor, J. Rudlin. Electromagnetic and eddy current NDT: a review, Insight: Non-Destr. Test. Condit. Monit. 43 (2001) 302-306.

[63] G. Davis et al. Laser ultrasonic inspection of additive manufactured components, Int. J. Adv. Manuf. Technol 102 (2019) 2571-2579. https://doi.org/10.1007/s00170-018-3046-y

[64] L.K. Shark, B.J. Matuszewski, J.P. Smith, M.R. Varley. Automatic feature-based fusion of ultrasonic, radiographic and shearographic images for aerospace NDT, Insight: Non-Destr. Test. Condit. Monit. 43 (2001) 607-615.

[65] A. Pacana, K. Czerwinska, L. Bednarova, Comprehensive improvement of the surface quality of the diesel engine piston, Metalurgija 58 (2019) 329-332.

[66] P. Zientek. Non-destructive testing methods for selected elements of small power turbogenerators, Napędy i Sterowanie 19(3) (2017) 114-119.

[67] M. Korzyński et al., Fatigue strength of chromium coated elements and possibility of its improvement with ball peening, Surface & Coatings Technology 204 (2009) 615-620, https://doi.org/10.1016/j.surfcoat.2009.08.049

Quality Production Improvement and System Safety: QPI 16 - CZOTO 10   Materials Research Forum LLC
Materials Research Proceedings 34 (2023) 199-206   https://doi.org/10.21741/9781644902691-24

Development of a Hybrid Method for Identifying the Causes of Product Incompatibility in Metallurgical Manufacturing

DWORNICKA Renata[1, a *], SIWIEC Dominika[2,b] and PACANA Andrzej[3,c]

[1]Cracow University of Technology, Kraków, Poland, ORCID: 0000-0002-2979-1614

[2]Rzeszow University of Technology, Rzeszów, Poland, ORCID: 0000-0002-6663-6621

[3]Rzeszow University of Technology, Rzeszów, Poland, ORCID: 0000-0003-1121-6352

[a]renata.dwornicka@pk.edu.pl, [b]d.siwiec@prz.edu.pl, [c]app@prz.edu.pl

Keywords: Mechanical Engineering, Quality Management, Making Decision, Non-Destructive Testing

Abstract. For the analysis of the quality of metallurgical products, it is important to use techniques that identify the internal and external unconformity of the product without destroying it. These techniques are non-destructive testing (NDT). Although these techniques identify the unconformity of the product, they do not indicate the source of their creation. The purpose of the study was to develop a hybrid method to make decisions about the causes of product incompatibility. This hybrid method was created as a combination of NDT and quality management techniques, i.e.: 5W2H method, Ishikawa diagram, 5Why method. The subject of the study was an unconformity detected in the tube made of the magnesium alloy AMS 4439. Research was carried out using the FPI method. In the analyzed case, its application allowed the detection of a linear indication in the product. To identify the root of the linear indication, the 5W2H method, the Ishikawa diagram, and the 5Why method were used sequentially. The main causes were bad casting and pollution. The root cause was defective supplier material. Integration of the FPI method, the 5W2H method, the Ishikawa diagram, and the 5Why method in the performance of a comprehensive qualitative analyze of products, after which it is possible to identify the unconformity and the root of its occurrence. The integration of FPI and quality management techniques can be practiced to analyze the quality of products (including metallurgical products) in manufacturing and service enterprises.

Introduction

Quality control of the product is a necessary stage of creation of the product. For the production industry, and aviation industry extremely useful are the non-destructive test (NDT), because it allows pointing out the non-conformity without damaging the product [1]. For non-destructive methods, the fluorescent method (i.e., penetrant method) is counted, which was used to quality test a selected subject of the study (tube). Fluorescent penetrant inspection (FPI) is the most commonly used method for the analysis of aviation components that has been applied, for example, in production and also in service inspection [2]. A review of selected elements of the literature indicates that the fluorescent method was used for example in quality analysis of product surface [3-5], detection and further processing of the unconformity by other methods to identify unconformity [1,4,6]. The FPI was improved or its effectiveness was assessed by, among others: the use of other substances or devices for penetration [7,8], the FPI performance rating [9]. A review of the literature shows that an important step for the qualitative analysis of products way shown that an effective is to integrate quality management techniques with the FPI method to identify the root of unconformity [10]. It is essential for enterprises, in which it is necessary not only to identify the unconformity but first of all point out the source of their occurrence.

Quality Production Improvement and System Safety: QPI 16 - CZOTO 10   Materials Research Forum LLC
Materials Research Proceedings 34 (2023) 199-206   https://doi.org/10.21741/9781644902691-24

The proposed hybrid method for identifying the causes of product nonconformities falls into the group of methods aimed at improving quality through organizational and technological changes [11-13]. In the organizational domain, these methods involve modifying management schemes [14-15] and pre-prepared scenarios of potential failures and their consequences [16-18]. In the technological domain, this influences the selection of materials with better technological properties [19], modification of already used materials through the production of special coatings [20-22], application of techniques such as electrospark deposition [23-25], or modification of the morphological characteristics of the surface layer [26]. This brings numerous benefits, including increased reliability of produced machine parts [27-29] and the expansion of the customer base to include demanding military recipients [30,31]. The implementation of such multidimensional technological changes inevitably involves the use of special computational [32,33] and statistical techniques [34-36], including non-parametric methods [37-39] and dimensionality reduction [40].

Method and Material
An attempt was made to solve the problem with linear indications on the AMS 4439 magnesium alloy cast tube in a selected production and service company located in south-eastern Poland, by extension of the fluorescent method about the quality management techniques. In the selected company, product analysis was carried out using NDT methods (i.e., magnetic powder and fluorescence methods). So far, after the unconformity because of their episodic character, no additional analyzes were not made, due to which it could be possible to identify the source of the unconformity. The fact that analysis of unconformity with use of quality management techniques could be made, for example, due to types of product and types of unconformity has been omitted. In turn, the types of unconformity were often repeated (for example, linear indication). Therefore, it was an order to implement the quality management techniques after the product analysis process by using NDT for types of identified unconformity. The analysis was carried out to show that it is effective to use NDT methods (based on the example of the FPI method) together with quality techniques to analyze product quality and identify the source of possible non-conformity. The purpose of the study was to develop a hybrid method to make decisions about the causes of product incompatibility. This hybrid method was created as a combination of NDT and quality management techniques, i.e.: 5W2H method, Ishikawa diagram, 5Why method. The subject of research was the so-called tube, applicable in the aviation industry. The choice of this product for analysis was determined by the type of unconformity identified on it (linear indication). The product was made of magnesium alloy AMS 4439 (SAE AMS 4439: 2012). AMS 4439 is a magnesium alloy, sand-cast 4.2 Zn - 1.2 rare earths - 0, 7 Zr. This material is used, among others, for products requiring uniform, medium strength up to 160 ° C, pressure tightness, good fatigue, and creep characteristics. Moreover, applicable to products that require welding during production.

The method was developed as a hybrid method to make decisions about the causes of product incompatibility. This hybrid method was created as a combination of NDT and quality management techniques, i.e.: 5W2H method, Ishikawa diagram, 5Why method. An analysis of the product quality used integrated one of method NDT, i.e. fluorescent method (FPI - fluorescent penetrant inspection) with these selected quality management techniques. The choice of the NDT method was conditioned by the type of material (magnesium alloy) from which the tube was made and also by the requirements of the customer ordering product quality control. The fluorescent method has applied to identify discontinuities on the surface free from pollution. However, it can be difficult to research the porous surface [41]. The way of conducting research using the fluorescent method has been characterized in the literature on the subject [42]. To identify the root of unconformity were implemented in sequential way methods, i.e. 5W2H, Ishikawa diagram, and 5Why. These techniques were selected because they were used sequentially to allow to analyze and define the problem (unconformity) (5W2H method), identifying the potential and main causes (Ishikawa diagram) and next to identify the root of the unconformity (5Why method) [43]. The choice of these techniques is an expert choice and, in other cases, it can be duplicated or the composition and order of the techniques used can be modified depending on the nature of the problem. The 5W2H has an application to analyze and characterize the problem by seven questions in a practical way (often in the form of a table). These questions relate to the most important information about the problem (in this case, it was a linear indication) [43]. The Ishikawa diagram, called causes and effects diagram, allows pointing the potential causes of problem. It was developed during a brainstorming with 7 employees [44]. To develop the diagram, from basic Ishikawa, categories (5M+E) were selected: man, method, machine, material, management, and environment [44,45]. For these categories, the potential causes of linear indication on the tube were noted. Of the indicated potential causes, three main causes were selected, which were further analyzed using the 5Why method. The 5Why method (that is, the Why-Why diagram) is used to identify the source of the problem [42,46]. The analysis of the linear indication on the tube was started from pointed main causes. Next, to each of the causes, the „Why?" question was asked. The method was completed when the source cause was indicated, that is, one after which improvement measures can be taken [43,47-49]. In the last stage, actions were proposed that could minimize or eliminate the formation of a linear indication r on the product.

Results

After analyzing the fluorescence cast tube of AMS 4439 magnesium alloy, unconformity was found, which was a linear indication, it is shown in Figure 1.

Figure 1. Linear indication identified on the tube from magnesium alloy AMS 4439.

Next, the analysis of the linear indications was performed using the 5W2H method (Table 1).

Table 1. The 5W2H method to linear indication problem on the product.

Question		Answer
Who?	Who has detected the problem?	An employee who checks the product using the FPI method
What?	What is the problem?	unconformity – linear indication
Why?	Why is this problem?	product disqualification
Where?	Where was the problem?	on the product surface
When?	When was the problem?	during quality control by FPI method
How?	How was the problem identified?	FPI method
How much?	What is the scale of this problem?	1 piece of the product

The analysis using the Ishikawa diagram shown in Figure 2. Potential causes of unconformity (linear indications) on the tube were analyzed according to selected Ishikawa categories, i.e., man, method, machine, material, management, and environment.

Quality Production Improvement and System Safety: QPI 16 - CZOTO 10   Materials Research Forum LLC
Materials Research Proceedings 34 (2023) 199-206   https://doi.org/10.21741/9781644902691-24

The identified source cause of the linear indication on the tube was a destructive material from the supplier. The supplier of the material was informed about the source of the cause of the irregularities. This action was taken in order to minimize or eliminate the cause of porosity cluster.

Conclusion

The product quality analysis for the aviation industry is one of the most demanding analyzes, and the fluorescent method (FPI) is one of the most practiced. Although its use allows one to assess the quality of products and indicate unconformity, it does not identify the source of their creation. This has consequences in the future, because according to the philosophy of continuous improvement, the cause of the unconformity must be resolved at the source, so that it does not occur in the future. Therefore, it was important to improve NDT analysis using qualitative techniques. The purpose of the study was to develop a hybrid method to make decisions about the causes of product incompatibility. This hybrid method was created as a combination of NDT and quality management techniques, i.e.: 5W2H method, Ishikawa diagram, 5Why method. This was done in a selected leading production and service company located in south-east Poland. The subject of the study was a tube made of the magnesium alloy AMS 4439.

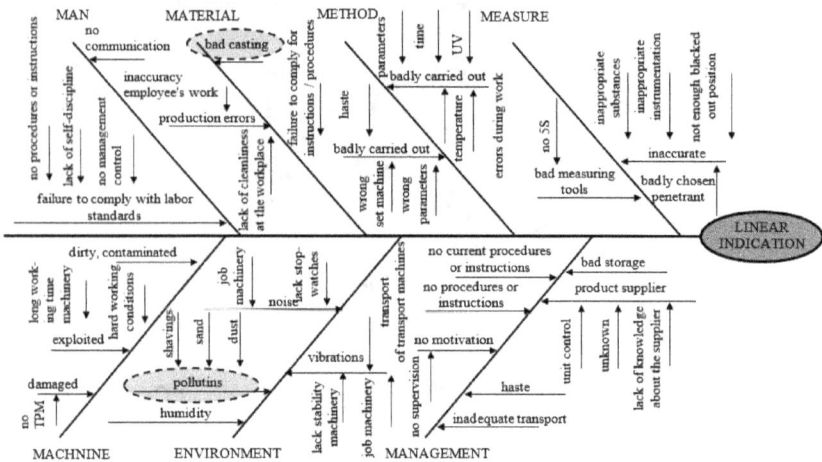

Fig.2. The Ishikawa diagram for the problem with linear indication identified on the tube.

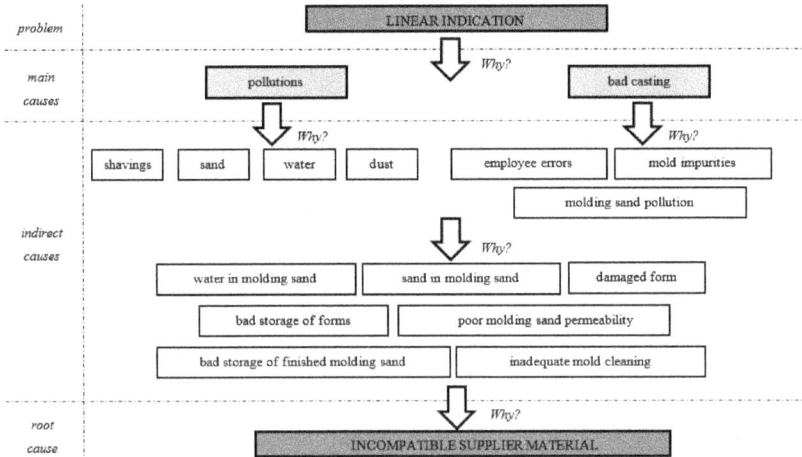

Fig.3. *The 5Why method for the problem with linear indication identified on the tube.*

Then, three main causes of the problem were selected. The main reasons were bad casting and pollutions. To identify the source of cause the 5Why method was performed, whose course is shown in Fig.3.

The analysis of the product was carried out using the FPI method by which the unconformity – linear identification - was detected. In the FPI method, the selected quality management techniques (5W2H method, Ishikawa diagram, and 5Why method) were implemented. These techniques were chosen because they are used sequentially to indicate the source of the problem, which was intended to improve the process of NDT tests (for example, using the FPI method). Using the 5W2H method, the linear indication on the tube was analyzed and characterized. Using the Ishikawa diagram, the main causes of the problem of the linear indication on the tube were identified, i.e., bad casting and pollutions. Next, the 5Why method was used, by which: the source of the cause of linear indication on the tube was identified, i.e., destructive material from supplier. Integrated NDT method with quality management techniques (5W2H method, Ishikawa diagram, and 5Why method) turned out to be effective in the analysis of the unconformity product (casting of magnesium alloy), so can be effective in the analysis of other types of unconformity or metallurgical products.

References

[1] J. Zheng et al. Design of an advanced automatic inspection system for aircraft parts based on fluorescent penetrant inspection analysis, Insight: Non-Destr. Test. Cond. Monit. 57 (2015) 18-34. https://doi.org/10.1784/insi.2014.57.1.18

[2] L.J.H. Brasche et al. Characterization of developer application methods used in fluorescent penetrant inspection, AIP Conf. Proc. 820 (2006) 598-605. https://doi.org/10.1063/1.2184582

[3] H. Fischer et al. Detection of microscopic cracks in dental ceramic materials by fluorescent penetrant method, J. Biomed. Mater. Res. 61(1) (2002) 153-158. https://doi.org/10.1002/jbm.10148

[4] N. J. Shipway et al. Performance Based Modifications of Random Forest to Perform Automated Defect Detection for Fluorescent Penetrant Inspection, J. Nondestructive Evaluation 38 (2019) art. 37. https://doi.org/10.1007/s10921-019-0574-9

[5] R. Wong, K. Cunningham. Mechanical processing of aluminum alloy components and its effect on fluorescent penetrant indications. Mater. Eval. 64 (2006) 1180-1186.

[6] N. J. Shipway et al. Automated defect detection for Fluorescent Penetrant Inspection using Random Forest, NDT & E International 101 (2019) 113-123. https://doi.org/10.1016/j.ndteint.2018.10.008

[7] K. Daneshvar, B. Dogan. Application of quantum dots as a fluorescent-penetrant for weld crack detection, Mater. at High Temp. 27 (2010) 179-182. https://doi.org/10.3184/096034010X12813744660988

[8] A.S. Bakunov et al. Increasing the reliability of magnetic-particle testing by means of a YMJIK-10 automated unit for magnetic fluorescent-penetrant inspection of pipe end faces, Russian J. Nondestruct. Test. 40 (2004) 311-316. https://doi.org/10.1023/B:RUNT.0000045935.55674.d3

[9] L. Brasche et al. Engineering studies for fluorescent penetrant inspection with a focus on developer application methods, Insight: Non-Destr. Test. Cond. Monit. 51 (2009) 88-91. https://doi.org/10.1784/insi.2009.51.2.88

[10] P. Ewert P., C.T. Kowalski. Detection of the damage of the submerged pump aggregates with induction motors. Prace Naukowe Instytutu Maszyn, Napędów i Pomiarów Elektrycznych Politechniki Wrocławskiej. Studia i Materiały 69 (2013) 181-193.

[11] S. Borkowski, R. Ulewicz, J. Selejdak, M. Konstanciak, D. Klimecka-Tatar. The use of 3x3 matrix to evaluation of ribbed wire manufacturing technology, METAL 2012 - 21st Int. Conf. Metallurgy and Materials (2012), Ostrava, Tanger 1722-1728.

[12] R. Ulewicz, F. Nový. Quality management systems in special processes, Transp. Res. Procedia 40 (2019) 113-118. https://doi.org/10.1016/j.trpro.2019.07.019

[13] T. Lipiński, R. Ulewicz. The effect of the impurities spaces on the quality of structural steel working at variable loads, Open Eng. 11 (2021) 233-238. https://doi.org/10.1515/eng-2021-0024

[14] R. Ulewicz, J. Selejdak, S. Borkowski, M. Jagusiak-Kocik. Process management in the cast iron foundry, METAL 2013 - 22nd Int. Conf. Metallurgy and Materials (2013), Ostrava, Tanger 1926-1931.

[15] R. Ulewicz, D. Jelonek, M. Mazur. Implementation of logic flow in planning and production control, Management and Production Engineering Review 7 (2016) 89-94. https://doi.org/10.1515/mper-2016-0010

[16] J. Fabiś-Domagała, G. Filo, H. Momeni, M. Domagała. Instruments of identification of hydraulic components potential failures, MATEC Web of Conf. 183 (2018) art.03008. https://doi.org/10.1051/matecconf/201818303008

[17] K. Knop et al. Evaluating and Improving the Effectiveness of Visual Inspection of Products from the Automotive Industry, Lecture Notes in Mechanical Engineering (2019) 231-243. https://doi.org/10.1007/978-3-030-17269-5_17

Quality Production Improvement and System Safety: QPI 16 - CZOTO 10   Materials Research Forum LLC
Materials Research Proceedings 34 (2023) 199-206   https://doi.org/10.21741/9781644902691-24

[18] P. Lempa, G. Filo. Analysis of Neural Network Training Algorithms for Implementation of the Prescriptive Maintenance Strategy, Mater. Res. Proc. 24 (2022) 281-287. https://doi.org/10.21741/9781644902059-41

[19] P. Jonšta et al. The effect of rare earth metals alloying on the internal quality of industrially produced heavy steel forgings, Materials 14 (2021) art.5160. https://doi.org/10.3390/ma14185160

[20] N. Radek et al. Technology and application of anti-graffiti coating systems for rolling stock, METAL 2019 28th Int. Conf. Metall. Mater. (2019) 1127-1132. ISBN 978-8087294925

[21] N. Radek et al. Influence of laser texturing on tribological properties of DLC coatings, Prod. Eng. Arch. 27 (2021) 119-123. https://doi.org/10.30657/pea.2021.27.15

[22] N. Radek et al. Formation of coatings with technologies using concentrated energy stream, Prod. Eng. Arch. 28 (2022) 117-122. https://doi.org/10.30657/pea.2022.28.13

[23] N. Radek, J. Pietraszek, A. Gadek-Moszczak, Ł.J. Orman, A. Szczotok. The morphology and mechanical properties of ESD coatings before and after laser beam machining, Materials 13 (2020) art. 2331. https://doi.org/10.3390/ma13102331

[24] N. Radek et al. The effect of laser treatment on operational properties of ESD coatings, METAL 2021 30th Ann. Int. Conf. Metall. Mater. (2021) 876-882. https://doi.org/10.37904/metal.2021.4212

[25] N. Radek et al. The impact of laser processing on the performance properties of electro-spark coatings, 14th World Congress in Computational Mechanics and ECCOMAS Congress 1000 (2021) 1-10. https://doi.org/10.23967/wccm-eccomas.2020.336

[26] N. Radek et al. The influence of plasma cutting parameters on the geometric structure of cut surfaces, Mater. Res. Proc. 17 (2020) 132-137. https://doi.org/10.21741/9781644901038-20

[27] A. Goroshko et al. Construction and practical application of hybrid statistically-determined models of multistage mechanical systems, Mechanika 20 (2014) 489-493. https://doi.org/10.5755/j01.mech.20.5.8221

[28] R. Ulewicz, M. Mazur. Economic aspects of robotization of production processes by example of a car semi-trailers manufacturer, Manufacturing Technology 19 (2019) 1054-1059. https://doi.org/10.21062/ujep/408.2019/a/1213-2489/MT/19/6/1054

[29] I. Drach et al. Design Principles of Horizontal Drum Machines with Low Vibration, Adv. Sci. Technol. Res. J. 15 (2021) 258-268. https://doi.org/10.12913/22998624/136441

[30] N. Radek et al. Operational tests of coating systems in military technology applications, Eksploat. i Niezawodn. 25 (2023) art.12. https://doi.org/10.17531/ein.2023.1.12

[31] W. Przybył et al. Microwave absorption properties of carbonyl iron-based paint coatings for military applications, Def. Technol. 22 (2023) 1-9. https://doi.org/10.1016/j.dt.2022.06.013

[32] S. Borkowski, R. Ulewicz, J. Selejdak, M. Konstanciak, D. Klimecka-Tatar. The use of 3x3 matrix to evaluation of ribbed wire manufacturing technology, METAL 2012 - 21st Int. Conf. Metallurgy and Materials (2012), Ostrava, Tanger 1722-1728.

[33] L. Cedro Model parameter on-line identification with nonlinear parametrization – manipulator model, Technical Transactions 119 (2022) art. e2022007. https://doi.org/10.37705/TechTrans/e2022007

[34] J. Pietraszek, A. Szczotok, N. Radek. The fixed-effects analysis of the relation between SDAS and carbides for the airfoil blade traces. Archives of Metallurgy and Materials 62 (2017) 235-239. https://doi.org/10.1515/amm-2017-0035

[35] J. Pietraszek, N. Radek, A.V. Goroshko. Challenges for the DOE methodology related to the introduction of Industry 4.0. Production Engineering Archives 26 (2020) 190-194. https://doi.org/10.30657/pea.2020.26.33

[36] B. Jasiewicz et al. Inter-observer and intra-observer reliability in the radiographic measurements of paediatric forefoot alignment, Foot Ankle Surg. 27 (2021) 371-376. https://doi.org/10.1016/j.fas.2020.04.015

[37] J. Pietraszek. The modified sequential-binary approach for fuzzy operations on correlated assessments, LNAI 7894 (2013) 353-364. https://doi.org/10.1007/978-3-642-38658-9_32

[38] J. Pietraszek et al. Non-parametric assessment of the uncertainty in the analysis of the airfoil blade traces, METAL 2017 26th Int. Conf. Metall. Mater. (2017) 1412-1418. ISBN 978-8087294796

[39] J. Pietraszek et al. The non-parametric approach to the quantification of the uncertainty in the design of experiments modelling, UNCECOMP 2017 Proc. 2nd Int. Conf. Uncert. Quant. Comput. Sci. Eng. (2017) 598-604. https://doi.org/10.7712/120217.5395.17225

[40] J. Pietraszek, E. Skrzypczak-Pietraszek. The uncertainty and robustness of the principal component analysis as a tool for the dimensionality reduction. Solid State Phenom. 235 (2015) 1-8. https://doi.org/10.4028/www.scientific.net/SSP.235.1

[41] A. Pacana, D. Siwiec. Universal Model to Support the Quality Improvement of Industrial Products, Materials 14 (2021) art. 7872. https://doi.org/10.3390/ma14247872

[42] A. Pacana et al. Analysis of the incompatibility of the product with fluorescent method, Metalurgija 58(3-4) (2019) 337-340.

[43] A. Pacana et al. Comprehensive improvement of the surface quality of the diesel engine piston, Metalurgija, 58(3-4) (2019), 329-332.

[44] D. Siwiec, A. Pacana. A New Model Supporting Stability Quality of Materials and Industrial Products, Materials 15 (2022) art. 4440. https://doi.org/10.3390/ma15134440

[45] M. Korzyński, A. Pacana. Centreless burnishing and influence of its parameters on machining effects, J. Mater. Process. Technol. 210 (2010) 1217-1223. https://doi.org/10.1016/j.jmatprotec.2010.03.008

[46] A. Pacana et al. Study on improving the quality of stretch film by Shainin method, Przemysl Chemiczny 93 (2014), 243-245. https://doi.org/10.12916/przemchem.2014.243

[47] D. Siwiec, A. Pacana. A Pro-Environmental Method of Sample Size Determination to Predict the Quality Level of Products Considering Current Customers' Expectations, Sustainability 13 (2021) art.5542. https://doi.org/10.3390/su13105542

[48] A. Gazda et al., Study on improving the quality of stretch film by Taguchi method, Przemysl Chemiczny, 92 6(2013), 980-982.

[49] R. Wolniak. Application methods for analysis car accident in industry on the example of power, Support Systems in Production Engineering 6 (2017) 34-40.

Quality Production Improvement and System Safety: QPI 16 - CZOTO 10   Materials Research Forum LLC
Materials Research Proceedings 34 (2023) 207-215   https://doi.org/10.21741/9781644902691-25

An Application of the Systematic Diagram in the Failure and Causes Analysis of a Vane Pump

FABIŚ-DOMAGAŁA Joanna[1,a] *, DOMAGAŁA Mariusz[1,b]
and PIETRASZEK Jacek[1,c]

[1]Cracow University of Technology, Al. Jana Pawła II 37, 31-864 Kraków, Poland

[a]joanna.fabis-domagala@pk.edu.pl, [b]mariusz.domagala@pk.edu.pl,
[c]jacek.pietraszek@pk.edu.pl

Keywords: Hydraulic Systems, Drive and Control Systems, Potential Failures, Causes, Quality Improvement Tools, Systematic Diagram, Preventive Measures, Hydraulic Vane Pump, Oil Contamination, Maintenance

Abstract. Hydraulic systems are widely spread among drive and control systems. They can play a crucial role in many applications; therefore, identifying potential failures and their causes might be required. Quality improvement tools and methods can be used to achieve this goal. This research attempts to apply one of the recently developed tools, which is a systematic diagram, to recognize possible failures and their causes and finally to define preventive measures for a typical hydraulic vane pump. The analysis of potential pump failures and their causes identified oil contamination as the primary source of pump failure or malfunction. Consequently, proper maintenance was found to be the proper preventive measure.

Introduction

Hydraulic systems, due to their advantages, are one of the top-rated drive systems in various industries, from agriculture, heavy machinery, mining, oil, and gas to the aerospace industry. Components of hydraulic systems have complex electro-hydro-mechanical structures. The most substantial and complicated structures among them are hydraulic pumps. Their main task is to convert mechanical energy into pressure energy and provide the required fluid flow rate. There are several main types of hydraulic pumps, such as gear, screw, piston, or vane. Regardless of the type, they might be one of the most expensive components in the system, and their failure might be catastrophic for the whole system. Therefore, it is crucial to identify symptoms of potential failures to take appropriate measures at an early stage of their occurrence or even during system design. In order to identify possible failures during operation, various diagnostic systems can be used. In contrast, quality improvement methods find application before the system is implemented. The most popular methods are Failure Modes and Effect Analysis (FMEA) [1-3], Quality Function Deployment (QFD), or methods using experimental data in designing products and processes. The abovementioned methods can be supported by quality improvement tools at various stages of their implementation. Those tools are used for collecting and processing data related to various quality aspects. They are instruments for monitoring and diagnosing design, manufacturing, control, or assembly processes throughout the product life cycle. These tools allow collecting information to define Total Quality Management (TQM) actions. Among the quality control tools, we can find the classical quality improvements tools such as the Ishikawa diagram, Pareto analysis, or correlation diagram. The "new" quality improvement tools are relationship diagrams or systematics diagrams [4,5]. The application of those tools is more and more extensive, particularly in the automotive industry. This paper attempts to implement a systematic diagram to identify the causes of potential failures of a vane pump.

Analogous methods for enhancing quality based on the development of potential failure scenarios are widely employed in the industry [6-8]. These methods allow for the optimization of preventive measures, minimizing them to the necessary minimum in line with lean principles [9,10]. This approach is particularly critical in the production of highly responsible products, such as in the machinery [11,12], railway [13], hydraulic power [14], and military [15-17] sectors.

Technological advancements play a significant role in improving reliability in these industries. This includes the selection of appropriate materials [18], their precise processing [19-21], and the shaping of desired functional characteristics in cooperating surface layers [22-24]. By implementing these techniques, several benefits are achieved. First, the corrosion resistance is significantly enhanced [25-27], leading to increased durability and longevity of the products. Second, the wear and tear rates during operation are reduced, resulting in prolonged lifespan and improved performance [28-30]. Lastly, the strength of welded joints is improved, ensuring structural integrity and safety [31-33].

An additional positive consequence of these improvements is the overall increase in product quality [34-37]. This has a direct impact on reducing the strain on the natural environment, contributing to sustainability efforts [38]. Moreover, these advancements inspire the application of image analysis methods for the identification and analysis of coating and surface layer characteristics [39,40]. By leveraging image analysis techniques, researchers and engineers can gain valuable insights into the performance and properties of surface coatings, facilitating further optimization and refinement of the manufacturing processes.

Design and Functionality of a Vane Pump

A vane pump is the positive displacement pump type. Its main features are high efficiency and reliability. An additional advantage is the low noise emission during operation and low operating costs. An example of the vane pump is presented in Fig.1. It has a relatively simple structure, where vanes are placed in a rotor socket and expanded to the stator during rotation, creating a pumping chamber [41]. Failure-free pump operation requires hydraulic oil to fulfill specific requirements for cleanliness, contamination level, and self-lubricating capability. The available research indicates that the cause of the majority of vane pump failures is improper operational conditions [42]. The knowledge about the causes of pump failures allows for defining preventive or/and corrective measures.

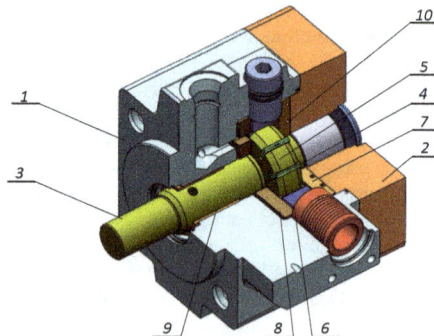

Fig.1. *Hydraulic vane pump (UPLV 32 type), where: 1 – pump body, 2 – cover, 3 – shaft, 4 – vanes, 5 – rotor, 6 – stator, 7,8 – plates, 9 – sleeve, 10 – plate sealings.*

Quality Production Improvement and System Safety: QPI 16 - CZOTO 10   Materials Research Forum LLC
Materials Research Proceedings 34 (2023) 207-215   https://doi.org/10.21741/9781644902691-25

Systematic Diagram

The Flow/Systematic diagram is one of the decent quality improvement tools known as a tree diagram or decision tree. It is most often used during the planning and managing processes in the organization to anticipate the consequences of decisions taken. It can also be used during concept development and the design of new products to identify possible failures or improvement actions for identified problems. It has a graphical form for presenting the ordering activities necessary for a given process or factors influencing the occurrence of a given failure. The diagram systematizes the causes of the problem in a chronological and logical order following the principle "*from general to detail*".

Furthermore, it can be used to arrange information in a relationship or dependency charts [43,44]. The main idea of the systematic diagram refers to the tree diagram, thanks to which it is possible to use specific techniques helpful in its preparation. One of these techniques is a systematic functional chart based on the FAST technique [45]. The procedure for developing a systematic diagram includes five steps:

-    stage 1 – level 0: defining the problem/effect (marked 000),
-    stage 2 – level 1: determining the main categories of causes for a given problem/effect (designation I00, II00),
-    stage 3 – level 2: determination of causes for a given effect (designation I10, II10)
-    stage 4 – level 3...n: determination of sub-causes for a given cause (designation I1a, II1a)
-    stage 5 – selection the cause that has the greatest impact on the problem/effect.

In the above-presented procedure, the systematic diagram logically presents cause and effect relationships for the problem/effect under consideration. Fig. 2 shows an example of a systematic diagram. The diagram is constructed from the general (left) to the detail (right).

The result of the analysis of the systematic diagram should be the determination of the leading cause category and the main cause that has the highest impact on the effect/problem. In the next step, the appropriate corrective measures should be specified so that the identified problem is resolved and there are no severe consequences due to its negligence.

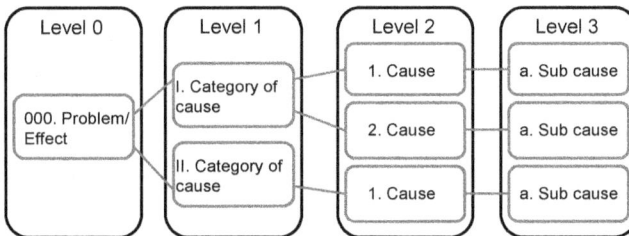

Fig.2. A graphical representation of the systematic diagram

Systematic Diagram in Failure Analysis of a Vane Pump

The operation of a vane pump is based on converting mechanical energy into the pressure energy of the working fluid. The failures that can occur during operation might cause pump malfunction or inability to operate. Some of those failures can be fixed directly onsite. The others may require sophisticated equipment or even component replacement. Therefore, the identification of the possible failure causes is a principal issue. This can be realized by a systematic diagram which is presented in Fig. 3. The diagram includes three main categories of failure causes: operation, design, and assembly/repair. For each of them, the causes that can be a potential source of the failure are specified. Consequently, the three sub-causes were assigned for each of them.

According to the literature review [42], the highest probability of failure of vane pumps arises during maintenance errors (I), which degrades the working fluid (IIa-d). Maintaining an appropriate quality of hydraulic oil is the primary cause affecting the proper operation of the vane pump and the entire hydraulic system. A change in the color of the working liquid indicates its property degradation or the initiation of contamination. Therefore, in the second step of the analysis, the systematic diagram was used to identify possible causes of failures of the vane pump during its operation, as presented in Fig.4.

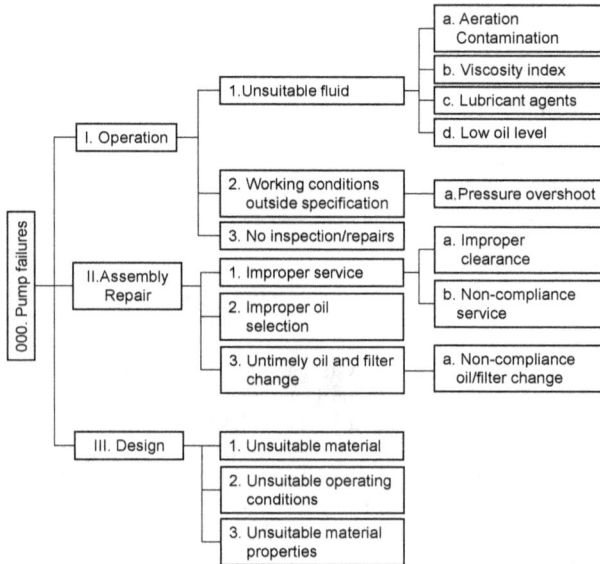

Fig.3. *A general systematic diagram of causes of vane pump failures*

000. Operational failures

I. Mechanical

1. Shaft axial/radial overload
2. Shaft bending
3. Cyclic pressure overshoot
4. Torsion fatigue

a. Lack of oil
b. Too high viscosity
c. Inproper air bleed-off
d. Clogged filter
e. Inlet velocity too low

II. Physical hydraulical chemical

1. The start-up failures
- a. Too high viscosity
- b. Too high viscosity
- c. Inlet tube too small
- d. Clogged oil filter
- e. Solid particles in oil

2. Erosion

3. Oil aeration
- a. Suction through shaft seal — aa. deteriorated shaft seal
- b. Oil velocity too high
- c. Oil level too low
- d. Bad deareation capabilities

4. Oil contamination
- a. Solid particles — aa. External sources
  - ab. Interal sources
  - ac. Insufiecient cleaning
  - ad. Contaminated fresh oil

5. Insuficient lubricant capabilities
- a. Water in oil
- b. Oil viscosity too low

6. Pressure failures
- a. Pressure overshoot — aa. Safety valve malfunction
- b. Cyclic pressure overshoot — ba. Safety valve malfunction
- c. Distance to safety valve too large

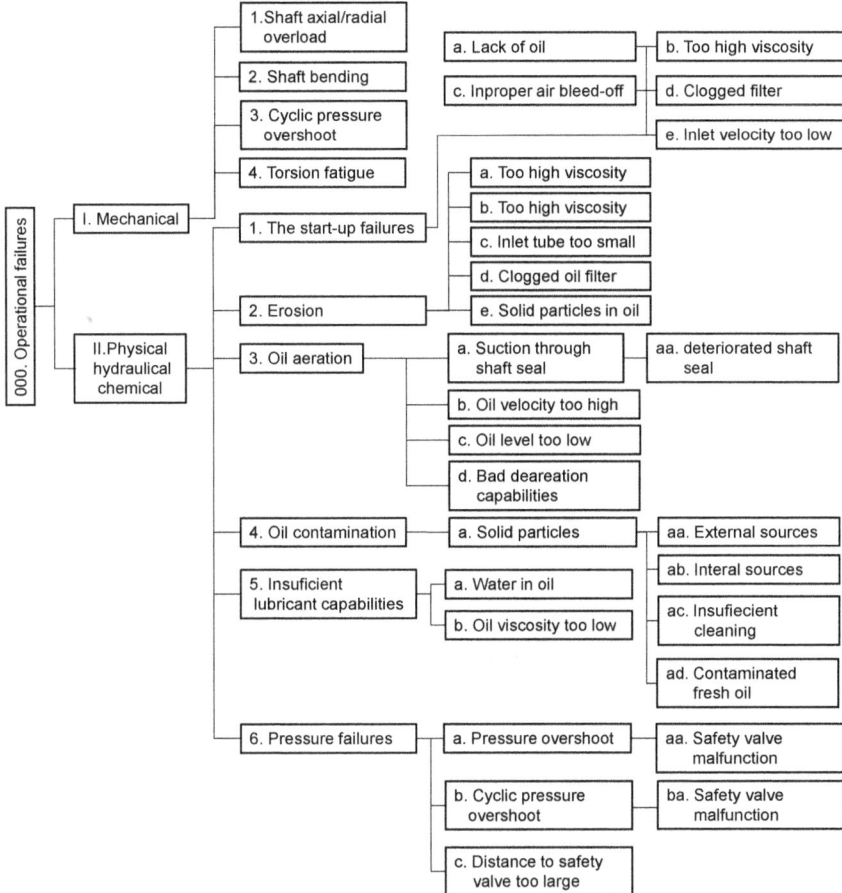

Fig.4. *A detailed systematic diagram for identification of vane pump failures*

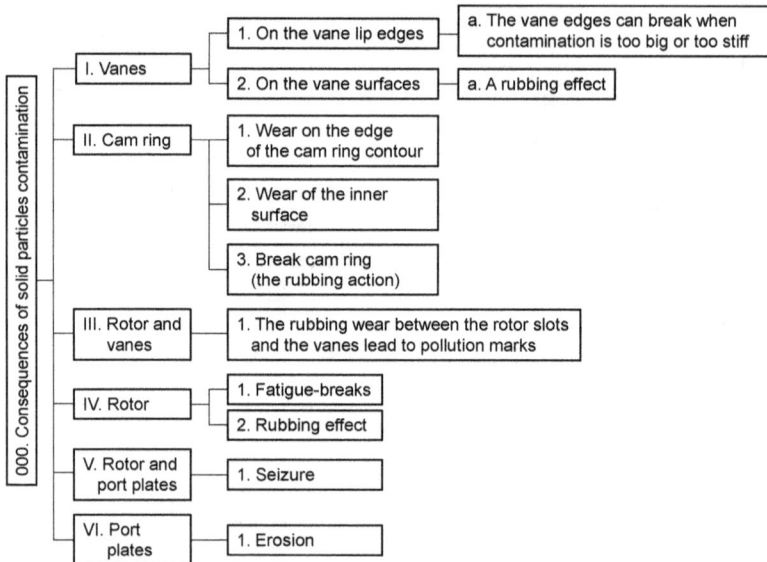

Fig.5. The systematic diagram for effects of solid particles contamination

The systematic diagram shows that maintenance errors are the sources of the vane pump failures. The most common errors are ineffective deaeration at the pump start-up, starting an operation without oil, or using oil with too high viscosity. Another related problem is cavitation, which may cause erosion and surface degradation. A widespread oil failure is the oil aeration caused by insufficient seals in the suction line or pump shaft. Dissolved air in oil affects oil properties, changes compressibility, disturbs the cooling and lubricating process, and, in the worst scenario, may cause pump components to be seized. The other typical oil contaminant is water which converts working fluid into a mixture with lower lubrication capabilities and viscosity. Under high pressure, such a mixture forms foam, significantly decreasing oil bulk modulus and disrupting proper vane operation. At this analysis stage, two main categories of causes were distinguished: I. mechanical and II. physical, hydraulic and chemical. Then, within these two categories, a total of 10 causes and 20 sub-causes were identified. The leading cause of the failure of the vane pump appeared to be contamination of oil with solid particles. It may cause cavitation corrosion (erosion), scratches on the surfaces of the pump components, and fluctuation in the pump operation. Oil contaminated with solid particles may also affect the entire hydraulic system and lead to faulty operation. In the last step of the pump failure analysis, the systematics diagram presented in Fig. 5 was created on which the consequences of solid particle contamination were presented [42].

A contaminated oil with solid particles and water may cause noisy operation, destruction of components, or excessive wear. Measures presented in Table 1 can be undertaken to prevent the failures caused by contaminated fluid.

Quality Production Improvement and System Safety: QPI 16 - CZOTO 10   Materials Research Forum LLC
Materials Research Proceedings 34 (2023) 207-215   https://doi.org/10.21741/9781644902691-25

Table 1. Preventing measures for contaminated oil

Cause	Symptom	Measures
Contaminated fluid	Noisy operation	Install an adequate filter or replace the oil more often.
	Breakage of parts inside the pump housing	
	excessive wear	Determine source of contaminants and correct

Conclusions

The implementation of a systematic diagram for vane pumps failure analysis allowed for identifying the causes of the most common failures. Regardless of the durability of pump components, it is very sensitive to the quality of working fluid. Presented in this research analysis indicates that oil contaminants are the significant sources of pump malfunction. They degrade oil properties, which may lead to pump malfunction or even inability to operate due to component destruction. Therefore, adequate maintenance procedures are key to the failure-free operation. Obtained results agree with other research, which defines oil as main source of failure for fluid power systems.

References

[1] J. Fabis-Domagala et al. A Matrix FMEA Analysis of Variable Delivery Vane Pumps, Energies 14 (2021) art.1741. https://doi.org/10.3390/en14061741

[2] J. Fabis-Domagala et al. A Concept of Risk Prioritization in FMEA Analysis for Fluid Power Systems, Energies 14 (2021) art.6482. https://doi.org/10.3390/en14206482

[3] J. Fabis-Domagala, M. Domagala. A Concept of Risk Prioritization in FMEA of Fluid Power Components, Energies 15 (2022) art.6180. https://doi.org/10.3390/en15176180

[4] A. Hamrol, W. Mantura. Zarządzanie jakością. Teoria i praktyka, Wydawnictwo Naukowe PWN, 2004. ISBN 978-8301167752

[5] H. Gołaś, A. Mazur. Zasady, metody i techniki wykorzystywane w zarządzaniu jakością, Wydawnictwo Politechniki Poznańskiej, Poznań, 2010. ISBN 978-8371439087

[6] K. Knop et al. Evaluating and Improving the Effectiveness of Visual Inspection of Products from the Automotive Industry, Lecture Notes in Mechanical Engineering (2019) 231-243. https://doi.org/10.1007/978-3-030-17269-5_17

[7] G. Filo, P. Lempa. Analysis of Neural Network Structure for Implementation of the Prescriptive Maintenance Strategy, Mater. Res. Proc. 24 (2022) 273-280. https://doi.org/10.21741/9781644902059-40

[8] P. Lempa, G. Filo. Analysis of Neural Network Training Algorithms for Implementation of the Prescriptive Maintenance Strategy, Mater. Res. Proc. 24 (2022) 281-287. https://doi.org/10.21741/9781644902059-41

[9] A. Maszke, R. Dwornicka, R. Ulewicz. Problems in the implementation of the lean concept at a steel works - Case study, MATEC Web of Conf. 183 (2018) art.01014. https://doi.org/10.1051/matecconf/201818301014

[10] R. Ulewicz, M. Ulewicz. Problems in the Implementation of the Lean Concept in the Construction Industries, LNCE 47 (2020) 495-500. https://doi.org/10.1007/978-3-030-27011-7_63

[11] R. Ulewicz, M. Mazur. Economic aspects of robotization of production processes by example of a car semi-trailers manufacturer, Manufacturing Technology 19 (2019) 1054-1059. https://doi.org/10.21062/ujep/408.2019/a/1213-2489/MT/19/6/1054

[12] I. Drach et al. Design Principles of Horizontal Drum Machines with Low Vibration, Adv. Sci. Technol. Res. J. 15 (2021) 258-268. https://doi.org/10.12913/22998624/136441

[13] N. Radek, R. Dwornicka. Fire properties of intumescent coating systems for the rolling stock, Commun. – Sci. Lett. Univ. Zilina 22 (2020) 90-96. https://doi.org/10.26552/com.C.2020.4.90-96

[14] G. Barucca et al. The potential of Λ and Ξ- studies with PANDA at FAIR, Europ. Phys. J. A 57 (2021) art.154 https://doi.org/10.1140/epja/s10050-021-00386-y

[15] W. Przybył et al. Virtual Methods of Testing Automatically Generated Camouflage Patterns Created Using Cellular Automata, Mater. Res. Proc. 24 (2022) 66-74. https://doi.org/10.21741/9781644902059-11

[16] N. Radek et al. Operational tests of coating systems in military technology applications, Eksploat. i Niezawodn. 25 (2023) art.12. https://doi.org/10.17531/ein.2023.1.12

[17] W. Przybył et al. Microwave absorption properties of carbonyl iron-based paint coatings for military applications, Def. Technol. 22 (2023) 1-9. https://doi.org/10.1016/j.dt.2022.06.013

[18] A. Dudek, B. Lisiecka, R. Ulewicz. The effect of alloying method on the structure and properties of sintered stainless steel, Archives of Metallurgy and Materials 62 (2017) 281-287. https://doi.org/10.1515/amm-2017-0042

[19] R. Ulewicz et al. Structure and mechanical properties of fine-grained steels, Period. Polytech. Transp. Eng. 41 (2013) 111-115. https://doi.org/10.3311/PPtr.7110

[20] D. Klimecka-Tatar, M. Ingaldi. Assessment of the technological position of a selected enterprise in the metallurgical industry, Mater. Res. Proc. 17 (2020) 72-78. https://doi.org/10.21741/9781644901038-11

[21] D. Siwiec et al. Improving the process of achieving required microstructure and mechanical properties of 38mnvs6 steel, METAL 2020 29th Int. Conf. Metall. Mater. (2020) 591-596. https://doi.org/10.37904/metal.2020.3525

[22] N. Radek et al. The WC-Co electrospark alloying coatings modified by laser treatment, Powder Metall. Met. Ceram. 47 (2008) 197-201. https://doi.org/10.1007/s11106-008-9005-7

[23] P. Kurp et al. The influence of treatment parameters on the microstructure, properties and bend angle of laser formed construction bars, Arch. Metall. Mater. 61 (2016) 1151-1156. https://doi.org/10.1515/amm-2016-0192

[24] P. Kurp, H. Danielewski Metal expansion joints manufacturing by a mechanically assisted laser forming hybrid method – concept, Technical Transactions 119 (2022) art. e2022008. https://doi.org/10.37705/TechTrans/e2022008

[25] K. Jagielska-Wiaderek, H. Bala, P. Wieczorek, J. Rudnicki, D. Klimecka-Tatar. Corrosion resistance depth profiles of nitrided layers on austenitic stainless steel produced at elevated temperatures, Arch. Metall. Mater. 54 (2009) 115-120.

[26] M. Scendo et al. Purine as an effective corrosion inhibitor for stainless steel in chloride acid solutions, Corr. Rev. 30 (2012) 33-45. https://doi.org/10.1515/CORRREV-2011-0039

[27] M. Scendo et al. Influence of laser treatment on the corrosive resistance of WC-Cu coating produced by electrospark deposition, Int. J. Electrochem. Sci. 8 (2013) 9264-9277.

[28] S. Marković et al. Exploitation characteristics of teeth flanks of gears regenerated by three hard-facing procedures, Materials 14 (20210 art. 4203. https://doi.org/10.3390/ma14154203

[29] M. Krynke et al. Maintenance management of large-size rolling bearings in heavy-duty machinery, Acta Montan. Slovaca 27 (2022) 327-341. https://doi.org/10.46544/AMS.v27i2.04

[30] P. Regulski, K.F. Abramek The application of neural networks for the life-cycle analysis of road and rail rolling stock during the operational phase, Technical Transactions 119 (2022) art. e2022002. https://doi.org/10.37705/TechTrans/e2022002

[31] M. Patek et al. Non-destructive testing of split sleeve welds by the ultrasonic TOFD method, Manuf. Technol. 14 (2014) 403-407. https://doi.org/10.21062/ujep/x.2014/a/1213-2489/MT/14/3/403

[32] P. Jonšta et al. Hydrogen embrittlement of welded joint made of supermartensitic stainless steel in environment containing sulfane, Arch. Metall. Mater. 61 (2016) 709-711. https://doi.org/10.1515/amm-2016-0121

[33] I. Miletić, A. Ilić, R.R. Nikolić, R. Ulewicz, L. Ivanović, N. Sczygiol. Analysis of selected properties of welded joints of the HSLA Steels, Materials 13 (2020) art.1301. https://doi.org/10.3390/ma13061301

[34] S. Borkowski, R. Ulewicz, J. Selejdak, M. Konstanciak, D. Klimecka-Tatar. The use of 3x3 matrix to evaluation of ribbed wire manufacturing technology, METAL 2012 – 21st Int. Conf. Metall. Mater. (2012), Ostrava, Tanger 1722-1728.

[35] R. Ulewicz. Outsorcing quality control in the automotive industry, MATEC Web of Conf. 183 (2018) art.03001. https://doi.org/10.1051/matecconf/201818303001

[36] R. Ulewicz, F. Nový. Quality management systems in special processes, Transp. Res. Procedia 40 (2019) 113-118. https://doi.org/10.1016/j.trpro.2019.07.019

[37] D. Siwiec, R. Dwornicka, A. Pacana. Improving the non-destructive test by initiating the quality management techniques on an example of the turbine nozzle outlet, Mater. Res. Proc. 17 (2020) 16-22. https://doi.org/10.21741/9781644901038-3

[38] A. Deja et al. Analysis and assessment of environmental threats in maritime transport, Transp. Res. Procedia 55 (2021) 1073-1080. https://doi.org/10.1016/j.trpro.2021.07.078

[39] A. Gadek-Moszczak, P. Matusiewicz. Polish stereology – A historical review, Image Analysis and Stereology 36 (2017) 207-221. https://doi.org/10.5566/ias.1808

[40] I. Jastrzębska, A. Piwowarczyk. Traditional vs. automated computer image analysis – A comparative assessment of use for analysis of digital SEM images of high-temperature ceramic material, Materials 16 (2023) art. 812. https://doi.org/10.3390/ma16020812

[41] S. Stryczek. Napęd hydrostatyczny T.1. WNT, Warszawa 1995. ISBN 978-8301189570

[42] Parker Hydraulic, Hydraulic Pumps & Motors Vane Troubleshooting Guide, Catalogue HY29-0022/UK.

[43] A. Pacana. Narzędzia zarządzania jakością, Wydawnictwo Politechnika Rzeszowska, 2022. ISBN 978-8379345601

[44] S. Wawak. Zarządzanie jakością. Podstawy, systemy i narzędzia, Wydawnictwo Helion, 2022. ISBN 83-73617876

[45] M. Ćwiklicki, H. Obora. Wprowadzenie do metod TQM, Wydawnictwo UEK, Kraków, 2011. ISBN 978-8372525277

Quality Production Improvement and System Safety: QPI 16 - CZOTO 10   Materials Research Forum LLC
Materials Research Proceedings 34 (2023) 216-225   https://doi.org/10.21741/9781644902691-26

Method for Analysing Quality Problems of Automotive Castings

PACANA Andrzej[1,a], CZERWIŃSKA Karolina[2,b] and DWORNICKA Renata[3,c] *

[1]Rzeszow University of Technology, Rzeszów, Poland; ORCID: 0000-0003-1121-6352

[2]Rzeszow University of Technology, Rzeszów, Poland; ORCID: 0000-0003-2150-0963

[3]Cracow University of Technology, Kraków, Poland; ORCID: 0000-0002-2979-1614

[a]app@prz.edu.pl, [b]k.czerwinska@prz.edu.pl, [c]*renata.dwornicka@pk.edu.pl

Keywords: Quality Management, Mechanical Engineering, Quality, Improving, Quality of Products

Abstract. The use of multifaceted quality analyses contributes to increasing the efficiency of production processes and quality control as part of maintaining competitiveness. The aim of this research was to develop an integrally configured method for the analysis of quality problems in automotive aluminium castings based on the methodological triangulation of NDT tests and quality management tools. The verification of the model was performed in one of the foundry companies against an engine block casting used in the automotive industry. The detection results obtained (VT and X-ray examination) allowed for the identification of a critical nonconformity - the presence of gas porosity, the description of the problem (5W2H method) and the identification of the root cause of the problem (5WHY method), which was the lack of up-to-date work instructions and a low level of employee competence. Corrective actions were proposed. The measures taken in the implementation of the quality problem analysis method were a new solution for the company – until now, in-depth analyses using a sequence of diagnostic methods and quality management technique tools had not been performed.

Introduction

When implementing a new technology or product, a manufacturing company should consider the quality level of the manufactured product, as it will have a significant impact on the organisation's market position and competitiveness. The quality strategy plays an important role here and should be taken into account when creating plans and objectives for the enterprise within the following areas: economy, organisation, finance or law [1,2]. Effective management of a manufacturing enterprise should be a well-thought-out and well-chosen group of activities enabling the organisation to overcome threats and unexpected events as well as obstacles and limitations, and to formulate original methods, solutions based on the desire to develop technologies that create and justify the necessary changes towards Industry 4.0 [3-5]. It should reinforce the company's focus on four correlative activities: quality, market, modernity and future.

The quality of a product starts already in the design phase. Even a correctly implemented technological process will not guarantee an exemplary level of quality if adequate quality is not ensured within the first phase of product creation. A significant number of non-conformities identified during the manufacturing process or operation of the product are initiated in the design phase [6,7]. However, design is not the only phase within the product quality life cycle. This cycle can be represented as a closed circle, which includes: the birth time of the product design, the stages of product development, the manufacturing and use of the product, and the decommissioning time [8,9].

Within the framework of improving the quality of the offered products, comprehensive methods of quality assurance by detecting inconsistencies and preventing their occurrence by detecting the sources of their origin and implementing corrective actions are constantly sought for. Methods enabling the implementation of the indicated activities are detection methods from the group of non-destructive methods [10] and quality management methods, which, skilfully applied, allow to increase the quality level of the offered products [11].

The aim of the research was to develop an integrally configured method for the analysis of quality problems of automotive aluminium castings based on the methodological triangulation of NDT tests and quality management tools. A clearly defined sequence of actions allows the problem to be specified, the cause of the drop in quality level to be identified and adequate improvement actions to be proposed.

The proposed method for addressing quality issues falls within the realm of organizational methods [12-14]. By enabling competence development, it modifies management frameworks [15,16] and influences changes in potential failure scenarios and their consequences [17-19]. Remedial actions are associated not only with purely organizational changes but also with interventions in material selection [20-22], the creation of special coatings [23-25], the methods employed for their application [26-28], and the execution of structural welds [29,30]. Improving the level of quality, which results in extended equipment lifespan and reduced failure occurrences, is also linked to a decrease in environmental pressure [31]. Furthermore, these changes may have implications in meeting the increased demand for military-grade production [32-34]. Given the complexity of production processes and the multidimensional interplay of significant process factors, the application of specialized analytical methods [35] and formal statistical techniques, such as Design of Experiments (DOE) methodology [36,37], including nonparametric variants [38,39], becomes essential.

Method for Analysing Qualitative Problems

In order to carry out a qualitative analysis of the castings, a study was carried out, in which the integration of detection methods and quality management tools was configured. Detection methods enable the identification of non-conformities and the determination of the basic characteristics of the non-conformity (type of non-conformity, and its size: depth, width and length of the non-conformity), while quality management tools will allow for the effective processing of the data on non-conformities and will make it possible to perform an analysis of potential causes of the occurrence of non-conformities. The execution of the indicated actions will enable the identification of the root cause of the decrease in the quality level of castings and the suggestion of adequate improvement actions. A diagram of the quality problem analysis method is shown in Fig.1.

The method for analysing quality problems is divided into three main areas (definition of research objectives, detection research and analysis of critical nonconformity) and into eight phases.

Phase 1. Selection of a test subject characterised by a decline in quality level

Due to the specific nature of the pro-quality analysis method developed (use of non-destructive methods - NDT), the selection of the test object should take into account the detection capabilities of the methods. The selection should concern a product characterised by a decrease in quality level and, at the same time, a product for which the detection of discontinuities of approximately (0.5% - 2.0%) of the object thickness represents a satisfactory result.

Phase 2: Selection of the expert team

The members of the expert team should have knowledge of the test object under analysis, the manufacturing process within which the product is manufactured and should have experience in the use of NDT methods used in the method to achieve the objective of the model realisation. The selection of the members of the expert team should be done according to the methods presented in the study, e.g. [40,41].

Quality Production Improvement and System Safety: QPI 16 - CZOTO 10   Materials Research Forum LLC
Materials Research Proceedings 34 (2023) 216-225   https://doi.org/10.21741/9781644902691-26

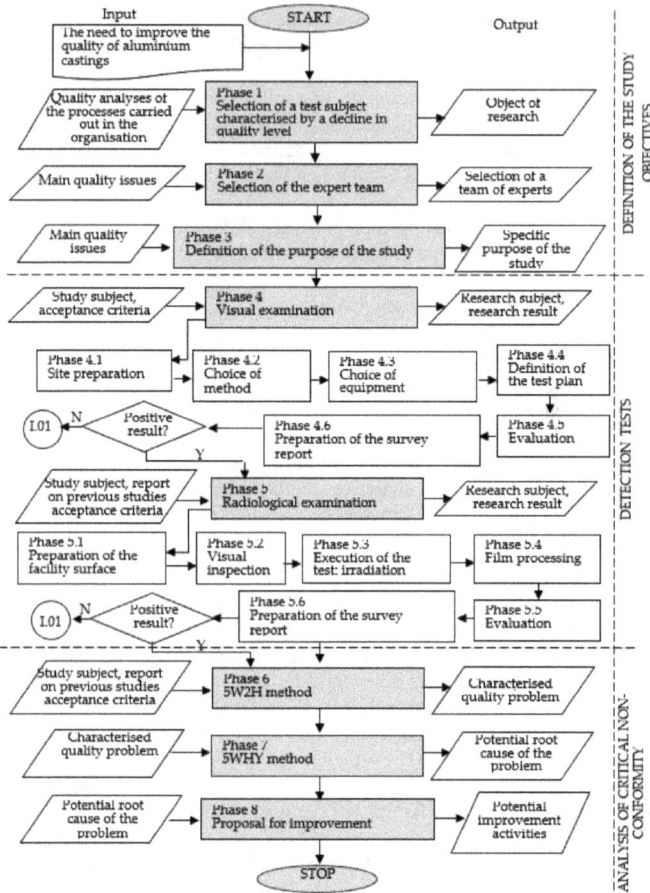

Fig.1. *Concept of a method for analysing qualitative problems*

Phase 3. Defining the research objective

The aim of the quality problem analysis method should be to identify the critical nonconformity and analyse its causes in order to identify its root causes and, ultimately, to propose appropriate corrective actions. In addition, the objective should take into account the customer's requirements for the subject of the study.

Phase 4: Visual examination

Visual examination of the test object is carried out as a preliminary visual inspection prior to X-ray examination. This examination includes preparation of the product surface and familiarisation with the specimen and specifications.

Depending on the accessibility of the surface to be inspected and the lighting conditions, the visual examination should be carried out using: endoscopes, videoscopes, periscopes and sets of mirrors, magnifying glasses and microscopes. The development of the test plan concerns the determination of the detection route. This survey allows the detection of any surface discontinuities, such as cracks. Discontinuities are classified and their number, type, severity and dimensions are determined. The results of the survey must be plotted on appropriate documentation [42,43].

Phase 5. Radiographic examination

Radiographic examination makes it possible to detect: spatial discontinuities, blisters, residual shrinkage cavities, planar discontinuities, shrinkage cracks, inclusions. With this test, it is also possible to detect and ocean changes in object thickness and coating thickness [44]. When preparing the surface for testing, residues of the gating system, moulding compound, reinforcement and cores must be removed. For the inspection of castings, the literature recommends that a visual inspection is carried out before the X-ray examination [45]. For this examination, a sending transducer is placed on one side, generating X-rays. The article being inspected absorbs radiation (gamma and X-rays), which is supplied from an external source. A detector (e.g. silver film) is placed on the other side of the device. The homogeneous beam of radiation passing through the device is partially absorbed, which depends on the variation of the internal structure and is noticeable in the final image. Radiographs show two-dimensional shadow images of three-dimensional discontinuities. The object's discrepancies take the form of darker areas, most often irregularly shaped located against a lighter object background. The evaluation and determination of dimensions is done by direct measurement on the radiographs. The test report should include key product information and identification of material discontinuities [46,47].

Phase 6. 5W2H method

The 5W2H method provides a detailed analysis of the problem. It is recognised that a correct and insightful description of the problem is a guarantee of reaching a correct conclusion, which is the basis for problem solving, so the 5W2H method is often used not only as a way of making a preliminary analysis, but also as a tool for characterising the identified problem. Answers to the seven questions should be gleaned from insightful interviews with employees [48,49].

Phase 7. The 5WHY method

The 5WHY method makes it possible to find the causes of the problem under analysis. The method addresses two aspects. The first relates to the causes of the problem and explaining why the problem arose, while the second aspect relates to the detectability of the problem and obtaining information on why the current quality control system did not detect the problem when it arose. The method is based on the assumption that each subsequent finding is determined by asking the question "why?" [50,51].

Phase 8. Suggesting improvement actions.

Once the main root causes of the quality problem have been identified, it is necessary to define adequate improvement actions, i.e. actions that, when properly implemented, will contribute to the elimination or significant reduction of the nonconformities that occur. These actions are determined by a team of experts.

The use of methodological triangulation in the model including sequential control (diagnostic-analytical) contributes to the validity of the collected data by including more sources and reducing measurement error.

Verification and Test of the Method

The test of the universal method for the analysis of the quality problem was carried out in a foundry company located in the south-eastern part of Poland. The test covered production data from 6 months of the year 2021.

Stage 1 – Defining the research assumptions (selection of the research subject, team of experts and definition of the research objective)

The method test was performed on an engine block casting that had lost quality stability. The loss of quality stability of the product was noticed after the implementation of design changes to the casting according to customer requirements. The product is used within engine and car technology. Fig.2a shows a model of the tested product. A team of experts was appointed, taking into account their experience, skills and knowledge of the manufacturing process and detection testing. The team consisted of: chief technologist, quality control manager, NDT specialist, claims specialist. The aim of applying the method to the engine block casting was to propose a course of action for the detection of the casting, characterise the quality problem, identify the root causes of the problem and finally indicate appropriate corrective actions.

Stage 2 – Detection tests (visual examination (VT), radiographic examination (X-ray))

Quality control of the engine block production process is performed on the basis of a control plan, which takes into account the relevant product parameters defined by standards and customer requirements. The tests carried out identified gas porosity as a critical nonconformity. In order to characterise the discontinuity areas, metallographic scrap and microscopic observations were made. An image of an example of a discontinuity is shown in Fig.2b. The detection results of the indicated methods within the control points were the input data of stage 3 of the method.

Stage 3 – Analysis of a critical nonconformity (method 5W2H, method 5WHY, proposal for improvement measures)

Based on the microstructure analysis of the non-conformity, gas porosities were identified in the engine block casting, which disqualified the product. In the next step, a group of experts, performed a gemba walk within the production process and then performed a 5W2H analysis to characterise the problem in detail (Table 1). After characterizing the quality problem in the subsequent step of the analysis, the 5WHY method was employed to identify the cause of the material discontinuity. Fig.4 displays the outcome of the analysis.

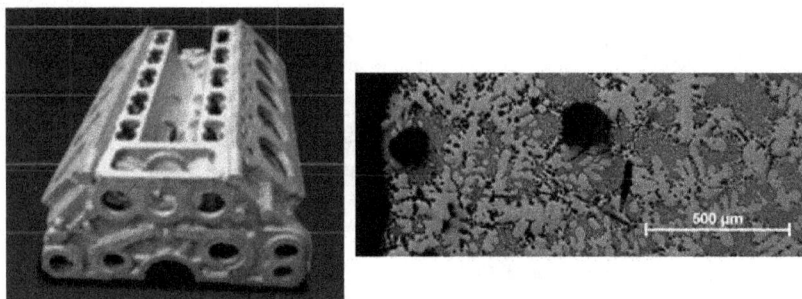

Fig.2. Left: model engine block, Right: Engine block area where material discontinuity has been identified

Quality Production Improvement and System Safety: QPI 16 - CZOTO 10   Materials Research Forum LLC
Materials Research Proceedings 34 (2023) 216-225   https://doi.org/10.21741/9781644902691-26

Table 1. 5W2H method for the engine block casting discontinuity problem

Question		Answer
Who?	Who has detected the problem?	The employee who performed the X-ray inspection
What?	What is the problem?	Gaseous porosity clusters in the engine block casting
Why?	Why is this a problem?	Failure to meet standards and customer requirements - disqualification of the product
Where?	Where was the problem detected?	In the area of engine block base
When?	When was the problem detected?	During X-ray inspection and nicroscopic observation
How?	How was the problem detected?	Observation of radiographic films, observation of computerised registration (in real-time radiography systems), observation of metallographic scans
How much?	How big is the problem?	9% of products manufactured within 6 months 2021

Based on the analysis conducted (as shown in Fig. 3), it was concluded that the root cause of gas porosity in the engine block castings was attributed to inadequate employee qualifications resulting from a lack of instructional training provided in the workplace. The observed situation may have resulted from insufficient employee training following the design modifications made to the product, and the root cause was identified in the human/management aspect.

As improvement measures, it was proposed to carry out a series of training courses for employees within the scope of which the introduced changes to the product are significant, as well as to conduct constant supervision over the performance of their duties and to develop adequate position instructions. The diagnostic and analytical activities carried out indicate the effectiveness of the developed method for analysing quality problems of automotive castings. The method is characterised by efficiency, universality, a broad approach to quality problems and a relatively fast implementation time.

Summary

In modern times, companies must prioritize meeting the demands of customers who expect high-quality products delivered promptly at the most competitive prices. Hence, companies are looking for solutions to increase the level of quality of the products offered. The aim of this research was to develop an integrally configured method for analysing the quality problems of automotive aluminium castings based on the methodological triangulation of NDT tests and quality management tools. A clearly defined sequence of actions allows the problem to be specified, the cause of the drop in quality level to be identified and adequate improvement actions to be proposed.

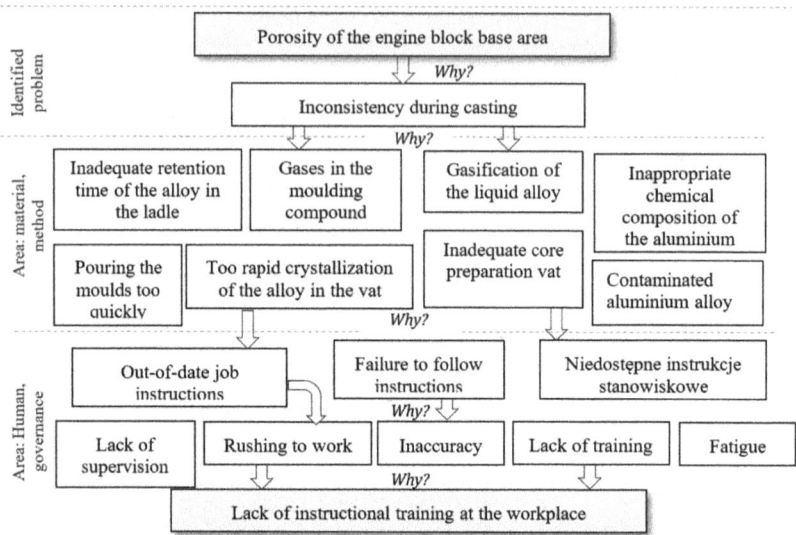

Fig.3 *5WHY method for the problem of the presence of clusters of porosity in the base area of an engine block.*

Visual and X-ray examinations detected the most serious type of non-conformity - the presence of gas porosity in the engine block casting. The presence of the indicated material discontinuity disqualifies the product. In order to recognise and characterise the problem, the expert team performed a 5W2H, while a 5WHY analysis was performed to identify the source of the problem. It was identified that the main cause of the casting quality deterioration was a worker's lack of relevant qualifications due to a lack of instructional training on the job. The implementation of training and the development of job instructions were proposed as corrective actions.

The application of the model does not require huge costs and its implication brings very accurate results. Further research directions will include the implication of the diagnostic-analytical method for the solution of possible quality problems within other casting processes in the company. This action is aimed at taking care of a high level of quality while optimising production costs.

References

[1] R. Wolniak. Main functions of operation management, Prod. Eng. Arch. 26 (2020) 11-14. https://doi.org/10.30657/pea.2020.26.03

[2] K. Hys. Tools and methods used by the Polish leading automotive companies in quality management system. Results of empirical research. Journal of Achievements in Materials and Manufacturing Engineering 63 (2014) 30-37.

[3] M.J. Ligarski, B. Rozalowska, K. Kalinowski. A study of the human factor in industry 4.0 based on the automotive industry, Energies 14 (2021) art. 6833. https://doi.org/10.3390/en14206833

[4] D. Klimecka-Tatar. Analysis and improvement of business processes management – based on value stream mapping (VSM) in manufacturing companies, Pol. J. Manag. Stud. 23 (2021) 213-231. https://doi.org/10.17512/pjms.2021.23.2.13

[5] J. Pietraszek, N. Radek, A.V. Goroshko. Challenges for the DOE methodology related to the introduction of Industry 4.0, Prod. Eng. Arch. 26 (2020) 190-194. https://doi.org/10.30657/pea.2020.26.33

[6] A. Hamrol, W. Mantura. Zarządzanie jakością – teoria i praktyka, Wydawnictwo Naukowe PWN, Warszawa 2005.

[7] A. Pacana, K. Czerwinska, R. Dwornicka. Analysis of non-compliance for the cast of the industrial robot basis, METAL 2019 28th Int. Conf. on Metallurgy and Materials (2019), Ostrava, Tanger, 644-650. https://doi.org/10.37904/metal.2019.869

[8] S.D. Shivankar, R. Deivanathan. Product design change propagation in automotive supply chain considering product life cycle. CIRP J. Manuf. Sci. Technol. 35 (2021) 390-399. https://doi.org/10.1016/j.cirpj.2021.07.001

[9] M. Korzyński, A. Pacana. Centreless burnishing and influence of its parameters on machining effects, J. Mater. Process. Technol. 210 (2010) 1217-1223. https://doi.org/10.1016/j.jmatprotec.2010.03.008.

[10] M. Gupta et al. Advances in applications of Non-Destructive Testing (NDT): A review, Adv. Mater. Process. Technol. 8 (2022) 2286-2307. https://doi.org/10.1080/2374068X.2021.1909332

[11] R. Ulewicz, D. Kleszcz, M. Ulewicz. Implementation of lean instruments in ceramics industries, Manag. Syst. Prod. Eng. 29 (2021) 203-207. https://doi.org/10.2478/mspe-2021-0025

[12] S. Borkowski, R. Ulewicz, J. Selejdak, M. Konstanciak, D. Klimecka-Tatar. The use of 3x3 matrix to evaluation of ribbed wire manufacturing technology, METAL 2012 - 21st Int. Conf. Metallurgy and Materials (2012), Ostrava, Tanger 1722-1728.

[13] R. Ulewicz. Outsorcing quality control in the automotive industry, MATEC Web of Conf. 183 (2018) art.03001. https://doi.org/10.1051/matecconf/201818303001

[14] R. Ulewicz, F. Nový. Quality management systems in special processes, Transp. Res. Procedia 40 (2019) 113-118. https://doi.org/10.1016/j.trpro.2019.07.019

[15] R. Ulewicz, J. Selejdak, S. Borkowski, M. Jagusiak-Kocik. Process management in the cast iron foundry, METAL 2013 - 22nd Int. Conf. Metallurgy and Materials (2013), Ostrava, Tanger 1926-1931.

[16] R. Ulewicz, D. Jelonek, M. Mazur. Implementation of logic flow in planning and production control, Management and Production Engineering Review 7 (2016) 89-94. https://doi.org/10.1515/mper-2016-0010

[17] J. Fabiś-Domagała, G. Filo, H. Momeni, M. Domagała. Instruments of identification of hydraulic components potential failures, MATEC Web of Conf. 183 (2018) art.03008. https://doi.org/10.1051/matecconf/201818303008

[18] K. Knop et al. Evaluating and Improving the Effectiveness of Visual Inspection of Products from the Automotive Industry, Lecture Notes in Mechanical Engineering (2019) 231-243. https://doi.org/10.1007/978-3-030-17269-5_17

[19] G. Filo, P. Lempa. Analysis of Neural Network Structure for Implementation of the Prescriptive Maintenance Strategy, Mater. Res. Proc. 24 (2022) 273-280. https://doi.org/10.21741/9781644902059-40

[20] R. Ulewicz et al. Structure and mechanical properties of fine-grained steels, Period. Polytech. Transp. Eng. 41 (2013) 111-115. https://doi.org/10.3311/PPtr.7110

[21] A. Szczotok et al. The Impact of the Thickness of the Ceramic Shell Mould on the $(\gamma + \gamma')$ Eutectic in the IN713C Superalloy Airfoil Blade Casting, Arch. Metall. Mater. 62 (2017) 587-593. https://doi.org/10.1515/amm-2017-0087

[22] P. Jonšta et al. The effect of rare earth metals alloying on the internal quality of industrially produced heavy steel forgings, Materials 14 (2021) art.5160. https://doi.org/10.3390/ma14185160

[23] N. Radek et al. Technology and application of anti-graffiti coating systems for rolling stock, METAL 2019 28th Int. Conf. Metall. Mater. (2019) 1127-1132. ISBN 978-8087294925

[24] N. Radek et al. Influence of laser texturing on tribological properties of DLC coatings, Prod. Eng. Arch. 27 (2021) 119-123. https://doi.org/10.30657/pea.2021.27.15

[25] N. Radek, J. Konstanty, J. Pietraszek, Ł.J. Orman, M. Szczepaniak, D. Przestacki. The effect of laser beam processing on the properties of WC-Co coatings deposited on steel. Materials 14 (2021) art. 538. https://doi.org/10.3390/ma14030538

[26] N. Radek, J. Konstanty. Cermet ESD coatings modified by laser treatment, Arch. Metall. Mater. 57 (2012) 665-670. https://doi.org/10.2478/v10172-012-0071-y

[27] N. Radek, J. Pietraszek, A. Gadek-Moszczak, Ł.J. Orman, A. Szczotok. The morphology and mechanical properties of ESD coatings before and after laser beam machining, Materials 13 (2020) art. 2331. https://doi.org/10.3390/ma13102331

[28] N. Radek et al. The effect of laser treatment on operational properties of ESD coatings, METAL 2021 30th Ann. Int. Conf. Metall. Mater. (2021) 876-882. https://doi.org/10.37904/metal.2021.4212

[29] M. Patek et al. Non-destructive testing of split sleeve welds by the ultrasonic TOFD method, Manuf. Technol. 14 (2014) 403-407. https://doi.org/10.21062/ujep/x.2014/a/1213-2489/MT/14/3/403

[30] N. Radek et al. Properties of Steel Welded with CO2 Laser, Lecture Notes in Mechanical Engineering (2020) 571-580. https://doi.org/10.1007/978-3-030-33146-7_65

[31] A. Deja et al. Analysis and assessment of environmental threats in maritime transport, Transp. Res. Procedia 55 (2021) 1073-1080. https://doi.org/10.1016/j.trpro.2021.07.078

[32] W. Przybył et al. Virtual Methods of Testing Automatically Generated Camouflage Patterns Created Using Cellular Automata, Mater. Res. Proc. 24 (2022) 66-74. https://doi.org/10.21741/9781644902059-11

[33] N. Radek et al. Operational tests of coating systems in military technology applications, Eksploat. i Niezawodn. 25 (2023) art.12. https://doi.org/10.17531/ein.2023.1.12

[34] W. Przybył et al. Microwave absorption properties of carbonyl iron-based paint coatings for military applications, Def. Technol. 22 (2023) 1-9. https://doi.org/10.1016/j.dt.2022.06.013

[35] J. Korzekwa et al. The influence of sample preparation on SEM measurements of anodic oxide layers, Pract. Metallogr. 53 (2016) 36-49. https://doi.org/10.3139/147.110367

[36] J. Pietraszek, A. Szczotok, N. Radek. The fixed-effects analysis of the relation between SDAS and carbides for the airfoil blade traces. Archives of Metallurgy and Materials 62 (2017) 235-239. https://doi.org/10.1515/amm-2017-0035

[37] B. Jasiewicz et al. Inter-observer and intra-observer reliability in the radiographic measurements of paediatric forefoot alignment, Foot Ankle Surg. 27 (2021) 371-376. https://doi.org/10.1016/j.fas.2020.04.015

[38] J. Pietraszek. The modified sequential-binary approach for fuzzy operations on correlated assessments, LNAI 7894 (2013) 353-364. https://doi.org/10.1007/978-3-642-38658-9_32

[39] J. Pietraszek et al. Non-parametric assessment of the uncertainty in the analysis of the airfoil blade traces, METAL 2017 26th Int. Conf. Metall. Mater. (2017) 1412-1418. ISBN 978-8087294796

[40] A. Pacana, K. Czerwińska. Analysis of the causes of control panel inconsistencies in the gravitational casting process by means of quality management instruments, Prod. Eng. Arch. 25 (2019) 12-16. https://doi.org/10.30657/pea.2019.25.03

[41] G. Ostasz, K. Czerwińska, A. Pacana. Quality management of aluminum pistons with the use of quality control points, Manag. Syst. Prod. Eng. 28 (2020) 29-33. https://doi.org/10.2478/mspe-2020-0005

[42] A. Sophian, G.Y. Tian, D. Taylor, J. Rudlin. Electromagnetic and eddy current NDT: a review, Insight: Non-Destr. Test. Condit. Monit. 43 (2001) 302-306.

[43] A. John. Intelligence Augmentation and Human-Machine Interface Best Practices for NDT 4.0 Reliability. Mater. Eval. 78 (2020) 869–879. https://doi.org/10.32548/2020.me-04133

[44] A. Pacana, K. Czerwinska, L. Bednarova. Comprehensive improvement of the surface quality of the diesel engine piston, Metalurgija, 58 (2019) 329-332.

[45] E. Denis et al. Automatic quality control of digitally reconstructed radiograph computation and comparison with standard methods, Progress in Biomedical Optics and Imaging – Proc. SPIE 6510, art. 65104. https://doi.org/10.1117/12.708193

[46] L.K. Shark, B.J. Matuszewski, J.P. Smith, M.R. Varley. Automatic feature-based fusion of ultrasonic, radiographic and shearographic images for aerospace NDT, Insight: Non-Destr. Test. Condit. Monit. 43 (2001) 607-615.

[47] A. Pacana et al., Study on improving the quality of stretch film by Shainin method, Przemysl Chemiczny, 93 (2014) 243-245. https://doi.org/10.12916/przemchem.2014.243

[48] M. Ingaldi, K. Nowakowska. Wykorzystanie metody 5W2H do doskonalenia produkcji wentylatorów. Archiwum Wiedzy Inżynierskiej, Vol. 1, nr 1, 2016, s. 39-41.

[49] K. Kowalik, D. Klimecka-Tatar. The service process quality improvement by using the 5W2H and 5Why methods. Arch. Eng. Knowl. 2 (2017) 24-26.

[50] A. Pacana, D. Siwiec. Method of Choice: A Fluorescent Penetrant Taking into Account Sustainability Criteria. Sustainability 12 (2020) art. 5854. https://doi.org/10.3390/su12145854

[51] J. Łuczak, A. Matuszak-Flejszman. Metody i techniki zarządzania jakością. Kompendium wiedzy, Quality Progress, Poznań 2007. ISBN 978-83-911169-1-3

Quality Production Improvement and System Safety: QPI 16 - CZOTO 10   Materials Research Forum LLC
Materials Research Proceedings 34 (2023) 226-236   https://doi.org/10.21741/9781644902691-27

Method Supporting Improving Products in Terms of Qualitative-Environmental

PACANA Andrzej[1,a], SIWIEC Dominika[2,b] * and DWORNICKA Renata[3,c] *

[1]Rzeszow University of Technology, Rzeszów, Poland, ORCID: 0000-0003-1121-6352

[2]Rzeszow University of Technology, Rzeszów, Poland, ORCID: 0000-0002-6663-6621

[3]Cracow University of Technology, Kraków, Poland, ORCID: 0000-0002-2979-1614

[a]app@prz.edu.pl, [b]d.siwiec@prz.edu.pl, [c]renata.dwornicka@pk.edu.pl

Keywords: Quality, Production Engineering, Multi-Criteria Decision Method, Sustainability Development, Importance Performance Analysis, Mechanical Engineering

Abstract. Problems with product quality during negative climate changes make the continuous improvement process difficult. The search for approaches consists of methodical and possible detailed analysis areas of these types of problem types. Therefore, the objective of the article was to develop a method to analyze problems with the quality of products considering the impact on the natural environment. The method was developed by a coherent combination of selected techniques, i.e.: brainstorming (BM), Ishikawa diagram, 5M rule, multiple voting, seven-point Likert scale, and the IPA method. A method was carried out for mechanical seal of the 410 alloy, on which a porosity cluster was detected. The originality of the study is that the proposed method supports a consistent methodical analysis of any problem with the quality of products and verifies these problems in view of their impact on the natural environment. Simultaneously, the result obtained from the method determines the main causes of the problem, from which to begin improving the quality of the product considering the impact on the natural environment. This method can be applied to any product and the incompatibilities detected on them. Therefore, the method can be used in service-production enterprises to improve products in terms of qualitative-environmental.

Introduction

Obtaining expected product quality remains a challenge. This is mainly due to dynamic changes in customer requirements and escalating negative climate change [1-3]. For this reason, it is important to adequately plan the process of product production in the design phase [4,5]. However, in view of the mentioned changes in customer requirements and the essence of caring for the natural environment, it is necessary to continuously improve the products. The popular action as part of achieving the expected quality of products is quality control, e.g. non-destructive testing (NDT) [6], which in view of the waste source are more beneficial than destructive testing (DT) [7]. It results from the effectiveness of NDT research in the identification of incompatibility of products without destructive elements, therefore these tests are more environmentally friendly. The quality control of the products after which incompatibility was detected is only the first stage in improving the process. This is due to the lack of application of non-destructive testing controls to identify the causes of these incompatibilities [8-10]. Later, it is necessary to use other techniques that support analysis of the causes of product incompatibility. Popular in use are, among other quality management techniques.

For example, the authors of articles [11-13] used the Ishikawa diagram to improve the quality of products. This diagram has application in the identification of potential causes of incompatibility. Other examples are studies [13-15], which the Pareto-Lorenz analysis and Ishikawa diagram were combined. Using these tools consists of determining potential causes and

Quality Production Improvement and System Safety: QPI 16 - CZOTO 10   Materials Research Forum LLC
Materials Research Proceedings 34 (2023) 226-236   https://doi.org/10.21741/9781644902691-27

then the main causes of the problem, that is, those that generate the greatest problem. The authors of the study [16] developed a universal model for improving the quality of industrial products, in which different quality management were used and combined sequentially. However, new approaches to analysis of the causes of problems with product quality problems were shown in studies [17,18]. These articles show the developed methods and models supporting the stability of the quality of materials and industrial products, where the innovation was the use simultaneously quality management tools and multi-criteria decision-making methods, e.g.: DEMATEL method, FAHP method (Fuzzy Analytic Hierarchy Process) or GRA method (Grey Relational Analysis). However, these analyzes were not aimed at improving the quality of products while caring for the natural environment. According to the authors of studies [19-21], in the era of negative climate change, customers pay more attention to the environmental friendliness of products. Therefore, it was concluded that a pro-ecological approach to product improvement is needed.

The purpose of this article was to develop a method to analyze problems with the quality of products considering the impact on the natural environment. The originality of the study is that the proposed method supports a consistent methodical analysis of any problem with the quality of products and verifies these problems in view of their impact on the natural environment. Simultaneously, the result obtained from the method determines the main causes of the problem, from which to begin improving the quality of the product considering the impact on the natural environment. A method was carried out for mechanical seal of the 410 alloy, on which a porosity cluster was detected.

New methods aimed at improving quality [22-24] in line with environmental protection inevitably bring about changes in both management systems [25-27] and associated production optimization systems [28]. Implementing these methods requires coordinated changes in various technological areas, such as special functional and protective coatings [29,30], hydraulic fluids in hydraulic systems [31-33], modification of surface layers [34] to reduce the rate of wear of interacting machine parts [35,36], and modifications of connections, including welding [37-39], to improve the integrity of structural components. These actions effectively reduce environmental pressure [40] by enhancing corrosion resistance [41-43], even in aggressive corrosive technological environments [44,45], and improving fatigue resistance [46]. Simultaneously, design changes enable increased levels of recycling of used parts and packaging materials [47]. Such multifactorial organizational and technological changes require a systematic approach to ensure the reliability and accuracy of the implemented modifications. In this regard, the methodology of experimental planning [48-50] is invaluable, as it allows for handling poorly defined variables in specific cases [51-53]. Ultimately, the introduced changes, by enhancing the quality and reliability of products and services, lead to simplified scenarios of potential failures and their consequences [54-56].

Method

The method was developed to improve the quality of products in terms of qualitative-environmental aspects. The concept of the method is based on the sequential and coherent analysis of the causes of problems with the quality of products and then the determination of the impact of these causes on the natural environment. The results of the method allow identification of the main causes of the problem, i.e.: the most contributing to the problem, and at the same time having the greatest negative impact on the environment. After testing different methods and making a literature review, it was possible to combine and use selected techniques, i.e. brainstorming (BM), Ishikawa diagram, 5M rule, multiple voting, seven-point Likert scale, and the IPA method (Importance Performance Analysis). The method was developed in five main stages, as shown in Fig.1. The characteristic of the method stage is shown in the next part of the study.

Stage 1. Choice of the subject of study and determine the purpose of the research. The subject of the study is selected by the entity (expert), where it can be, e.g. product unstable in quality or

Quality Production Improvement and System Safety: QPI 16 - CZOTO 10   Materials Research Forum LLC
Materials Research Proceedings 34 (2023) 226-236   https://doi.org/10.21741/9781644902691-27

incompatibility, most occurred on products. In the proposed approach, the main incompatibility, so most often occurred. The choice of this incompatibility can be done on catalogue of incompatibility of products, which is often realized by the entity. In case of a large number of incompatibilities of different types, it is necessary to use tools supporting, e.g. Pareto-Lorenz analysis (20/80) [15]. Then, depending on the selected subject of study, it is necessary to determine the purpose of the research. The purpose is determined by the entity (expert) using SMARTER method [57]. It was assumed that purpose should refer to identifying the main causes of the problem, which will have the largest probability impact on the problem, and simultaneously will have a negative impact on the natural environment.

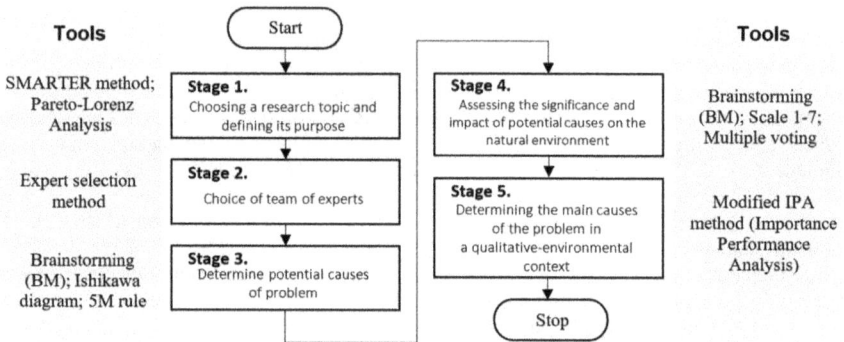

Fig. 1. *Algorithm of Method supporting improving products in terms of qualitative-environmental.*

Stage 2. Choice of the team of experts. The proposed method should be performed with the participation of a selected team of experts, so that the members who will be responsible for achieving the purpose of the research. A team of experts is selected by the entity (expert) according to the competent members and their skills in dealing with the analyzed problem. The method that supports the choice of experts is shown in studies, e.g. [16,18].

Stage 3. Determine potential causes of the problem. A team of experts identified all potential causes, so causes which probably caused the problem. It is necessary to generate possibly the most possible causes. The tool supporting this process is brainstorming (BM) [58]. A leader of the team of experts should note all causes in a place visible to the team, e.g. table. Additionally, it is appropriate to visualize and group all potential causes according to the main causes. The Ishikawa diagram (causes-and-effect diagram) with the 5M rule (man, material, method, machine, and measure) is an effective tool for it [11].

Stage 4. Assessment of the degree of significance of potential causes and their impact on the natural environment. All potential causes of the problem should be assessed in view of: a) the degree of significance of these causes for the emergence of the problem, and b) the impact of causes on the natural environment. Assessments admit experts on a scale from 1 to 7, where 1 – the cause causes the problem to a negligible extent (or the cause has a negligible impact on the environment), 7 – the cause causes the problem to a very significant extent (or the cause has a very significant negative impact on the environment) [59]. Assessments can be given during brainstorming (BM) and multiple voting [60]. Assessments can be noted in the Ishikawa diagram or in the summary table of assessments.

Stage 5. Determining the main causes of the problem in a qualitative-environmental context. In the proposed approach, the main causes of the problem have a simultaneous important impact

Quality Production Improvement and System Safety: QPI 16 - CZOTO 10   Materials Research Forum LLC
Materials Research Proceedings 34 (2023) 226-236   https://doi.org/10.21741/9781644902691-27

on the occurrence of the problem and on the natural environment. To determine these causes, it was assumed that the IPA diagram (Importance Performance Analysis) [61-63]. The traditional form of the IPA diagram was modified to combine the importance of causes and impact on the natural environment, as shown in Fig.2.

The diagram is created according to the assessments of potential causes. The potential causes in the area "concentrate here" are the main causes, i.e., causes that have a significant impact on the emergence of the problem and simultaneously a significant impact on the natural environment. For the main causes, improvement actions should be proposed in the first place, i.e., those that will ensure the improvement of product quality and minimize the negative impact on the natural environment. This is the last step of the proposed method.

Fig.2. Modified IPA diagram to qualitative-environmental analysis. Own study based on [26].

Results

Test of the proposed method performed based on the incompatibility of the porosity cluster on the mechanical seal of the 410 alloy. Incompatibility was identified in a service-production enterprise localized in Poland. The test was carried out according to developed algorithm, i.e., in the five main stages.

Firstly, the study of research was selected and then the purpose of the research was determined. The subject of the research was selected by the entity (expert) according to the catalogue of the incompatibilities of products. One of the most frequent incompatibilities was the porosity cluster in the mechanical seal. This incompatibility was identified after non-destructive testing (NDT), i.e. fluorescent method (FPI). An example of incompatibility is shown in Fig.3.

Subsequently, the purpose of the research was defined according to the selected research subject. The goal was defined by the entity (an expert) using the SMARTER method. The purpose was to identify the main causes of the cluster of porosity on the mechanical seal from 410 alloy, where these causes will have the largest impact on the occurrence the of incompatibility, and simultaneously will have the most negative impact on the natural environment.

Quality Production Improvement and System Safety: QPI 16 - CZOTO 10   Materials Research Forum LLC
Materials Research Proceedings 34 (2023) 226-236   https://doi.org/10.21741/9781644902691-27

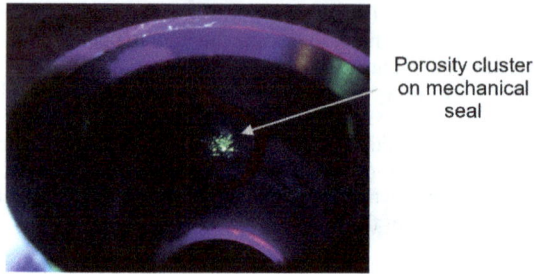

Porosity cluster on mechanical seal

Fig.3. *Example of porosity cluster on mechanical seal.*

Then, the team of experts was selected. The team consisted of the authors of the article and the product quality control manager. A team of experts realized other stages of the method. Later, the team determined the possible causes of the porosity cluster. The causes generated during brainstorming (BM) and then noted on the Ishikawa diagram with the 5M rule (man, material, method, machine, and measure). The result of this stage is shown in Fig.4.

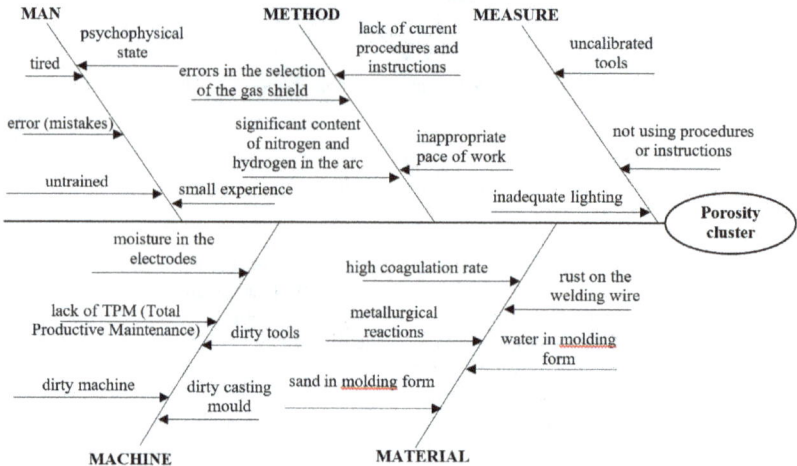

Fig.4. *Ishikawa diagram for problem of porosity cluster on the mechanical seal.*

Quality Production Improvement and System Safety: QPI 16 - CZOTO 10   Materials Research Forum LLC
Materials Research Proceedings 34 (2023) 226-236   https://doi.org/10.21741/9781644902691-27

Table 1. Assessments of the degree of importance of the potential causes and their impact on the natural environment.

No.	Potential causes	assessment of the degree of influence of the potential cause on	
		the emergence of the problem	the natural environment
C1	tired	2	1
C2	error (mistakes)	3	1
C3	untrained	5	3
C4	psychophysical state	5	3
C5	small experience	4	5
C6	significant content of nitrogen and hydrogen in the arc	7	7
C7	errors in the selection of the gas shield	2	2
C8	lack of current procedures and instructions	1	2
C9	inappropriate pace of work	1	2
C10	uncalibrated tools	1	1
C11	not using procedures or instructions	1	3
C12	inadequate lighting	1	4
C13	moisture in the electrodes	2	5
C14	lack of TPM (Total Productive Maintenance)	3	4
C15	dirty machine	2	5
C16	dirty tools	2	5
C17	dirty casting mould	2	5
C18	high coagulation rate	4	2
C19	metallurgical reactions	6	5
C20	sand in molding form	6	4
C21	rust on the welding wire	5	6
C22	water in molding form	2	3

Then, a team of experts assessed the degree of influence of the potential causes on the emergence of the porosity cluster and their impact on the natural environment. As assumed by the method, all potential causes from the Ishikawa diagram were assessed on a Likert scale from 1 to 7. Potential causes were assessed as part of the brainstorming method (BM) implemented among of team of experts. The results of this process are shown in Table 1.

In the last stage, the main causes of the porosity cluster in the mechanical seal were determined in the qualitative-environmental context. For this purpose, a modified IPA diagram was used, which included assessments of the degree of impact of potential causes on the occurrence of the problem and the natural environment, as in Fig.5.

It was shown that the main causes of porosity cluster are causes belonging to the area of "concentrate here". Therefore, the main causes in this case were as follows: C12 – inadequate lighting, C13 – moisture in the electrodes, C14 – lack of TPM (Total Productive Maintenance), C15 – dirty machine, C16 – dirty tools, and C17 – dirty casting mould. For these causes, improvement actions should be applied in the first place. These activities should aim at improving the quality of the product and, at the same time, at reducing the negative impact on the natural environment.

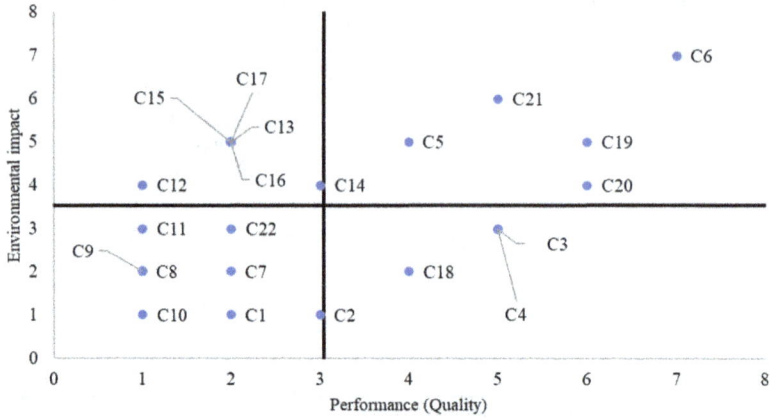

Fig.5. *Results from qualitative-environmental analysis on modified IPA diagram
for the problem of porosity cluster.*

Summary

Changes in customers' expectations cause the need for continuous improvement of products. It is a complicated process, also in view of the need for environmental protection. Therefore, the objective of the article was to develop a method to analyze problems with the quality of products considering the impact on the natural environment. The method was developed by a coherent combination of selected techniques, i.e.: brainstorming (BM), Ishikawa diagram, 5M rule, multiple voting, seven-point Likert scale, and the IPA method. A method was carried out for mechanical seal of the 410 alloy, on which porosity cluster was detected. Initially, during the test of the method, the purpose of the research was determined, i.e.: identify the main causes of porosity cluster on the mechanical seal from 410 alloy, where these causes will have the largest impact on occurrence the of incompatibility and simultaneously will have the most negative impact on the natural environment. Then, a team of experts was selected, with whose participation an Ishikawa diagram was developed for the potential causes of this problem. Next, on a scale of 1-7 the degree of importance of potential causes in the porosity cluster and its impact on the natural environment. On the basis of these assessments, an IPA chart was developed, after which it was shown that the main causes of the porosity cluster were inadequate lighting, moisture in the electrodes, lack of TPM (Total Productive Maintenance), dirty machine, dirty tools, and dirty casting mould. For these causes, improvement actions should be applied in the first place. These activities should aim to improve the quality of the product and simultaneously reduce the negative impact on the natural environment. This method can be applied to any product and incompatibilities detected on them. Therefore, the method can be used in service-production enterprises to improve products in terms of qualitative-environmental.

References

[1] D. Siwiec, A. Pacana. Model Supporting Development Decisions by Considering Qualitative–Environmental Aspects, Sustainability 13 (2021) art. 9067. https://doi.org/10.3390/su13169067

Quality Production Improvement and System Safety: QPI 16 - CZOTO 10   Materials Research Forum LLC
Materials Research Proceedings 34 (2023) 226-236   https://doi.org/10.21741/9781644902691-27

[2] D. Siwiec, A. Pacana. A Pro-Environmental Method of Sample Size Determination to Predict the Quality Level of Products Considering Current Customers' Expectations, Sustainability 13 (2021) art. 5542. https://doi.org/10.3390/su13105542

[3] KC. Tan, XX. Shen. Integrating Kano's Model in the Planning Matrix of Quality Function Deployment, Total Quality Management 11 (2000) 1141-1151. https://doi.org/10.1080/095441200440395

[4] MF. Elmorshedy, MR. Elkadeem, KM. Kotb, IB. Taha, D. Mazzeo. Optimal Design and Energy Management of an Isolated Fully Renewable Energy System Integrating Batteries and Supercapacitors, Energy Convers. Manag. 245 (2021) art. 114584. https://doi.org/10.1016/j.enconman.2021.114584

[5] R. Wolniak. Downtime in the Automotive Industry Production Process – Cause Analysis, Quality Innovation Prosperity 23 (2019) art. 101. https://doi.org/10.12776/qip.v23i2.1259

[6] A. Naskar, S. Paul. Non-Destructive Measurement of Grinding-Induced Deformation-Depth Using Grazing Incidence X-Ray Diffraction Technique, NDT & E International 126 (2022) 102592. https://doi.org/10.1016/j.ndteint.2021.102592

[7] M. Brown, D. Wright, R. M'Saoubi, J. McGourlay, M. Wallis, A. Mantle, P. Crawforth, H. Ghadbeigi. Destructive and Non-Destructive Testing Methods for Characterization and Detection of Machining-Induced White Layer: A Review Paper, CIRP J. Manuf. Sci. Technol. 23 (2018) 39–53. https://doi.org/10.1016/j.cirpj.2018.10.001

[8] M. Ingaldi, R. Ulewicz. How to Make E-Commerce More Successful by Use of Kano's Model to Assess Customer Satisfaction in Terms of Sustainable Development, Sustainability 11 (2019) art. 4830. https://doi.org/10.3390/su11184830

[9] P. Jonšta, Z. Jonšta, S. Brožová, M. Ingaldi, J. Pietraszek, D. Klimecka-Tatar. The Effect of Rare Earth Metals Alloying on the Internal Quality of Industrially Produced Heavy Steel Forgings, Materials 14 (2021) art. 5160. https://doi.org/10.3390/ma14185160

[10] Ł. J. Orman, G. Majewski, N. Radek, J. Pietraszek. Analysis of Thermal Comfort in Intelligent and Traditional Buildings, Energies (Basel) 15 (2022) art. 6522. https://doi.org/10.3390/en15186522

[11] S. Ishikawa. The Gear Geometry of Tooth Engagement in Harmonic Drive, Proc. of the JSME Semi-International Symposium, Tokyo (1967) 94–104.

[12] L. Liliana. A New Model of Ishikawa Diagram for Quality Assessment, IOP Conf Ser Mater Sci Eng 161 (2016) 012099. https://doi.org/10.1088/1757-899X/161/1/012099

[13] D.D. Shinde, S. Ahirrao, R. Prasad. Fishbone Diagram: Application to Identify the Root Causes of Student-Staff Problems in Technical Education, Wireless Pers. Commun. 100 (2018) 653–664. https://doi.org/10.1007/s11277-018-5344-y

[14] R. Raman, Y. Basavaraj. Quality Improvement of Capacitors Through Fishbone and Pareto Techniques, Int. J. Recent Eng. Technol. 8 (2019) 2248-2252. https://doi.org/10.3940/ijrte.B2444.078219

[15] A. Hoła, M. Sawicki, M. Szóstak. Methodology of Classifying the Causes of Occupational Accidents Involving Construction Scaffolding Using Pareto-Lorenz Analysis, Applied Sciences 8 (2018) art. 48. https://doi.org/10.3390/app8010048

[16] A. Pacana, D. Siwiec. Universal Model to Support the Quality Improvement of Industrial Products, Materials 14 (2021) art. 7872. https://doi.org/10.3390/ma14247872

[17] D. Siwiec, A. Pacana. A New Model Supporting Stability Quality of Materials and Industrial Products, Materials 15 (2022) art. 4440. https://doi.org/10.3390/ma15134440

[18] A. Pacana, D. Siwiec. Method of Choice: A Fluorescent Penetrant Taking into Account Sustainability Criteria. Sustainability 12 (2020) art. 5854. https://doi.org/10.3390/su12145854

[19] Y. Yi, M. Yang, C. Fu, Y. Li. Gaming Strategies within a Green Supply Chain Considering Consumers' Concern about the Greenness and Conformance Quality of Products, Environ. Sci. Pollut. Res. 29 (2022) 69082–69100. https://doi.org/10.1007/s11356-022-20318-7

[20] G. Li, H. Wu, S. P. Sethi, X. Zhang. Contracting green product supply chains considering marketing efforts in the circular economy era, Int. J. Prod. Econ. 234 (2021) art. 108041. https://doi.org/10.1016/j.ijpe.2021.108041

[21] Z. Hong, X. Guo. Green Product Supply Chain Contracts Considering Environmental Responsibilities, Omega 83 (2019) 155-166. https://doi.org/10.1016/j.omega.2018.02.010

[22] S. Borkowski, R. Ulewicz, J. Selejdak, M. Konstanciak, D. Klimecka-Tatar. The use of 3x3 matrix to evaluation of ribbed wire manufacturing technology, METAL 2012 - 21st Int. Conf. Metallurgy and Materials (2012), Ostrava, Tanger 1722-1728.

[23] R. Ulewicz, F. Nový. Quality management systems in special processes, Transp. Res. Procedia 40 (2019) 113-118. https://doi.org/10.1016/j.trpro.2019.07.019

[24] T. Lipiński, R. Ulewicz. The effect of the impurities spaces on the quality of structural steel working at variable loads, Open Eng. 11 (2021) 233-238. https://doi.org/10.1515/eng-2021-0024

[25] R. Ulewicz, J. Selejdak, S. Borkowski, M. Jagusiak-Kocik. Process management in the cast iron foundry, METAL 2013 - 22nd Int. Conf. Metallurgy and Materials (2013), Ostrava, Tanger 1926-1931.

[26] R. Ulewicz, D. Jelonek, M. Mazur. Implementation of logic flow in planning and production control, Management and Production Engineering Review 7 (2016) 89-94. https://doi.org/10.1515/mper-2016-0010

[27] N. Baryshnikova et al. Management approach on food export expansion in the conditions of limited internal demand, Pol. J. Manag. Stud. 21 (2020) 101-114. https://doi.org/10.17512/pjms.2020.21.2.08

[28] R. Ulewicz, M. Ulewicz. Problems in the Implementation of the Lean Concept in the Construction Industries, LNCE 47 (2020) 495-500. https://doi.org/10.1007/978-3-030-27011-7_63

[29] N. Radek et al. Technology and application of anti-graffiti coating systems for rolling stock, METAL 2019 28th Int. Conf. Metall. Mater. (2019) 1127-1132. ISBN 978-8087294925

[30] N. Radek et al. Influence of laser texturing on tribological properties of DLC coatings, Prod. Eng. Arch. 27 (2021) 119-123. https://doi.org/10.30657/pea.2021.27.15

[31] G. Filo, E. Lisowski, M. Domagała, J. Fabiś-Domagała, H. Momeni. Modelling of pressure pulse generator with the use of a flow control valve and a fuzzy logic controller, AIP Conf. Proc. 2029 (2018) art.20015. https://doi.org/10.1063/1.5066477

[32] G. Barucca et al. The potential of Λ and Ξ- studies with PANDA at FAIR, Europ. Phys. J. A 57 (2021) art.154 https://doi.org/10.1140/epja/s10050-021-00386-y

[33] M. Domagala et al. CFD Estimation of a Resistance Coefficient for an Egg-Shaped Geometric Dome, Appl. Sci. 12 (2022) art.10780. https://doi.org/10.3390/app122110780

Quality Production Improvement and System Safety: QPI 16 - CZOTO 10   Materials Research Forum LLC
Materials Research Proceedings 34 (2023) 226-236   https://doi.org/10.21741/9781644902691-27

[34] N. Radek et al. The influence of plasma cutting parameters on the geometric structure of cut surfaces, Mater. Res. Proc. 17 (2020) 132-137. https://doi.org/10.21741/9781644901038-20

[35] M. Krynke et al. Maintenance management of large-size rolling bearings in heavy-duty machinery, Acta Montan. Slovaca 27 (2022) 327-341. https://doi.org/10.46544/AMS.v27i2.04

[36] P. Regulski, K.F. Abramek The application of neural networks for the life-cycle analysis of road and rail rolling stock during the operational phase, Technical Transactions 119 (2022) art. e2022002. https://doi.org/10.37705/TechTrans/e2022002

[37] M. Patek et al. Non-destructive testing of split sleeve welds by the ultrasonic TOFD method, Manuf. Technol. 14 (2014) 403-407. https://doi.org/10.21062/ujep/x.2014/a/1213-2489/MT/14/3/403

[38] I. Miletić, A. Ilić, R.R. Nikolić, R. Ulewicz, L. Ivanović, N. Sczygiol. Analysis of selected properties of welded joints of the HSLA Steels, Materials 13 (2020) art.1301. https://doi.org/10.3390/ma13061301

[39] N. Radek et al. Properties of Steel Welded with CO2 Laser, Lecture Notes in Mechanical Engineering (2020) 571-580. https://doi.org/10.1007/978-3-030-33146-7_65

[40] A. Deja et al. Analysis and assessment of environmental threats in maritime transport, Transp. Res. Procedia 55 (2021) 1073-1080. https://doi.org/10.1016/j.trpro.2021.07.078

[41] M. Scendo et al. Influence of laser treatment on the corrosive resistance of WC-Cu coating produced by electrospark deposition, Int. J. Electrochem. Sci. 8 (2013) 9264-9277.

[42] T. Lipinski, J. Pietraszek. Influence of animal slurry on carbon C35 steel with different microstructure at room temperature, Engineering for Rural Development 21 (2022) 344-350. https://doi.org/10.22616/ERDev.2022.21.TF115

[43] T. Lipiński, J. Pietraszek. Corrosion of the S235JR Carbon Steel after Normalizing and Overheating Annealing in 2.5% Sulphuric Acid at Room Temperature, Mater. Res. Proc. 24 (2022) 102-108.

[44] E. Skrzypczak-Pietraszek, J. Pietraszek. Seasonal changes of flavonoid content in Melittis melissophyllum L. (Lamiaceae), Chem. Biodiv. 11 (2014) 562-570. https://doi.org/10.1002/cbdv.201300148

[45] E. Skrzypczak-Pietraszek et al. Enhanced accumulation of harpagide and 8-O-acetyl-harpagide in Melittis melissophyllum L. agitated shoot cultures analyzed by UPLC-MS/MS. PLoS ONE 13 (2018) art. e0202556. https://doi.org/10.1371/journal.pone.0202556

[46] R. Ulewicz et al. Fatigue strength of ductile iron in ultra-high cycle regime, Adv. Mater. Res. 874 (2014) 43-48. https://doi.org/10.4028/www.scientific.net/AMR.874.43

[47] S.T. Dziuba, M. Ingaldi. Segragation and recycling of packaging waste by individual consumers in Poland, Int. Multidisciplinary Scientific GeoConference Surveying Geology and Mining Ecology Management, SGEM 3 (2015) 545-552.

[48] J. Pietraszek, A. Szczotok, N. Radek. The fixed-effects analysis of the relation between SDAS and carbides for the airfoil blade traces. Archives of Metallurgy and Materials 62 (2017) 235-239. https://doi.org/10.1515/amm-2017-0035

[49] J. Pietraszek, N. Radek, A.V. Goroshko. Challenges for the DOE methodology related to the introduction of Industry 4.0. Production Engineering Archives 26 (2020) 190-194. https://doi.org/10.30657/pea.2020.26.33

[50] B. Jasiewicz et al. Inter-observer and intra-observer reliability in the radiographic measurements of paediatric forefoot alignment, Foot Ankle Surg. 27 (2021) 371-376. https://doi.org/10.1016/j.fas.2020.04.015

[51] J. Pietraszek. The modified sequential-binary approach for fuzzy operations on correlated assessments, LNAI 7894 (2013) 353-364. https://doi.org/10.1007/978-3-642-38658-9_32

[52] J. Pietraszek et al. Non-parametric assessment of the uncertainty in the analysis of the airfoil blade traces, METAL 2017 26th Int. Conf. Metall. Mater. (2017) 1412-1418. ISBN 978-8087294796

[53] J. Pietraszek et al. A. Advantages and disadvantages of various uncertainty assessment methods in material and technological predictive models, World Congress in Comput. Mech. and ECCOMAS Congress 1000 (2021) 1-8. https://doi.org/10.23967/wccm-eccomas.2020.053

[54] J. Fabiś-Domagała, G. Filo, H. Momeni, M. Domagała. Instruments of identification of hydraulic components potential failures, MATEC Web of Conf. 183 (2018) art.03008. https://doi.org/10.1051/matecconf/201818303008

[55] K. Knop et al. Evaluating and Improving the Effectiveness of Visual Inspection of Products from the Automotive Industry, Lecture Notes in Mechanical Engineering (2019) 231-243. https://doi.org/10.1007/978-3-030-17269-5_17

[56] G. Filo, P. Lempa. Analysis of Neural Network Structure for Implementation of the Prescriptive Maintenance Strategy, Mater. Res. Proc. 24 (2022) 273-280. https://doi.org/10.21741/9781644902059-40

[57] A. Pacana et al., Study on improving the quality of stretch film by Shainin method, Przemysl Chemiczny, 93 (2014) 243-245. https://doi.org/10.12916/przemchem.2014.243

[58] V.L. Putman, P.B. Paulus. Brainstorming, Brainstorming Rules and Decision Making, J. Creat. Behav. 43 (2009) 29-40. https://doi.org/10.1002/j.2162-6057.2009.tb01304.x

[59] G.M. Sullivan, A.R. Artino. Analyzing and Interpreting Data From Likert-Type Scales, J. Grad. Med. Educ. 5 (2013) 541–542. https://doi.org/10.4300/JGME-5-4-18

[60] J.M. Crosson, G. Tsebelis. Multiple Vote Electoral Systems: A Remedy for Political Polarization, J. Eur. Public Policy 29 (2022) 932–952. https://doi.org/10.1080/13501763.2021.1901962

[61] A. Gazda et al. Study on improving the quality of stretch film by Taguchi method, Przemysl Chemiczny, 92 6(2013) 980-982.

[62] J.J. Choi, C.A. Boger Jr. Association Planners' Satisfaction: An Application of Importance-Performance Analysis, J. Convention & Exhibition Manag. 2 (2000) 113-129. https://doi.org/10.1300/J143v02n02_10

[63] BY. Kim, H. Oh. An Extended Application of Importance-Performance Analysis, J. Hosp. Leisure Mark. 9 (2001) 107-125. https://doi.org/10.1300/J150v09n03_08

Quality Production Improvement and System Safety: QPI 16 - CZOTO 10   Materials Research Forum LLC
Materials Research Proceedings 34 (2023) 237-245   https://doi.org/10.21741/9781644902691-28

Numerical Analysis of a Pneumatic Cushion Operation Safety, including the Airflow Conditions

FILO Grzegorz[1, a *] and LEMPA Paweł[1,b]

[1]Cracow University of Technology, Kraków, Poland

[a]grzegorz.filo@pk.edu.pl, [b]pawel.lempa@pk.edu.pl

Keywords: Pneumatic Cushion, Heavy Load Movement, Air Velocity, Critical Air Flow, Numerical Modelling, Matlab, Simulink

Abstract. This work concerns studying the characteristics of a pneumatic cushion in a system for moving heavy loads. Particular emphasis has been placed on the airflow parameters through the main supply nozzle, including the average flow velocity. Critical flow at the speed of sound can occur under certain circumstances. In the subsequent steps, a mathematical model of a pneumatic cushion was formulated, next, a simulation model was built in the Matlab/Simulink system, and then numerical simulations were carried out. As a result, it was estimated whether there is a flow with a safe, appropriately low speed in a given range of load and supply pressure.

Introduction

The indoor transport of heavy loads is a common problem in many industrial companies. The movement must be carried out quickly and efficiently. Various transportation devices, such as cranes, forklifts, or conveyors, can be used for this purpose. Sometimes a beneficial alternative may be using transport platforms on pneumatic cushions [1, 2]. They are particularly convenient under the safety requirements regarding the inability to use combustion engines and electrical devices. The simple and compact design allows the cushions to be easily inserted under bulky loads or devices, even heavy ones or of enormous dimensions. The handy feature is their small height of approximately $30 - 40$ mm. However, using air cushion systems requires specific requirements to be met. Firstly, access to an air supply system with a certain pressure and flow rate must be provided. In addition, the operating floor must be suitably smooth and level, including the inclination $\alpha \leq 0.1^0$ and the roughness $R_a \leq 12.5$ μm [3, 4]. When the air supply is turned on, a thin film is formed between the lower plane of the cushion and the ground. The film is created by the air radially flowing out of the chamber located in the central part. It reduces the friction coefficient between the cushion and the ground to approximately $\mu = 0.001$.

In the standard design of a pneumatic cushion [2, 5, 6], the air flows through a nozzle located centrally in the cushion axis. An important issue is ensuring proper outflow uniformity and estimating the required parameters of the air supply system. The phenomena accompanying the flow of gases are complex; therefore, they are the subject of research in many research centres. The issue of the work includes modelling and numerical analyses of elements and systems such as pneumatic cylinders [7], pneumatic vibration isolation systems with multiple air chambers [8] or air springs [9]. Another crucial issue is the appropriate control of the airflow rate. The manufacturers usually provide manual control systems. Nevertheless, the possibility of using automatic controllers with advanced algorithms, such as fuzzy logic [10] or pulse width modulation (PWM) technique [11], is also being investigated. The aim of the research is often a reduction in energy consumption [12] or an increase in the efficiency of the system [13].

This paper deals with the issue of ensuring the safety of a pneumatic cushion operation by analysing the speed of airflow through the supply nozzle. The built model and numerical simulations carried out in the Matlab/Simulink system were aimed at estimating whether, given the power supply parameters and load values, there would appear conditions for the formation of a critical flow with the speed of sound.

Reliability and safety of pneumatic devices are of paramount importance as potential failures can pose a threat to the health and lives of workers, especially in high-pressure conditions [14-16]. Therefore, it is necessary to develop scenarios for potential failures [17-19], prevent them, and mitigate their consequences when they occur. This places a requirement on the devices and their individual working components to exhibit high-quality characteristics [20-22].

The main risk factors for failures are equipment wear [23-25], failures of structural joints, including welded joints [26-28], and unwanted gas and liquids ingress [29-31]. Preventing failures requires the use of materials with desired technological and functional properties [32-34], and in specific cases, complementing them with coatings that modify their characteristics [35], including coatings with special functional properties [36-38] and those that alter mechanical properties [39-41]. Additionally, the morphology of the surface layer can be modified by changing frictional conditions [42-44]. The complexity of methods and the multifactorial nature of processes necessitate the application of dimensionality reduction techniques [45,46] and process optimization methods [47-49], including nonparametric techniques [50,51]. Ultimately, in conjunction with FMEA scenarios, this enables a reduction in the required resources [52].

In addition to the primary benefits of increased reliability and safety, it also leads to lower energy consumption [53], thereby increasing the applicability of pneumatic cushion transporters, both as standalone systems and as part of larger devices [54-56], posing competition to typical wheeled or conveyor-based transportation [57]. The military has long recognized the advantages of such transportation, especially in the case of air-cushion landing crafts [58-60].

Working Principle of a Pneumatic Cushion

The analysed pneumatic cushion is based on a DELU 4LTM-200-1 type (fig. 1). It consists of a rigid aluminium plate (2) with four landing pads (3). In the lower part, there are rubber bellows (4) of a shape similar to a torus. The air flows through the inlet port (1) and the nozzle (5) to the central chamber of the V_1 volume. The cushion has the following nominal operational parameters: input pressure $p_0 = 2.4$ bar, and the air consumption $Q_0 = 180$ dm^3/min.

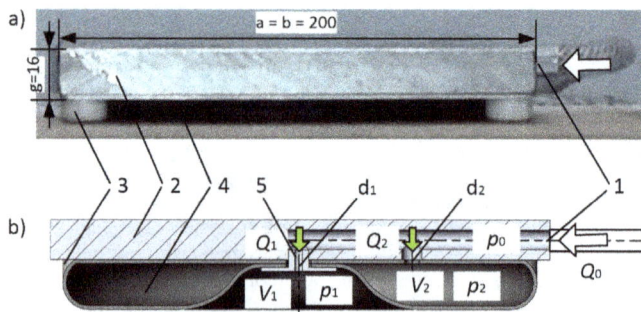

Fig. 1 Pneumatic cushion: (a) real cushion, b) cross-section through the 3D model;
1 – inlet port, 2 – aluminium plate, 3 – landing pad, 4 – rubber bellows, 5 – nozzle

After opening the airflow, the p_1 pressure appears at the inlet. Air flows into the central chamber of V_1 volume through the d_1 nozzle and simultaneously into the bellows through the d_2 nozzle. After some time, the bellows fill up, and the flow occurs through the d_1 nozzle. When the p_1 pressure reaches a sufficiently high value, the gap between the lower surface of the bellows and the ground is formed, and the outflow to the atmosphere begins. The pressure difference

Quality Production Improvement and System Safety: QPI 16 - CZOTO 10   Materials Research Forum LLC
Materials Research Proceedings 34 (2023) 237-245   https://doi.org/10.21741/9781644902691-28

$\Delta p = p_0 - p_1$ is established, which determines the Q_1 flow rate through the nozzle and, thus, the average velocity of the air stream.

The mass stream $\dot{m}_1(t)$ of air flowing into the central chamber through the d_1 nozzle can be determined by taking into account the critical velocity of the flow [5, 6]. The air velocity depends on the pressure ratio p_1 / p_0. In the case of assuming an adiabatic change during the flow (omission heat exchange with the environment), the threshold value of this coefficient is constant and equal to $\beta_0 = 0.53$. If the value of the pressure ratio $p_1 / p_0 \leq \beta_0$, which means $p_1 \leq 0.53 \cdot p_0$, the supercritical flow occurs at the speed of sound, otherwise the flow velocity is lower. Hence, the equation of the air inflow to the V_1 volume can be written as follows:

$$\frac{dm_1(t)}{dt} = \frac{\pi \cdot d_1^2}{4} \cdot p_0(t) \cdot \sqrt{\frac{2}{R \cdot T_0}} \cdot \Psi(t),\tag{1}$$

where T_0 is air temperature at the inlet, the specific air constant $R = 287 \ J/(kg \cdot K)$ and the $\psi(t)$ coefficient for the subcritical flow (heat capacity ratio $\kappa = 1.4$) is determined from the relation ([6]):

$$\psi(t) = \sqrt{\frac{\kappa}{\kappa - 1} \left[\left(\frac{p_1(t)}{p_0(t)} \right)^{2/\kappa} - \left(\frac{p_1(t)}{p_0(t)} \right)^{(\kappa+1)/\kappa} \right]}.\tag{2}$$

In the case of supercritical flow, the coefficient ψ has a constant value of $\psi_{crit} = 0.684$. Hence, the equation (1) is simplified to the following form:

$$\frac{dm_1(t)}{dt} = 0.7594 \cdot \frac{d_1^2 \cdot p_0(t)}{\sqrt{R \cdot T_0}}.\tag{3}$$

The volumetric flow rate can be determined by dividing the mass flow by the air density:

$$Q_1(t) = \frac{1}{\rho} \cdot \frac{dm_1(t)}{dt},\tag{4}$$

and thus the average velocity of the air stream is:

$$v_{av}(t) = \frac{4 Q_1(t)}{\pi \cdot d_1^2}.\tag{5}$$

The mathematical model also includes equations related to the V_2 chamber flow, bottom gap formation, payload, equation of motion in the vertical direction, etc. The complete model the one can find in the publication [1].

Simulink Model and Numerical Analysis
The created simulation model in the form of a Simulink block diagram is shown in fig. 2. It includes subsystems for an input step function U_in, a relief valve INLET_VALVE, two nozzles d1_nozzle, d2_nozzle and two chambers V1_chamber, V2_chamber, respectively. Signals are related to p_0, p_1, p_2 pressures, dm1/dt, dm2/dt flow rates, z_f, z_2 heights of the air gap and the bellows, V_2 chamber volume and v_1_avg flow velocity through the nozzle. The results are stored in a file using a Results_to_file block.

The test plan assumed checking the flow velocity through the nozzle d1 for the load values in the $F_{load} = 100, 500, 1500, 2500$ N and the supply pressure $p_0 = 1.2, 1.8, 2.4$ bar. The first step determined the velocity for different load values at a fixed supply pressure. Next, simulations were carried out at different pressure values for the fixed load force. Fig. 3 and fig. 4 show the results obtained in the first and the second step, respectively.

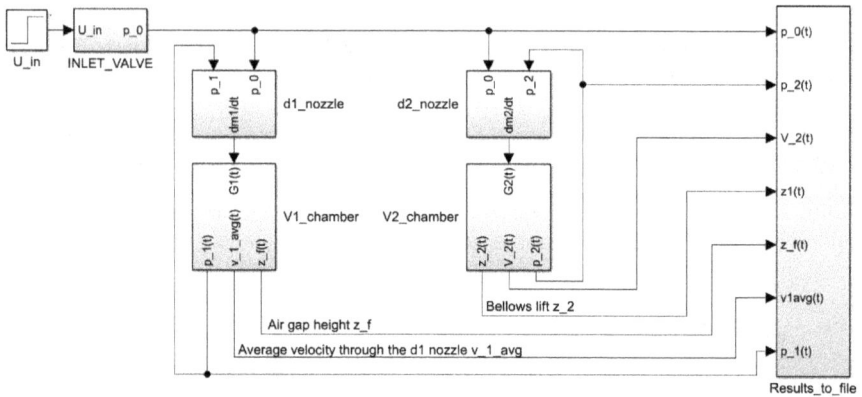

Fig. 2 Block diagram of a Simulink model

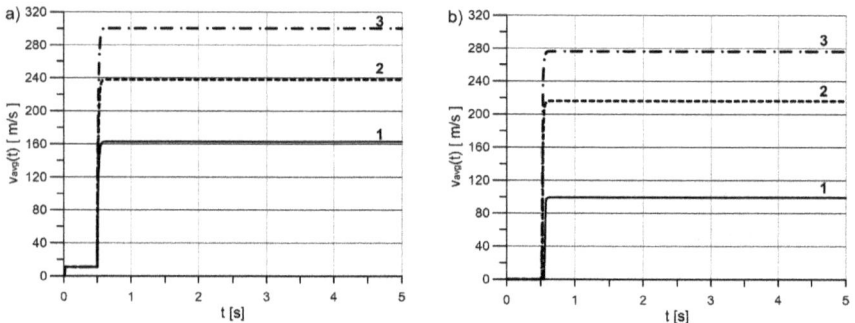

Fig. 3 Air velocity for fixed load force value: a) F_{load} = 100 N, b) F_{load} = 1500 N; $1 - p_0$ = 1.2 bar, $2 - p_0$ = 1.8 bar, $3 - p_0$ = 2.4 bar

Quality Production Improvement and System Safety: QPI 16 - CZOTO 10   Materials Research Forum LLC
Materials Research Proceedings 34 (2023) 237-245   https://doi.org/10.21741/9781644902691-28

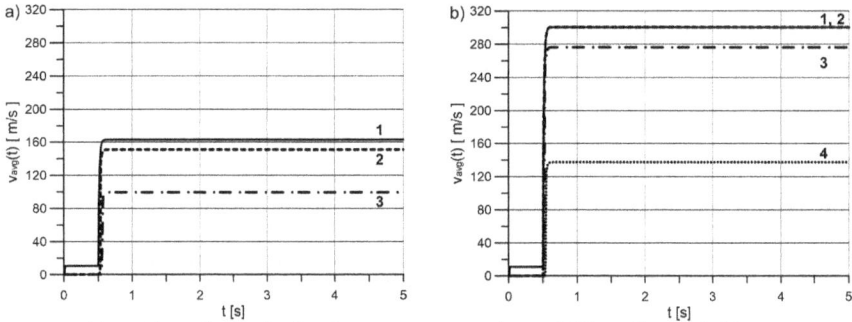

Fig. 4 Air velocity for fixed input pressure: a) p_0 = 1.2 bar, *b)* p_0 = 2.4 bar;
1 – F_{load} = 100 N, 2 – F_{load} = 500 N, 3 – F_{load} = 1500 N, 4 – F_{load} = 2500 N

Fig. 3 shows that the velocity value for F_{load} =100 N is in the range between 160 and 295 m/s, while for F_{load} = 1500 N, between 100 and 275 m/s, respectively. In the case of F_{load} =2500 N, the air film was formed only at the pressure p_0 = 2.4 bar, and in this case, the velocity value 137 m/s was obtained. As arises from fig. 4, at a supply pressure of 1.2 bar, the velocity value ranges from 100 to 160 m/s, and the maximum load force is 1500 N. When the supply pressure is 2.4 bar, the obtained air velocity varies from 140 to 295 m/s, respectively.

Summary
The article aimed to conduct a numerical analysis of the parameters of a pneumatic cushion intended for use in heavy load handling systems. In particular, it was checked whether the conditions of critical flow formation with the speed of sound could occur in the air supply nozzle. The simulations in Simulink showed that the highest air flow velocity occurs when the maximum supply pressure is set at minimum load. However, the maximum flow velocity obtained is less than 300 m/s. In practice, putting the full airflow at minimum payload can occur with manual pressure setting and may be prevented by applying an automatic control system.

References
[1] E. Lisowski, G. Filo. Automated heavy load lifting and moving system using pneumatic cushions, Autom. Constr. 50 (2015) 91-101. https://doi.org/10.1016/j.autcon.2014.12.004.

[2] E. Lisowski, D. Kwiatkowski. Determination of basic parameters of pneumatic transport platform system with pneumatic bags, Przegląd Mechaniczny 12(2) (2012) 45-48.

[3] K. Kaya, O. Özcan. A numerical investigation on aerodynamic characteristics of an air-cushion vehicle. J. Wind Eng. Ind. Aerodyn. 120 (2013) 70-80.
https://doi.org/10.1016/j.jweia.2013.06.012.

[4] E. Lisowski, D. Kwiatkowski. Nonlinear static analysis of air cushion in SolidWorks Simulation 2016. Technical Transactions 115 (2018) 211-218.
https://doi.org/10.4467/2353737XCT.18.030.8003

[5] J. Wołkow, R. Dindorf. Teoria i obliczenia układów pneumatycznych. Wydawnictwo Politechniki Krakowskiej, Kraków, 1994.

[6] W. Szejnach. Napęd i sterowanie pneumatyczne. WNT, Warszawa, 2003. ISBN 8320429188

Quality Production Improvement and System Safety: QPI 16 - CZOTO 10   Materials Research Forum LLC
Materials Research Proceedings 34 (2023) 237-245   https://doi.org/10.21741/9781644902691-28

[7] D.N. Dihovicni, M. Medenica. Simulation, Analyze and Program Support for Pneumatic Cylinder System. Proc. WCE 2009 World Cong. Engin., Vol I, 149-152. ISBN: 9789881701251

[8] J.-H. Moon, B.-G. Lee. Modeling and sensitivity analysis of a pneumatic vibration isolation system with two air chambers. Mechanism and Machine Theory 45 (2010) 1828-1850. https://doi.org/10.1016/j.mechmachtheory.2010.08.006.

[9] T. Bešter, M. Fajdiga, M. Nagode. Application of Constant Amplitude Dynamic Tests for Life Prediction of Air Springs at Various Control Parameters. Strojniški Vestnik/Journal of Mechanical Engineering 60 (2014) 241-249. https://doi.org/10.5545/sv-jme.2013.1348

[10]E. Lisowski, G. Filo. Pressure control in air cushions of the mobile platform. Journal of KONES Powertrain and Transport 18 (2011) 261-270.

[11]A. Messina, N. Giannoccaro, A. Gentile. Experimenting and modeling the dynamics of pneumatic actuators controlled by the pulse width modulation (PWM) technique. Mechatronics 15 (2005) 859-881. https://doi.org/10.1016/j.mechatronics.2005.01.003

[12]R. Dindorf. Estimating Potential Energy Savings in Compressed Air Systems. Procedia Engineering 39 (2012) 204-211. https://doi.org/10.1016/j.proeng.2012.07.026

[13]E. Richer, Y. Hurmuzlu. A high performance pneumatic force actuator system :part 1 - nonlinear mathematical model. J. Dyn. Syst. Meas. Control 122 (2000) 416-425. https://doi.org/10.1115/1.1286336

[14] G. Barucca et al. The potential of Λ and Ξ- studies with PANDA at FAIR, Europ. Phys. J. A 57 (2021) art.154 https://doi.org/10.1140/epja/s10050-021-00386-y

[15] M. Domagala et al. CFD Estimation of a Resistance Coefficient for an Egg-Shaped Geometric Dome, Appl. Sci. 12 (2022) art.10780. https://doi.org/10.3390/app122110780

[16] M. Domagala et al. The Influence of Oil Contamination on Flow Control Valve Operation, Mater. Res. Proc. 24 (2022) 1-8. https://doi.org/10.21741/9781644902059-1

[17] J. Fabiś-Domagała et al. Instruments of identification of hydraulic components potential failures, MATEC Web of Conf. 183 (2018) art.03008. https://doi.org/10.1051/matecconf/201818303008

[18] K. Knop et al. Evaluating and Improving the Effectiveness of Visual Inspection of Products from the Automotive Industry, Lecture Notes in Mechanical Engineering (2019) 231-243. https://doi.org/10.1007/978-3-030-17269-5_17

[19] J. Fabis-Domagala et al. A concept of risk prioritization in FMEA analysis for fluid power systems, Energies 14 (2021) art. 6482. https://doi.org/10.3390/en14206482

[20] R. Ulewicz, F. Nový. Quality management systems in special processes, Transp. Res. Procedia 40 (2019) 113-118. https://doi.org/10.1016/j.trpro.2019.07.019

[21] D. Siwiec et al. Improving the non-destructive test by initiating the quality management techniques on an example of the turbine nozzle outlet, Materials Research Proceedings 17 (2020) 16-22. https://doi.org/10.21741/9781644901038-3

[22] K. Czerwinska et al. Improving quality control of siluminial castings used in the automotive industry, METAL 2020 29th Int. Conf. Metall. Mater. (2020) 1382-1387. https://doi.org/10.37904/metal.2020.3661

Quality Production Improvement and System Safety: QPI 16 - CZOTO 10   Materials Research Forum LLC
Materials Research Proceedings 34 (2023) 237-245   https://doi.org/10.21741/9781644902691-28

[23] S. Marković et al. Exploitation characteristics of teeth flanks of gears regenerated by three hard-facing procedures, Materials 14 (20210 art. 4203. https://doi.org/10.3390/ma14154203

[24] M. Krynke et al. Maintenance management of large-size rolling bearings in heavy-duty machinery, Acta Montan. Slovaca 27 (2022) 327-341. https://doi.org/10.46544/AMS.v27i2.04

[25] P. Regulski, K.F. Abramek. The application of neural networks for the life-cycle analysis of road and rail rolling stock during the operational phase, Technical Transactions 119 (2022) art. e2022002. https://doi.org/10.37705/TechTrans/e2022002

[26] M. Patek et al. Non-destructive testing of split sleeve welds by the ultrasonic TOFD method, Manuf. Technol. 14 (2014) 403-407. https://doi.org/10.21062/ujep/x.2014/a/1213-2489/MT/14/3/403

[27] N. Radek, J. Pietraszek, A. Goroshko. The impact of laser welding parameters on the mechanical properties of the weld, AIP Conf. Proc. 2017 (2018) art.20025. https://doi.org/10.1063/1.5056288"

[28] N. Radek et al. Properties of Steel Welded with CO2 Laser, Lecture Notes in Mechanical Engineering (2020) 571-580. https://doi.org/10.1007/978-3-030-33146-7_65

[29] M. Ulewicz et al. Ion flotation of zinc(II) and cadmium(II) in the presence of side-armed diphosphaza-16-crown-6 ethers, Sep. Sci. Technol. 38 (2003) 633-645. https://doi.org/10.1081/SS-120016655

[30] M. Zenkiewicz, T. Zuk, J. Pietraszek, P. Rytlewski, K. Moraczewski, M. Stepczyńska. Electrostatic separation of binary mixtures of some biodegradable polymers and poly(vinyl chloride) or poly(ethylene terephthalate), Polimery/Polymers 61 (2016) 835-843. https://doi.org/10.14314/polimery.2016.835

[31] T. Zuk et al. Modeling of electrostatic separation process for some polymer mixtures, Polymers 61 (2016) 519-527. https://doi.org/10.14314/polimery.2016.519

[32] P. Szataniak, F. Novy, R. Ulewicz. HSLA steels - Comparison of cutting techniques, METAL 2014 - 23rd Int. Conf. Metallurgy and Materials (2014), Ostrava, Tanger, 778-783.

[33] D. Klimecka-Tatar, M. Ingaldi. Assessment of the technological position of a selected enterprise in the metallurgical industry, Mater. Res. Proc. 17 (2020) 72-78. https://doi.org/10.21741/9781644901038-11

[34] P. Jonšta et al. The effect of rare earth metals alloying on the internal quality of industrially produced heavy steel forgings, Materials 14 (2021) art.5160. https://doi.org/10.3390/ma14185160

[35] W. Zórawski et al. Plasma-sprayed composite coatings with reduced friction coefficient, Surf. Coat. Technol. 202 (2008) 4578-4582. https://doi.org/10.1016/j.surfcoat.2008.04.026

[36] N. Radek et al. Microstructure and tribological properties of DLC coatings, Mater. Res. Proc. 17 (2020) 171-176. https://doi.org/10.21741/9781644901038-26

[37] N. Radek et al. Influence of laser texturing on tribological properties of DLC coatings, Prod. Eng. Arch. 27 (2021) 119-123. https://doi.org/10.30657/pea.2021.27.15

[38] N. Radek et al. Operational properties of DLC coatings and their potential application, METAL 2022 31st Int. Conf. Metall. Mater. (2022) 531-536. https://doi.org/10.37904/metal.2022.4491

[39] N. Radek, J. Konstanty. Cermet ESD coatings modified by laser treatment, Arch. Metall. Mater. 57 (2012) 665-670. https://doi.org/10.2478/v10172-012-0071-y

[40] N. Radek et al. The effect of laser treatment on operational properties of ESD coatings, METAL 2021 30th Ann. Int. Conf. Metall. Mater. (2021) 876-882. https://doi.org/10.37904/metal.2021.4212

[41] N. Radek et al. The impact of laser processing on the performance properties of electro-spark coatings, 14th World Congress in Computational Mechanics and ECCOMAS Congress 1000 (2021) 1-10. https://doi.org/10.23967/wccm-eccomas.2020.336

[42] N. Radek et al. The WC-Co electrospark alloying coatings modified by laser treatment, Powder Metall. Met. Ceram. 47 (2008) 197-201. https://doi.org/10.1007/s11106-008-9005-7

[43] N. Radek et al. Laser Processing of WC-Co Coatings, Mater. Res. Proc. 24 (2022) 34-38. https:10.21741/9781644902059-6

[44] P. Kurp, H. Danielewski. Metal expansion joints manufacturing by a mechanically assisted laser forming hybrid method – concept, Technical Transactions 119 (2022) art. e2022008. https://doi.org/10.37705/TechTrans/e2022008

[45] J. Pietraszek et al. The principal component analysis of tribological tests of surface layers modified with IF-WS2 nanoparticles, Solid State Phenom. 235 (2015) 9-15. https://doi.org/10.4028/www.scientific.net/SSP.235.9

[46] J. Pietraszek, E. Skrzypczak-Pietraszek. The uncertainty and robustness of the principal component analysis as a tool for the dimensionality reduction. Solid State Phenom. 235 (2015) 1-8. https://doi.org/10.4028/www.scientific.net/SSP.235.1

[47] R. Dwornicka, J. Pietraszek. The outline of the expert system for the design of experiment, Prod. Eng. Arch. 20 (2018) 43-48. https://doi.org/10.30657/pea.2018.20.09

[48] J. Pietraszek, N. Radek, A.V. Goroshko. Challenges for the DOE methodology related to the introduction of Industry 4.0. Prod. Eng. Arch. 26 (2020) 190-194. https://doi.org/10.30657/pea.2020.26.33

[49] B. Jasiewicz et al. Inter-observer and intra-observer reliability in the radiographic measurements of paediatric forefoot alignment, Foot Ankle Surg. 27 (2021) 371-376. https://doi.org/10.1016/j.fas.2020.04.015

[50] J. Pietraszek. The modified sequential-binary approach for fuzzy operations on correlated assessments, LNAI 7894 (2013) 353-364. https://doi.org/10.1007/978-3-642-38658-9_32

[51] J. Pietraszek et al. Non-parametric assessment of the uncertainty in the analysis of the airfoil blade traces, METAL 2017 26th Int. Conf. Metall. Mater. (2017) 1412-1418. ISBN 978-8087294796

[52] A. Maszke, R. Dwornicka, R. Ulewicz. Problems in the implementation of the lean concept at a steel works - Case study, MATEC Web of Conf. 183 (2018) art.01014. https://doi.org/10.1051/matecconf/201818301014

[53] Ł.J. Orman. Enhancementof pool boiling heat transfer with pin-fin microstructures, J. Enhanc. Heat Transf. 23 (2016) 137-153. https://doi.org/10.1615/JEnhHeatTransf.2017019452

Quality Production Improvement and System Safety: QPI 16 - CZOTO 10   Materials Research Forum LLC
Materials Research Proceedings 34 (2023) 237-245   https://doi.org/10.21741/9781644902691-28

[54] A. Goroshko et al. Construction and practical application of hybrid statistically-determined models of multistage mechanical systems, Mechanika 20 (2014) 489-493. https://doi.org/10.5755/j01.mech.20.5.8221

[55] R. Ulewicz, M. Mazur. Economic aspects of robotization of production processes by example of a car semi-trailers manufacturer, Manuf. Technol. 19 (2019) 1054-1059. https://doi.org/10.21062/ujep/408.2019/a/1213-2489/MT/19/6/1054

[56] I. Drach et al. Design Principles of Horizontal Drum Machines with Low Vibration, Adv. Sci. Technol. Res. J. 15 (2021) 258-268. https://doi.org/10.12913/22998624/136441

[57] N. Radek, R. Dwornicka. Fire properties of intumescent coating systems for the rolling stock, Commun. - Sci. Lett. Univ. Zilina 22 (2020) 90-96. https://doi.org/10.26552/com.C.2020.4.90-96

[58] W. Przybył et al. Virtual Methods of Testing Automatically Generated Camouflage Patterns Created Using Cellular Automata, Mater. Res. Proc. 24 (2022) 66-74. https://doi.org/10.21741/9781644902059-11

[59] N. Radek et al. Operational tests of coating systems in military technology applications, Eksploat. i Niezawodn. 25 (2023) art.12. https://doi.org/10.17531/ein.2023.1.12

[60] W. Przybył et al. Microwave absorption properties of carbonyl iron-based paint coatings for military applications, Def. Technol. 22 (2023) 1-9. https://doi.org/10.1016/j.dt.2022.06.013

Quality Production Improvement and System Safety: QPI 16 - CZOTO 10   Materials Research Forum LLC
Materials Research Proceedings 34 (2023) 246-251   https://doi.org/10.21741/9781644902691-29

Static Mechanical Force Amplifier on the Example of a Fastener with an Electromagnetic Bolt

JASIŃSKI Wiesław [1] *, KRYSIAK Piotr [1] and PICHLAK Cezary [1]

[1]Military Institute of Engineer Technology, ul. Obornicka 136, 50-961 Wrocław

Key words: Strength Amplifier, Fastener, Bolt, Construction Optimization

Abstract. Modern mechanical constructions consist of purchased elements and workshop-made elements. Undoubtedly, purchased components – made in specialized factories – have a definite advantage in quality and durability. Unfortunately, sometimes the operating parameters of a component for a given application are insufficient. Then it remains to design and manufacture a given element in the workshop – losing quality and costs. This paper presents a different approach. Application requirements can be customized just include an additional (simple) element, e.g. a force amplifier, and a purchased component can be used.

Introduction

In the construction of mechanisms, fasteners are used to hold stored energy and then release it to perform the necessary work. These types of fasteners should hold the given load when fastened, and then easily unfasten under load. Therefore, the design of such mechanisms should be specific. It is particularly important that at the time of opening the clasp there are no abrasions and mechanical jerks.

The above-mentioned requirements are met by electric strikes available on the market, commonly used in door locks. Unfortunately, they are not designed to transfer too much force when closed. It is therefore necessary to limit the force to an acceptable value. This can be done with a static force amplifier which is based on the principle of equilibrium of mechanical moments.

This paper presents an exemplary construction of a mechanical force amplifier mechanism with the use of an electromagnetic lock.

Materials and Methods

Autodesk Inventor Professional 2021 parametric software version 2021.3.3 was used for the design (Fig.1 and Fig.2.). The program contains the necessary mechanical standards and a number of useful calculation modules and wizards.

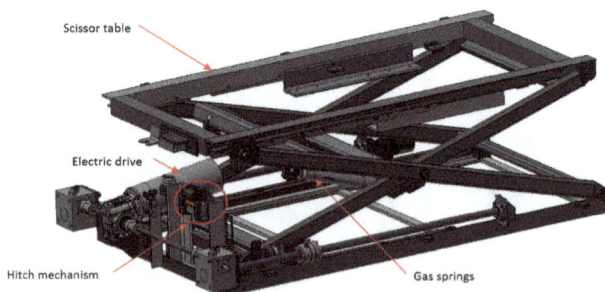

Fig.1. *View of the device containing the considered mechanism.*

Quality Production Improvement and System Safety: QPI 16 - CZOTO 10   Materials Research Forum LLC
Materials Research Proceedings 34 (2023) 246-251   https://doi.org/10.21741/9781644902691-29

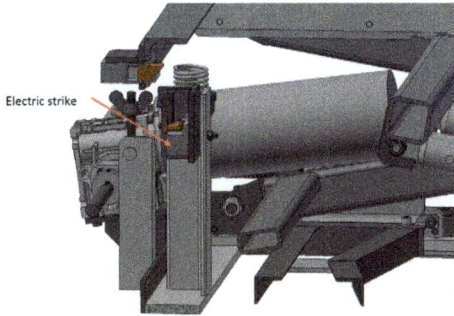

Fig.2. *Cross-section of the working fastening mechanism.*

The process of creating the mechanism required the initial preparation of pre-designed components. All the parts were then interconnected by mutual ties/connections. This procedure allowed to verify the matching of components and interconnections.

Results and Discussion

The purpose of this work was to present a specific approach to purchasing components commonly used in the industry, taking into account the permissible nominal parameters. If the permissible values are exceeded in a specific application, it is possible to use an additional mechanism, e.g. a static mechanical force amplifier, which will adjust these values.

Design of the hitch mechanism. The ASSA ABLOY shopping component Electric strike 143.U Cat. No. 143.Q34 was used to construct the fastening mechanism. The cam with eccentrically made teeth is an element that strengthens the force of the impact on the electric strike. One of the teeth is supported on the latch of the electric strike and the other of the teeth is used to fasten the lock (door) tongue carrying the payload. The diagram of the mechanism's operation is shown in Fig.3.

Quality Production Improvement and System Safety: QPI 16 - CZOTO 10   Materials Research Forum LLC
Materials Research Proceedings 34 (2023) 246-251   https://doi.org/10.21741/9781644902691-29

b)

a)

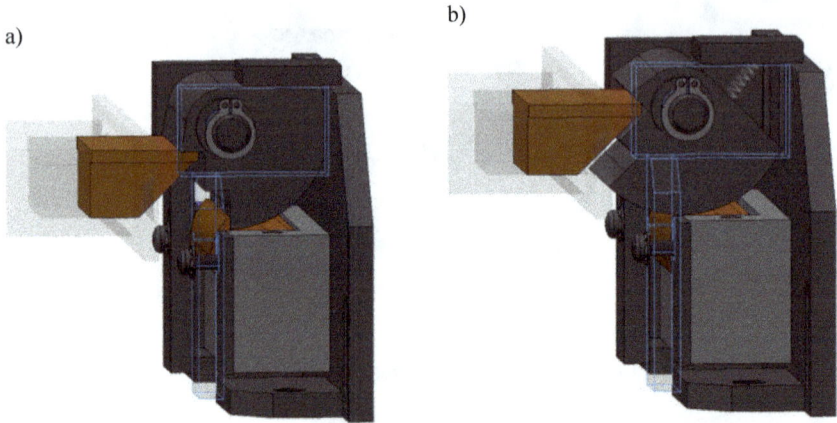

Fig.3. *The essence of the operation of the hitch mechanism: a) locked, b) unlocked*

Such a mechanism should effectively enhance the ability of the catch to transfer static forces. This means that the load may exceed the ratings specified by the manufacturer of the electric strike. At the same time, the catch can be opened and the load placed on it can be released without using much force and almost instantaneously. The instantaneous operation of such a mechanism is sometimes a particularly important feature.

Calculations for the hitch mechanism. Calculation of the fastening mechanism was done in a classic way. Fig.4 shows the load diagram of the fastener.

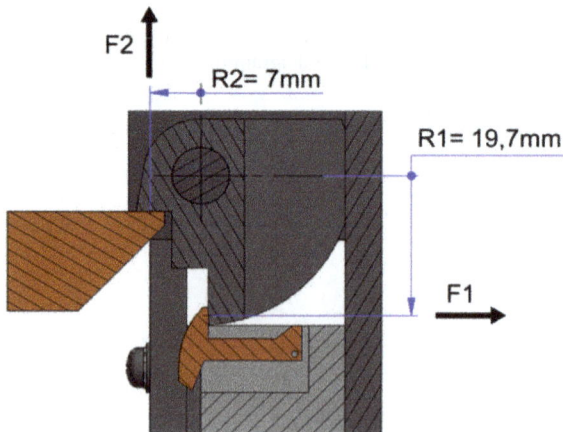

Fig.4. *Clasp load diagram*

Quality Production Improvement and System Safety: QPI 16 - CZOTO 10   Materials Research Forum LLC
Materials Research Proceedings 34 (2023) 246-251   https://doi.org/10.21741/9781644902691-29

Calculation formula from the equation of moments:

$$F2 = \frac{F1 \cdot R1}{R2}$$

where F1 acc. manufacturer's data is 8,000 N

$$F2 = \frac{8\,000\ N \cdot 19.7\ mm}{7\ mm} = 22\,514.3\ N$$

so the mechanism can carry 22,514.3 N. Reinforcement of mechanism strength:

$$w = \frac{F2}{F1}$$

totals:

$$w = \frac{22\,514.3}{8\,000} = 2.8$$

Thus, the design of the mechanism can be considered correct.

Conclusions

Contemporary designed mechanical structures should consist of as many purchased components as possible, rather than workshop-made components. These principal results from the fact that purchased components – made in specialized factories – have a definite advantage in quality and durability. In a situation where it is difficult to select a component with the right parameters, the problem can be solved by reversing the situation – lowering the requirements for the component by using, for example, a static force amplifier that will reduce these requirements.

The presented approach, which compensates for parametric or functional deficiencies of a factory-prepared component by adding an additional mechanical or structural adapter, can be successfully applied in analogous situations in other industries to reduce costs for unit or small-batch productions. Such adapters can be used in construction, particularly in the field of "smart" building information modeling (BIM), where effectors need to be widely utilized [11], sometimes based on composite materials [12]. Another example could be the deposition of additional special coatings [13, 14] to enhance wear resistance. However, in each of these cases, a thorough analysis of capabilities [15] is necessary beforehand to ensure that the desired modifications will be achieved. Stereology methods [16] and industrial statistics, both in classical [17-19] and nonparametric approaches [20-22], even supported by an expert system [23], can be helpful in this regard.

References

[1] A. Arian et al. Kinematic and dynamic analysis of the Gantry-Tau, a 3-DoF translational parallel manipulator. Appl. Math. Model. 51 (2017) 217-231. https://doi.org/10.1016/j.apm.2017.06.012

[2] J. Frączek, M. Wojtyra. Kinematyka układów wieloczłonowych – metody obliczeniowe. WNT, Warszawa, 2008.

[3] M. Gawrysiak. Mechatronika i projektowanie mechatroniczne. Wyd. Politechniki Białostockiej, Białystok, 1997.

[4] A. Gronowicz. Podstawy analizy układów kinematycznych. Oficyna Wydawnicza Politechniki Wrocławskiej, Wrocław, 2003.

[5] P. Hejma et al. Analytic Analysis of a Cam Mechanism. Procedia Eng. 177 (2017) 3-10. https://doi.org/10.1016/j.proeng.2017.02.175

[6] S. Miller. Teoria maszyn i mechanizmów – analiza układów kinematycznych. Oficyna Wydawnicza Politechniki Wrocławskiej, Wrocław, 1996.

[7] J. Ormezowski. Analiza dynamiczna mechanizmu hamującego. Archiwum Motoryzacji 1 (2010) 27-34.

[8] A. Sapietova et al. Analysis and Implementation of Input Load Effects on an Air Compressor Piston in MSC.ADAMS. Procedia Eng. 177 (2017) 554-561. https://doi.org/10.1016/j.proeng.2017.02.260

[9] P. Sperzyński, J. Szrek, A. Gronowicz. Synthesis of a mechanism for generating straight line indexing trajectory. Acta Mechanica et Automatica 4 (2010) 124-129.

[10] J. Vavro Jr. et al. Kinematic and dynamic analysis of planar mechanisms by means of the SolidWorks software. Procedia Eng. 177 (2017) 476-481. https://doi.org/10.1016/j.proeng.2017.02.248

[11] J.M. Djoković et al. Selection of the Optimal Window Type and Orientation for the Two Cities in Serbia and One in Slovakia, Energies 15 (2022) art.323. https://doi.org/10.3390/en15010323

[12] J. Korzekwa et al. Tribological behaviour of Al_2O_3/inorganic fullerene-like WS_2 composite layer sliding against plastic, Int. J. Surf. Sci. Eng. 10 (2016) 570-584. https://doi.org/10.1504/IJSURFSE.2016.081035

[13] N. Radek et al. Microstructure and tribological properties of DLC coatings, Mater. Res. Proc. 17 (2020) 171-176. https://doi.org/10.21741/9781644901038-26

[14] N. Radek et al. Influence of laser texturing on tribological properties of DLC coatings, Prod. Eng. Arch. 27 (2021) 119-123. https://doi.org/10.30657/pea.2021.27.15

[15] B. Jasiewicz et al. Inter-observer and intra-observer reliability in the radiographic measurements of paediatric forefoot alignment, Foot Ankle Surg. 27 (2021) 371-376. https://doi.org/10.1016/j.fas.2020.04.015

[16] J. Pietraszek, A. Szczotok, N. Radek. The fixed-effects analysis of the relation between SDAS and carbides for the airfoil blade traces. Arch. Metall. Mater. 62 (2017) 235-239. https://doi.org/10.1515/amm-2017-0035

[17] J. Pietraszek et al. Factorial approach to assessment of GPU computational efficiency in surrogate models, Adv. Mater. Res. 874 (2014) 157-162. https://doi.org/10.4028/www.scientific.net/AMR.874.157

[18] J. Pietraszek et al. The parametric RSM model with higher order terms for the meat tumbler machine process, Solid State Phenom. 235 (2015) 37-44. https://doi.org/10.4028/www.scientific.net/SSP.235.37

[19] J. Pietraszek et al. Challenges for the DOE methodology related to the introduction of Industry 4.0. Prod. Eng. Arch. 26 (2020) 190-194. https://doi.org/10.30657/pea.2020.26.33

[20] J. Pietraszek. Fuzzy regression compared to classical experimental design in the case of flywheel assembly, LNAI 7267 LNAI (2012) 310-317. https://doi.org/10.1007/978-3-642-29347-4_36

Quality Production Improvement and System Safety: QPI 16 - CZOTO 10   Materials Research Forum LLC
Materials Research Proceedings 34 (2023) 246-251   https://doi.org/10.21741/9781644902691-29

[21] J. Pietraszek. The modified sequential-binary approach for fuzzy operations on correlated assessments, LNAI 7894 (2013) 353-364. https://doi.org/10.1007/978-3-642-38658-9_32

[22] J. Pietraszek et al. The fuzzy approach to assessment of ANOVA results, LNAI 9875 (2016) 260-268. https://doi.org/10.1007/978-3-319-45243-2_24

[23] R. Dwornicka, J. Pietraszek. The outline of the expert system for the design of experiment, Prod. Eng. Arch. 20 (2018) 43-48. https://doi.org/10.30657/pea.2018.20.09

Quality Production Improvement and System Safety: QPI 16 - CZOTO 10   Materials Research Forum LLC
Materials Research Proceedings 34 (2023) 252-261   https://doi.org/10.21741/9781644902691-30

An Importance of the Roof of the Toyota House Factors in the Food Industry

MIELCZAREK Krzysztof[1, a],

[1]Department of Production Engineering and Safety, Faculty of Management, Czestochowa University of Technology, Al. Armii Krajowej 19b, 42-218 Czestochowa, Poland

[a] krzysztof.mielczarek@pcz.pl

Keywords: BOST Method, Toyotarity, Statistical Analysis, Production Management

Abstract. The article presents a case study of the practical use of BOST surveys to identify the most important areas in the functioning of enterprise in perception of the company mission by their employees. This is an important element in the consciousness of the worker and is part of the delegation and self-organization to the lowest level. The research object is company from food industry. Some production workers of the company with the help of BOST questionnaire survey showed, which factors are the most important. Based on the survey results of carried out on the population of production workers, a series of importance areas for improvement was formulated. The aim of the analysis is to present which factors are the most important by building the significance sequences of obtained results. The results obtained for the type of small and medium-sized enterprises overlap with the results of tests verified in other enterprises.

Introduction

Toyota Production System (TPS) is based on scientific principles and assumes that all separate elements work well for the benefit of the entirety [1]. The Toyota's management style has its origins in textile industry. Management in reference to automotive industry has elements of an American management with consideration of a Japanese culture. Toyotarity is a concept that is legally protected by confirming the date. This document contains the following definition of Toyotarity: "*Toyotarity is a field of scientific research dealing with human-machine relations and human and human taking into account the process approach, Japanese culture, particularly Toyota, oriented towards continuous improvement with the use of knowledge*" [2]. The primary research tool of toyotarity is the BOST method. It presents Toyota management principles in the form of distinctive sets of issues, describing a particular principle, these sets are called areas. Survey and research method determined as BOST was formed as a result of author's fascination in Toyota Motor Company [3]. This method describes Toyota's management principles with its characteristic factors [4]. The presented questionnaire has a ranking scale. Respondents may assess the significance of a given factor by placing one of the numbers within the range of scale in an appropriate box. After the description of the main part of this method its further elements will be outlined briefly. The BOST method allows assessing the significance of factors describing the 14 Toyota management principles [5]. Complement of carried out research there is interpretation of BOST questionnaire results. It lets better look on the enterprise by eyes of their workers. In the purpose to form an opinion it is essential to know the judgement of workers from different ranks in enterprise. BOST is survey where the questions are matched to judge enterprise and its immaterial stores are possible [6].

Methodology

The research company is a medium-sized enterprise from the food industry located in south part of the Silesian region and operating in the industry since 1990. The company produces a wide range of food products, among the most important, are: concentrates, dinner accessories vinegar,

Quality Production Improvement and System Safety: QPI 16 - CZOTO 10   Materials Research Forum LLC
Materials Research Proceedings 34 (2023) 252-261   https://doi.org/10.21741/9781644902691-30

spices, sauces, juices and fruit preserves. The company currently employs over 100 employees. Company is competitive with foreign countries – exporting its products to many countries. An organizational structure of the company can determine as linear, flat. There are not a large number of managerial ranks what quicken the flow of information and the procedure of deciding. Production managers are responsible for the course of an entire production process and for creating and correct implementing of new ideas or the technology. New workers before working on full-time must have an internal examination in the company. Thanks to this procedure the enterprise has the guarantee that a new worker will be performed tasks competently. The quality control is present at every stage of the production, having conceived from careful selection of sub-suppliers, through the control of materials delivered by them and sub-assemblies, the control of half-finished products processing in every phase, and on final goods finishing. It is supposed to assure that the product is safe and is characterized by good quality. The department manager may have supervision over a certain number of employees. The department employee has a strictly defined supervisor. Significant decisions are made by the owners who consider the beliefs and suggestions of the crew during the consideration.

In selected enterprise the population of respondents was chosen, which consisted from production workers of the examined enterprise, having a contact with manufacturing process in the workplace [7]. Stability of the basic production process is crucial for continuous manufacture of the product consistent with the highest quality standards. The control of its particular elements and the awareness of their significance among employees is the key factor to optimization of the whole process. This article presents an analysis of the answers given to the question contained in the BOST questionnaire, referring to the roof of Toyota's house – called mission of enterprises.

Employees have answered the following question: *"Which factor is the most important in your enterprise? Fill in the blanks with 1; 2; 3; 4; 5 (where 5 the most important factor)"* (Table 1).

Table 1. *Questionnaire form*

JA		Quality
KO		Cost
CR		Execution time
BP		Work safety
MZ		Attitude of the crew

Statement of responses obtained in studies that concerned the company mission is presented in Table 2. The presented results are part of a research BOST covering a range of issues relating to Toyota management principles which were carried out in a company from food industry. The questionnaire survey was carried out amongst 60% production workers. i.e. more than half of workers. Such a large research group of directly production workers will allow to precise identification the most important areas in the surveyed enterprise.

Table 2. *Roof of the Toyota house: evaluation structure (%) of factors importance for E1 area (it concerns production of food products)*

Evaluation	Factor indication				
	JA	KO	CR	BP	MZ
1	0	10.0	13.3	20.0	56.6
2	3.3	63.3	13.3	13.3	6.7
3	677	6.7	40.0	16.7	30.0
4	26.7	3.3	33.3	30.0	6.7
5	63.3	16.7	0	20.0	0

On the basis of Table 2 was presented importance series of factors for individual evaluations. Summing up, a range of important factors in examined enterprise is following:

For evaluation "1" the importance series is: MZ > BP > CR > KO > JA. It proves that the factor of *attitude of the crew* (MZ) has received the biggest number of rates "1" – 56.6% and takes the first place in the significance sequence for this rate. For evaluation "2" the importance series is: KO > BP > CR > MZ > JA. In the case of rate "3" the following significance sequence of analysed factors has been developed: CR > MZ > BP > JA > KO. For a rate "4" respondents declared that in the analysed enterprise the following significance sequence describing importance of factors: CR > BP > JA > MZ > KO. For a maximum rate "5" respondents declared that in the analysed enterprise the following significance sequence has been achieved: JA > BP > KO > (CR; MZ). The results of the study were detailed in the analysis. As is clear from the date presented in Table 2 it can be concluded that the largest number of respondents (63.3%) indicated *quality* (JA) as the most important factor in a company. As a supplement, Fig.1 is presenting radar graphs made for evaluations of importance the research factors.

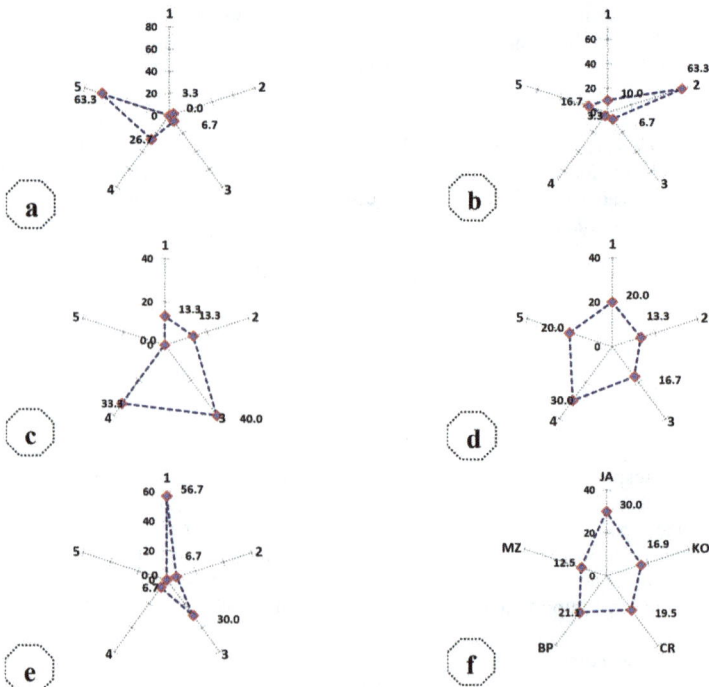

Fig. 1. Elements of the roof of the Toyota house. Circle graphs – structure of evaluations for: a) JA, b) KO, c) CR, d) BP, e) MZ, f) average – it concerns the enterprise producing food products (source: own study)

On the basis of carried out analysis it was state, that *quality* (JA) is the most important factor for 63.3% respondents. *Costs* (KO), as well as *attitude of the crew* (MZ) are the most important only for 16.7% and 20% of respondent's respondents.). From Fig.1f results that in the enterprise a

Quality Production Improvement and System Safety: QPI 16 - CZOTO 10   Materials Research Forum LLC
Materials Research Proceedings 34 (2023) 252-261   https://doi.org/10.21741/9781644902691-30

quality (JA) is the most important element - 30.0%. *Work safety* (BP) were on a next place 21.1% and farther *execution time* (CR) 19.5%. Factor *attitude of the crew* (MZ) turned out to be the least crucial factor 12.5%.

Statistical Analysis of the Results from the BOST Questionnaire

Making statistical analysis of studied area six statistical tools were used: arithmetic average, variance, standard deviation, the coefficient of variation, skewness and kurtosis. The aim of application of this statistical tool is to show distribution of evaluation for individual factors [8].

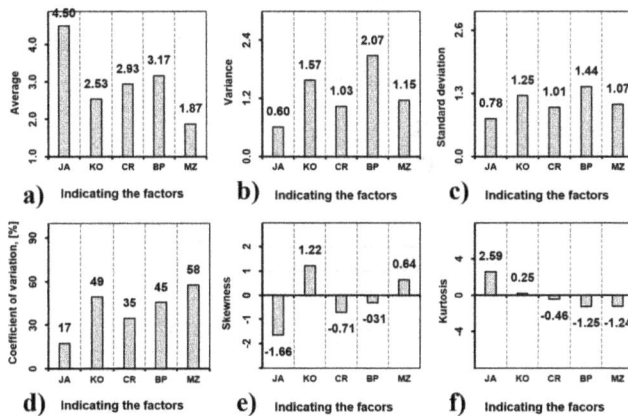

Fig.2. *Elements of the roof of the Toyota house. Comparison: a) averages, b) variance, c) standard deviation, d) coefficient of variation, e) skewness, f) kurtosis for factors in E1 area (source: own study).*

The average level of the measurable feature was presented with the help of the average. Analyzing the result concerning average it was taken the conclusion that the majority of respondents judged the *quality* (JA) on the level 4.50. The smallest value of the average amounting 1.87 fell for *attitude of the crew* (MZ). In case of variance there are small differences between factors. The highest level was achieved by the factor *work safety* (BP) 2.07, and the lowest – 0.60 – by the factor *quality* (JA). On Fig.2c was determined a standard deviation. It described provided all examined individuals features are different from average. It results from them that *work safety* (BP) having the biggest standard deviation. The next analyzed statistical measure determining the area of the changeability being a difference between greatest and smallest value is coefficient of variation [9]. The biggest diversity it is possible to observe for *attitude of the crew* (MZ). It is providing about the nonconformity of polled workers in this topic. The analysis of skewness of the factor importance rates distribution describing roof the Toyota house management principle, comes down to the following facts that the greatest asymmetry force occurred for the distribution of importance rates for *quality* (JA) and amounted to -1.66. The distribution of rates for the rest of factors indicates weak and moderate skewness [10]. The last factor for analyzing is kurtosis. It determines the measure of distribution and concentrating the results in surroundings of the average [11]. It is possible to conclude that all factors are characterized by bigger flattening in the attitude to the normal distribution. For appropriate interpretation of results the following statement is necessary: We < 0 – distribution is characterized by lower than standard peakedness, We = 0

Quality Production Improvement and System Safety: QPI 16 - CZOTO 10   Materials Research Forum LLC
Materials Research Proceedings 34 (2023) 252-261   https://doi.org/10.21741/9781644902691-30

distribution is characterized by standard peakedness, We > 0 – distribution is characterized by peakedness higher than standard [12]. For the factor *quality* (JA) and *cost* (KO) kurtosis that is measure the concentration of the disintegration, is positive. For remaining factors kurtosis is negative, i.e. flatter, and value of individual factors are less concentrated, than at the normal distribution. This statistical tool confirm that distribution of results is logical and can be helpful for evaluation actual state in enterprises.

The starting point for changes (improvement) is recording the existing condition. The present situation is known best by participants of the processes implemented in a given enterprise. Data obtained from BOST analysis allowed to know the opinions of the representative group of workers in the topic of functioning of the enterprises concerning the most important elements in the company [13]. Fig.3 show the structure of ratings of the importance of factors related to the construction of the roof of a Toyota house. Division of drawings into two parts (according to the structure of assessments - Fig.3a and according to the importance of factors– Fig.3b.

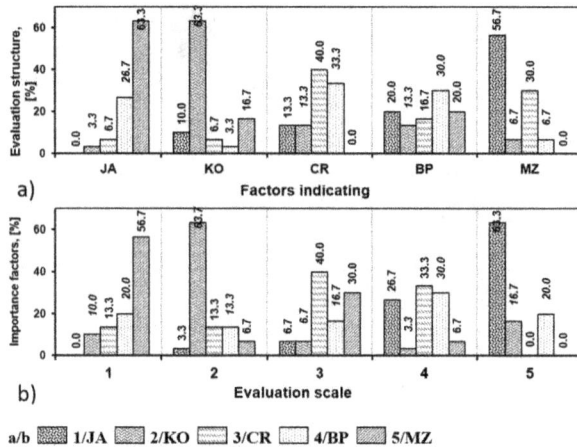

Fig.3. *Elements of the roof of the Toyota house. Comparison: a) structure of factor ratings, b) the importance of factors in rating scales (source: own study).*

Based on the analysis of Fig.3a, the statement was based that the distribution of ratings of the factors that were rated the highest is characterized by the concentration of the number of occurrences of a given rating on the right. It can therefore be concluded that the most important area of the company's activity among the elements of the roof of the Toyota house was considered by the respondents is *quality* (JA), as it received 63.3% of "5" ratings, i.e. the majority of employees rated this factor as the most important. None of the participants in the BOST study ranked quality as the least important aspect. It also shows that the *cost* (KO) received the most "2" ratings. 63.3% of the respondents identified it as the second most important area of the company's operation. The *execution time* (CR) factor can be described as a factor of medium importance from the point of view of the respondents, as it received the highest scores of "3" (40%) and "4" (33.3%), and no one considered it the most important factor (assessed "5" - 0%). The structure of *work safety* (BP) ratings is relatively uniform and ranges from 16.7% for a '3' to 30% for a '4'. The respondents considered the *attitude of the crew* (MZ) as the least important area of company activity, as evidenced by the highest number of "1" ratings – 56.7%.

Quality Production Improvement and System Safety: QPI 16 - CZOTO 10   Materials Research Forum LLC
Materials Research Proceedings 34 (2023) 252-261   https://doi.org/10.21741/9781644902691-30

Analysis Fig 3b revealed the following facts that the respondents most willingly gave a "1" rating to *attitude of the crew* (MZ) – 56.7%, considering it the least important area, while *quality* (JA) did not receive a single such rating. Employees considered the *quality* (JA) the most important, as evidenced by the largest number of "5" ratings given to this factor.

Influence of Respondents' Features on Factor Importance Evaluation – Correlation Graphs
To the purpose of a wider statistical analysis of respondent's replies it was carried out correlation analysis [14]. On the base of Fig.4a it is possible to state that the gender is feature of respondents which is entering correlations *cost* (KO) and *work safety* (BP). Examining the relation between education of respondents we can notice that in this case she appeared correlation for work safety (BP) on all α level.

Age has no significant impact on the assessment of any of the analyzed factors. Analyzing the correlation between the factors and job seniority it is possible to notice that this feature is entering correlation with the *attitude of the crew* (MZ) on all α level. In the case of the relation between the mobility and factors it was noticed that the mobility correlating with only one factor execution time (CR). This correlation is only on two α level. Analysis the relation between factors and the way of employment it was showed that the way the employment was entering correlations with only one factor cost (KO) on all α level.

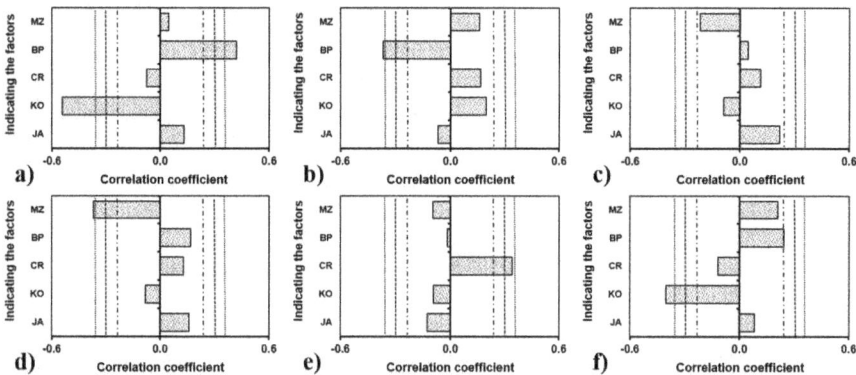

Fig.4. *Correlation graphs for factors in E1 area depending on the respondents features: a) gender, b) education, c) age, d) job seniority, e) mobility, f) way of employment. It concerns the enterprise producing food products [own study].*

The next stage of results analysis includes interpretation of graphs for voices distribution. Fig.5 is presenting interpretation of evaluation for the factor *quality* (JA) depending on individual features of the respondent.

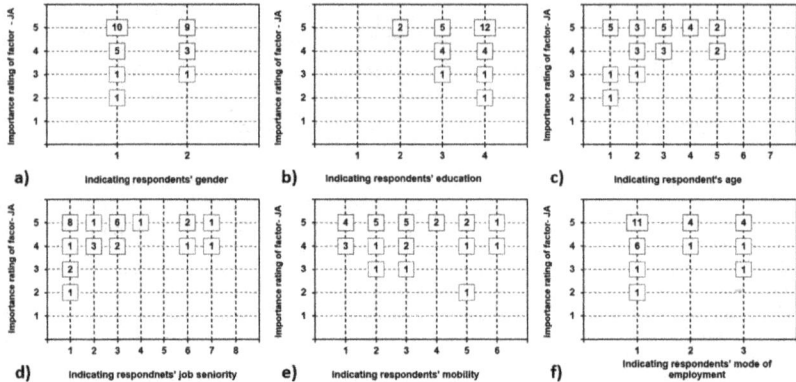

Fig.5. *Roof of the Toyota house – the map of evaluation number for quality (JA) factors depending on the respondent's feature: a) gender, b) education, c) age, d) job seniority, e) mobility, f) mode of the employment; it concerns the enterprise producing food products (source: own study).*

Analyzing the graph of the evaluation of the importance the factor *quality* (JA) depending on the respondent gender we notice that the most of men (10) granted this factor the evaluation "5" and 9 women granted the same evaluation. Analysing the respondents' education, it is possible to state, that persons with the higher education on "5" gave 12 votes, on "4" 4 voices. Analysing replies from the work experience point of view of, the most often been a granted assessment "5" from persons with work experience 1-5 years The lowest evaluation for factor *quality* (JA) granted 8 persons which researched enterprise is the first place of employment. The analysis of the dependence of grades on the mode of employment shows that 11 people employed on a regular basis, 4 employed on a transfer basis and 4 employed due to finances assess quality (JA) as the highest.

Summary

Surveyed studies have shown the effectiveness of the selection of specific factors in the purpose of determining a series of validity. The use of this knowledge will enable effective use of the company's resources in improving the indicated areas. As a result of the research work, it was found that the most important area requiring improvement is quality. It is an important element of research for small and medium enterprises. The results of research are consistent with the research carried out in other such enterprises. Innovative BOST questionnaire survey, which are an attempt to convert Toyota's management principles into questions was described. Data obtained from BOST analysis allowed to know the opinions of the representative group. Research was carried out amongst production workers of the company from food industry. It allowed detailing factors which in the greatest degree can contribute for improvement processes in the company and the ones which have this smallest contribution. The results of research are consistent with the research carried out in other such enterprises. The above fragment of analysis has revealed diversity in the significance of factors describing the roof of the Toyota house. In this way the usefulness of the presented BOST method has been proved for assessing a production process of goods of high-quality requirements. In the respondents' opinion the proposed set of factors has been arranged in a way characteristic for the enterprise producing different type of groceries. The acquired significance sequence of factors describing this management principle is logical, thus confirming the correctness of their selection and the research results can be used in another small and medium-sized enterprises. Factor *quality* (JA) has been judged with the highest evaluation by men than by women. In the researched company a quality of products, as well as a tidiness in production sectors, and in the entire company. Functioning of the production system, although now is working perfectly, it is possible still to be improved. The quality of offered products can be improved through applying tools of the quality management. The practical use of BOST surveys gives the opportunity to benefit from the practical knowledge of employees at the company's production level. This may contribute to the identification of key areas for the functioning of the enterprise.

Toyotarity aims at continuous improvement of quality [15]. It is a specifically Japanese approach, but it yields beneficial effects wherever the processing of challenging materials such as metals [16-18], alloys [19], and special coatings [20, 21] like DLC [22] is involved, significantly modifying the properties of surface layers [23]. Quality improvement is particularly desirable in areas with critical requirements, such as the military [24], structural welding [25], and construction [26, 27]. To effectively implement this approach, sophisticated design tools [28] and analytical tools [29] are necessary, which reduce the complexity of the problem and allow for an understanding of the structure of controlled factor relationships [30-32], including the use of non-parametric methods [33-35]. Resampling techniques [36] enable the estimation of result uncertainties, even in the case of relatively small-scale measurements.

References

[1]S. Gao, S.P. Low. Toyota way style human resource management in large Chinese construction firms, A qualitative study, Int. J. Constr. Manag. 15 (2015) 17-32. https://doi.org/10.1080/15623599.2015.1012139

[2]S. Borkowski. Scientific Potential of Toyotarity and BOST Method, Polish Quality Institute, Warszawa, 2016.

[3]S. Borkowski. BOST Method as the Instrument of Assessment Process Functioning according to Toyota Principles, University of Maribor, Maribor, 2012.

[4]K. Knop. Indicating and analysis the interrelation between terms – visual: management, control, inspection and testing, Prod. Eng. Arch. 26 (2020) 110-120. https://doi.org/10.30657/pea.2020.26.22

[5]K. Knop. Importance of visual management in metal and automotive branch and its influence in building a competitive advantage, Pol. J. Manag. Stud. 22 (2020) 263-278. https://doi.org/10.17512/pjms.2020.22.1.17

[6]J.K. Liker, J. Franz. The Toyota Way to continuous improvement, McGraw Hill, New York, 2011. ISBN 978-0071477468

[7]M. Matuszny. Building decision trees based on production knowledge as support in decision-making process, Prod. Eng. Arch. 26 (2020) 36-40. https://doi.org/10.30657/pea.2020.26.08

[8]K. Knop. Statistical Control of the Production Process of Rolled Products, Prod. Eng. Arch. 20 (2018) 26-31. https://doi.org/10.30657/pea.2018.20.06

[9]M. Krynke. Risk Management in the Process of Personnel Allocation to Jobs, System Safety: Human-Technical Facility-Environment (2020) 82-90. https://doi.org/10.2478/czoto-2020-0011

[10]M. Ingaldi. A new approach to quality management: conceptual matrix of service attributes, Pol. J. Manag. Stud. 22 (2020) 187-200. https://doi.org/10.17512/pjms.2020.22.2.13

[11] D. Klimecka-Tatar. Context of production engineering in management model of Value Stream Flow according to manufacturing industry, Prod. Eng. Arch. 21 (2018) 32-35. https://doi.org/10.30657/pea.2018.21.07

[12] M. Krynke et al. Cost Optimization and Risk Minimization During Teamwork Organization, Manag. Sys. Prod. Eng. 29 (2021) 145-150. https://doi.org/10.2478/mspe-2021-0019

[13] J. Selejdak. Use of the Toyota Management Principles for Evaluation of the Company's Mission, Prod. Eng. Arch. 1 (2013) 13-15. https://doi.org/10.30657/pea.2013.01.04

[14] K. Knop, K. Mielczarek. Assessment of Production Processes Functioning in the Case of Air Bag Production, MATEC Web Conf. 183 (2018) art.04009. https://doi.org/10.1051/matecconf/201818304009

[15] T. Lipiński, R. Ulewicz. The effect of the impurities spaces on the quality of structural steel working at variable loads, Open Eng. 11 (2021) 233-238. https://doi.org/10.1515/eng-2021-0024

[16] M. Mazur et al. The structure and mechanical properties of domex 700 MC steel, Comm. - Sci. Lett. Univ. Žilina 15 (2013) 54-57. https://doi.org/10.26552/com.C.2013.4.54-57

[17] R. Ulewicz et al. Structure and mechanical properties of fine-grained steels, Period. Polytech. Transp. Eng. 41 (2013) 111-115. https://doi.org/10.3311/PPtr.7110

[18] P. Jonšta et al. The effect of rare earth metals alloying on the internal quality of industrially produced heavy steel forgings, Materials 14 (2021) art.5160. https://doi.org/10.3390/ma14185160

[19] A. Dudek et al. The effect of alloying method on the structure and properties of sintered stainless steel, Archives of Metallurgy and Materials 62 (2017) 281-287. https://doi.org/10.1515/amm-2017-0042

[20] N. Radek et al. Technology and application of anti-graffiti coating systems for rolling stock, METAL 2019 – 28th Int. Conf. Metall. Mater. (2019) 1127-1132. ISBN 978-8087294925

[21] N. Radek et al. Formation of coatings with technologies using concentrated energy stream, Prod. Eng. Arch. 28 (2022) 117-122. https://doi.org/10.30657/pea.2022.28.13

[22] N. Radek et al. Microstructure and tribological properties of DLC coatings, Mater. Res. Proc. 17 (2020) 171-176. https://doi.org/10.21741/9781644901038-26

[23] N. Radek et al. The influence of plasma cutting parameters on the geometric structure of cut surfaces, Mater. Res. Proc. 17 (2020) 132-137. https://doi.org/10.21741/9781644901038-20

[24] W. Przybył et al. Microwave absorption properties of carbonyl iron-based paint coatings for military applications, Def. Technol. 22 (2023) 1-9. https://doi.org/10.1016/j.dt.2022.06.013

[25] N. Radek et al. The impact of laser welding parameters on the mechanical properties of the weld, AIP Conf. Proc. 2017 (2018) art.20025. https://doi.org/10.1063/1.5056288

[26] A. Bakowski et al. Frequency analysis of urban traffic noise, ICCC 2019 20th Int. Carpathian Contr. Conf. (2019) 1660-1670. https://doi.org/10.1109/CarpathianCC.2019.8766012

[27] J.M. Djoković et al. Selection of the Optimal Window Type and Orientation for the Two Cities in Serbia and One in Slovakia, Energies 15 (2022) art.323. https://doi.org/10.3390/en15010323

Quality Production Improvement and System Safety: QPI 16 - CZOTO 10   Materials Research Forum LLC
Materials Research Proceedings 34 (2023) 252-261   https://doi.org/10.21741/9781644902691-30

[28] L. Cedro. Model parameter on-line identification with nonlinear parametrization – manipulator model, Technical Transactions 119 (2022) art. e2022007. https://doi.org/10.37705/TechTrans/e2022007

[29] J. Pietraszek, E. Skrzypczak-Pietraszek. The uncertainty and robustness of the principal component analysis as a tool for the dimensionality reduction. Solid State Phenom. 235 (2015) 1-8. https://doi.org/10.4028/www.scientific.net/SSP.235.1

[30] J. Pietraszek et al. The parametric RSM model with higher order terms for the meat tumbler machine process, Solid State Phenom. 235 (2015) 37-44. https://doi.org/10.4028/www.scientific.net/SSP.235.37

[31] R. Dwornicka, J. Pietraszek. The outline of the expert system for the design of experiment, Prod. Eng. Arch. 20 (2018) 43-48. https://doi.org/10.30657/pea.2018.20.09

[32] J. Pietraszek et al. Challenges for the DOE methodology related to the introduction of Industry 4.0. Prod. Eng. Arch. 26 (2020) 190-194. https://doi.org/10.30657/pea.2020.26.33

[33] J. Pietraszek. Fuzzy regression compared to classical experimental design in the case of flywheel assembly, LNAI 7267 LNAI (2012) 310-317. https://doi.org/10.1007/978-3-642-29347-4_36

[34] J. Pietraszek. The modified sequential-binary approach for fuzzy operations on correlated assessments, LNAI 7894 (2013) 353-364. https://doi.org/10.1007/978-3-642-38658-9_32

[35] J. Pietraszek et al. The fuzzy approach to assessment of ANOVA results, LNAI 9875 (2016) 260-268. https://doi.org/10.1007/978-3-319-45243-2_24

[36] J. Pietraszek, L. Wojnar. The bootstrap approach to the statistical significance of parameters in RSM model, ECCOMAS Congress 2016 Proc. 7th Europ. Congr. Comput. Methods in Appl. Sci. Eng. 1 (2016) 2003-2009. https://doi.org/10.7712/100016.1937.9138

Quality Production Improvement and System Safety: QPI 16 - CZOTO 10   Materials Research Forum LLC
Materials Research Proceedings 34 (2023) 262-267   https://doi.org/10.21741/9781644902691-31

Building Competitiveness through Improvement of Quality Process in Paper Industry Enterprises

KONIECZNA Monika[1,a] * and MRUGALSKA Beata[2,b]

[1]Faculty of Engineering Management, Poznan University of Technology, 60-965 Poznan, Poland

[2]Faculty of Engineering Management, Poznan University of Technology, 60-965 Poznan, Poland

[a]monika.t.konieczna@doctorate.put.poznan.pl, [b]beata.mrugalska@put.poznan.pl

Keywords: Quality, Quality Management, Competitiveness, Paper Industry

Abstract. Nowadays, organizations continuously compete with each other and other market competitors. It is clearly visible in paper production market in Poland which is constantly developing taking over even the plastic product market due to the regulation of the reduction of the impact of certain *plastic* products. Therefore, for a single company there is a need to search and take competitive advantage. In order to build competitiveness it is possible to refer to improvement of quality process which can lead to additional benefits for organizations or customers. It can concern various aspects of the company such as structures, processes, activities or products. The aim of this paper is to propose methods and tools of quality management at particular stages of paper production process to build or improve market competitiveness in paper industry.

Introduction

Competition is perceived as one of the exclusive components of a free market [2]. It is a phenomenon that describes mutual relations that appear in rivalry environment between entities that have been covered by this phenomenon. Enterprises that compete effectively and despite the obstacles created by competitors, achieve their goals, are called competitive [8]. Competitive advantage is limited by changes taking place in a given sector and actions taken by competitors. Presently, the enterprises in Polish paper industry are constantly looking for new sources of building a competitive advantage. It is important to identify the factors affecting market competitiveness that should be part of planned strategic goals. Despite strong foreign competition, Polish enterprises are characterized by intensive growth on the local and international market. This may be due to special features, modern technology, innovation or a high level of know-how [9].

Improvement of quality process may be one of the ideas of building competitiveness of paper industry enterprises. It consists of reducing its volatility through the optimal use of resources, the use of various concepts, methods, techniques and tools, the elimination of waste, shortening the production time and reducing costs. The aim of improvement is to obtain appropriate economic and production results, affecting the increase in quality and customer satisfaction [1]. Continuous improvement should be an integral part of the functioning of any organization, regardless of the stage it is in [4, 5, 6]. The process approach is used in management more and more often, due to the fact that identification and understanding of processes in organization has an impact on increasing overall efficiency [10, 3]. Chosen concepts, methods, techniques and tool implemented in the particular stages of production process in paper industry may help to improve of quality process and by this to build or improve the market competitiveness.

Quality Production Improvement and System Safety: QPI 16 - CZOTO 10   Materials Research Forum LLC
Materials Research Proceedings 34 (2023) 262-267   https://doi.org/10.21741/9781644902691-31

Dynamics of Polish Paper Industry

Polish paper market is one of the fastest growing branches of industrial processing in Poland. The production (that includes paper and cardboard) in 2017 was equal to 6 million tons. It gives an increase of 105% compared to 2000. Despite the changing environment, the sector is developing dynamically and has good perspectives. It may seem that in the world dominated by digitalization, there is no place for development of paper production. However, it may observe that that demand for packaging segment and sanitary products is growing rapidly, while less and less important are the graphics paper. The value of sold production of paper products is growing at 8% per year which is one of the best industry performances during recent years [11]. Each stage of paper production process takes place in Poland. From obtaining raw materials (pulp), waste paper, production of various types of paper and paper products. The first stage of production is performed mainly is only a few large enterprises in Poland.

The Polish paper market is characterized by high international competitiveness due to the high quality of products while maintaining low production costs. The introduction of new EU regulations, emphasizing the importance of ecology and caring for the natural environment, resulted in an increase in the importance of paper as a natural raw material with less harmful effects. The paper sector also developed during the COVID-19 pandemic, when the demand for household hygiene and sanitary products and various types of packaging paper increased. Paper sector, despite the crisis, recorded an increase in the value of sold production. It can be assumed that the paper industry will continue to develop along with the growing importance of e-commerce [11].

There are a lot of opportunities for development in paper industry, for instance to increase the range of manufactured products. Due to the huge problem regarding the amount of plastic in the world, the European Parliament has decided to implement The Single-Use Plastics Directive. The organizations are obliged to stop producing and using disposable plastics products such as polystyrene cups, food containers, cutlery, straws [7]. It may result in increasing the range of assortment of other sectors. According to forecasts, the role of the paper industry will continue to grow, especially in the context of domestic consumption and exports.

Material and Methods

Nowadays, many difficulties may be encountered when conducting empirical researches among entrepreneurs in Poland. These difficulties appear mainly due to entrepreneurs reluctance to participate in surveys, because of lack of time, lack of knowledge or fear of disclosing sensitive information. The main criterion qualifying for participation in the survey was the affiliation of enterprise to the paper sector which is defined in Polish Classification of Activities in Section C – Industrial processing and in Division 17 – Manufacturing of paper and paper products. The second qualifying criterion was the position of the respondent, defined as the director, the manager, the representative of Quality Management Systems or an employee of quality department. Another criterion to participate in the survey was the size of enterprise.

Quantitative research conducted by the author in February-December 2021 were part of a research project conducted at the Faculty of Management Engineering at Poznań University of Technology. The research was conducted by using CAWI (Computer-Assisted Web Interview) method which is a technique of collecting information in which the respondent is asked to complete a questionnaire in electronic form. The research was conducted mainly through the international social network LinkedIn, specializing in business contacts. The initial stage included pilot studies, which resulted in 15 completed questionnaires. A total of 686 employees were invited to participate in the study. As a result of the conducted research, 90 correctly completed questionnaires were received.

Quality Production Improvement and System Safety: QPI 16 - CZOTO 10   Materials Research Forum LLC
Materials Research Proceedings 34 (2023) 262-267   https://doi.org/10.21741/9781644902691-31

The respondents were asked to answer the questions:

Q1: Please indicate which concepts, methods, techniques and tools of quality management are used in your company?

The respondents had 25 different concepts, methods, techniques and tools of quality management to choose from. There was a possibility to mark more than one answer.

Q2: Please indicate to what extent the presented concepts, methods, techniques and tools of quality management are important for continuous improvement of production processes (where 1 means not important and 5 means very important)?

In this question the respondents were asked to assess each of concept, method, technique or tool of quality management in Likert scale (from 1 to 5).

Results and Discussion

As a result, in response to the question about the use of concepts, methods, techniques and tools for quality management, 2.2% of respondents did not indicate any of them. The rest gave from 1 to 17 answers. According to the respondents, the most frequently used in the company were:

- brainstorming – 77.8% of people,
- check sheet – 67.8% of people,
- 5S – 58.9% of people,
- 5WHY – 40.0% of people.

The results of the first question from the survey may be found in Fig.1.

However, according to the respondents, the most important concepts, methods, techniques and tools of continuous improvement of production processes in the company were:

- brainstorming – average 3.84,
- check sheet – average 3.40,
- 5S – average 3.14,
- 5WHY – average 2.80.

The results of the second question from the survey may be found in Fig.2.

Quality Production Improvement and System Safety: QPI 16 - CZOTO 10   Materials Research Forum LLC
Materials Research Proceedings 34 (2023) 262-267   https://doi.org/10.21741/9781644902691-31

Table 1. *Proposition of implementation chosen concept, method, technique or tool of quality management at particular stages of paper production [Based on results from survey]*

Concept, ethod, technique or tool	Description	Stage of production
FMEA **QFD** **DoE**	• elimination of weak points in the product design process, • improvement of the product design process, • determination of the influence of input parameters on the process output	Production planning and design
Check sheet	• collecting and organizing data from measurements and observations	Preparation of proof Preparation of ozalid
Matrix diagram	• determination of the dependencies of various analyzed objects	Technical preparation of production
Matrix data analysis	• systematizing the priorities set in the matrix diagram	Technical preparation of production
Brainstorming **5 WHY**	• group analysis of solving the problems that may occur in production process	Whole production process
5S	• elimination of non-value-added activities	Whole production process
Statistical Process Control	• monitoring of production process	Interior printing Folding Cover printing
Report 8D **Ishikawa diagram**	• identifying the causes of repetitive problems in the production process	Whole production process
Statistical acceptance control	• declaration of compliance with technical standards	Finished product

Quality Production Improvement and System Safety: QPI 16 - CZOTO 10   Materials Research Forum LLC
Materials Research Proceedings 34 (2023) 262-267   https://doi.org/10.21741/9781644902691-31

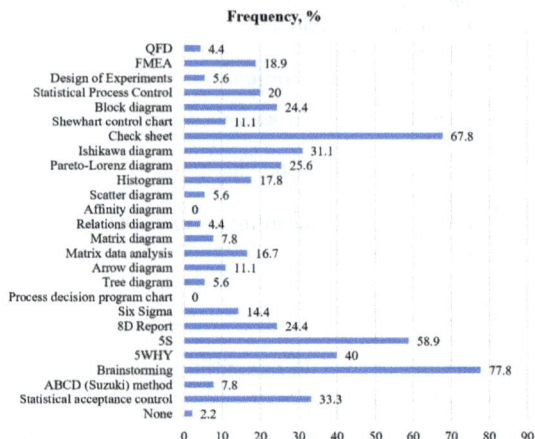

Fig. 1. *Concepts, methods tools and techniques implemented in investigated paper enterprises (source: based on results from survey).*

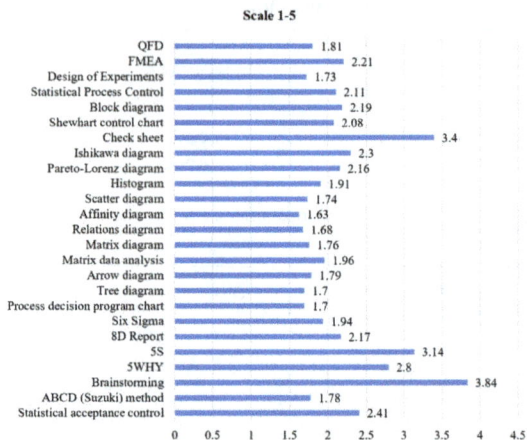

Fig.2. *Concepts, methods, techniques and tools contributing to build competitiveness (source: based on results from survey)*

On the basis of collected answers there was created a proposition of implementation chosen concept, method, technique or tool of quality management at the particular stages of paper production process which is presented in Table 1. There were chosen some elements of quality management that implemented may contribute to building competitiveness through improvement of quality process in paper industry enterprises.

Quality Production Improvement and System Safety: QPI 16 - CZOTO 10   Materials Research Forum LLC
Materials Research Proceedings 34 (2023) 262-267   https://doi.org/10.21741/9781644902691-31

Summary

Nowadays, the companies in Polish paper industry are continuously competing, so the new sources of competitive advantage are needed. It may be observed dynamic and steady increase of production every year. Paper products are perceived as very attractive and essential objects while paper itself as a key raw material in Polish economy today. Improvement of quality process may be one of the ideas of building competitiveness of paper industry enterprises. The research conducted with the representatives of paper industry showed that different concepts, methods, techniques and tools are widely used in enterprises of this sector. Respondents also perceived many of them as being useful in building competitiveness. Proposed solution of implementation specific concepts, methods, techniques and tools at particular stages of paper production process may contribute to improvement of quality process and building competitiveness.

References

[1] A. Hamrol. Zarządzanie jakością z przykładami, Wydawnictwo Naukowe PWN, Warszawa, 2013. ISBN 978-83-01-17466-8

[2] R. Kolman et al. Wybrane zagadnienia zarządzania jakością, Wyższa Szkoła Administracji i Biznesu w Gdyni, Gdynia, 1996.

[3] M. Konieczna, B. Mrugalska, M. K. Wyrwicka. Application of Single Minute Exchange of Die in Production Process Improvement: A Review, Logistics & Transport 39 (2018). https://doi.org/10.26411/83-1734-2015-3-39-7-18

[4] M. Smętkowska, B. Mrugalska. Using Six Sigma DMAIC to Improve the Quality of the Production Process: A Case Study, Procedia Soc. Behav. Sci. 238 (2018) 590-596. https://doi.org/10.1016/j.sbspro.2018.04.039

[5] M. Konieczna. Quality Improvement in Production Enterprise: Case study, Scientific Papers of Silesian University of Technology. Organization and Management Series (2018) 151-162. http://doi.org/10.29119/1641-3466.2018.119.10

[6] M. Konieczna, B. Mrugalska, M. K. Wyrwicka. How to Improve Manufacturing Process Implementing 5S Practices: A Case Study, IHSI 2020: Intelligent Human Systems Integration 2020 (2020) 1225–1232. http://doi.org/10.1007/978-3-030-39512-4_187

[7] M. Niemir, B. Mrugalska. Identifying the Cognitive Gap in the Causes of Product Name Ambiguity in E-Commerce, Logforum 18 (2022) 257-364. https://doi.org/10.17270/J.LOG.2022.738

[8] M. Strużycki. Podstawy zarządzania przedsiębiorstwem, SGH, Warszawa, 1998. ISBN 8386689978

[9] R. Śliwiński. Resource Based View and the Key Factors of International Competitiveness of Polish Fast-Growing Firms, Studia Ekonomiczne/Uniwersytet Ekonomiczny w Katowicach 116 (2012) 241-252.

[10] M. Trenkner. Process Improvement – Social Perspective, J. Manag. Finance 14(2) (2016) 429-438.

[11] Bank Pekao report: Polish paper industry and its role in the modern economy. [online] 2021 [Viewed: 31-01-2023] Available from: https://media.pekao.com.pl/pr/659692/raport-banku-pekao-polski-przemysl-papierniczy-i-jego-rola-w-nowoczesnej-gospodarce

Quality Production Improvement and System Safety: QPI 16 - CZOTO 10   Materials Research Forum LLC
Materials Research Proceedings 34 (2023) 268-277   https://doi.org/10.21741/9781644902691-32

Commercialization of Research Results – Overview of Assumptions and General Definitions

JAGUSIAK-KOCIK Marta [1,a] * and JANASIK Michał [2,b]

[1] Czestochowa University of Technology, Department of Production Engineering and Safety, Armii Krajowej 19B, 42-201 Czestochowa, Poland

[2] Łukasiewicz Centre, 19 Poleczki street, 02-822 Warsaw, Poland

[a]m.jagusiak-kocik@pcz.pl*

Keywords: Scientific Network, Commercialization of Research Results, Indirect Commercialization, Direct Commercialization, Financing of Commercialization

Abstract. The work is based on a literature review in the field of commercialization of scientific research – it presents various definitions of commercialization, the concept of indirect and direct commercialization, as well as ways of financing the commercialization of research results. The aim of the work is to present and learn about various approaches to the concept of commercialization, to show what forms the process of commercialization of research results takes place and how it can be financed. The aim is to identify and analyze the research area, because the work is an introduction to practical research and the creation of a model for the commercialization of scientific research results in the largest research network in Poland, dealing with i.e. providing attractive and competitive technological solutions.

Introduction

The commercialization of research results is the foundation of cooperation between science and business/industry [1]. It is necessary to properly structure it (based on international experience in the field of best practices) in order to successfully implement projects that will involve both representatives from the world of science and the world of industry. Important from the point of view of proper planning of all activities preceding successful commercialization, and then its proper implementation, is the choice of the method of commercialization and, among others, defining the form of intellectual property protection [1]. The subject of commercialization of scientific research results is any product or service produced in a scientific unit as a result of research conducted there [2]. This is a very important issue in the realities of the knowledge economy, because it translates into the competitive position of the national economy in the era of globalization [3].

The aim of the work is to present the concept of commercialization according to various authors, to present the forms of indirect and direct commercialization as well as to present the method of financing commercialization. By recognizing the topic and analyzing the available literature, research related to the analysis of commercialization models will be possible and an attempt to create a model for the commercialization of research results in the largest research network in Poland.

The Concept of Commercialization

One of the first definitions of the term commercialization in the Polish legal system appeared in the Act of August 30th, 1996 on commercialization and privatization of enterprises (UKPP). According to this definition, commercialization consists in transforming a state-owned enterprise into a company; unless the provisions of the Act provide otherwise, this company enters into all legal relationships of which the state-owned enterprise was the subject, regardless of the legal nature of these relationships [4]. After 2000, when the importance of innovation policy and the use

Quality Production Improvement and System Safety: QPI 16 - CZOTO 10   Materials Research Forum LLC
Materials Research Proceedings 34 (2023) 268-277   https://doi.org/10.21741/9781644902691-32

of the potential of Polish science to increase the competitiveness of enterprises began to grow, and the privatization of state-owned enterprises was practically completed, the concept of commercialization in the Polish legal system changed and was linked with the commercialization of the results of research activity [4]. In 2014, in the Law Act on higher education (UPSW), the concept of indirect and direct commercialization appeared. These concepts and the division were maintained in Act 2.0 Law on Higher Education and science, in Chapter 6, on Commercialization of the results of scientific activity and know-how. According to the Dictionary of Foreign Words and Foreign Language Phrases, "commercialization" (from Middle Latin *commercialis* – commercial, from Latin *commercium* – trade in goods, from English commerce – trade or trade exchange) is basing something on commercial, mercantile principles [4, 5]. According to this definition, commercialization is making something marketable, making it possible to buy it on the market and that someone has to pay for it [5-7]. Commercialization is defined as the transfer (transfer) of knowledge, ideas, research results from research laboratories to the market [5, 6, 8-10] – that is, it is supplying the market with new solutions. According to the National Center for Research and Science, commercialization is all activities related to the transfer of research results to economic and social practice, i.e., the transfer and sale by science institutions of a given technical or organizational knowledge and related know-how to the business sphere [4, 11, 12]. Commercialization is the process of creating added value for ideas, research results, technologies and new products, the spread of innovations within economies and industry sectors. It is building a business model for a current or future organization based on new technologies or new products. It results from patterns adopted and shaped by technological and innovation policy [13-16].

Other notions of commercialization can also be found in the literature, such as [4-6, 12, 17, 18]:

- Webster's Third New International Dictionary: Commercialization is to ensure that a product or technology with the potential to be produced, presented, sold, and used brings added value, profits and increases the company's capital.
- J.R. Meredith, S.M. Shafer: Commercialization is the process of transferring ideas into a new product or service from the conception of the product to its launch on the market.
- D.R. Prebble, G. A. de Waal, C. de Groot: Commercialization is something between innovation and entrepreneurship. It includes processes and activities that fill the gap between the creation of economic added value and the realization of economic added value.
- M.J. DeGeeter: Successful commercialization is a complex process dependent on many factors (inputs to the process) skilfully managed in order to create appropriate and prudent strategic and tactical plans to be implemented effectively and efficiently.
- K.B. Matusiak: Commercialization is all activities related to the transfer of a given technical or organizational knowledge and related know-how to business practice. Technology commercialization can be defined as the process of supplying the market with new technologies.

Methods of Commercialization of Research Results
Commercialization can take many forms. These forms are not established once and for all, which is why new types of transferring knowledge to the economy are often found in economic practice [19, 20]. Commercialization may take the form of indirect or direct commercialization [4, 7, 19].

According to the National Center for Research and Development, direct commercialization is a process in which the right to R&D results grants a license directly to the entity implementing these results, e.g., for production according to a patent and using know-how, or sells rights [4]. In direct commercialization, the creator of a new solution makes it available to other people or companies (i.e., he gives ownership rights to the innovation that is protected by a patent). Then, these companies or people introduce the new idea into practice on their own (the market is exploited and benefits from the competitive advantage offered by this innovation are taken). The process of direct commercialization may consist in the sale of exclusive rights or granting licenses and collecting license fees. Sieńczyło-Chlabicz [21] defines direct commercialization as based on the legislator listing exemplary forms of making research results available to a third party. In direct commercialization, this form is an agreement, which is connected with a regulation or authorization to exploit an intangible asset. Direct commercialization is the sale of the results of scientific research, development work or know-how related to these results, or making these results or know-how available for use, in particular on the basis of a license, rental or lease agreement [4].

The consequence of direct commercialization is that the results of scientific research gain contact with the market - they are offered to third parties (buyers, licensees) for use or purchase. Through this form of commercialization, the entity that exclusively uses the results of scientific research is not only the scientific unit that produced them.

Direct commercialization can take place in two ways:

- sale of research and development results,
- granting a license for the results of research and development works.

Direct commercialization has many benefits, but also carries certain risks. The formula of this form of commercialization enables the "cross" benefit from different specializations that are held by both parties. And so, the creator of innovation (innovator) gains access to future income, which is generated by the entity undertaking the market exploitation of innovation (licensee), and the licensee has access to innovation, which he would not be able to produce himself or such production would require disproportionately large expenditures.

Licensing [22] consists in authorizing another entity to use the results of R&D works in a specific scope and time of use. The license agreement specifies in detail and in writing, the scope of the authorization to use the invention, which is granted by the patent holder to another entity. They make it possible for the relationship between the entity that grants the license and the licensee to be shaped in a variety of ways. For the licensor, the main source of profit (benefits) are license fees. There are at least six types of licenses [12] – different types of intellectual property rights are associated with them. They are not separable and may occur in several variants together, e.g. a non-exclusive license may be full or limited.

The sale of property rights [22] is a common form of commercialization of developed technologies. One of its forms is the sale of only know-how regarding a given technology and the right to use it by the buyer. In this case, the concept itself is subject to sale. Intellectual property rights can in most cases be freely traded. Acquisition and disposal (e.g. sale) of proprietary copyrights requires the conclusion of a contract.

Indirect commercialization is a more complex and advanced process than direct commercialization. In this commercialization, the creator of the innovation (originator) is involved in the process of implementing this innovation on the market by creating a business entity (the originator launches it alone or with partners), through which the given solution will be implemented in business practice [21]. The goal is to start market production, which is based on an innovation that is the intellectual property of its creator. In this way, the "intermediary" (licensee) is avoided, which stands in the case of direct commercialization between the creator of the innovation and the market. The creator of the innovation takes responsibility for the course of all stages of the commercialization process, which results in successfully introducing the innovation to the market, developing it and maintaining it on this market for a certain period of time. The creator of innovation should have new skills and qualifications and be aware that risk is a feature of commercialization. Indirect commercialization takes place through established companies [4]. Shares in companies are subscribed for or purchased, or subscription warrants are subscribed for, which entitle to subscribe for shares in companies. The goal is to implement or prepare for implementation the results of scientific research, development works or know-how related to these results.

Quality Production Improvement and System Safety: QPI 16 - CZOTO 10   Materials Research Forum LLC
Materials Research Proceedings 34 (2023) 268-277   https://doi.org/10.21741/9781644902691-32

A company is a business enterprise which is carried out by two or more persons and managed by them. Companies are one of the organizational and legal forms under which enterprises operate. Contrary to direct commercialization, in this form of commercialization there is no real contact between research results and the market. According to the definition of the National Center for Research and Development [4], indirect commercialization takes the following forms:

- formation of a company,
- transfer of intellectual property rights to the company,
- another form of indirect commercialization.

Among the commercial companies (they are divided into partnerships and capital companies), which are established by the State Research Units (PJB), there are:

- general partnership,
- professional partnership,
- limited partnership,
- limited joint-stock partnership,
- limited liability company,
- joint-stock company,
- a simple joint-stock company.

General partnerships, professional partnerships, limited partnerships and limited joint-stock partnerships are partnerships, while limited liability companies, joint-stock companies and simple joint-stock companies are capital companies. Commercial companies are registered in the Register of Entrepreneurs in the National Court Register (KRS). There are two methods of submitting an application for registration in the National Court Register:

- via the S24 system - in the case of companies whose agreement was concluded in this system (this system can be used to establish a general partnership, limited partnership, limited liability company and a simple joint-stock company),
- through the Court Registers Portal - in the case of companies whose agreement has been concluded in traditional form.

At the moment of registration in the National Court Register, capital companies acquire legal personality and partnerships are established. State Research Units (PJB), by co-creating a given company, derive financial and corporate benefits from it, but it is impossible for them to create partnerships, because such companies can only be established by natural persons to practice a freelance profession. Each capital company has characteristics such as:

- full legal personality,
- share capital held,
- liability borne by the company (management board and supervisory board), and not by its shareholders,
- the accumulated assets are separate from the personal assets of partners or shareholders,
- separation of the ownership sphere and the management sphere (this means that the company is run by its governing bodies, not partners or shareholders).

The governing bodies of capital companies include:

- management board - it implements decisions made by shareholders and persons representing the company,
- supervisory board or audit committee – its task is to supervise the company's operations,
- general meeting or shareholders' meeting – makes the most important decisions in the company.

Quality Production Improvement and System Safety: QPI 16 - CZOTO 10   Materials Research Forum LLC
Materials Research Proceedings 34 (2023) 268-277   https://doi.org/10.21741/9781644902691-32

Financing the Commercialization of Research Results

The financing process consists of all undertakings in the company that provide the company with capital, and also shape the entire structure of financing sources in given market conditions. The financing strategy, including the acquisition of specific sources of financing for the company, is an expression of the entire financing process. The elements of the financing process are:

- financing objectives,
- principles of financing business operations,
- methods and tools used to achieve particular goals and principles,
- stages of the financing process.

The commercialization process of research results encompasses various stages, and while they may vary in practice, they share certain common characteristics from a financial perspective. The research and development stage [12] is characterized by expenses without generating any revenues. During this stage, companies invest in research projects, testing developed solutions, and prototyping, but there is no income generated. This stage poses a double burden for companies as it involves significant financial outlays for laboratory and research equipment, materials, and staffing expenses. Additionally, it prolongs the time between project initiation and the generation of sales revenue. As a result, entrepreneurs, whenever possible, tend to avoid this stage and explore alternative options such as acquiring technology from external sources. The stage of implementation and introduction to the market [12] is a stage that includes activities such as:

- undertaking an investment – to produce a new product or provide a new service involves significant expenses, including factors such as production costs,
- introducing a product or service to the market – involves financial resources for manufacturing, promotion, and establishing distribution channels.

The stage of sales development [12] refers to the period when a new product or service gains acceptance in the market, resulting in increasing revenue. During this stage, the sales cover the operating costs associated with producing the product or delivering the service. For projects that involve further development and expansion into new markets, additional financial resources are required. However, projects that do not aim for dynamic growth can reach a satisfactory operational scale and generate profits. Obtaining external financing for a project related to the commercialization of new solutions requires significant commitment from the implementing entity. It can be challenging due to factors such as the skepticism of capital providers. In the early stages of implementing technical solutions, external financing options include *business angels* (individual investors) and seed capital. Business angels are experienced individuals who have successfully developed their own businesses and are willing to invest in the ideas of new creators, start-ups, and early-stage enterprises. They not only provide financial support but also offer mentorship, scientific guidance, and valuable networking opportunities. In return, business angels become shareholders in the company. *Seed funds* are initial-stage investment funds that focus on investing in early implementation or planning phase ideas and receive equity in return. These funds not only provide capital but also offer valuable knowledge and mentorship. They are managed by experienced investors who possess industry contacts and expertise in entering the market and gaining a competitive edge. When it comes to financing innovative projects based on new technical solutions, industrial enterprises can serve as potential sources of investment. They can provide capital as strategic investors or corporate venture capitalists. The banking sector is the main provider of capital to enterprises, however, when it comes to financing innovative projects, banks are cautious and their financing can only be used for advanced projects. Banks assess the company's creditworthiness and a sufficient level of collateral – these conditions are usually met by companies with a relatively long market history and good financial standing [24].

Quality Production Improvement and System Safety: QPI 16 - CZOTO 10   Materials Research Forum LLC
Materials Research Proceedings 34 (2023) 268-277   https://doi.org/10.21741/9781644902691-32

Areas of Commercialization in STEM Sciences

The commercialization of scientific research results must have economic value for stakeholders, meaning it must either significantly improve the quality of products and services [25-27] or reduce costs, including the management of production and service processes [28]. Scientific outcomes in the STEM field are typically associated with either typical material-related problems within the scope of materials engineering [29-31], especially in the case of special alloys [32, 33], or machine-related issues [34, 35], which encounter specific problems beyond the competence and cognition of an average engineer, particularly in the railway industry [36], issues related to wear [37], including fatigue of welded joints [38, 39] and hydraulic power systems [40]. The demand for scientific support also arises in the case of particularly advanced technologies associated with coatings [41-43] and surface layer [44]. This primarily includes DLC coatings [45-47] and ESD coatings [48-50], as well as separation issues [51]. In any of these cases, analytical and statistical support is necessary [52], including experimental design methodology [53-55] and non-classical approaches [56-58]. In the case of a large set of controlled factors, prior dimensionality reduction [59] is also necessary.

Summary

In the process of effective commercialization that brings measurable financial benefits (taking the form of direct and indirect commercialization), the basis is sales, implementation on the market using market principles and its practical use in economic turnover. Therefore, already at an early stage of this process, it is necessary to precisely define the commercialization strategy and carefully and thoroughly identify barriers and risks that may delay this process (thus reducing the entire commercialization potential or economic value of intellectual property), or in extreme cases prevent it. These barriers may relate to technological, legal and economic areas [23, 24]. This will make intellectual property an innovative and perhaps even groundbreaking technology.

References

[1] M. Makowiec. Start-upy technologiczne generujące innowacje w gospodarce jako efekt komercjalizacji badań naukowych. Nierówności Społeczne a Wzrost Gospodarczy 4 (2017) 424-441. https://doi.org/10.15584/nsawg.2017.4.31

[2] B. Flisiuk, A. Gołąbek. Commercialization of research results – models, procedures, barriers and best practices, Zeszyty Naukowe. Organizacja i Zarządzanie / Politechnika Śląska 77 (2015) 63-73.

[3] U. Mikiewicz. Problems of management of special purpose vehicles set up by the local government and scientific institutions to commercialize research – case study, Prace Naukowe Uniwersytetu Ekonomicznego we Wrocławiu 543 (2018) 98-109.

[4] M.J. Radło, M. Baranowski, T. Napiórkowski, J. Chojecki. Komercjalizacja, wdrożenia i transfer technologii. Definicje i pomiar. Dobre praktyki wybranych krajów, Oficyna Wydawnicza SGH, Narodowe Centrum Badań i Rozwoju, Warszawa, 2020. ISBN 978-8380303942

[5] E. Gwarda-Gruszczyńska. Strategie przedsiębiorstw a ochrona własności intelektualnej, in: D.M. Trzmielak (Ed.), Komercjalizacja wiedzy i technologii a własność intelektualna, Centrum Transferu Technologii Uniwersytetu Łódzkiego, Łódź, 2010.

[6] T.B. Kalinowski. Modele komercjalizacji i transferu technologii, in: D.M. Trzmielak (Ed.), Komercjalizacja wiedzy i technologii a własność intelektualna, Centrum Transferu Technologii Uniwersytetu Łódzkiego, Łódź, 2010.

[7] D. Markiewicz. Komercjalizacja wyników badań naukowych – krok po kroku. Centrum Transferu Technologii Politechnika Krakowska, Kraków, 2009.

[8] A. Białek-Jaworska, R. Gabryelczyk, A. Pugacewicz. Czy komercjalizacja wyników badań naukowych wpływa na dojrzałość modelu biznesowego? Przedsiębiorczość i Zarządzanie 16 (2015) 91-108.

[9] S. Thore. Technology Commercialization: DEA and Related Analytical Methods for Evaluating the Use and Implementation of Technical Innovation, Boston-Dordrecht-London, Kluwer Academic Publishers, 2002. https://doi.org/10.1007/978-1-4615-1001-7

[10] B. Poteralska, M. Walasik. Commercialisation Models for R&D Organisations, ECIE 2021 Proc. 16th Eur. Conf. Innov. Entrepreneurship 2 (2021) 782-290. https://doi.org/10.34190/EIE.21.118

[11] W. Gaweł. Komercjalizacja badań naukowych pomiędzy powinnością a niezależnością nauki, in: K. Karpińska, A. Protasiewicz (Eds.), Współczesne problemy ekonomiczne w badaniach młodych naukowców. T.1, Wzrost, rozwój i polityka gospodarcza, Polskie Towarzystwo Ekonomiczne, 196-206, 2018.

[12] S. Łobejko, A. Sosnowska (Eds.). Komercjalizacja wyników badań naukowych praktyczny poradnik dla naukowców. Urząd Marszałkowski Województwa Mazowieckiego w Warszawie, Departament Rozwoju Regionalnego i Funduszy Europejskich, Wydział Innowacyjności, Warszawa, 2013.

[13] D.M. Trzmielak. Współpraca ośrodków naukowych i przedsiębiorstw we wdrażaniu wyników badan, in: W. Wiśniowski (Ed.), Marketing instytucji naukowych i badawczych, Prace Instytutu Lotnictwa, Warszawa, 2013.

[14] D.M. Trzmielak. Komercjalizacja wiedzy i technologii – determinanty i strategie. Wydawnictwo Uniwersytetu Łódzkiego, Łódź, 2013. ISBN 978-8379691401

[15] D. Avimanyu, M. Debmalya, J. Len. Understanding commercialization of technological innovation: Taking stock and moving forward. R&D Manag. 45 (2015) 215-249. https://doi.org/10.1111/radm.12068

[16] R. Svensson. Commercialization, renewal, and quality of patents, Econ. Innov. New Technol. 21 (2012) 175-201. https://doi.org/10.1080/10438599.2011.561996

[17] V.Virchenko et al. Commercialization of intellectual property: innovative impact on global competitiveness of national economies. Market. Manag. Innov. 5 (2021) 25-39. https://doi.org/10.21272/mmi.2021.2-02

[18] B.M. Frischmann. Commercializing University Research System in Economic Perspective: a View from the Demand Side, in: G.D. Libecap (Ed.) University Entrepreneurship and Technology Transfer: Process, Design, and Intellectual Property, Emerald, 2005, 155-186. http://doi.org/10.1016/S1048-4736(05)16006-8

[19] P. Głodek. Akademicki spin off. Wiedza, zasoby i ścieżki rozwoju. Wydawnictwo Uniwersytetu Łódzkiego, Łódź, 2019. ISBN 978-8381425605

[20] T. Dehghani. Technology commercialization: From generating ideas to creating economic value. Int. J. Organ. Leadersh. 4 (2015) 192-199. https://doi.org/10.33844/IJOL.2015.60449

Quality Production Improvement and System Safety: QPI 16 - CZOTO 10   Materials Research Forum LLC
Materials Research Proceedings 34 (2023) 268-277   https://doi.org/10.21741/9781644902691-32

[21] D. Dec, K. Dobrowolska (Eds.). Badanie w zakresie przedsiębiorczości akademickiej i spin-off, w tym programów akademickich dotyczących przedsiębiorczości i funkcjonowania uczelnianych jednostek promujących przedsiębiorczość akademicką, Raport końcowy z badania, Urząd Marszałkowski Województwa Mazowieckiego, Warszawa 2014.

[22] G.D. Markman, D.S. Siegel, M. Wright. Research and Technology Commercialization. J. Manag. Stud. 45 (2008) 1401-1423. https://doi.org/10.1111/j.1467-6486.2008.00803.x

[23] T. Caulfield, U. Ogbogu. The commercialization of university-based research: Balancing risks and benefits, BMC Med. Ethics 16 (2015) art.70. https://doi.org/10.1186/s12910-015-0064-2

[24] K.B. Matusiak, J. Guliński (Eds.) System transferu i komercjalizacji wiedzy w Polsce – siły motoryczne i bariery. PARP, Warszawa, 2010. ISBN 978-8376330013

[25] S. Borkowski et al. The use of 3x3 matrix to evaluation of ribbed wire manufacturing technology, METAL 2012 – 21st Int. Conf. Metallurgy and Materials (2012), Ostrava, Tanger 1722-1728.

[26] D. Siwiec et al. Improving the non-destructive test by initiating the quality management techniques on an example of the turbine nozzle outlet, Materials Research Proceedings 17 (2020) 16-22. https://doi.org/10.21741/9781644901038-3

[27] T. Lipiński, R. Ulewicz. The effect of the impurities spaces on the quality of structural steel working at variable loads, Open Eng. 11 (2021) 233-238. https://doi.org/10.1515/eng-2021-0024

[28] P. Fobel, A. Kuzior. The future (Industry 4.0) is closer than we think. Will it also be ethical? AIP Conf. Proc. 2186 (2019) art. 80003. https://doi.org/10.1063/1.5137987

[29] P. Szataniak et al. HSLA steels - Comparison of cutting techniques, METAL 2014 - 23rd Int. Conf. Metallurgy and Materials (2014), Ostrava, Tanger, 778-783.

[30] A. Szczotok et al. The Impact of the Thickness of the Ceramic Shell Mould on the $(\gamma + \gamma')$ Eutectic in the IN713C Superalloy Airfoil Blade Casting, Arch. Metall. Mater. 62 (2017) 587-593. https://doi.org/10.1515/amm-2017-0087

[31] D. Klimecka-Tatar, R. Dwornicka. The assembly process stability assessment based on the strength parameters statistical control of complex metal products, METAL 2019 – 28th Int. Conf. Metall. Mater. (2019) 709-714. ISBN 978-808729492-5

[32] P. Jonšta et al. The effect of rare earth metals alloying on the internal quality of industrially produced heavy steel forgings, Materials 14 (2021) art.5160. https://doi.org/10.3390/ma14185160

[33] A. Dudek et al. The effect of alloying method on the structure and properties of sintered stainless steel, Arch. Metall. Mater. 62 (2017) 281-287. https://doi.org/10.1515/amm-2017-0042

[34] R. Ulewicz, M. Mazur. Economic aspects of robotization of production processes by example of a car semi-trailers manufacturer, Manuf. Technol. 19 (2019) 1054-1059. https://doi.org/10.21062/ujep/408.2019/a/1213-2489/MT/19/6/1054

[35] S. Blasiak et al. Rapid prototyping of pneumatic directional control valves, Polymers 13 (2021) art. 1458. https://doi.org/10.3390/polym13091458

[36] N. Radek, R. Dwornicka. Fire properties of intumescent coating systems for the rolling stock, Commun. – Sci. Lett. Univ. Zilina 22 (2020) 90-96. https://doi.org/10.26552/com.C.2020.4.90-96

[37] S. Marković et al. Exploitation characteristics of teeth flanks of gears regenerated by three hard-facing procedures, Materials 14 (2021) art. 4203. https://doi.org/10.3390/ma14154203

[38] N. Radek et al. The impact of laser welding parameters on the mechanical properties of the weld, AIP Conf. Proc. 2017 (2018) art. 20025. https://doi.org/10.1063/1.5056288

[39] N. Radek et al. Properties of Steel Welded with CO_2 Laser, Lecture Notes in Mechanical Engineering (2020) 571-580. https://doi.org/10.1007/978-3-030-33146-7_65

[40] G. Filo et al. Modelling of pressure pulse generator with the use of a flow control valve and a fuzzy logic controller, AIP Conf. Proc. 2029 (2018) art.20015. https://doi.org/10.1063/1.5066477

[41] N. Radek et al. Technology and application of anti-graffiti coating systems for rolling stock, METAL 2019 28th Int. Conf. Metall. Mater. (2019) 1127-1132. ISBN 978-8087294925

[42] N. Radek et al. The effect of laser beam processing on the properties of WC-Co coatings deposited on steel. Materials 14 (2021) art. 538. https://doi.org/10.3390/ma14030538

[43] N. Radek et al. Formation of coatings with technologies using concentrated energy stream, Prod. Eng. Arch. 28 (2022) 117-122. https://doi.org/10.30657/pea.2022.28.13

[44] N. Radek et al. The influence of plasma cutting parameters on the geometric structure of cut surfaces, Mater. Res. Proc. 17 (2020) 132-137. https://doi.org/10.21741/9781644901038-20

[45] N. Radek et al. Microstructure and tribological properties of DLC coatings, Mater. Res. Proc. 17 (2020) 171-176. https://doi.org/10.21741/9781644901038-26

[46] N. Radek et al. Influence of laser texturing on tribological properties of DLC coatings, Prod. Eng. Arch. 27 (2021) 119-123. https://doi.org/10.30657/pea.2021.27.15

[47] N. Radek et al. Operational properties of DLC coatings and their potential application, METAL 2022 – 31st Int. Conf. Metall. Mater. (2022) 531-536. https://doi.org/10.37904/metal.2022.4491

[48] N. Radek, K. Bartkowiak. Laser Treatment of Electro-Spark Coatings Deposited in the Carbon Steel Substrate with using Nanostructured WC-Cu Electrodes, Physics Procedia 39 (2012) 295-301. https://doi.org/10.1016/j.phpro.2012.10.041

[49] N. Radek et al. The morphology and mechanical properties of ESD coatings before and after laser beam machining, Materials 13 (2020) art. 2331. https://doi.org/10.3390/ma13102331

[50] N. Radek et al. The impact of laser processing on the performance properties of electro-spark coatings, 14th World Congr. Comput. Mech. and ECCOMAS Congr. 1000 (2021) 1-10. https://doi.org/10.23967/wccm-eccomas.2020.336

[51] M. Zenkiewicz et al. Electrostatic separation of binary mixtures of some biodegradable polymers and poly(vinyl chloride) or poly(ethylene terephthalate), Polimery 61 (2016) 835-843. https://doi.org/10.14314/polimery.2016.835

Quality Production Improvement and System Safety: QPI 16 - CZOTO 10   Materials Research Forum LLC
Materials Research Proceedings 34 (2023) 268-277   https://doi.org/10.21741/9781644902691-32

[52] L. Cedro. Model parameter on-line identification with nonlinear parametrization – manipulator model, Technical Transactions 119 (2022) art. e2022007. https://doi.org/10.37705/TechTrans/e2022007

[53] J. Pietraszek et al. The fixed-effects analysis of the relation between SDAS and carbides for the airfoil blade traces. Arch. Metall. Mater. 62 (2017) 235-239. https://doi.org/10.1515/amm-2017-0035

[54] R. Dwornicka, J. Pietraszek. The outline of the expert system for the design of experiment, Prod. Eng. Arch. 20 (2018) 43-48. https://doi.org/10.30657/pea.2018.20.09

[55] J. Pietraszek et al. Challenges for the DOE methodology related to the introduction of Industry 4.0. Prod. Eng. Arch. 26 (2020) 190-194. https://doi.org/10.30657/pea.2020.26.33

[56] J. Pietraszek. The modified sequential-binary approach for fuzzy operations on correlated assessments, LNAI 7894 (2013) 353-364. https://doi.org/10.1007/978-3-642-38658-9_32

[57] J. Pietraszek et al. Non-parametric assessment of the uncertainty in the analysis of the airfoil blade traces, METAL 2017 – 26th Int. Conf. Metall. Mater. (2017) 1412-1418. ISBN 978-8087294796

[58] J. Pietraszek et al. The non-parametric approach to the quantification of the uncertainty in the design of experiments modelling, UNCECOMP 2017 Proc. 2nd Int. Conf. Uncert. Quant. Comput. Sci. Eng. (2017) 598-604. https://doi.org/10.7712/120217.5395.17225

[59] J. Pietraszek, E. Skrzypczak-Pietraszek. The uncertainty and robustness of the principal component analysis as a tool for the dimensionality reduction. Solid State Phenom. 235 (2015) 1-8. https://doi.org/10.4028/www.scientific.net/SSP.235.1

Quality Production Improvement and System Safety: QPI 16 - CZOTO 10   Materials Research Forum LLC
Materials Research Proceedings 34 (2023) 278-287   https://doi.org/10.21741/9781644902691-33

Complaint Analysis as Part of Service Quality and Safety Management

INGALDI Manuela[1, a *]

[1]Department of Production Engineering and Safety, Faculty of Management, Czestochowa University of Technology, Al. Armii Krajowej 19b, 42-218 Czestochowa, Poland

[a]manuela.ingaldi@pcz.pl

Keywords: Service Quality, Quality Management, Service Safety, Complaints

Abstract. Quality and safety of products and services are very important factors influencing customers' decisions about the purchase. They affect the level of customer satisfaction, but also their safety during consumption. A great deal of information about customers' perceptions of products can be inferred from the complaints they file. Complaint indicators can easily be used to analyze the quality of services available on the market. The purpose of the research was to analyze complaints filed by customers of a chosen e-shop from Poland. The analysis was based on data from January 2017 to December 2021. An analysis was made of the number of complaints filed in relation to the number of orders placed. The percentage of complaints recognized by the manufacturer and the reasons for filing complaints were presented. This analysis should be a starting point for the researched e-shop for improvement measures to raise the level of customer satisfaction.

Introduction

Product quality and safety are important aspects considered by customers. Many scientists have conducted researches to evaluate the quality of products [1,2], the technology of their production, which affects the final result [3,4], the use of materials during their production [5,6], production machinery and equipment [7,8], or safety during their use [9,10]. Only such an approach allows to offer products of the highest quality and influence customers' satisfaction level.

A peculiar situation in terms of quality can be seen in the case of services due to the lack of strictly defined service parameters that can be easily measured. In this case, completely different methods of quality assessment are used, which take into account different types of attributes related to the service in question. However, complaint analysis can be used for both tangible products and services. A complaint can be defined as a directed expression of dissatisfaction to an organization, related to its products or the complaint handling process itself, where a response or solution is expected or required [11,12]. Complaints are a way for a customer to communicate with an enterprise to notify of inadequate quality of products or services, and are an important element of management in an organization [13].

A complaint is an expression of customer's dissatisfaction directed to a service provider, third parties or consumer protection authorities [14]. It is a set of behavioral and non-behavioral reactions, some or all of which are triggered by the customer's dissatisfaction with the purchase of a service or product [15]. Complaints can also be viewed as actions that directly express dissatisfaction with the performance of services that do not meet acceptable or tolerable standards [16]. Customers complain when they experience service performance that falls below their expectations and the dissatisfaction, they feel stems from this [17,18]. Thinking of complaints, people often have a negative attitude toward this term because of the claim-compensation overtones it carries. However, it should be acknowledged that it is one of the most important sources of information about the quality of a product or service that a company can use. If an enterprise treats complaints as a mere problem to be gotten rid of as quickly as possible, it has little awareness of the role of complaints in quality assessment. Often the way complaints are reported

Quality Production Improvement and System Safety: QPI 16 - CZOTO 10   Materials Research Forum LLC
Materials Research Proceedings 34 (2023) 278-287   https://doi.org/10.21741/9781644902691-33

and handled by the enterprise also does not have a positive effect on its image in the eyes of the customer.

The challenge is to use complaints data to make decisions that result in substantive action. By using complaint data to address design, marketing, distribution, and after-sales use and maintenance issues, a basic understanding of customer preferences and market behavior is gained. Moreover, a complaint should not be viewed as something very negative. Complaints allow to point out weaknesses in the product, reasons for customer's dissatisfaction, so that a response from the company is possible. A worse situation is when a customer is dissatisfied, tells other customers about it, describes it on various online forums, but does not report it to the enterprise. The result is bad publicity for the enterprise and the inability to improve, as well as the inability to influence the dissatisfied customer.

Understanding the need for such analysis will provide a framework for interpreting the data and extrapolate them to the entire customer base. The framework will allow organizations not only to quantify the consequences of these complaints, but also to prioritize and allocate limited quality assurance resources to mitigate problems. Some argue that complaints made at the time of a problem are less costly than systematic sampling and inspection, and provide more timely information than what is typically available in warranty data [19,20].

Complaint handling encompasses the first contact with the customer during the complaint, as well as a set of company actions during the complaint handling process. Since service and communication with customers have a direct impact on customer satisfaction and loyalty, for a company this part of complaint management is a very important part of quality management [21,22].

The reported complaints make it possible to ascertain not only the existence of problems, but additional analysis of them makes it possible to indicate their causes as well. This is a clue to determining the location of the problem or other, less obvious reasons for its appearance. Due to in-depth analysis, it is possible to indicate where such an error occurred and on which side (e.g., manufacturer, carrier, vendor, etc.), who or what was at fault. This makes it possible to analyze the entire production process of a given product [23].

Many companies use various types of indicators to analyze complaints, which are regularly calculated and compared. Indicators can be, for example, the fraction of product units returned to the manufacturer in the form of quality complaints, the causes of quality complaints (complaints relating to a specific cause), or the cause of complaint recognition (positive handling of complaints due to a specific cause). These indicators can be very useful, but companies using them must be very careful, especially when it comes to the reasons for advertising complaints or the reasons for recognition of complaints, because the same service can be advertised several times, in the same service the same reason can be recognized several times and the same problem can be rectified.

Complaints should be handled and resolved as quickly as possible, even if they may seem annoying, time-consuming and costly, otherwise they can lead to reputational damage caused by bad publicity. Dealing with customer complaints is often the last chance that an organization has to change customer attitudes and offset customer dissatisfaction [24]. Researchers emphasized the potential of complaint management and service improvement systems to increase customers' satisfaction [25]. All information from customer complaints must be analyzed to enable strategic planning to improve the quality of offered services [26]. Considered and resolved, they should be a driver for improving customer satisfaction rather than a consequence of dissatisfaction [20], but also a motivator for improving one's own operations.

An important aspect of a complaint is to analyze it and notify the customer of its outcome. It should be borne in mind that a properly handled complaint process can positively affect customer satisfaction, despite the fact that the complaint itself implies customer dissatisfaction, and that the outcome of the complaint was negative. As a result, it can prevent the customer from running away

Quality Production Improvement and System Safety: QPI 16 - CZOTO 10   Materials Research Forum LLC
Materials Research Proceedings 34 (2023) 278-287   https://doi.org/10.21741/9781644902691-33

from the particular company whose product he complained about. Most customers pass on their dissatisfaction to relatives or friends, often changing companies immediately before expressing their dissatisfaction with the company [27,28]. And this is something that companies cannot afford to do.

Complaint indicators can easily be used to analyze the quality of services available on the market. The purpose of the research was to analyze complaints filed by customers of a chosen e-shop from Poland. The analysis was based on data from January 2017 to December 2021. An analysis was made of the number of complaints filed in relation to the number of orders placed. This analysis should be a starting point for the researched e-shop for improvement measures to raise the level of customer satisfaction.

The methodology presented in this article can find wide application both for external customers and internal recipients within the company. Quality issues [29-31], as well as the consideration of potential failure scenarios [32-34], prevention methods, or mitigating their consequences, are among the most important tools for managers at all levels. The main risk factors can be attributed to failures resulting from both external factors [35-37], the ingress of unwanted gases or liquids [38-40], as well as wear [41-43] and material defects, especially in joints [44,45]. The main methods of preventing failures include the proper technological and operational selection of materials [46-48], the use of appropriate protective coatings [49,50], and special coatings [51-53], as well as modifying their properties by influencing the morphology of the surface layer [54-56]. The complexity of preventive measures requires, in accordance with the Taguchi methodology, the consideration of interfering factors already in the design stage of the product or service [57,58]. Design of Experiments (DOE) [59-61] approach is useful in identifying the hierarchy of interfering factors, including nonparametric methods [62] and resampling techniques [63]. Properly modified products achieve high reliability, enhance user comfort, and thereby reduce the number of reported complaints. This also applies to products in the machinery industry [64-66], including those with specific requirements for military recipients [67-69]. Ultimately, the correct identification of risk factors enables the reduction of excessive resources, leading to cost reduction [70].

Methodology
The research was conducted using data from one e-shop in Poland. This e-shop sells clothing. It ships its products to both domestic and international customers. As a source of data for the research, a summary of customer complaints from January 2017 to December 2021 was used. Importantly, this period takes into account the time of the pandemic, when the number of orders increased sharply.

Data analysis was carried out in terms of the number of complaints filed and the number of complaints lodged, in addition, the percentage of complaints accepted to complaints lodged was calculated. Then the ratio of complaints lodged to the number of orders was calculated. The next step was to analyze the causes of complaints, in this case the analysis was carried out for the entire period combined. The results of the study were presented overall and for each year of the research period.

Results and Discussion
The results of the total number of complaints lodged and complaints accepted are presented first (Table 1).

Table 1. Number of complaints in individual years of the research period [own study]

Factor	Period (year)					
	2017	**2018**	**2019**	**2020**	**2021**	**2017-2021**
Number of complaints lodged [pc]	84	71	76	102	86	419
Number of complaints accepted [pc]	76	68	72	79	77	372
Ratio of accepted to lodged complaints [%]	90.48	95.77	94.74	77.45	89.53	88.78

It can be seen some variability in the number of complaints lodged and complaints accepted by the research e-shop. The largest number of complaints lodged was in 2022, and the largest number were also accepted, but the percentage of accepted complaints to those lodged in 2022 was the smallest. This means that many complaints were unfounded. Perhaps due to the prevailing Covid-19 pandemic and the increase in online shopping, the number of complaints increased.

Table 2. Complaint ratio compared to number of orders for individual years [own study]

Factor	Period (year)					
	2017	**2018**	**2019**	**2020**	**2021**	**2017-2021**
Number of orders [%]	5671	5821	5793	6540	6205	30030
Ratio of complaints lodged [%]	1.48	1.22	1.31	1.56	1.39	1.40
Ratio of complaints accepted [%]	1.34	1.17	1.24	1.21	1.24	1.24

Table 3. The causes of complaints during the research period [own study]

No	Cause of the complaint	Number of complaints lodged [pc]	Number of complaints accepted [pc]	Ratio of accepted to lodged complaints [%]
1.	No search engine	1	1	100.00
2.	Sparse product information on the e-hop site	3	3	100.00
3.	Lack of goods in stock	5	5	100.00
4.	Problems with placing an order	3	3	100.00
5.	Virus after visiting the e-shop site	2	2	100.00
6.	Problems with online payment	6	6	100.00
7.	Leakage of customer data	1	1	100.00
8.	Untimely delivery	15	15	100.00
9.	Lost shipment	3	3	100.00
10.	Goods was not delivered	5	5	100.00
11.	Evidence of opening the shipment	6	6	100.00
12.	Damaged packaging	14	14	100.00
13.	Price does not match the one on the e-shop's website	39	38	97.44

14.	Goods damaged during transport	27	27	100.00
15.	Inappropriate color of the goods	19	16	84.21
16.	Inappropriate size of the goods	34	31	91.18
17.	Customer changed his mind	29	22	75.86
18.	Goods do not agree with the order	34	27	79.41
19.	Goods do not match the description	21	17	80.95
20.	Low quality of the ordered goods	44	38	86.36
21.	Goods do not meet expectations	59	47	79.66
22.	Fake goods	18	14	77.78
23.	Problems with returning the goods	24	24	100.00
24.	No cash refund for returning the goods	7	7	100.00
	Sum	**419**	**372**	**88.78**

But this was also due to the fact that the lockdown did not allow people to leave their homes, and so customers were able to buy and send back ordered products as complaints, instead of normal returns, because they do not fully know their rights as e-commerce customers. In such cases, the customer has an extended possibility of returns, since they can't touch or try on the product, and photos often don't reflect reality. The fewest complaints were lodged in 2018, and that's when the fewest complaints were accepted, but the percentage of accepted complaints to those lodged was the highest. Customers made more than 400 different complaints throughout the entire research period, almost 89% of them were accepted by the research e-shop.

The ratio of complaints accepted was then calculated by dividing the number of complaints by the number of orders and giving the value as a percentage. On a similar basis, the rate of accepted complaints was calculated. The results of these calculations are shown in Table 2.

The most orders were placed in 2022, and the following year the volume was also high. This was due to the change in distribution channels due to the Covid-19 pandemic. Customers, despite being confined to their homes, still wanted to shop online. Many people, skeptical until then, decided to shop online for the first time. The 2020 habit, but also the uncertainty of the pandemic in 2021, caused customers to continue shopping online. An interesting result can be seen in the case of the number of complaints lodged versus the number of orders. The year the pandemic began was not only a year of increased online shopping, but the ratio also increased significantly, with customers complaining more often about the products they purchased. However, the ratio of accepted complaints to the number of orders was not the highest at that time.

In Table 3 the basic reasons for customer complaints are shown. The frequency of their submission by customers and recognition by the company is presented. In addition, the ratio of complaints accepted to those lodged was calculated.

When the causes of complaints were analyzed, they were classified, first grouped together, and then the general name of the cause was determined. 24 different causes of complaints were identified. The various causes of complaints were divided into four groups: site operation and order placement (causes 1-7); delivery (8-10); condition of shipment and goods (11-22); return of goods (23-24). The most common reasons for complaints would be: Goods do not meet expectations, Low quality of the ordered goods, Price does not match the one on the e-shop's website, and Inappropriate size of the goods. Thus, one can conclude that not all customers are satisfied with the products ordered from the research e-shop. Perhaps a more thorough analysis should be carried out from this angle to determine why this is happening. The problem of inadequately sized goods should be resolved as soon as possible so that there are as few complaints of this type as possible. However, the e-shop is not the manufacturer of the goods it sells, they come from different manufacturers. Therefore, it would be necessary to take this into account somehow, for example, in the size tables for individual goods. As previously highlighted, not all complaints were accepted.

Quality Production Improvement and System Safety: QPI 16 - CZOTO 10   Materials Research Forum LLC
Materials Research Proceedings 34 (2023) 278-287   https://doi.org/10.21741/9781644902691-33

As the main reasons why complaints were not accepted in the documentation were cited: complaint submitted after the deadline, inappropriate use of the advertised goods.

In conclusion, it can be said that the number of complaints is not large and does not affect the overall assessment of the quality of services offered by the research e-shop. It should be added that a large part of the complaints lodged by customers are legitimate.

Conclusions

An inseparable part of producing products and providing services is the process of customer complaints. Complaints are the customer's reaction to their level of quality, even more low level of quality.

The purpose of the research was to analyze complaints lodged by customers of one of the e-shops in Poland. Such analysis made it possible to identify the main causes of complaints, which should be a source of improvement, but also allowed to show the level of complaints in relation to the number of orders placed.

The number of complaints lodged by customers and the number of complaints accepted by the e-shop fluctuated during the period under review. However, their percentage in relation to the number of orders placed was not high. Most often, customers complained about Goods do not meet expectations, Low quality of the ordered goods, Price does not match the one on the e-shop's website, and Inappropriate size of the goods. These are areas of potential improvement for the e-shop. Not all complaints were accepted. As the main reasons why complaints were not accepted in the documentation were cited: complaint submitted after the deadline, inappropriate use of the advertised goods.

The occurrence of complaints can result in a decrease of the customer's satisfaction, and ultimately in customers leaving. This, too, will affect the profit for the company, but can also threaten its failure. Hence the importance of the results of such an analysis for the company's management, but also because of this the need to implement improvement measures in these areas.

Among the limitations of the research is the research period. Perhaps extending the research period would show other causes of complaints. In addition, the research was conducted in a clothing e-shop. This is a very specific type of shop. Certainly, many of the indicated causes of complaints could not be included in research of other types of e-commerce stores, hence the limited interpretive possibilities. It should also be noted that the shop's assortment of products affects the structure of customers, and thus the structure of claimants.

However, it should be noted that the analysis of complaints provides a lot of interesting information about the services provided by the e-shop or the reaction of customers to these services. In addition, it gives information about the level of quality without the need for additional research requiring, in the case of services, customer participation. Therefore, such research is worth continuing and drawing conclusions from it in order to improve the organization's operations.

References

[1] R. Stasiak-Betlejewska, A. Czajkowska. Quantification of the quality problems in the construction machinery production, MATEC Web of Conf. 94 (2017) art. 04011. https://doi.org/10.1051/matecconf/20179404011

[2] B. Nurakhova, G. Ilyashova, U. Torekulova. Quality control in dairy supply chain management, Pol. J. Manag. Stud. 21(1) (2020) 236-250. https://doi.org/10.17512/pjms.2020.21.1.18

[3] T. Wróbel, N., Przyszlak, A. Dulska, Technology of Alloy Layers on Surface of Castings, Int. J. Metal Cast. 13(3) (2019) 604–610. https://doi.org/10.1007/s40962-018-00304-x

[4] Y. Blikharskyy et al. Study of concrete under combined action of aggressive environment and long-term loading, Materials 14 (2021) art.6612. https://doi.org/10.3390/ma14216612

[5] P. Jonšta et al. The Effect of Rare Earth Metals Alloying on the Internal Quality of Industrially Produced Heavy Steel Forgings, Materials 14 (2021) art.5160. https://doi.org/10.3390/ma14185160

[6] Y. Nikitin, P. Božek, A. Turygin. Vibration Diagnostics of Spiroid Gear, Manag. Sys. Prod. Eng. 30 (2022) 69-73. https://doi.org/10.2478/mspe-2022-0009

[7] R. Ulewicz, K. Mielczarek. Machine Operation Efficiency in the Production of Car Equipment, AIP Conf. Proc. 2503 (2022) art.050070. https://doi.org/10.1063/5.0099981

[8] M. Krynke, K. Knop, M. Mazur. Maintenance management of large-size rolling bearings in heavy-duty machinery, Acta Montan. Slovaca 27(2) (2022) 327-341. https://doi.org/10.46544/AMS.v27i2.04

[9] D. Palka et al. The role, importance and impact of the methane hazard on the safety and efficiency of mining production, Prod. Eng. Arch. 28 (2022) 390-397. https://doi.org/10.30657/pea.2022.28.48

[10] E. Nedeliaková, M.P. Hranický, M. Valla. Risk identification methodology regarding the safety and quality of railway services, Prod. Eng. Arch. 28 (2022) 21-29. https://doi.org/10.30657/pea.2022.28.03

[11] PN-ISO 10002:2006. Quality management – Customer satisfaction – Guidelines for dealing with complaints in organizations.

[12] B. Olszewska, P. Szewczyk. Effective and Efficient Management of Reclamations Based on a Chosen Enterprise, Zeszyty Naukowe Politechniki Śląskiej – Organizacja i Zarządzanie 63a (1891) (2012) 277-289.

[13] U. Balon, Zachowania młodych konsumentów wobec reklamacji a koszty w przedsiębiorstwie, Zeszyty Naukowe Uniwersytetu Ekonomicznego w Katowicach. Studia Ekonomiczne 255 (2016) 239-249.

[14] B.W. Ateke, I.F. Asiegbu, C.S. Nwulu. Customer Complaint Handling and Relationship Quality: Any Correlation? Ilorin Journal of Marketing 2(2) (2015) 16-34.

[15] J. Singh. Determinants of Consumer Decision to Seek Redress: An Empirical Study of Dissatisfied Patients, J. Consum. Aff. 23 (1989) 329-363. https://doi.org/10.1111/j.1745-6606.1989.tb00251.x

[16] D. Halstead, C. Droge. Consumer Attitude Towards Complaining and the Prediction of Multiple Complaint Responses, Adv. Consumer Res. 18 (1991) 210-216.

[17] T.L. Keiningham et al. A Five-Component Customer Commitment Model: Implications for Repurchase Intentions in Goods and Services Industries, J. Serv. Res. 18 (2015) 433-450. https://doi.org/10.1177/1094670515578823

[18] B.W. Ateke, S.E. Kalu. Complaint Handling and Post-Complaint Satisfaction of Customers of Eateries in Port Harcourt, Nigeria, International J. Res. Bus. Stud. Manag. 3(12) (2016) 16-26.

[19] J. Goodman, S. Newman. Understand Customer Behavior and Complaints. Eight Areas of Quantifiable Data Can Be Integrated into Quality Assurance Decisions, Quality Progress. January 2003 (2003) 51-55.

[20] R.R. Ramphal. A Complaints Handling System for the Hospitality Industry. African Journal of Hospitality, Tourism and Leisure. 5(2) (2016) 1-15.

[21] H. Estelami. Competitive and Procedural Determinants of Delight and Disappointment in Consumer Complaint Outcomes, J. Serv. Res. 42 (2000) 285-300. https://doi.org/10.1177/109467050023006

[22] T. Gruber, I. Szmigin, R. Voss. Developing a Deeper Understanding of the Attributes of Effective Customer Contact Employees in Personal Complaint-Handling Encounters, J. Serv. Market. 23 (2009) 422-435. https://doi.org/10.1108/08876040910985889

[23] A. Juszczak, B. Białecka. Complain process as supporting tool for production process. Zeszyty Naukowe Politechniki Śląskiej. Organizacja i Zarządzanie 63a(1891) (2012) 33-49.

[24] N.A. Vincent, M. Webster. Emotions and Response Actions in Consumer Complaint Behaviour, ANZMAC2005 Conf.: Consumer Behaviour, Australian and New Zealand Marketing Academy Conference, Fremantle, WA, (2005) 352-358.

[25] A.K. Smith, R.N. Bolton, J. Wagner. A Model of Customer Satisfaction with Service Encounters Involving Failure and Recovery, J. Market. Res. 36 (1999) 356-372. https://doi.org/10.2307/3152082

[26] P. Merrill. Do It Right the Second Time: Benchmarking Best Practices in the Quality Change Process, QualityPress, Milwaukee, 2009. ISBN 978-0873897334

[27] C. Homburg, A. Fürst. Complaint Behaviour and Management, Die Betriebswirtschaft 67(1) (2007) 45-58.

[28] S. Swanson, S. Kelly. Service Recovery Attributions and Word-of-Mouth Intentions, European J. Market. 35 (2001) 194-211. http://doi.org/10.1108/03090560110363463

[29] S. Borkowski et al. The use of 3x3 matrix to evaluation of ribbed wire manufacturing technology, METAL 2012 – 21st Int. Conf. Metall. Mater. (2012), Ostrava, Tanger 1722-1728.

[30] R. Ulewicz, F. Nový. Quality management systems in special processes, Transp. Res. Procedia 40 (2019) 113-118. https://doi.org/10.1016/j.trpro.2019.07.019

[31] A. Pacana et al. Analysis of quality control efficiency in the automotive industry, Transp. Res. Procedia 55 (2021) 691-698. https://doi.org/10.1016/j.trpro.2021.07.037

[32] J. Fabiś-Domagała et al. Instruments of identification of hydraulic components potential failures, MATEC Web of Conf. 183 (2018) art.03008. https://doi.org/10.1051/matecconf/201818303008

[33] K. Knop et al. Evaluating and Improving the Effectiveness of Visual Inspection of Products from the Automotive Industry, Lecture Notes in Mechanical Engineering (2019) 231-243. https://doi.org/10.1007/978-3-030-17269-5_17

[34] J. Fabis-Domagala et al. A concept of risk prioritization in FMEA analysis for fluid power systems, Energies 14 (2021) art. 6482. https://doi.org/10.3390/en14206482

[35] M. Scendo et al. Purine as an effective corrosion inhibitor for stainless steel in chloride acid solutions, Corr. Rev. 30 (2012) 33-45. https://doi.org/10.1515/CORRREV-2011-0039

[36] M. Scendo et al. Influence of laser treatment on the corrosive resistance of WC-Cu coating produced by electrospark deposition, Int. J. Electrochem. Sci. 8 (2013) 9264-9277.

[37] D. Klimecka-Tatar et al. The effect of powder particle biencapsulation with Ni-P layer on local corrosion of bonded Nd-(Fe, Co)-B magnetic material, Arch. Metall. Mater. 60 (2015) 154-157. https://doi.org/10.1515/amm-2015-0024

[38] M. Ulewicz et al. Ion flotation of zinc(II) and cadmium(II) in the presence of side-armed diphosphaza-16-crown-6 ethers, Sep. Sci. Technol. 38 (2003) 633-645. https://doi.org/10.1081/SS-120016655

[39] M. Ulewicz et al. Transport of lead across polymer inclusion membrane with p-tert - butylcalix[4]arene derivative , Physicochem. Probl. Miner. Process. 44 (2010) 245-256.

[40] M. Zenkiewicz et al. Electrostatic separation of binary mixtures of some biodegradable polymers and poly(vinyl chloride) or poly(ethylene terephthalate), Polimery/Polymers 61 (2016) 835-843. https://doi.org/10.14314/polimery.2016.835

[41] S. Marković et al. Exploitation characteristics of teeth flanks of gears regenerated by three hard-facing procedures, Materials 14 (20210 art. 4203. https://doi.org/10.3390/ma14154203"

[42] M. Krynke et al. Maintenance management of large-size rolling bearings in heavy-duty machinery, Acta Montan. Slovaca 27 (2022) 327-341. https://doi.org/10.46544/AMS.v27i2.04

[43] P. Regulski, K.F. Abramek. The application of neural networks for the life-cycle analysis of road and rail rolling stock during the operational phase, Technical Transactions 119 (2022) art. e2022002. https://doi.org/10.37705/TechTrans/e2022002

[44] M. Patek et al. Non-destructive testing of split sleeve welds by the ultrasonic TOFD method, Manuf. Technol. 14 (2014) 403-407. https://doi.org/10.21062/ujep/x.2014/a/1213-2489/MT/14/3/403

[45] N. Radek, J. Pietraszek, A. Goroshko. The impact of laser welding parameters on the mechanical properties of the weld, AIP Conf. Proc. 2017 (2018) art.20025. https://doi.org/10.1063/1.5056288

[46] A. Szczotok et al. The Impact of the Thickness of the Ceramic Shell Mould on the $(\gamma + \gamma')$ Eutectic in the IN713C Superalloy Airfoil Blade Casting, Arch. Metall. Mater. 62 (2017) 587-593. https://doi.org/10.1515/amm-2017-0087

[47] D. Klimecka-Tatar, R. Dwornicka. The assembly process stability assessment based on the strength parameters statistical control of complex metal products, METAL 2019 28th Int. Conf. Metall. Mater. (2019) 709-714. ISBN 978-808729492-5

[48] D. Siwiec et al. Improving the process of achieving required microstructure and mechanical properties of 38mnvs6 steel, METAL 2020 29th Int. Conf. Metall. Mater. (2020) 591-596. https://doi.org/10.37904/metal.2020.3525

[49] W. Zórawski et al. Plasma-sprayed composite coatings with reduced friction coefficient, Surf. Coat. Technol. 202 (2008) 4578-4582. https://doi.org/10.1016/j.surfcoat.2008.04.026

[50] N. Radek et al. Formation of coatings with technologies using concentrated energy stream, Prod. Eng. Arch. 28 (2022) 117-122. https://doi.org/10.30657/pea.2022.28.13

[51] N. Radek et al. Microstructure and tribological properties of DLC coatings, Mater. Res. Proc. 17 (2020) 171-176. https://doi.org/10.21741/9781644901038-26

[52] N. Radek et al. Influence of laser texturing on tribological properties of DLC coatings, Prod. Eng. Arch. 27 (2021) 119-123. https://doi.org/10.30657/pea.2021.27.15

[53] N. Radek et al. Operational properties of DLC coatings and their potential application, METAL 2022 31st Int. Conf. Metall. Mater. (2022) 531-536. https://doi.org/10.37904/metal.2022.4491

[54] N. Radek et al. The WC-Co electrospark alloying coatings modified by laser treatment, Powder Metall. Met. Ceram. 47 (2008) 197-201. https://doi.org/10.1007/s11106-008-9005-7

[55] N. Radek et al. Laser Processing of WC-Co Coatings, Mater. Res. Proc. 24 (2022) 34-38. https:10.21741/9781644902059-6

[56] P. Kurp, H. Danielewski. Metal expansion joints manufacturing by a mechanically assisted laser forming hybrid method – concept, Technical Transactions 119 (2022) art. e2022008. https://doi.org/10.37705/TechTrans/e2022008

[57] R. Dwornicka et al. The laser textured surfaces of the silicon carbide analyzed with the bootstrapped tribology model, METAL 2017 26th Int. Conf. Metall. Mater. (2017) 1252-1257. ISBN 978-8087294796

[58] L. Cedro. Model parameter on-line identification with nonlinear parametrization – manipulator model, Technical Transactions 119 (2022) art. e2022007. https://doi.org/10.37705/TechTrans/e2022007

[59] R. Dwornicka, J. Pietraszek. The outline of the expert system for the design of experiment, Prod. Eng. Arch. 20 (2018) 43-48. https://doi.org/10.30657/pea.2018.20.09

[60] J. Pietraszek, N. Radek, A.V. Goroshko. Challenges for the DOE methodology related to the introduction of Industry 4.0. Prod. Eng. Arch. 26 (2020) 190-194. https://doi.org/10.30657/pea.2020.26.33

[61] B. Jasiewicz et al. Inter-observer and intra-observer reliability in the radiographic measurements of paediatric forefoot alignment, Foot Ankle Surg. 27 (2021) 371-376. https://doi.org/10.1016/j.fas.2020.04.015

[62] J. Pietraszek. The modified sequential-binary approach for fuzzy operations on correlated assessments, LNAI 7894 (2013) 353-364. https://doi.org/10.1007/978-3-642-38658-9_32

[63] J. Pietraszek, L. Wojnar. The bootstrap approach to the statistical significance of parameters in RSM model, ECCOMAS Congress 2016 Proc. 7th Europ. Congr. Comput. Methods in Appl. Sci. Eng. 1 (2016) 2003-2009. https://doi.org/10.7712/100016.1937.9138

[64] A. Goroshko et al. Construction and practical application of hybrid statistically-determined models of multistage mechanical systems, Mechanika 20 (2014) 489-493. https://doi.org/10.5755/j01.mech.20.5.8221

[65] R. Ulewicz, M. Mazur. Economic aspects of robotization of production processes by example of a car semi-trailers manufacturer, Manufacturing Technology 19 (2019) 1054-1059. https://doi.org/10.21062/ujep/408.2019/a/1213-2489/MT/19/6/1054

[66] I. Drach et al. Design Principles of Horizontal Drum Machines with Low Vibration, Adv. Sci. Technol. Res. J. 15 (2021) 258-268. https://doi.org/10.12913/22998624/136441

[67] W. Przybył et al. Virtual Methods of Testing Automatically Generated Camouflage Patterns Created Using Cellular Automata, Mater. Res. Proc. 24 (2022) 66-74. https://doi.org/10.21741/9781644902059-11

[68] N. Radek et al. Operational tests of coating systems in military technology applications, Eksploat. i Niezawodn. 25 (2023) art.12. https://doi.org/10.17531/ein.2023.1.12

[69] W. Przybył et al. Microwave absorption properties of carbonyl iron-based paint coatings for military applications, Def. Technol. 22 (2023) 1-9. https://doi.org/10.1016/j.dt.2022.06.013

[70] A. Maszke, R. Dwornicka, R. Ulewicz. Problems in the implementation of the lean concept at a steel works – Case study, MATEC Web of Conf. 183 (2018) art.01014. https://doi.org/10.1051/matecconf/201818301014

Quality Production Improvement and System Safety: QPI 16 - CZOTO 10   Materials Research Forum LLC
Materials Research Proceedings 34 (2023) 288-297   https://doi.org/10.21741/9781644902691-34

Good Manufacturing Practices for Quality and Safety Management in the Food Industry

SZCZYRBA Anna [1,a] * and DZIUBA Szymon [2,b]

[1] Medical University of Silesia Faculty of Health Sciences in Bytom (student research club), Czestochowa University of Technology, Faculty of Management (student research club - Quality and safety promoter), Czestochowa, Poland

[2] Wroclaw University of Economics and Business, Faculty of Business and Management, Wroclaw, Poland

[a]anna.szczyrba.09@gmail.com, [b] szymon.dziuba@ue.wroc.pl

Keywords: Quality, HACCP, FMEA, Food Industry

Abstract. The article presents a description of the design and implementation of the FMEA methodology in a food company dealing with the processing of vegetables and fruits, where it is used in combination with the HACCP system as a tool for ensuring product quality and as a means of improving the operational efficiency of the production cycle. The use of the FMEA method as support for HACCP in terms of safety and quality allowed for step-by-step analysis of the production cycle, provided additional knowledge about processes and ultimately positively influenced both the process and the product itself.

Introduction

In today's world, the quality of food products is a key factor in a company's success. Customers are paying more attention to the composition and origin of products, so the food industry must adapt to their needs and expectations [1,2]. To achieve this, companies use various tools and quality management methods to ensure the safety and compliance of products with requirements. Quality is defined as the set of product characteristics that meet both stated and implied customer needs [3]. Adopting predefined standards enables Quality Assurance to provide customers with the assurance that the company is operating in accordance with these requirements. In the Food Industry, there are two distinct aspects of product quality: food safety and sanitary integrity, which are mandatory for selling food products, and other factors such as appearance, functionality, and nutritional characteristics that appeal to customers [4]. The use of Failure Mode and Effects Analysis (FMEA) methodology and incorporating its findings into an already established HACCP (Hazard Analysis and Critical Control Points) system allows for a comprehensive examination and analysis of both aspects of food quality. In the food industry, HACCP system is considered one of the good practices, other include [5, 6]:

- Quality control – a process of controlling the quality of products and services to ensure customer requirements are met.
- Proper identification and labelling of products – a process of identifying and labelling products, which allows for easy recognition of products and information about them.
- Cleanliness and sanitation – a process of maintaining cleanliness and sanitation to ensure food safety and the health of workers.
- Compliance with regulations and laws related to food safety, environmental protection and worker health.
- Employee training – training employees in food safety, compliance with regulations and good manufacturing practices.

Quality Production Improvement and System Safety: QPI 16 - CZOTO 10   Materials Research Forum LLC
Materials Research Proceedings 34 (2023) 288-297   https://doi.org/10.21741/9781644902691-34

- Monitoring and oversight – regular monitoring and oversight of production processes and food safety to ensure their effectiveness.

The HACCP system is counted among so called good practices in the food industry, and includes:

- Quality control – the process of controlling the quality of products and services to ensure that customer requirements are met.
- Correct product identification and labelling – the process of identifying and labelling products so that products and information about them can be easily identified.
- Cleanliness and sanitation – the process of maintaining cleanliness and sanitation to ensure food safety and employee health.
- Compliance with food safety, environmental and worker health laws and regulations.
- Worker training – training of workers in food safety, compliance with regulations and good production practices.
- Monitoring and supervision – regular monitoring and supervision of production processes and food safety to ensure their effectiveness.

Hazard Analysis and Critical Control Points (HACCP)

HACCP was first introduced to food service by the Minnesota Foodservice Quality Assurance program in 1974. The HACCP program can monitor food production from raw materials to end products and even as far as serving, based on controlling factors such as time, temperature, and specific factors that contribute to foodborne disease outbreaks.

End-point testing is not an effective way to guarantee food safety because by the time the results are available, the food has already been sent to the customer. A more effective approach is to implement additional procedures during the processing stage and monitor the process using a HACCP system, which has been shown to be a reliable method. HACCP is designed to prevent problems from occurring in the first place, rather than addressing them after they have happened. By following the guidelines of safe food production with the HACCP system, the risk of foodborne illnesses will be reduced. It should be emphasized that the HACCP system is a food supervision system and is aimed at eliminating direct causes of health risks, directly at the place of their origin. This system cannot be treated as a constant, once developed for a given facility procedure, including documents resulting from it [7]. The HACCP system should ensure food safety, through constant monitoring of biological, chemical and physical hazards in the process of production, storage and food distribution. Although the HACCP system is effective in ensuring food safety, it has several limitations, particularly for vegetable and fruit processing companies. One of the limitations of the HACCP system is limited effectiveness. The HACCP system is focused on identifying and controlling specific hazards, such as pathogenic bacteria, but does not take into account other potential problems, such as production errors. In the case of vegetable and fruit processing companies, this may lead to a lack of control over other hazards, such as chemical contamination or microbiological pollution. Another limitation is the dependence on the accuracy and reliability of the analysis [8, 9] The effectiveness of the HACCP system depends on the accuracy and reliability of the analysis of hazards, so it is important to conduct it in an appropriate and objective manner. In the case of vegetable and fruit processing companies, difficulties related to identifying all hazards may lead to a lack of effective control over product safety. HACCP focuses on identifying and controlling critical control points to prevent food safety hazards and ignores the aspect of production. In this case, the FMEA method, which focuses on identifying and assessing potential errors and their effects, is appropriate. The use of the FMEA method can help identify potential problems in the production process, allowing for preventative actions to be taken, which in turn can reduce the risk of food safety hazards. However, FMEA and HACCP are complementary methods and both should be used together to ensure the highest level of food safety.

Failure Mode and Effects Analysis (FMEA)

FMEA involves conducting a risk analysis for each stage of the production process, in order to identify potential errors and their effects. These analyses include criteria such as the frequency of error occurrence, its impact, and the level of difficulty in detecting it [12-14]. Based on these criteria, the risk of a given error is evaluated and preventive actions such as modifying the production process or adding quality control are taken. In the food industry, FMEA is particularly useful in identifying potential food safety hazards such as microorganisms, chemical contamination, or physical contamination. This analysis allows for the identification of potential sources of hazards and preventive actions such as changing the production process or adding quality control, which can reduce the risk of these hazards occurring. Furthermore, FMEA can also help in identifying potential issues related to product quality, such as lack of taste or smell, which allows for preventative actions such as changing ingredients or production processes. As with HACCP, FMEA has certain limitations related to costs, both in terms of time and human resources. FMEA analysis is based on expert opinions, which can lead to subjectivity and a lack of a consistent approach to risk assessment. FMEA is effective in identifying potential hazards, but it does not always allow for effective risk management, as it does not take into account the actual occurrence of errors. FMEA analysis should be regularly updated to reflect changes in the production process, but this is often ignored. The method itself focuses on identifying errors and potential hazards, which can lead to a lack of attention on other important aspects of safety and product quality

Results and Discussion

The research was carried out in an SME company involved in the production of: canned and jarred vegetables, concentrates, purees, dinner additives, spices, fruit preparations and juices. For the case study, the production of canned cucumber in one-letter glass packaging was proposed (Fig.1).

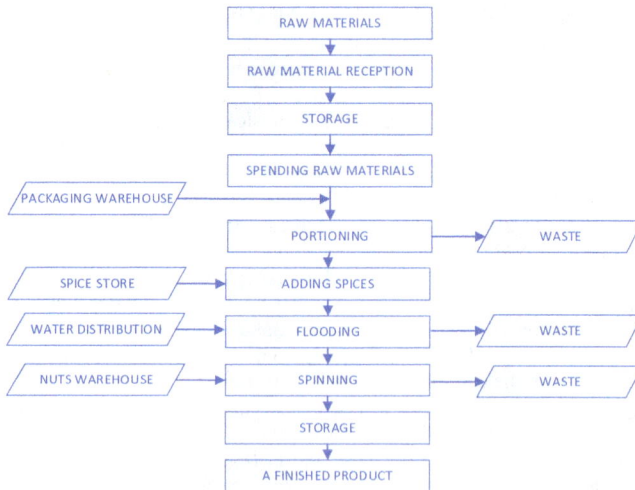

Fig.1. *Flowchart of the production process of pickled cucumbers in a 1lir jar.*

Quality Production Improvement and System Safety: QPI 16 - CZOTO 10   Materials Research Forum LLC
Materials Research Proceedings 34 (2023) 288-297   https://doi.org/10.21741/9781644902691-34

According to the principles of the HACCP system, a list of all expected biological, chemical, and physical hazards should be made at each stage specified on the diagram. These hazards should be specified as precisely as possible. In this case, the following should be taken into account: an assessment of the significance and impact of a given hazard on human health, the probability of a given hazard occurring, the possibility of survival or reproduction of microorganisms that may infect the health of the consumer, as well as the possibility of eliminating hazards. Table 1 presents a summary of hazards and identification of CCPs.

***Table 1.** Hazard identification – CCPs*

Process stage	Type and source of the threat	Control	CCP
Raw material reception	Infected raw materials	approvals, visual assessment of packaging tightness	CCP 1
	Incorrect composition of raw materials	approvals, System procedure	
	Foreign bodies in raw materials	approvals, System procedure	
Storage	Development of microorganisms caused by pest activity	System procedure	CCP 2
	Growth of microorganisms caused by inappropriate temperature and humidity	System procedure	CCP 3
	Pests and their droppings	System procedure	CCP 4
Spending raw materials	Undefined threat	-	-
Portioning	Secondary human infection	GHP Code	-
	Incorrect amount/weight of cucumbers	Weighing instructions	CCP 5
Adding spices	Secondary human infection	GHP Code	-
Spinning	Secondary human infection	GHP Code	-
Storage	Development of microorganisms caused by pest activity	System procedure	-
	Growth of microorganisms caused by inappropriate temperature and humidity	System procedure	CCP 6
	Pests and their droppings	System procedure	CCP 7

The most commonly identified critical control points are the receipt of products into the warehouse and the storage of food. All potential hazards should be examined and the products accepted should be evaluated in terms of health quality. For each critical control point, so-called target values along with allowable tolerances and boundary values of specified parameters, i.e. critical values, should be established. These values should guarantee the effective elimination of the hazard. These criteria usually include indicators such as time, temperature, humidity, water activity, etc., and sometimes sensory characteristics (e.g. smell, taste, color).

The monitoring method must be strictly defined. It should specify:

- what parameters are measured,
- what are the critical limits for the measurements made,
- how, when, and with what frequency monitoring is carried out.
- whether the monitoring procedure is reliable.

Figure 2 shows the hazard identification decision tree.

For each critical control point, corrective actions must be established. These actions should allow for the immediate removal of any deviations in the values of the accepted parameters and ensure that the critical point is under control. Corrective actions should be taken after exceeding the critical value for a given point. Corrective actions should include: ways to regain control over the critical control point, methods of dealing with uncertain product (what to do with a product that was poorly stored - throw it away, test it, shorten the shelf life for consumption?).

Corrective actions should be as simple and easy to implement as possible and the personnel responsible for them should be appropriately trained. Verification aims to determine whether the procedures introduced within the HACCP system give the desired result. It should also detect any shortcomings. Activities may include, for example, reviews of records with identified deviations, determining tolerance levels, etc. Verification should be carried out by persons other than those responsible for the activities being verified. Verification should be carried out periodically in a planned manner to ensure the effective implementation of HACCP principles.

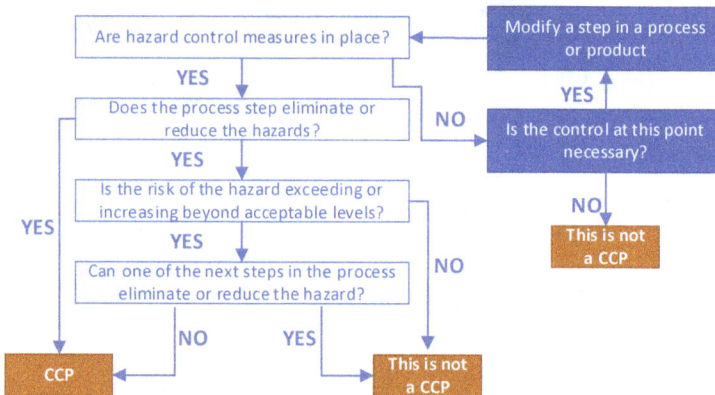

Fig.2. *Identification of critical points of CCPs.*

The FMEA method is a widely used tool in the field of reliability engineering. It is used to identify potential failure modes in products or systems, and to analyze the effects and causes of those failures. The FMEA process begins by identifying potential failure modes based on existing information and then uses statistical methods to determine the severity, occurrence, and detectability of each failure mode (Fig.3).

The risk priority number (RPN) is calculated for each failure mode using a formula (RPN=S×O×D), and the failure modes are ranked according to their RPN value. Finally, measures are put in place to improve the reliability of the product or system by addressing the most significant failure modes. The severity (S) of a failure mode is a measure of its impact on the system, product, or customer, and is typically rated on a scale from 1 to 10. The occurrence (O) of a failure mode is a measure of its likelihood and is also rated on a scale from 1 to 10. The detectability (D) of a failure mode refers to the ease with which it can be detected and is rated on a scale from 1 to 10. Based on the results of the FEMA analysis obtained for vegetable processing products, potential product failure modes have been established. Failures with the highest number of RPNs are: Pollution of sources of vegetable (RPN=320), Lack of cold storage protection (RPN=196) and Cross-contamination (RPN=120). The potential failure factors need to be combined with the flow of vegetable products from the growing site to distribution to the customer in order to identify critical CCP control points. As shown in Table 2, the critical control points of the meat product supply chain are cultivation, distribution and transport and sales.

Table 2. Analysis of critical control points in the supply chain of plant products

Potential failure modes	Link(s) of the occurrence of the failure	RPN	CCP
Pollution of sources of vegetable	Cultivation	320	YES
Lack of cold storage protection	Distribution and transportation and sales	196	YES
Cross-contamination	Cultivation, transport, processing and sales	120	NO
False label marks	Processing and sales	56	NO
Pollution of packaging materials	Storage and distribution and transportation	36	NO

Functional requirements of vegetable processing products	Potential failure modes	Consequences of potential failures	S	Potential failure mechanism	O	Current process control	D	RPN	The link of responsibility supply chain
Preservation. The product must be able to effectively preserve the freshness and nutritional value of the vegetables for an extended period of time. / Shelf life. The product should have a long shelf life to ensure that it can be stored for a reasonable period of time before consumption.	Pollution of vegetable sources	Vegetables products are polluted by bacteria such as E. coli, Salmonella, and Listeria, pesticides, heavy metals, mycotoxins, and microplastics.	8	Use of pesticides by farmers, soil contamination, water contamination, air pollution, animal waste, lack of proper sanitation and hygiene practices.	5	Strictly implement the relevant access approval standards, check the "phytosanitary certificate" and the "official certificate of non-epizootic area" Testing. Regularly testing the vegetables for contaminants. Traceability: implementing traceability systems.	8	320	Cultivation
Food safety. The product must be safe for consumption and should not pose a risk of food poisoning or other health hazards.	Cross-contamination	Microbes and pathogenic bacteria contaminate vegetables products	6	Cross-contamination of equipment, inadequate cleaning of equipment, the transmission of harmful microorganisms due to non-compliance with sanitary and hygienic procedures.	5	Avoid workers and customers having direct contact with vegetables; implement operational prerequisite programs	4	120	cultivation, transport, processing and sales
Taste and texture. The product should retain the taste and texture of fresh vegetables as much as possible, without being overly processed or altered.	Pollution of packaging materials	Vegetables products suffer chemical pollution	4	Packing materials contain harmful components as well as mobility and diffusivity	3	Strictly implement the national technical standards and regulations of packaging materials	3	36	Storage, distribution and transportation
Nutritional value. The product should retain as much of the nutritional value of the fresh vegetables as possible, and should not introduce any harmful additives or preservatives.	Lack of cold storage protection	Vegetable products suffer spoilage and are polluted by spoilage organisms and pathogenic bacteria	7	The storage of fresh vegetable is incomplete, no attention is paid to securing vegetables in the warehouse	7	professional distribution teams; strengthening of temperature monitoring and records	4	196	Distribution and transportation, sales
Convenience. The product should be easy to use and store, and should not require extensive preparation or cooking time.	False label marks	Expired and bad vegetable products	7	Use exaggerated and false packaging to change the expired vegetables products and sell them again	4	Authenticity test, information query, monitoring and recording of expiration dates	2	56	Processing and sales

Fig.3. *FMEA analysis form of vegetable products*

Summary

In conclusion, this research suggests that the use of the Failure Modes and Effects Analysis (FMEA) method as support for the Hazard Analysis and Critical Control Points (HACCP) system in the food industry provides better benefits than using HACCP alone. By identifying potential failures using FMEA, the research found that the most significant risks included pollution of sources of vegetables, lack of cold storage protection, cross-contamination, and false label marks. These failures were then classified according to their Risk Priority Number (RPN) value, with the highest RPN values assigned to pollution of sources of vegetables and lack of cold storage protection. Both of these failures were defined as Critical Control Points (CCPs) and are therefore supervised and monitored. However, cross-contamination and false label marks were not defined as CCPs and may require additional attention. Overall, this research suggests that the combination of HACCP and FMEA can be an effective way to identify and mitigate risks in the food industry because:

- Together, these methods provide a comprehensive approach to identifying and assessing risks in the food industry.
- The combination of HACCP and FMEA allows for a more detailed and thorough assessment of risks, leading to more effective risk management. By identifying potential hazards and analyzing their potential effects, companies can take proactive measures to prevent or mitigate these risks.
- Using FMEA as support for HACCP can lead to more efficient risk management. By identifying potential hazards early on in the process, companies can take steps to prevent them from occurring, rather than having to react to problems after they have already occurred.
- By using both HACCP and FMEA, companies can continuously monitor and improve their food safety processes. This helps ensure that risks are identified and addressed in a timely manner, leading to safer products for consumers.

Safety in the food processing industry is associated with both the safety of workers and the sanitary safety of the produced food, and proper management procedures are crucial here [15, 16]. Regardless of that, technological measures related to proper preparation of surfaces [17], protective coatings [18, 19], special coatings [20, 21], and applied joints [22] will be helpful. The appropriate design of buildings [23] and the organization of internal and external traffic [24] also have a significant impact. Useful tools will be proper mathematical procedures [25, 26] that enable effective reduction of a very large number of process factors, often mutually correlated, to a much smaller but easier-to-analyze set [27]. Then, known statistical analysis procedures of DOE methodology [28-30] can be applied, including non-parametric approaches [31-33] and those specific to small sample sizes [34-36].

References

[1] A. Pacana, R., Ulewicz. Analysis of causes and effects of implementation of the quality management system compliant with ISO 9001. Pol. J. Manag. Stud. 21 (2020) 283-296. https://doi.org/10.17512/pjms.2020.21.1.21

[2] J.C Lee et al. Implementation of Food Safety Management Systems along with Other Management Tools (HAZOP, FMEA, Ishikawa, Pareto). The Case Study of Listeria monocytogenes and Correlation with Microbiological Criteria. Foods 10 (2021) art.2169. https://doi.org/10.3390/foods10092169

[3] D. Siwiec et al. Concept of a model to predict the qualitative-cost level considering customers' expectations. Pol. J. Manag. Stud. 26 (2022) 330-340. https://doi.org/10.17512/pjms.2022.26.2.20

[4] J. Rosak-Szyrocka, A.A. Abbase. Quality management and safety of food in HACCP system aspect. Prod. Eng. Arch., 26 (2020) 50-53. https://doi.org/10.30657/pea.2020.26.11

[5] Y.-M., Sun, H.W. Ockerman. A review of the needs and current applications of hazard analysis and critical control point (HACCP) system in foodservice areas. Food Control 16 (2005) 325-332. https://doi.org/10.1016/j.foodcont.2004.03.012

[6] K.V. Kotsanopoulos, I.S. Arvanitoyannis. The Role of Auditing, Food Safety, and Food Quality Standards in the Food Industry: A Review. Compr. Rev. Food Sci. Food Saf. 16 (2017) 760-775. https://doi.org/10.1111/1541-4337.12293

[7] P.J. Panisello, P.C. Quantick. Technical barriers to Hazard Analysis Critical Control Point (HACCP). Food Control 12 (2001) 165-173. https://doi.org/10.1016/S0956-7135(00)00035-9

[8] V. Dadi et al. Agri-Food 4.0 and Innovations: Revamping the Supply Chain Operations. Prod. Eng. Arch. 27 (2021) 75-89. https://doi.org/10.30657/pea.2021.27.10

[9] M. Ingaldi, S.T. Dziuba. Market of the organic products in Poland according to potential customers, In: 16th Int. Multidiscip. Sci. GeoConf. Surv. Geol. Mining Ecol. Manag. SGEM 3 (2016), 341-348. https://doi.org/10.5593/SGEM2016/B53/S21.044

[10] S.T. Dziuba, M. Ingaldi, Systems providing food safety and its perception by polish customers – introduction, In: 17th Int. Multidiscip. Sci. GeoConf. Surv. Geol. Mining Ecol. Manag. SGEM 17 (2017) 854-860. https://doi.org/10.5593/sgem2017/53/S21.104

[11] N. Baryshnikova et al. Management approach on food export expansion in the conditions of limited internal demand. Pol. J. Manag. Stud. 21 (2020) 101-114. https://doi.org/10.17512/pjms.2020.21.2.08

[12] R Wolniak. Problems of use of FMEA method in industrial enterprise. Prod. Eng. Arch. 23 (2019) 12-17. https://doi.org/10.30657/pea.2019.23.02

[13] R. Ulewicz. Practical Application of Quality Tools in the Cast Iron Foundry. Manuf. Technol. 14 (2014) 104-111. https://doi.org/10.21062/ujep/x.2014/a/1213-2489/MT/14/1/104

[14] R. Ulewicz et al. Implementation of logic flow in planning and production control, Manag. Prod. Eng. Rev. 7 (2016) 89-94. https://doi.org/10.1515/mper-2016-0010

[15] R. Ulewicz et al. Implementation of logic flow in planning and production control, Manag. Prod. Eng. Rev. 7 (2016) 89-94. https://doi.org/10.1515/mper-2016-0010

[16] N. Baryshnikova et al. Management approach on food export expansion in the conditions of limited internal demand, Pol. J. Manag. Stud. 21 (2020) 101-114. https://doi.org/10.17512/pjms.2020.21.2.08

[17] N. Radek et al. The influence of plasma cutting parameters on the geometric structure of cut surfaces, Mater. Res. Proc. 17 (2020) 132-137. https://doi.org/10.21741/9781644901038-20

[18] N. Radek et al. Technology and application of anti-graffiti coating systems for rolling stock, METAL 2019 – 28th Int. Conf. Metall. Mater. (2019) 1127-1132. ISBN 978-8087294925

[19] N. Radek et al. Formation of coatings with technologies using concentrated energy stream, Prod. Eng. Arch. 28 (2022) 117-122. https://doi.org/10.30657/pea.2022.28.13

[20] N. Radek et al. Microstructure and tribological properties of DLC coatings, Mater. Res. Proc. 17 (2020) 171-176. https://doi.org/10.21741/9781644901038-26

[21] N. Radek et al. Influence of laser texturing on tribological properties of DLC coatings, Prod. Eng. Arch. 27 (2021) 119-123. https://doi.org/10.30657/pea.2021.27.15

[22] N. Radek et al. The impact of laser welding parameters on the mechanical properties of the weld, AIP Conf. Proc. 2017 (2018) art.20025. https://doi.org/10.1063/1.5056288

[23] J.M. Djoković et al. Selection of the Optimal Window Type and Orientation for the Two Cities in Serbia and One in Slovakia, Energies 15 (2022) art.323. https://doi.org/10.3390/en15010323

[24] A. Bąkowski et al. Frequency analysis of urban traffic noise, ICCC 2019 20th Int. Carpathian Contr. Conf. (2019) 1660-1670. https://doi.org/10.1109/CarpathianCC.2019.8766012

[25] B. Jasiewicz et al. Inter-observer and intra-observer reliability in the radiographic measurements of paediatric forefoot alignment, Foot Ankle Surg. 27 (2021) 371-376. https://doi.org/10.1016/j.fas.2020.04.015

[26] L. Cedro. Model parameter on-line identification with nonlinear parametrization – manipulator model, Technical Transactions 119 (2022) art. e2022007. https://doi.org/10.37705/TechTrans/e2022007

[27] J. Pietraszek, E. Skrzypczak-Pietraszek. The uncertainty and robustness of the principal component analysis as a tool for the dimensionality reduction. Solid State Phenom. 235 (2015) 1-8. https://doi.org/10.4028/www.scientific.net/SSP.235.1

[28] J. Pietraszek et al. The parametric RSM model with higher order terms for the meat tumbler machine process, Solid State Phenom. 235 (2015) 37-44. https://doi.org/10.4028/www.scientific.net/SSP.235.37

[29] J. Pietraszek, A. Szczotok, N. Radek. The fixed-effects analysis of the relation between SDAS and carbides for the airfoil blade traces. Archives of Metallurgy and Materials 62 (2017) 235-239. https://doi.org/10.1515/amm-2017-0035

[30] R. Dwornicka, J. Pietraszek. The outline of the expert system for the design of experiment, Prod. Eng. Arch. 20 (2018) 43-48. https://doi.org/10.30657/pea.2018.20.09

[31] J. Pietraszek. Fuzzy regression compared to classical experimental design in the case of flywheel assembly, LNAI 7267 LNAI (2012) 310-317. https://doi.org/10.1007/978-3-642-29347-4_36

[32] J. Pietraszek. The modified sequential-binary approach for fuzzy operations on correlated assessments, LNAI 7894 (2013) 353-364. https://doi.org/10.1007/978-3-642-38658-9_32

[33] J. Pietraszek et al. The fuzzy approach to assessment of ANOVA results, LNAI 9875 (2016) 260-268. https://doi.org/10.1007/978-3-319-45243-2_24

[34] A. Gądek-Moszczak et al. The bootstrap approach to the comparison of two methods applied to the evaluation of the growth index in the analysis of the digital X-ray image of a bone regenerate, Studies in Computational Intelligence 572 (2015) 127-136. https://doi.org/10.1007/978-3-319-10774-5_12

[35] J. Pietraszek, L. Wojnar. The bootstrap approach to the statistical significance of parameters in RSM model, ECCOMAS Congress 2016 Proc. 7th Europ. Cong. Comput. Methods in Appl. Sci. Eng. 1 (2016) 2003-2009. https://doi.org/10.7712/100016.1937.9138

[36] J. Pietraszek et al. Challenges for the DOE methodology related to the introduction of Industry 4.0. Prod. Eng. Arch. 26 (2020) 190-194. https://doi.org/10.30657/pea.2020.26.33

Quality Production Improvement and System Safety: QPI 16 - CZOTO 10   Materials Research Forum LLC
Materials Research Proceedings 34 (2023) 298-304   https://doi.org/10.21741/9781644902691-35

Key Projects in the Area of Occupational Health and Safety Management – Case of Polish Company

STASIAK-BETLEJEWSKA Renata[1, a],

[1]Department of Production Engineering and Safety, Faculty of Management, Czestochowa University of Technology, Al. Armii Krajowej 19b, 42-218 Częstochowa, Poland

[a] renata.stasiak-betlejewska@pcz.pl

Keywords: Occupational Health and Safety, Company, Management, Project

Abstract. Project management in the area of occupational health and safety management is related to implementing conditions that affect the safety of industrial practices. The goal of the paper is the review of the systematic projects in the area of the occupational health and safety management system in the chosen Polish large company that developed and adopted practices enforcing work safety system.

Introduction

The Fourth Industrial Revolution requires the health and safety of workers to be adapted to new economic and corporate environment. This process of strategic and organizational adaptation requires research, development and innovation (RDI) in occupational health and safety as a key tool to anticipate undesirable events and facilitate preventive decision-making in advance (Rubio-Romero, 2004). As a result, research into new occupational health and safety solutions become a necessity that is closely linked to companies' competitiveness [1, 2].

Occupational health and safety are generally defined as the science of the anticipation, recognition, evaluation and control of hazards arising in or from the workplace that could impair the health and well-being of workers, taking into account the possible impact on the surrounding communities and the general environment. This domain is necessarily vast, encompassing a large number of disciplines and numerous workplace and environmental hazards [3-5].

The literature on industrial safety focuses upon numerous themes related to the activities of different players such as the errors made at operational level, communication between teams, the issues of management and of leadership and the impact of strategic decisions and organizational dimensions [6]. Some studies show the difficulties and the limitations encountered by industrial safety specialists when they seek to establish some counter power in relation to the orientations or decisions that can jeopardize safety [7-9]. Some research confirms, that industry uses rigorous project management, modern and safe facilities and robust rules of occupational health and safety but accidents continue to cause human and social problems [10-12].

According to Recommended Practices for Safety & Health Programs in Construction elaborated by Occupational Safety and Health Administration establishing a safety and health program at the job site is one of the most effective ways of protecting company most valuable asset: workers. Losing workers to injury or illness, even for a short time, can cause significant disruption and cost to company as well as the workers and their families. It can also damage workplace morale, productivity, turnover, and reputation. Safety and health programs foster a proactive approach to "finding and fixing" job site hazards before they can cause injury or illness. Rather than reacting to an incident, management and workers collaborate to identify and solve issues before they occur. This collaboration builds trust, enhances communication, and often leads to other business improvements [13].

EU-OSHA's foresight studies and review projects aim to anticipate and prioritize risks to help develop OSH practices and policies in areas such as digitalization and green jobs, and stress and psychosocial risks. EU-OSHA also provides easy-to-use resources to help put prevention in the workplace in practice, as well as a range of guidelines being developed to keep workers safe during the pandemic, both on the front lines and adapting to teleworking. The Agency's involvement in the development of the Carcinogens Action Plan and its Healthy Workplaces campaigns demonstrate the Agency's commitment to promoting a culture of prevention across Europe and beyond, which is at the heart of EU OSH policy [14].

The paper presents review of the key projects in the area of occupational and safety management in chosen large Polish company that let also indicate solutions to improve the health and safety conditions on the chosen work placement in companies' group from SME sector.

Characteristic of the Analyzed Company

The analyzed company is an integrated multi-energy concern operating in Central Europe and Canada. It provides energy and fuel to over 100 million Europeans, and our advanced products are available in nearly 100 countries on 6 continents. ORLEN Group companies operate:

- production in the energy segment (generation of electricity and heat), and in the refining and petrochemical segments, including crude oil processing and production of refinery, petrochemical and chemical products and semi-finished products;
- commercial: distribution and sale of electricity and heat, electricity trading, wholesale and retail sale of fuels, petrochemicals, chemicals and other products;
- services: storage of crude oil and fuels, transport, maintenance and repair, laboratory, security, design, administration, insurance and financial services;
- related to exploration, appraisal and production of hydrocarbons.

In accordance with the Strategy of the ORLEN Group adopted in November 2020, the key goal of the Company is to be a regional leader in the field of energy transformation, to build new energy capacities from renewable sources and to implement the decarbonization process, while maintaining operational efficiency and financial strength in traditional business segments.

The ORLEN2030 strategy is based on renewable energy and modern petrochemistry, which pave the way for the transformation of the ORLEN Group until 2030. The development of new areas of activity takes into account technological changes, environmental and consumer trends. The implementation of the strategy will lead to further diversification of the Company's revenue sources, in line with the long-term goal of achieving emission neutrality in 2050.

Ensuring safe working conditions for the employees of the ORLEN Group, employees of external companies performing work on its behalf, as well as the safety of operational and production processes is treated as a priority. Within the scope of this objective, multi-directional activities are carried out in the field of continuous improvement of the work safety culture and maintenance and development of the existing safety standards. In accordance with the assumptions, the implementation of strategic goals was focused, among others, on:

- continued implementation of unified safety standards in the ORLEN Group as part of the Safety Plus+ project;
- improving the contractor management system and achieving the involvement of contractors at the level of identifying with the vision of co-creating a safe workplace;
- initiating, designing and implementing projects and solutions in the aspect of counteracting the COVID-19 pandemic;
- implementation of various activities to improve the awareness of safe work performance and to create proactive attitudes among employees and contractors.

Key Projects in the Area of the Occupational Health and Safety

In the year 2021, *"Directions for the development of the personal and process safety area of the ORLEN Group"* were developed for the years 2022-2026. One of the important undertakings will be to support the processes of taking over capital control over companies and their groups in the area of occupational safety. As part of the implementation of these activities, a coherent work safety management system will be created within the ORLEN Group, mainly in terms of incorporated companies. A common occupational safety policy of the Concern will be defined and uniform safety standards will be implemented in the ORLEN Group. One of the main objectives will be to maintain the values of key KPIs in the area of occupational safety TRR1) ≤ 1.70 and T1PSER2) ≤ 0.3 while implementing the processes of taking capital control over companies / groups of companies.

The area of occupational safety of the ORLEN Group is regulated by national regulations applicable to a given area, European Union regulations, harmonized national standards and rules resulting from good practices identified in the fuel and energy industry.

In the area of personal, process and fire safety as well as the safety of operations related to dangerous goods, uniform Safety Standards, Technical Standards and Guidelines are issued for the ORLEN Group companies. They contain good practices identified in the Group's companies, as well as standards applied in the considered areas in the leading companies in the industry. In 2021, the ORLEN Group completed the process of implementing uniform safety standards under the Safety Plus+ project, which includes the implementation of 15 safety standards constituting the highest functioning safety standards identified in the fuel and energy industry. In 2021, technical standards were developed and issued for the LOGISTYKA+ and DETAL+ projects. The aim of the projects is to standardize and increase the level of work safety in the area of logistics and petrol stations in the ORLEN Group. The technical standards were developed in cooperation with representatives of the occupational safety areas of the companies covered by the projects and constitute a set of guidelines based on the best practices applied in the ORLEN Group.

The vast majority of ORLEN Group companies operate an Occupational Health and Safety Management System in accordance with PN-N-18001 / OHSAS 18001 (until March 11, 2021) / ISO 45001 - the scope of certification covers their individual activities, for which only in a single case an exemption for processes. The scope in question applies to all employees of organizations in which this system operates. The Process Safety Management System based on the US OSHA 1910.119 standard has been implemented in the production companies of the ORLEN Group. The system allows for the effective implementation of activities in the area of operational excellence by improving the technical safety of process, storage and auxiliary installations, and thus preventing the occurrence of any undesirable events that may potentially affect the safety of employees and/or production processes. The Process Safety Management System (SZBP) as an element of the general management and organization system of PKN ORLEN was introduced in order to ensure the highest safety standards in workplaces and to meet the national requirements of Art. 252 of the Environmental Protection Law of April 27, 2001 (consolidated text, Journal of Laws of 2021, item 1973, as amended) in the field of systemic process safety management and further increasing the effectiveness and efficiency of actions to prevent major industrial accidents.

The concept of the safety area management in the ORLEN Group, is based on building and developing a uniform safety standard within the ORLEN Group, while increasing operational excellence and preventive excellence in personal and process safety. The strategy includes the following strategic areas: Management and leadership, Personal security, Process security. In the following years, these assumptions will be continued as part of the implementation of the "Development directions for the area of personal and process safety of the ORLEN Group 2022-2026".

The Comprehensive Prevention System is the basic element of the Occupational Health and Safety Management System, which consists of internal organizational acts in the field of occupational health and safety, fire and chemical safety, radiological protection, technical safety, process safety and safety of operations related to dangerous goods.

The PKN ORLEN Safety Units are a set of basic rules that must be followed along with all applicable regulations and standards, and the Safety Units for PKN ORLEN's own petrol stations, which are a set of basic rules to be applied at PKN ORLEN's own service stations.

The Safety Standards, Technical Standards and Guidelines of the ORLEN Group are unified requirements in the field of occupational safety, process safety and fire safety, safety of operations related to dangerous goods, containing good practices identified in the ORLEN Group companies, as well as standards applied in the considered areas in the leading concerns in industry.

In the ORLEN Group, legal regulations regarding occupational safety have been identified, taking into account the aspect of the company's operations. The implementation of these requirements has been ensured and ongoing supervision of legislative changes in the area of occupational safety is carried out. In 2021, an important element of the process was ensuring compliance in the implementation of preventive measures regarding the COVID-19 pandemic in the aspect of changing legal requirements, ongoing supervision of the changes in question and responding to them.

More than 2,300 on-the-job training courses in the field of operations with dangerous goods have been conducted. The correctness of the preparation of transport documentation was verified, the drivers' licenses and the compliance with the regulations of vehicles that were dispatched to transport dangerous goods were checked. The correct marking of the vehicles, the marking of the packages, the completeness and correctness of the equipment that should be in the vehicle were checked.

Projects in the Area of Safety Standards and Technical Standards
The ORLEN Group conducts various activities to continuously improve the area of occupational safety management. These include, among others: setting and implementing strategic goals at the level of both the ORLEN Group and individually for individual companies, monitoring KPIs in the area of occupational safety of the companies and the ORLEN Group, developing plans to improve health and safety conditions, performing annual health and safety analyses, implementing to improve the work safety culture of employees and contractors (e.g. "Safe Contractor", "ANWIL Safe Program", "Report a Threat", Behavior-Based Safety Programs, "Occupational Safety and Health Protection Days in the ORLEN Group", etc.), the functioning of control and audit systems, both internal and external, an extensive training system for employees and contractors, incentive systems for employees and contractors (e.g. "Health and Safety Incentives Programme", "Security Eagles" competitions, "Good Personal and Process Safety Practices" ORLEN Group", "Millions of man-hours without an accident at work of an employee and contractor in the ORLEN Group"), implementation of post-inspection assessments, preventive and corrective actions identified in the area of occupational safety, implementation of good practices identified in the fuel and energy industry.

In the scope of improving the safety management system, consultations with external advisors and cooperation with universities are carried out. In 2021, the following tests were carried out in selected companies: consultations on the Occupational Health and Safety Management System in accordance with the ISO 45001:2018 standard, while PKN ORLEN continued the implementation of the strategy of cooperation with universities. Objectives and tasks in the area of occupational safety are implemented in the ORLEN Group companies, taking into account their own OHS Service, with the support of the resources of ORLEN Eko Sp. z o. o. or with a small share of outsourcing of health and safety services.

The best practices are constantly identified, both through the exchange of experience between the ORLEN Group companies and the use of the experience of global oil and gas companies. Among the implemented projects, we can mention the LOTO System and the Employee Support System.

Quality Production Improvement and System Safety: QPI 16 - CZOTO 10   Materials Research Forum LLC
Materials Research Proceedings 34 (2023) 298-304   https://doi.org/10.21741/9781644902691-35

The Employee Support System is a behavioral program aimed at strengthening safe behaviors' and eliminating risky behaviors. The implementation of this project is one of the multi-directional activities aimed at improving the safety culture of the ORLEN Group. In 2021, development workshops were organized for Advisors of the Employee Support System. The aim of the workshops was to improve the skills of effective training, sharing knowledge and experience with PKN ORLEN employees and contractors, using various methods of knowledge transfer. One of the most important elements of the Employee Support System are conversations between SWP Advisors and employees of PKN ORLEN and external companies. Therefore, SWP advisers should communicate effectively so that the message reflects their intentions as faithfully as possible and employees understand this message correctly.

The LOTO system is a planned action that includes disconnecting the power supply from industrial equipment and machines for the duration of maintenance or repair work on them. The purpose of the LOTO system is to eliminate accidental and uncontrolled switching on of machines or the release of dangerous energy during operation, investment and repair works as well as service and maintenance works, so as to prevent accidents and events resulting from unexpected start-up or re-supply of energy to machines, devices or installations.

Cooperation with universities is one of the elements of implementing OHS standards.

In 2021, the 5th edition of lectures on the subject "Technical Safety" was conducted in the field of full-time and part-time studies in the field of Chemical Technology. It was a continuation of the project implemented in 2017-2020 by the Occupational Health and Safety Office, the PKN ORLEN Technology Office and the Warsaw University of Technology, Branch in Płock. The initiative concerned a series of expert lectures, conducted in accordance with the idea adopted by PKN ORLEN to educate the future engineering staff of the company in areas relevant to the oil processing and energy industry and as part of the adopted strategy of cooperation with universities. In addition, cooperation with the Lodz University of Technology (through the Education Centre in Płock) was continued in the field of post-graduate studies on Industrial Process Safety.

The ORLEN Group conducts a competition entitled "Good practices of personal and process safety, in which the companies of the Concern take part". The aim of the competition is to spread good practices in security that promote a pragmatic and modern approach to security, including all technical, quality, efficiency, communication, organizational and infrastructural solutions. The implementation of the competition supports the process of improving and streamlining the elements of work safety management and increasing the motivation of employees. Participation in it gives the opportunity to exchange knowledge and experience in the field of shaping the safety culture between the ORLEN Group companies. In addition, the competition "Millions of man-hours without an accident at work of an employee and contractor in the ORLEN Group" is being carried out, also intended for the companies of the Concern. In this case, the aim of the competition is to promote accident-free work.

The ORLEN Group companies carry out ongoing identification of threats, risk assessment and analysis of accident events. The process in question is carried out in the scope of developing and updating the Occupational Risk Assessment for individual work positions of employees, performing work environment tests, identifying hazards in work instructions and monitoring working conditions. An important aspect of this process is the implementation of the Report a Work Safety Threat program in the companies. The process of hazard identification, risk assessment and analysis of accident events is carried out individually by the companies and at the level of the ORLEN Group. In this case, it is carried out using the following tools: Safety Alerts, Lesson Learned, analysis of accidents at work using the "event tree" method, Safety Meeting. Accidents and emergencies occurring in the ORLEN Group are subject to analysis and assessment of the possibility of their recurrence.

An element that is constantly improved and necessary for further development on the scale of the ORLEN Group are interdisciplinary meetings of representatives of the areas of personal, process and fire safety, the Technology Office and other areas of the organization, as part of the so-called safety meeting.

Summary

In order to constantly supervise the state of process safety of the company's facilities in accordance with legal requirements, emphasis was placed not only on adapting activities to standards and requirements, but also on hazard analyses as part of investment projects and periodic reviews and updates. Knowledge of the level of risk of a potential adverse event is an important element when making decisions regarding the effectiveness of security measures used to prevent failures. Training in the field of the process safety management system was conducted for newly recruited employees as part of trainings conducted at the Training Centre and for employees of fuel terminals. The conducted training confirmed that knowledge sharing is one of the practical organizational activities aimed at minimizing the risk of emergencies.

References

[1] R. Romero, J. Carlos. Métodos de evaluación de riesgos laborales, Díaz de Santos, Madrid (2004). ISBN 978-8479781354

[2] J.A. Torrecilla-García et al. Assessment of research, development and innovation in occupational health and safety in Spain, Saf. Sci. 141 (2021) art.105321. https://doi.org/10.1016/j.ssci.2021.105321

[3] B.O. Alli. Fundamental Principles of Occupational Health and Safety, 2nd. Edition. Geneva, International Labour Office, 2008. ISBN 978-9221204541

[4] OSH management system: A tool for continual improvement. Geneva, International Labour Organization, 2011. ISBN 978-9221247401

[5] O. Costa et al. Occupational health and safety in construction projects: a case study on chemical industry sector, Int. J. Occup. Environ. Health 5 (2021) 14-21. https://doi.org/10.24840/2184-0954_005.002_0002

[6] F. Guennoc et al. The activities of occupational health and safety specialists in a high-risk industry, Saf. Sci. 112 (2019) 71-80. https://doi.org/10.1016/j.ssci.2018.10.004

[7] A. Hopkins. Failure to Learn: The BP Texas City Refinery Disaster, CCH Australia Limited, 2008. ISBN 978-1921322440

[8] D. Vaughan. The Challenger Launch Decision: Risky Technology, Culture and Deviance at NASA, University of Chicago Press, Chicago, 1996. ISBN 978-0226851761

[9] J.C. Le Coze. Accident in a French dynamite factory: An example of an organisational investigation, Saf. Sci. 48 (2010) 80-90. https://doi.org/10.1016/j.ssci.2009.06.007

[10] A.A. Shikdar, N.M. Sawaqed. Worker productivity and occupational health and safety issues in selected industries, Comput. Ind. Eng. 45 (2003) 563-572. https://doi.org/10.1016/S0360-8352(03)00074-3

[11] J.J. Smallwood. The Influence of engineering designers on health and safety during construction, J. S. Afr. Inst. Civ. Eng. 46 (2004) 2-8.

[12] A. Badri et al. Occupational health and safety risks: Towards the integration into project management, Saf. Sci. 50 (2012) 190-198. https://doi.org/10.1016/j.ssci.2011.08.008

[13] Occupational Safety and Health Administration. Recommended Practices for Safety & Health Programs in Construction, (2016). [online] [Viewed: 31-01-2023] Available from : https://www.osha.gov/sites/default/files/publications/OSHA3885.pdf

[14] G. Dharmayanti et al. Improving Occupational Health and Safety (OHS) implementation in construction project in Bali. MATEC Web of Conf. 276 (2022) art.02022. https://doi.org/10.1051/matecconf/201927602022

Quality Production Improvement and System Safety: QPI 16 - CZOTO 10   Materials Research Forum LLC
Materials Research Proceedings 34 (2023) 305-314   https://doi.org/10.21741/9781644902691-36

Public Safety and Information Obligations for Upper-Tier Establishments

Monika OSYRA *

Czestochowa University of Technology, Poland

monika.osyra@pcz.pl

Keywords: Upper-Tier Establishment, Seveso Directive, Public Information

Abstract. The operation of the upper-tier establishments, which are prone to a major accident hazard, poses a real threat to the public in terms of health and potential environmental pollution. Therefore, the norms of EU directives seek to introduce as much transparency as possible to the operation of upper-tier establishments. The purpose of this article is to give an overview of the rules concerning the issue of information provision by the establishment to both the public and the relevant state and EU authorities. The following part of the article will provide a brief analysis of how the posting of public information on the websites of the upper-tier establishments looks like in practice.

Introduction

The issue of establishments posing a risk of a major industrial accident due to the fact that they possess hazardous substances is an important part of each country's social safety regulations. The idea of unified EU regulations was born after the 1976 industrial disaster in Seveso, Italy. The release of large quantities of dioxin into the air at the time caused an environmental disaster in the area adjacent to the site. In particular, animals suffered as a result of this major industrial accident, dying *en masse* after coming into contact with the poisonous substance [1,2]. The measures taken by the European The operation of the upper-tier establishments, which are prone to a major accident hazard, poses a real threat to the public in terms of health and potential environmental pollution. Therefore, the norms of EU directives seek to introduce as much transparency as possible to the operation of upper-tier establishments. The purpose of this article is to give an overview of the rules concerning the issue of information provision by the establishment to both the public and the relevant state and EU authorities. The following part of the article will provide a brief analysis of how the posting of public information on the websites of the upper-tier establishments looks like in practice. Economic Community culminated in the enactment of the so-called Seveso I Directive in 1982 [3]. Unfortunately, subsequent tragic industrial disasters that occurred in the following years, such as Bhopal 1984, Chernobyl 1986 and Schweizerhalle 1986, meant that EU regulations needed to be clarified and amended. The result was the Seveso II Directive of 1996 [4], and the one currently in force is the Seveso III Directive of 2012, which regulates in detail the activities of upper-tier establishments. The Directive takes the subject of providing information to the public very seriously. The Seveso III Directive [5] implements the provisions of Council Decision of 17 February 2005 on the conclusion of the Convention on access to information, public participation in decision-making and access to justice in environmental matters [6]. It results in introducing obligations for the operator of a high-risk facility to communicate information to the public.

One of the goals of the Seveso III Directive is to improve the level and quality of information that is made available to the public.

Information obligations for upper-tier establishment

In order to ensure safety, upper-tier establishments are required to provide information on both issues related to the production process, the possession of hazardous substances and the establishment's fulfillment of its organizational responsibilities [7].

Public information relating to the operation of the upper-tier establishments can be divided into three types:

- preventive information regarding actions to prevent the occurrence of a major accident - known as "pre-pre" information,
- information concerning behavior in the event of a major accident - called "pre-post" information,
- information provided after the occurrence of the accident - called "post-post" information [8].

Information on the upper-tier establishment is directed to three audiences:

- the public,
- the relevant authorities (in Poland these will be: The State Fire Service and The Environmental Protection Inspectorate),
- the relevant institutions of the European Union.

Both types of information, the one directly for the public and the one for authorities and institutions, have an impact on public safety.

Information directed to the public

Information that goes directly to the public is regulated in Article 19 and Annex V of the Seveso III Directive. The changes in the Seveso Directive show how the importance of informing the public about the operation of upper-tier establishments has increased over the years. Table 1 shows how the scope of the information obligation has changed over the last 30 years.

Table 1. *Changes in the scope of the information obligation between 1982 and 2012 in the Seveso Directives (source: own study based on Seveso Directives)*

Year	The legal act	Meaning of the provision
1982	Council Directive 82/501/EEC of 24 June 1982 on the major-accident hazards of certain industrial activities – Seveso I	Art. 8(1) "Member states shall ensure that persons liable to be affected by a major accident originating in a notified industrial activity within the meaning of article 5 are informed in an appropriate manner of the safety measures and of the correct behavior to adopt in the event of an accident".
1988	Council Directive 88/610/EEC of 24 November 1988 amending Directive 82/501/EEC on the major-accident hazards of certain industrial activities [9]	The 1988 amendment to the Seveso I Directive clarified what should be included in public information. A list of information elements is included in Annex VII to the directive - "items of information to be communicated to the public in application of article 8 (1). The information should include the following: – the name and address of the enterprise, – who is obliged to provide further information, – confirmation that the establishment is subject to the provisions of the Directive with the notification submitted to the competent authority, – an explanation in simple words of the activities of the establishment,

Year	The legal act	Meaning of the provision
		– the common names of the hazardous substances and preparations on the site with an indication of their main characteristics, – general information on major accident hazards and their potential consequences, – information on how to warn and inform the population in the event of a major accident, – Information on actions and behavior that the population should take in case of an accident, – Confirmation that the plant is obliged to cooperate with emergency services to cope with accidents and minimize their effects, – information on the off-site emergency plan, – information on where the further relevant information can be obtained.
1996	Council Directive 96/82/EC of 9 December 1996 on the control of major-accident hazards involving dangerous substances – Seveso II	In the Seveso II Directive, the issues related to informing the interested public were included in Article 13. It added an obligation to verify the information provided to the public at least once every 3 years. The list of elements that public information should contain is included in Annex V of the directive. The only difference from the previous version was the change of name from the off-site emergency plan to the external emergency plan.
2012	Directive 2012/18/Eu Of The European Parliament And Of The Council of 4 July 2012 on the control of major-accident hazards involving dangerous substances, amending and subsequently repealing Council Directive 96/82/EC – Seveso III	In the Seveso III Directive, issues of public information are included in Article 14 and its elements in Annex V. According to Article 14, information prepared under Annex V should be available electronically at all times. In addition, the safety report and the inventory of hazardous substances should be available upon request, subject to information covered by company confidentiality. The list of information contained in Annex V is divided into two parts, the first of which applies to all establishments covered by the provisions of the Seveso III Directive, while the second refers to the upper-tier establishments. Information common to all establishments should include: – the data of the operator of the establishment or the name of the establishment, – confirmation that the establishment is subject to the directive, – an explanation in simple terms of the plant's activities, – providing the names of hazardous substances present on the site, – general information on how the population should behave in the event of a major accident, – information about the date of the inspection that will be carried out by the relevant authority, and where additional information can be obtained.

Year	The legal act	Meaning of the provision
		Upper-tier establishments additionally inform about: − the general nature of major accidents and their potential danger to people and the environment, along with scenarios for the course of the accident, − confirmation of cooperation with emergency services to minimize the effects of the accident, − the relevant information from the external emergency plan − the possibility of transboundary effects of major accidents if the plant is located near the border with another country.

The first Seveso Directive stated that an establishment that may pose a risk of a major accident should inform those exposed about safety measures and how to proceed in the event of an accident. This provision provided very general guidelines in this regard. The breakthrough came in 1988, when the amendment resulted in an annex to the directive that included items of information. The Seveso II Directive clarified certain terms and indicated the obligation to update the data contained in the information. On the other hand, the Seveso III Directive introduced the division of establishments into the lower-tier and upper-tier establishments and assigned corresponding information obligations to them.

In addition, the Seveso III Directive regulates issues related to: providing information on documentation stored by the upper-tier establishment, access to the safety report, external emergency plan or inventory of hazardous substances [10].

Information for the Authorities

Another duty for the upper-tier establishment is to provide relevant information to the services and the authorities related to the site where the establishment operates. The operator of the upper-tier establishment is obliged to:

1. report the fact of meeting the criterion of an upper-tier establishment to the relevant authority of the State Fire Service. This information should include the following data:
   − information on the operator of the establishment with the address of residence or registered office,
   − the address of the establishment and its website
   − legal title data of the establishment,
   − information about the nature of the establishment's activities and the type of installation and safety systems on site,
   − the nature, quantity and the characteristics of the dangerous substances, including the dangerous substance stored, taken into consideration for the classification of the establishment as an upper-tier establishment,
   − a description of the terrain in the immediate vicinity of the establishment (of particular relevance if there are factors which could increase the risk of an industrial accident).
2. provide information on major accidents, including:
   − produce immediate notification of the fact of a major accident to the competent authority of the State Fire Service and the provincial inspector for environmental protection,
   − inform the above-mentioned authorities without undue delay, and updating it as the situation changes:
   − produce information on the circumstances of the industrial accident and the hazardous substances involved,

- publish data making it possible to assess the human, material and environmental consequences of the accident,
- produce information on all actions taken to limit the effects of the industrial accident and prevent its recurrence.

The competent authorities of the State Fire Service are obliged to publish all data obtained from the operator of the establishment, in the form of submitted information and documentation, on the website of the Public Information Bulletin. The list of this information is contained in Article 267 of the Environmental Protection Act and these include:

- Information on approved, accepted or positively assessed documents such as safety report, external emergency plans or major accident prevention policy,
- information on notifications of the upper-tier establishments,
- information on planned inspections of the upper-tier establishments,
- information on the possibility for the public to participate in the procedure for drawing up the external emergency plan,
- emergency instructions for the public,
- information on the inventory of dangerous substances contained in the upper-tier establishments.

The inclusion of this information is an additional source of information for the public, concerning the operation of upper-tier establishments.

Information for the European Commission
One of the aims of the Seveso III Directive is to provide the public with the best possible information on the risks to life and limb and to the environment arising from the operation of the upper-tier establishments. To ensure this and to be able to control the correct implementation of the Directive, the European Commission, collects information from the Member States. The information is provided by the Minister for Climate Affairs, who obtains it from the operators of the plant through the authorities of the State Fire Service and the Environmental Inspectorate. The European Commission is informed of:

- industrial accidents, with a description and analysis of their consequences;
- national experiences with accident prevention measures;
- lower-tier and upper-tier establishments;
- the recognition of hazardous substances as not posing an industrial accident hazard.

The Major Accident Hazards Bureau of the Joint Research Centre (JRC) of the European Commission has created the Minerva portal, which contains a collection of technical information and tools to support policy on major accident hazards. This portal also contains two databases:

- eMARS - the Major Accident Reporting System, established by the Seveso I Directive, (currently operating electronically). Its purpose is to facilitate the exchange of lessons learned from accidents and incidents involving dangerous substances. Entry into this database is mandatory for the upper-tier establishments operating in EU Member States [11].
- eSPIRS - is an electronic version of the SPIRS (Seveso Plants Information Retrieval System); the database was created in 2001 by the JRC at the request of the European Commission's Directorate-General for the Environment (DG ENV). The Seveso III Directive required EU Member States to register the upper-tier and lower-tier establishments in this database[12].

Information obligation - Research results
According to Article 261a of the Environmental Law [13], the information made public should include 7 points:
1) the designation of the operator of the upper-tier establishment;

2) information about the notification of the establishment to the relevant authorities, the communication of the major accident prevention policy to them and information that establishment is subject to Seveso Directive;
3) a description of the activity of the establishment;
4) a summary of the substances stored, including their names or categories and the hazards they may cause;
5) information on how to warn and instruct the public to behave in the event of an accident, which has been agreed with the competent authorities of the State Fire Service;
6) information on the preparation of the safety report;
7) a list of the main potential industrial accident scenarios with the safety measures that can be taken.

This information should be posted and made permanently available on the websites of the establishments. And they should also be updated in a way that is understandable to the average citizen.

On the basis of those guidelines, 40 websites of establishments with a high risk of a major industrial accident were analyzed, which constitutes 1/5 of all the establishments of that category situated in the territory of Poland.

The websites of all the establishments contained obligatory information. However, a frequent problem was their location in the structure of the website, causing difficulties in finding them. Sometimes it was necessary to use a web browser to find a file with information on the prevention of major accidents. Table 2 shows examples of establishments with the places on the website where the above information was located.

In the absence of legislation standardizing where information on an upper-tier establishment should be located on a website, each establishment approaches this issue differently. The result is that it is difficult for an average citizen to access this information.

Table 2. *The examples of data on the location of the public information on the potential risk posed by the upper-tier establishment (source: own study)*

No.	Name of the establishment	Website tab	File name
1.	3M Wrocław Sp. z o. o. ul. Kowalska 143 51-424 Wrocław	Information on security measures	PDF file - Information on safety measures and how to deal with major accidents
2.	ADAMA Manufacturing Poland S.A., ul. Sienkiewicza 4 56-120 Brzeg Dolny	Quality and Environmental > Safety measures	PDF file - Information on safety measures and how to deal with major accidents.
3.	Guotai - Huarong (Poland) Sp. z o.o. ul. Powstańców Śląskich 2-4, 53-333 Wrocław	About us > Upper-Tier Establishment	PDF file – Information of the Upper-Tier Establishment
4	Anwil S.A., ul. Toruńska 222, 87-805 Włocławek	Responsible business > Environmental protection > Accident protection	PDF file Information on hazards occurring, preventive measures applied and actions to be taken in the event of an accident on the premises of ANWIL S.A.
5	Bałtyk-Gaz Sp. z o.o. , 84-230 Rumia, ulica Sobieskiego 5 Magazyn Gazu w Bydgoszczy	Safety measures	Information on safety measures and how to proceed in the event of a major

	85-461 Bydgoszcz, ul. Ołowiana 41,		industrial accident at the gas storage facility of Bałtyk- Gaz in Bydgoszcz
6	Solino S.A. Inowrocławskie Kopalnie, ul. Świętego Ducha 26a, 88-100 Inowrocław, Województwo kujawsko-pomorskie	About company > Our standards > Process safety	PDF file - Information on safety measures and on how to deal with a major industrial accident at the upper-tier establishment
7	TRANSGAZ S.A. 21- 512 Zalesie Wólka Dobryńska 15	Information on risks	PDF file - Information on safety measures and how to deal with a major industrial accident at the TRANSGAZ S.A. transhipment terminal.
8.	Bialchem Group Sp. z o.o. 15-062 Białystok, ul. Warszawska 39 adres zakładu 21-512 Zalesie Wólka Dobryńska 15	Terminal	PDF file - Information on the existing hazards, the anticipated effects of these hazards, the preventive measures applied and the actions that are being taken in the event of an industrial accident at the Bialchem Group Transshipment Terminal in Wólka Dobryńska.
9.	Grupa Azoty Zakłady Azotowe PUŁAWY S.A., 24-100 Puławy, Al. Tysiąclecia Państwa Polskiego 13	Responsible business > For safety	PDF file Safety in the Azoty Pulawy Group
10.	BATERPOL S.A. ul. Obr. Westerplatte 108, 40-335 Katowice,	About company > Upper-Tier Establishment	Text on website – About The Upper-Tier Establishment

Another problem is the arbitrary description of the information file and its form. For the most part, the information is provided by the establishments in a separate PDF file which can be downloaded from the website, but there are also situations in which the information is merely the content of a web page.

As a rule, the data contained in the separate file are carefully prepared and contain detailed and comprehensive information on the functioning of the establishment and on how to warn the public in the event of a major accident. In most cases, they are written in a way that is understandable to the average citizen. On the other hand, there were rare cases of establishments posting only brief information, aiming only to meet the statutory minimum.

Summary
The aim of the provisions of the Seveso III Directive is to prevent a major industrial accident or to minimize the negative effects, should the accident occur. One instrument for achieving this objective is to provide as much information as possible on the risks associated with the operation of the upper-tier establishment. In order to reach all persons concerned, this information must be provided in an accessible form and be available at all times. And although all analyzed websites of the upper-tier establishments contained the mandatory information, it was often very difficult to reach to access it.

The data presented show that establishments take their obligations seriously; unfortunately, the lack of standardization in terms of posting public information on the website means that it is often not accessible to the public. A solution to this problem would be to create uniform rules for the

Quality Production Improvement and System Safety: QPI 16 - CZOTO 10   Materials Research Forum LLC
Materials Research Proceedings 34 (2023) 305-314   https://doi.org/10.21741/9781644902691-36

publication of information related to the safety of citizens and to implement them, at least in the form of recommendations or good practices for the upper-tier establishments.

In the past, public safety has been repeatedly compromised by both the regular operations of businesses and emergency situations. Nowadays, safety systems are much more sophisticated and reliable; however, there is always a risk of accidents. A particular case is represented by relatively small metal processing establishments [14] that, due to the technologies used, handle toxic substances or generate such substances as waste during their operations. Welding facilities [15], surface treatment operations [16], various types of coatings application [17, 18], including special coatings [19, 20], and window manufacturing [21] can be mentioned in this regard. Given the substantial resources involved in the construction and maintenance of protective systems, a thorough identification of hazards is essential. Comparative analyses [22], complexity reduction methods [23], methods supporting analysis of small sample sizes [24, 25], as well as traditional [26-28] or non-classical [29-31] methods of industrial statistics, come to the aid in this regard.

References

[1] B. Eskenazi et al. The Seveso accident: A look at 40 years of health research and beyond, Environ. Int. 121 (2018) 71-84. https://doi.org/10.1016/j.envint.2018.08.051

[2] C.Nerin et al. Seveso Disaster and the European Seveso Directives, encyclopedia of Toxicology (Third Edition), Academic Press, 2014, 244-247. http://doi.org/10.1016/B978-0-12-386454-3.00461-9

[3] Council Directive 82/501/EEC of 24 June 1982 on the major-accident hazards of certain industrial activities.

[4] Council Directive 96/82/EC of 9 December 1996 on the control of major-accident hazards involving dangerous substances.

[5] Council Directive 2012 Directive 2012/18/EU of the European Parliament and of the Council of 4 July 2012 on the control of major-accident hazards involving dangerous substances.

[6] Council Decision 2005/370/EC on the conclusion, on behalf of the European Community, of the Convention on access to information, public participation in decision-making and access to justice in environmental matters.

[7] B. De Marchi, S. Funtowicz. General Guidelines for. Content of Information to the Public (Directive 82/501/EEC – Annex VII), European Commission, Luxembourg, 1994.

[8] B. Wynne. Implementation of Article 8 of Directive 82/501/EWG: A Study of Public Information. European Commission, 1987, Contract 86-B-6641-11-006-llN.

[9] Council Directive 88/610/EEC of 24 November 1988 amending Directive 82/501/EEC on the major-accident hazards of certain industrial activities.

[10] M. Micińska. Udział społeczeństwa w zapobieganiu poważnym awariom przemysłowym – możliwości prawne. Bezpieczeństwo Pracy: Nauka i Praktyka 11 (2005) 8-10.

[11] The Major Accident Reporting System – eMARS website. [online] Viewed: 31-01-2023. Available from: https://emars.jrc.ec.europa.eu/en/emars/content

[12] The SPIRS (Seveso Plants Information Retrieval System). [online] Viewed: 31-01-2023. Available from: https://espirs.jrc.ec.europa.eu/en/espirs/content

Quality Production Improvement and System Safety: QPI 16 - CZOTO 10   Materials Research Forum LLC
Materials Research Proceedings 34 (2023) 305-314   https://doi.org/10.21741/9781644902691-36

[13] Ustawa z dnia 27 kwietnia 2001 r. Prawo ochrony środowiska Dz.U. z 2022 r. poz. 2556. [online] Viewed: 31-01-2023. Available from: https://isap.sejm.gov.pl/isap.nsf/download.xsp/WDU20220002556/U/D20222556Lj.pdf

[14] P. Jonšta et al. The effect of rare earth metals alloying on the internal quality of industrially produced heavy steel forgings, Materials 14 (2021) art.5160. https://doi.org/10.3390/ma14185160

[15] N. Radek et al. The impact of laser welding parameters on the mechanical properties of the weld, AIP Conf. Proc. 2017 (2018) art.20025. https://doi.org/10.1063/1.5056288

[16] N. Radek et al. The influence of plasma cutting parameters on the geometric structure of cut surfaces, Mater. Res. Proc. 17 (2020) 132-137. https://doi.org/10.21741/9781644901038-20

[17] N. Radek et al. Technology and application of anti-graffiti coating systems for rolling stock, METAL 2019 28th Int. Conf. Metall. Mater. (2019) 1127-1132. ISBN 978-8087294925

[18] N. Radek et al. Formation of coatings with technologies using concentrated energy stream, Prod. Eng. Arch. 28 (2022) 117-122. https://doi.org/10.30657/pea.2022.28.13

[19] N. Radek et al. Microstructure and tribological properties of DLC coatings, Mater. Res. Proc. 17 (2020) 171-176. https://doi.org/10.21741/9781644901038-26

[20] N. Radek et al. Influence of laser texturing on tribological properties of DLC coatings, Prod. Eng. Arch. 27 (2021) 119-123. https://doi.org/10.30657/pea.2021.27.15

[21] J.M. Djoković et al. Selection of the Optimal Window Type and Orientation for the Two Cities in Serbia and One in Slovakia, Energies 15 (2022) art.323. https://doi.org/10.3390/en15010323

[22] B. Jasiewicz et al. Inter-observer and intra-observer reliability in the radiographic measurements of paediatric forefoot alignment, Foot Ankle Surg. 27 (2021) 371-376. https://doi.org/10.1016/j.fas.2020.04.015

[23] J. Pietraszek, E. Skrzypczak-Pietraszek. The uncertainty and robustness of the principal component analysis as a tool for the dimensionality reduction. Solid State Phenom. 235 (2015) 1-8. https://doi.org/10.4028/www.scientific.net/SSP.235.1

[24] A. Gądek-Moszczak et al. The bootstrap approach to the comparison of two methods applied to the evaluation of the growth index in the analysis of the digital X-ray image of a bone regenerate, Studies in Computational Intelligence 572 (2015) 127-136. https://doi.org/10.1007/978-3-319-10774-5_12

[25] J. Pietraszek, L. Wojnar. The bootstrap approach to the statistical significance of parameters in RSM model, ECCOMAS Congress 2016 Proc. 7th Europ. Congr. Comput. Methods in Appl. Sci. Eng. 1 (2016) 2003-2009. https://doi.org/10.7712/100016.1937.9138

[26] J. Pietraszek et al. The parametric RSM model with higher order terms for the meat tumbler machine process, Solid State Phenom. 235 (2015) 37-44. https://doi.org/10.4028/www.scientific.net/SSP.235.37

[27] J. Pietraszek, A. Szczotok, N. Radek. The fixed-effects analysis of the relation between SDAS and carbides for the airfoil blade traces. Arch. Metall. Mater. 62 (2017) 235-239. https://doi.org/10.1515/amm-2017-0035

[28] R. Dwornicka, J. Pietraszek. The outline of the expert system for the design of experiment, Prod. Eng. Arch. 20 (2018) 43-48. https://doi.org/10.30657/pea.2018.20.09

[29] J. Pietraszek. The modified sequential-binary approach for fuzzy operations on correlated assessments, LNAI 7894 (2013) 353-364. https://doi.org/10.1007/978-3-642-38658-9_32

[30] J. Pietraszek et al. Non-parametric assessment of the uncertainty in the analysis of the airfoil blade traces, METAL 2017 – 26th Int. Conf. Metall. Mater. (2017) 1412-1418. ISBN 978-8087294796

[31] J. Pietraszek et al. The non-parametric approach to the quantification of the uncertainty in the design of experiments modelling, UNCECOMP 2017 Proc. 2nd Int. Conf. Uncert. Quant. Comput. Sci. Eng. (2017) 598-604. https://doi.org/10.7712/120217.5395.17225

Quality Production Improvement and System Safety: QPI 16 - CZOTO 10   Materials Research Forum LLC
Materials Research Proceedings 34 (2023) 315-322   https://doi.org/10.21741/9781644902691-37

The Importance of the Diet Catering Sector During the COVID-19 Pandemic and Its Impact on Families Eating Together

ROWIECKA Jacqueline Adriana [1, a *], DZIUBA Szymon [1, b],
CIERNIAK-EMERYCH Anna [1, c] and JAROSSOVÁ Malgorzata A. [2d]

[1] Department of Labour, Capital and Innovation, Wroclaw University of Economics and Business, ul. Komandorska 118/120, 53-345 Wrocław, Poland

[2] Department of Marketing, University of Economics in Bratislava, ul. Dolnozemská cesta 1, 852-35 Bratislava, Slovakia

[a]jacqueline.rowiecka@ue.wroc.pl*, [b]szymon.dziuba@ue.wroc.pl*, [c]anna.cierniak-emerych@ue.wroc.pl, [d]malgorzata.jarossova@euba.sk

Keywords: Pandemic COVID-19, Diet Catering Sector, Families Eating

Abstract. Proper nutrition, physical activity, positive interpersonal relations, sleeping, and resting promote health and well-being. A varied and well-balanced diet is one of the fundamental factors influencing human mental and physical balance. Diet catering services involve providing customers with individualized sets of meals chosen based on their preferences and needs. The purpose of this paper is to analyze the importance of the diet catering sector during the COVID-19 pandemic, whose activities were not limited by government restrictions, and the impact of ordering a healthy boxed diet on eating together of all family members. The empirical study included customers of diet catering enterprises. Individual in-depth interview was used as a quantitative research technique. The analysis of the empirical study proved the extremely different behavior of the respondents, who had completely different views on the basic issues related to the measures that limit social and economic life. They often had no knowledge of the current pandemic and often performed a variety of mutually exclusive actions.

Introduction

Eating habits are one of the key issues of physical and mental health. World Health Organization defines health as "complete, physical, mental and social well-being, and not just the absence of disease or disability" [1]. An individual's dietary choices depend on a variety of factors. By forming the most important environment in a child's life, parents shape children's food preferences, eating behaviors, and reduce caloric intake [2]. Based on the values instilled at home, children learn what foods to eat, how to eat them, and at what times of the day [3]. In addition, the manner of meals offered and the products eaten by family members are important [4]. Determinants of learning proper eating habits include exposure and availability (the presence of fruit and vegetables in a visible place increases the likelihood of eating them more often) [5], trying new tastes, and eating meals together [6]. The last element was included in the empirical field of the present study and the literature considerations at the level of an increasingly popular form of nutrition, i.e. boxed diets offered by external catering firms.

Analyses of international empirical studies indicate [6] that families that eat meals together (≥ 5 meals per week) are less likely to have children's nutrition-related health problems by 25% (compared to children who eat ≤ 1 meal with their family members). Furthermore, children and youth raised in families that share at least 3 family meals per week are less likely to be overweight (12%), consume unhealthy foods (20%), and have eating disorders (35%), and have increased chances of eating healthy foods (24%). A correlation is also indicated between better parental education/higher socioeconomic status, and healthy eating patterns [7,8] Increased frequency of family meals correlates with higher intake of fruits and vegetables, fiber and micronutrients, less fried foods and carbonated beverages, saturated fats, trans fats [9], sweets, and fast foods [8]. Shared family meals consumed during adolescence may have lasting positive effects on diet quality and meal patterns in young adulthood [10].

Lifestyle Changes in Families During the COVID-19 Pandemic
On January 31, 2020, the World Health Organization declared a public health emergency of international concern [11] and on March 11, 2020, a pandemic state. Awareness of the morbidity risk of COVID-19, with the limited medical capacity to treat it, where non-pharmaceutical interventions were the main strategy to contain the pandemic [12], prompted both individuals and national governments to take preventive action and introduce restrictions. In addition to the massive health impacts, the COVID-19 pandemic carries significant economic losses for households, businesses, and governments, with massive social, economic, and livelihood disruptions as a result of closures, lockdowns, supply chain disturbances, and a sharp decline in commercial activity [13]. The economic losses generated by the pandemic caused by the SARS-CoV-2 virus will affect the entire world for a long time while the uncertainty about the further course of the pandemic continues to worry people and economies [14].

A large percentage of the population has been forced to switch from the traditional form of on-site work to remote, home-based work. Polish Economic Institute indicates that in Poland, 27% of employees have the opportunity to work remotely. Instability and uncertainty of the situation generate anxiety, amplified by the dynamic changes as to the prognosis of further progress of the epidemic. Serious mental disorders (caused by stress and anxiety) have emerged in society associated with decreased physical activity and reduced social interaction. People who stay at home for a long time (both those who work online and those who take care of children) are characterized by disturbed emotional reactions, with their intensity individually variable and depending mainly on personality traits, psychiatric and somatic co-morbidities, and the environment in which the individual stays. Changing the circadian cycle often results in overeating, with a particular focus on convenience foods, which are rich in sugar, high fats, and simple carbohydrates [15], which stimulate serotonin production and positively affect mood [16]. Furthermore, less frequent shopping due to government restrictions is indicated, resulting in the consumption of fewer fresh fruits and vegetables relative to highly processed foods [17].

Having to stay at home also changes the priorities and lifestyles of family members, including meal preparation and eating. The involvement of family members and the way meals are prepared can influence the establishment of relationships, learning proper eating patterns, building a sense of security, self-esteem, and identity. Meals offered by diet catering companies requiring only a final heat treatment (if the meal is to be eaten hot) do not provide an opportunity to spend time together while preparing the meals. Very often, these meals are eaten hurriedly and alone.

The Situation of the Catering Market During the COVID-19 Pandemic
The country's socio-economic situation in 2020 was determined by the restrictions implemented to counteract the COVID-19 pandemic, which affected performance in core business areas with varying degrees of intensity. The dynamics in most basic areas of economic activity in 2020 was much weaker than in previous years, which contributed to the first decline in the gross domestic product since the transformation period in Poland. Gross domestic product is estimated to have reduced by 2.8% in real terms in 2020, compared to the growth of 4.5% in 2019. The Polish Economic Institute forecasts that the year 2021 will be irregular: poor first months will be accompanied by a strong rebound in the second half of the year. In 2022, growth rates are expected to be similar to those observed in the years preceding the epidemic [18] (Fig. 1).

The most unfavorable business climate assessment concerned entities from the accommodation and catering section (Table 1), whose activity has been limited since November due to the

Quality Production Improvement and System Safety: QPI 16 - CZOTO 10   Materials Research Forum LLC
Materials Research Proceedings 34 (2023) 315-322   https://doi.org/10.21741/9781644902691-37

introduced restrictions. Of entrepreneurs, 60% declared that their businesses would not survive in such conditions for more than 3 months. These entities reported a reduction in employment in January by around 9% [19]. Food services directly impact the lives of individuals. Food services, including the activities of diet catering businesses, have seen increasing growth rates in recent years, but their situation has changed significantly following the introduction of restrictions.

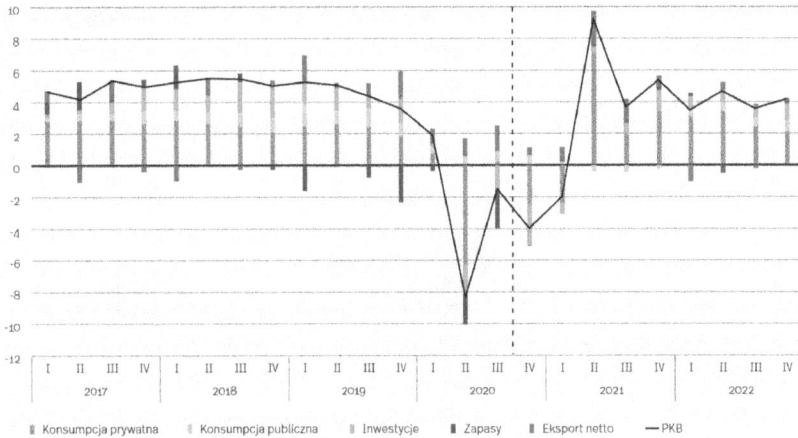

Fig.1. Structure of GDP growth: PIE forecast; gray – private consumption, yellow – public consumption, green – investments, blue – stocks, red – net export, black – GDP [18].

Table 1. Economic trends in January 2021 [19]

General business climate indicator	
industrial processing	-9.9
construction	-18.7
retail commerce	-13.9
transport and storage	-7.8
accommodation and catering	**-52.1**

The catering industry has suffered significantly from the loss of customers and regulatory uncertainty associated with the COVID-19 crisis. Diet catering provides services based on delivering (to the address indicated by the customer) meal kits personalized in terms of quantity, calories, and type of diet. These meals are prepared by a team of cooks based on menus compliant with the requirements of IŻŻ (Institute of Food and Nutrition) and FAO/WHO, prepared by dieticians taking into account the balanced diet. During the pandemic, this form of meals turned out to be one of the most desirable for customers, both because of the awareness of the importance of a varied and well-balanced diet and the possibility to maintain the disinfection regime and social distancing (non-contact delivery). Enterprises commit to: accept cashless payments, the mandatory wearing of gloves and protective masks, delivery of meals in a non-contact manner (so that they are also recommended for people in quarantine and isolation), constant internal checks on workers' body temperature and health status, and disinfection of all facilities, machinery, and equipment.

Methodology of Empirical Research

The study examined 147 people. The respondents were a group of customers of diet catering enterprises. The selection criterion was the form of home-office work duties performed and having children. The research was conducted in the second half of February 2019. The research method was a diagnostic survey and the research tool was a focus group interview. Based on the interviews, the respondents identified factors affecting the demand for services offered by boxed diet enterprises during the COVID-19 pandemic. The study group was selected using a non-probabilistic distribution with a network nature. The aim of the study was to outline the effect of ordering diet catering services on the meals families eat together.

Analysis of Empirical Research

The group of respondents consisted of customers of diet catering enterprises in the south-western part of Poland, with 77% of respondents being female. Furthermore, 73% of the respondents declared a master's degree, 22% – a bachelor's degree, 4% – secondary education, and 1% – primary education. The largest percentage of people were respondents aged 25-35 (38%), 29% were those aged 25-35, 20% – 45-55, 3% – over 50, and 9% – respondents under 25. The majority of respondents, specifically 55%, resided in big cities with a population exceeding 300,000 inhabitants. Additionally, 22% of the respondents lived in cities with a population of up to 300,000 inhabitants, while 14% resided in towns with a population of up to 50,000 inhabitants. Lastly, 9% of the respondents lived in villages.

The situation of enterprises offering diet catering changed after the Polish government introduced restrictions on their operation. The shift from on-site to remote work often led to customers no longer ordering meals. This was influenced by economic conditions and the greater amount of time spent at home associated with the possibility to prepare meals for the entire family. Clients indicated that they needed to allocate the amount they were previously spending on boxed diet meals to other purposes. This situation has continued from the time the restrictions were introduced (March 2020) until May 2021. Only in the period from May to October 2020, there were small fluctuations in order quantities with an upward trend.

A large part of respondents indicated ordering a boxed diet on a regular basis regardless of the ongoing epidemiological situation, due to lack of time to prepare meals. In many cases, remote work involves even more time, because in the vast majority of enterprises (as indicated by clients) there is no rigid time frame allocated to work duties, as was the case when individuals worked on the company's premises, and the employee is paid for work done. These respondents indicated smaller financial resources due to the need to stay at home, and therefore less opportunity for travel and in-store shopping. An additional determinant is the presence of children at home. Those in infancy, preschool, and early school age need their parent's care and attention throughout the day.

The ongoing COVID-19 pandemic has also affected customers' perception of the meals delivered. In this case, customers fall into two main ideological groups. The first group is convinced of the hygienic nature of meal preparation and distribution. These people indicated the advantages of ordering diet catering such as no need for meeting other people during shopping, maintaining social distancing, the health-promoting nature of the diet which increases the human body's immunity, the observance of personal protection rules by the enterprise's employees, including drivers, and several safeguards guaranteed in the catering enterprise.

The second group, who perceived the situation from a completely different perspective, is convinced that the use of diet catering services during the pandemic is not epidemiologically safe enough. These people declare that meals should be isolated for at least 24 hours to eliminate the likelihood of SARS- CoV-2 virus transmission through physical things, which is not feasible with boxed diet delivery. They declare that they are afraid of ordering diet catering meals in order not to expose themselves and their relatives to possible infection with the virus.

At the same time, the respondents indicated that the manner of eating meals has not changed, despite the fact that they perform most of their professional duties at home, and not, as before, on the company's premises. Although respondents worked at home, they did not eat more meals with their families than when they had worked away from home. When asked to indicate the reasons why, despite the change in the style of performing professional duties at home, they eat meals alone, they mostly indicated the need to focus on work and do not treat meals as a break during work (they eat while working at the same time), or, on the contrary, they explain that during meals they want to rest/relax in silence.

Table 2. Opinions declared by the respondents on the reasons for ordering/ceasing to order diet catering services [own study]

Opinions of respondents declared according to groups		
Free time criterion	Lack of time to prepare meals (more time devoted to remote work due to higher demands of employers/ problems with combining remote work with family life)	More free time (remote work, no need to spend time commuting, parental leave taken due to closure of nurseries, kindergartens, and schools)
Economic criterion	Reduced financial resources (job loss, reduced salaries due to restrictions imposed on various sectors of the economy)	Greater financial resources (anti-crisis shields, limited travel, and traditional in-store shopping)
Epidemiological safety criterion	The conviction that epidemiological safety measures are insufficient (lack of meal isolation for 24 hours, uncertainty whether all individual safeguards are met)	The conviction that the sterility of meals delivered is ensured (accepting only non-cash payments, obligation to wear gloves and protective masks, delivery of meals in a non-contact manner, constant internal checks on employees' body temperature and health status, disinfection of all equipment, machines, and devices.
Replacement criterion	No impact of the inability to use restaurant services on the number of orders in diet catering firms (people using restaurant services require, in addition to the food served in an appropriate form, the celebration of the food eaten: social meetings, a way to spend free time, show love, celebrate)	Lack of possibility to use restaurant services led to the increase in orders of diet catering firms (substitution of services of a similar nature of the benefits)

An important criterion is to distinguish between families where diet catering is ordered by only one person, both parents, and the whole family (parents and children). This has a determining effect on the perception of eating meals together (Table 3). Families in which meals are ordered by only one person relatively less frequently consume them together with the family (19%), whereas in the situation when meal kits are provided for all family members, the percentage of shared consumption is the highest (93%). When analyzing the way how diet catering foods are consumed, it is important to consider the reason for ordering such products. One of the most common (occurring primarily among people who order boxed diets as the only family member) is the need or desire to lose weight. In this case, eating alone may have an effect on increasing feelings of control over eating. When meals are eaten together and the other family members are eating meals high in calories and tastier, the individual needs higher self-control, and this situation can arouse negative emotions. When the same meals are eaten together (when the whole family orders and eats the boxed diet meals), there is no need to monitor the quality and quantity of food portions, while adherence to the group with whom the meal is eaten is reinforced. On the other hand, it is worth noting the effect of social facilitation: an increase in the effectiveness of performed tasks caused by the presence of other people.

Another conclusion that can be drawn from the study is that people who order boxed diets as the only ones in their families decide most often to purchase five meals, whereas when there are more orders in the family, the percentage of three- and one-meal kits increases, which may confirm the thesis that the reason for ordering meals for one person is the desire/need to change the diet, whereas for whole families, the reasons include the unwillingness to cook, no need to do shopping or the need to devote time to other duties or hobbies.

In addition, it should be recalled that often customer behavior changes depending on the market situation [20-22]. The COVID-19 pandemic is a good example. It is important to analyze such behavior, learn from it and use it in future development plans [23-24].

Table 3. Number of ordered diet catering kits per family and percentage distribution of meals consumed together [own study]

1 person		2 people (parents)		family (parents and children)	
47 people	32%	72 people	49%	28 people	19%
the number of meals ordered per day:					
5 meals	91%	5 meals	85%	5 meals	32%
3 meals	6%	3 meals	11%	3 meals	57%
1 meal	3%	1 meal	4%	1 meal	11%
percentage of people who eat all meals with their family during the day					
19%		68%		93%	

Conclusion

The analysis of the empirical study proved the extremely different behavior of the respondents, who had completely different views on the basic issues related to the measures that limit social and economic life. They often had no knowledge of the current pandemic and often performed a variety of mutually exclusive actions. There was no correlation between government restrictions of the food service industry and the activities of boxed diet companies. It can be argued that diet catering does not affect the operation of restaurants and other food service operators as they meet different customer needs. Eating together in families depends on the number of boxed diet kits ordered per family. In the case of ordering boxed diets for the whole family, the percentage of eating together is 93%, whereas when ordering by only one member of the family, this percentage is 19%. This is undoubtedly related to ordering boxed diets. Predictors of ordering diet catering meals by individuals are primarily the desire or need to lose weight, while in the case of whole families, the reasons are different (lack of time or willingness to cook).

In terms of future research and limitations, it is important to address certain issues. Specifically, one area that requires further examination is the relationship between individuals who discontinued using dietary catering services during the COVID-19 pandemic and those who consistently ordered meals, with regard to their practice of eating meals together with their families. It is also worth noting that this study has certain limitations in terms of its geographic scope and duration.

Acknowledgement
The paper is the output of the project VEGA no. 1/0398/22 – The current status and perspectives of the development of the market of healthy, environmentally friendly and carbon-neutral products in Slovakia and the European Union.

The authors would like to express their sincere gratitude to Ms. Małgorzata Wecko from Wrocław University of Economics for her invaluable assistance in preparing the bibliometric data.

References

[1] World Health Organization, Basic documents, Forty-eighth edition, 2014.

[2] J.S. Savage et al. Parental influence on eating behavior: Conception to adolescence, J. Law Med. Ethics 35 (2007) 22-34. https://doi.org/10.1111/j.1748-720x.2007.00111.x

[3] E. Adessi et al. Specific social influences on the acceptance of novel food in 2-5 years-old children, Appetite 45 (2005) 264-271. https://doi.org/10.1016/j.appet.2005.07.007

[4] J.O. Fisher et al. Parental influences on young girls'fruit and vegetable, micronutrient, and fat intakes, J. Am. Diet. Assoc. 102 (2002) 58-64. https://doi.org/10.1016/s0002-8223(02)90017-9

[5] T. Baranowski et al. Psychosocial correlates of dietary intake: Advancing dietary intervention, Ann. Rev. Nutr. 19 (1999) 17-40. https://doi.org/10.1146/annurev.nutr.19.1.17

[6] J.A. Hammsons, B.H. Fiese. Is Frequency of Shared Family Meals Related to the Nutritional Health of Children and Adolescents? Pediatrics 127 (2011) e1565–e1574. https://doi.org/10.1542/peds.2010-1440

[7] T.M. Videon, C.K. Manning. Influences on adolescent eating patterns: the importance of family meals, J. Adolesc. Health 32 (2003) 365-73. https://doi.org/10.1016/S1054-139X(02)00711-5

[8] M. Haapalahti et al. Meal patterns and food use in 10-to 11-year-old Finnish children, Public. Health Nutr. 6 (2003) 365-370. https://doi.org/10.1079/PHN2002433

[9] M.W. Gillman et al. Family dinner and diet quality among older children and adolescents, Arch. Fam. Med. 9 (2000) 235-240. https://doi.org/10.1001/archfami.9.3.235

[10] N.I. Larson et al. Family meals during adolescence are associated with higher diet quality and healthful meal patterns during young adulthood, Am. Diet. Assoc. 107 (2007) 1502-1510. https://doi.org/10.1016/j.jada.2007.06.012

[11] J. Wen et al. Effects of misleading media coverage on public health crisis: a case of the 2019 novel coronavirus outbreak in China, Anatolia 31 (2020) 331-336. https://doi.org/10.1080/13032917.2020.1730621

[12] S. Gössling et al. Pandemics, tourism and global change: a rapid assessment of COVID-1, J. Sustain. Tour. 29 (2020) 1-20. https://doi.org/10.1080/09669582.2020.1758708

[13] R.E. Caraka et al. Impact of COVID-19 large scale restriction on environment and economy in Indonesia, Glob. J. Environ. Sci. Manag. 6 (2020) 65-84. https://doi.org/10.22034/GJESM.2019.06.SI.07

[14] O. Gulcin et al. A comparative evaluation between the impact of previous out breaks and COVID-19 on the tourism industry, Int. Hosp. Rev. 36 (2021) 2516-8142. https://doi.org/10.1108/IHR-05-2020-0015

[15] C. Yılmaz, V. Gökmen, Neuroactive compounds in foods: occurrence, mechanism and potential health effects. Food Res. 128 (2020) art.108744. https://doi.org/10.1016/j.foodres.2019.108744

Quality Production Improvement and System Safety: QPI 16 - CZOTO 10   Materials Research Forum LLC
Materials Research Proceedings 34 (2023) 315-322   https://doi.org/10.21741/9781644902691-37

[16] Y. Ma et al. Carbohydrate craving: not everything is sweet. Curr. Opin. Clin. Nutr. Metab. Care 20 (2017) 261-265. https://doi.org/10.1097/MCO.0000000000000374

[17] L. Renzo et al. Eating habits and lifestyle changes during COVID-19 lockdown: an Italian survey, J. Transl. Med. 18 (2020) art.229. https://doi.org/10.1186/s12967-020-02399-5

[18] M. Gniazdowski, J. Rybacki, J. Sawulski. Polski Instytut Ekonomiczny 2020, Przegląd Gospodarczy PIE. 11 (2020) 20-27.

[19] Główny Urząd Statystyczny. Sytuacja społeczno-gospodarcza kraju w 2020, GUS, Warszawa, 2021.

[20] M. Ingaldi. E-service quality assessment according to hierarchical service quality models. Manag. Sys. Prod. Eng. 30 (2022) 311-318. https://doi.org/10.2478/mspe-2022-0040

[21] D. Klimecka-Tatar, M. Ingaldi. Service Quality Management in Term of IHRM Concept and the Employee Internationalization. Int. J. Qual. Res. 15 (2021) 753-772. https://doi.org/10.24874/IJQR15.03-05

[22] M. Jagusiak-Kocik, R. Ulewicz. Implementation of the QFD Method in a Construction Industry Company, LNCE 290 (2023) 416-423. https://doi.org/10.1007/978-3-031-14141-6_42

[23] M. Grebski, M. Mazur. Social climate of support for innovativeness, Prod. Eng. Arch. 28 (2022) 110-116. https://doi.org/10.30657/pea.2022.28.12

[24] M. Ingaldi. A new approach to quality management: conceptual matrix of service attributes. Pol. J. Manag. Stud. 22 (2020) 187-200. https://doi.org/10.17512/pjms.2020.22.2.13

The Investigation of Safety Behavior in Logistic Companies of Malaysia

IMRAN Muhammad [1, a *], ZULKIFLY Syazwan Syah bin [2, b]
and KOT Sebastian [3, c]

[1]School of Business Management, Universiti Utara Malaysia, Kedah, Malaysia & Department of Project and Operation Management, The Islamia University of Bahawalpur, Pakistan

[2] School of Business Management, Universiti Utara Malaysia, Kedah, Malaysia

[3] The Management Faculty, Czestochowa University of Technology, Częstochowa, Poland and North-West University, Faculty of Economics and Management Sciences, South Africa

[a]muhammad.imran@uum.edu.my, [b] syazwan.syah@uum.edu.my, [c] sebastian.kot@pcz.pl

Keywords: Safety Leadership, Supervisor Safety Roles, Safety Behavior, Logistics, Malaysia

Abstract. In Malaysia, industrialization and population growth have both contributed to an increase in workplace occupational injuries. The most recent statistics on reported accidents show that out of 10,000 employees, 99 have been directly involved in work-related accidents. Accidents do not occur by chance, thus there are multiple factors which are contributing to workplace accidents. It is important to spread safety awareness to employees inside the organization. The main objective of the current study is to investigate the relationship between safety leadership, supervisor safety roles and safety behavior in logistics companies of Malaysia. The study respondents included 160 employees from logistics companies of Malaysia. A partial least square equation modeling "Higher Order Two-Stage Approach" analysis was performed to assess the measurement and structural model involving variables of safety leadership, supervisor safety roles and safety behavior to draw the results. The results of the study revealed that safety leadership roles imposed by the managers have a significant effect on safety behavior, mediated by supervisors' safety roles. Therefore, the findings of the study suggest that safety leadership and supervisor safety roles be implemented in logistics companies. These make employees more inclined to take part in safety initiatives which can make the workplace safer.

Introduction

Positive safety behavior at the workplace is one of the most essential elements that cannot be taken lightly in an organization [1]. The negligence of positive safety behavior could lead into high accident case at workplace [2, 3]. Accident, even if minor, can cause serious adverse effect to the organization [4]. Kamardeen [5] also affirmed that workplace accident inflicts direct and indirect costs on a business. The direct costs include investigation cost, equipment loss, legal fine, and property damages, while the indirect costs are skillful manpower loss, medical cost, and others [6].

The workplace accidents are possibly caused by behavior of taking shortcuts and bypassing the standard operating procedures [4]. Other than that, safety behavior is among the critical factors of safety performance that contributes towards work-related accidents [7]. Accident at workplaces has also recently become main issue in Malaysia. Based on the data shown in the Social Security Organization (SOCSO), the average number for industrial accidents in Malaysia for the period between 2014 and 2018 were reported at 35,791 cases per year. In other words, it is equivalent to 98 cases per day. More particularly, the numbers of accident storage and transportation sector in Malaysia recorded from 3,600 cases to 4,200 cases, which is alarming [8]. Anyway, hit by flying objects while performing loading-unloading activity, struck by moving loader vehicle while

Quality Production Improvement and System Safety: QPI 16 - CZOTO 10   Materials Research Forum LLC
Materials Research Proceedings 34 (2023) 323-333   https://doi.org/10.21741/9781644902691-38

walking, and paper cut injury while performing packaging task are the common types of accidents in logistics companies [9].

However, the factors which cause accidents in the workplace need to be seriously addressed. Henceforth, the researcher decided to embark on studying the contributing factors such as safety leadership and supervisor safety management roles to safety performance such as safety behavior [2, 10]. However, the primary cause of accident has been identified as safety behavior [11-13]. Therefore, factors towards safety behavior had been intensively studied [14-16], where most of the studies measured safety performance by safety behavior. Furthermore, past studies have also proven that crucial role of supervisors in terms of safety has significant influence on safety performance (safety behavior) and supervisors in safety management has a significant effect on safety performance concerning accident and injury reduction [17-19].

However, in Malaysian settings, supervisor safety roles is found to have insignificant effect towards safety performance in terms of accident and injury reduction [20]. On one hand, Khoo, Surienty and Selamat [21] determined a significant relationship between supervisor's support on safety and safety behavior in manufacturing sector. In realizing this empirical gap, this study proposes to examine the effect of safety management practices including supervisor safety roles towards safety performance. The variables for safety leadership consisting the safety concern, and safety motivation dimensions which adapted from Shang, Yang and Lu [10] study. However, based on the consideration of the most proactive approach, safety behavior is selected to be the measurement of safety performance for this study [2, 22] and consisting two dimensions such as safety compliance and safety participation, which is different from the measurement used by Shang, Yang and Lu [10], Zulkifly, Baharudin, Mahadi, Ismail and Hasan [23]. Safety behavior is found to be the leading factor in industrial accidents. Thus, it is vital for this research to study the level of safety behavior and its predictors to overcome accident issues in Logistic sector of Malaysia.

Literature Review

Safety Behavior. Generally speaking, safety performance is the degree of safety as determined by workplace mistakes, injuries, and fatalities [24]. The several earlier researchers promoted the proactive method of measuring safety performance by measuring safety behavior [25-27]. Furthermore, researchers have recently used four dimensions to evaluate the safety performance such as perceived accident reduction, perceived equipment failure, perceived goods defect & damage, and reported personal injury reduction [28].

In respect of Malaysia's transportation sector is performing poorly in terms of safety [29]. According to Malaysian industrial accident statistics, 60–70% ratio of accidents are recorded in manufacturing and transport sector simultaneously [30]. The purpose of this research is to investigate how safety management techniques improve safety performance. More specifically, the logistics companies working to enhance their workplace safety and control the accidents rations actively [31, 32]. In this regard, the current research is to investigate that how safety leadership and supervisor roles can affect positively the safety performance (behavior safety) in logistics sector of Malaysia.

Safety leadership, supervisor safety roles and safety behavior. As per past studies, most of researchers are agreed that safety management practices played the important role to ensuring the effective safety performance [33, 34]. Currently, a few important studies have been conducted to look at how safety management practices affect safety performance considering safety concern, and safety motivation factors. For instance, in order to predict the safety behavior, Shang, Yang and Lu [10] evaluated the relationship between safety concern, safety motivation as the major practices of safety management and safety behavior. Further, he suggested that effective safety

Quality Production Improvement and System Safety: QPI 16 - CZOTO 10   Materials Research Forum LLC
Materials Research Proceedings 34 (2023) 323-333   https://doi.org/10.21741/9781644902691-38

management such as safety leadership (safety motivation and safety concern) could reduce the accidents, injuries, product loss value, and equipment failure value in logistics sector.

Furthermore, the importance of the supervisor's role in occupational safety and health is also investigated [35]. Moreover, the Noe [36] stated that supervisors are the first management level where they are given key duties and obligations to establish and direct workgroups in organizations to guarantee that the workforce satisfies all corporate objectives including safe workplace [37]. According to Vinodkumar and Bhasi [2] that supervisor has a positive role to build the positive working relationship between management and employees, this could bring the higher safety awareness and effective implementation of safety practices at workplace which also can boost the safety behavior. Hence, Subramaniam, Shamsudin and Alshuaibi [38] recommended that supervisors play a significant role in improving the safety behavior which can improve the safety performance more specifically in Malaysian context. Hence, proposed the following hypothesis:

H_1: safety leadership has a significant positive effect on safety behavior.

H_2: safety leadership has significant positive effect on supervisor safety roles.

Furthermore, Shang, Yang and Lu [39] found the strong link between supervisor safety roles and safety behavior. Moreover, Yanar, Lay and Smith [40] revealed that missing link of supervisor between management and safety implementation can decrease the safety performance which also can increase the injury rate up to 3.5 percent. Additionally, some studies also found the partial impact of employers' and supervisors' safety management on safety performance, particularly in the transportation industry. Besides, there is very little research that has been noticed in the transportation sector as well. Hence, it's very important to evaluate the intervening role of supervisor safety role between leadership safety and safety behavior.

H_3: supervisor safety roles mediated the effect of safety leadership and safety behavior.

Methodology

The respondents of this study were middle management staff from the logistics companies of Malaysia. Furthermore, they should have enough knowledge about safety management, and they are also responsible directly and indirectly for the implementation of safety practices inside the organization. The thirty-four major logistics companies were identified from the website of stock exchange of Malaysia. Sampling is a statistical analysis procedure in which a preset number of observations are collected from a wider population or in which a suitable number of elements are chosen from the population by choosing the appropriate sample. Anyway, purposive sampling techniques have been used to approach the respondents. Basically, purposive sampling is defined as a sampling that is limited to people/employees who can give the necessary information, in other words, those are only person who keeping the required information. However, the questionnaire has been distributed among the middle management of logistics companies.

We have given one month to respondents to respond to the questionnaire as per their convenience. At the same time, questionnaires were also sent to respondents through e-mail and google survey, because some of respondents requested online surveys. A 5-point Likert scale was used in the questionnaire to get the response. However, the 300 questionnaires were distributed among the middle management of logistics companies of Malaysia, the only 160 out of 300 questionnaires have received, most of companies from Knag Valley participated in the survey. Only 162 of the 300 surveys that were sent out were returned. The response rate of study is recorded 53.33% and other 46.66 % refused to participate in current study, maybe due to busy schedule and reluctance to give the answer.

Furthermore, the current study makes sure that the 160-sample size is enough to justify the population and secondly is this enough to run the analysis. In this regard, we calculate the sample size using G*Power software (Fig.1). The minimum 160 respondents are needed for the present study to draw the results, you may see the sample calculation in the figure below. The snowball sampling technique has been used to approach respondents.

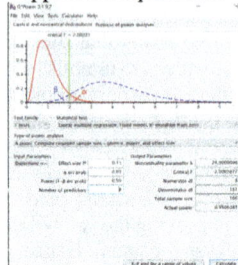

Fig. 1. *Evaluation of sample size with G*Power software*

Measurement

In this study, safety leadership comprise two dimensions: safety concern and safety motivation [26, 41]. Basically, the items are adapted the questions from [10]. Anyway, safety motivation and safety concern are measured with 3 and 5 items respectively. On one hand, supervisor safety roles are measured through three items which are adapted from previous research [10]. The current study, the safety behavior scale has two dimensions such as safety compliance and safety participation which is adopted from the study of Vinodkumar and Bhasi [2], Griffin and Neal [42]. The safety behavior has two dimensions such as safety compliance behavior and safety participation behavior. However, the safety compliance is measured with 7 items in this study. On the other side, safety participation is measured with 5 items. The current study scale used the 5-point Likert Scale (1 = strongly disagree, 5 = strongly agree) [43].

Analysis

Measurement Model. This study assessed the two-stage higher order technique. First, the reliability, convergent validity, and discriminant validity of the first order reflective measurement model were evaluated. If the loadings are more than 0.5, it is possible to determine the convergent validity [44]. As mentioned in Table 1, composite dependability is established when the average variance extracted (AVE) is at least 0.5 [45] and the values are larger than 0.7 [46].

Subsequently, the discriminant validity test was performed using the evaluation of the Heterotrait-Monotrait (HTMT) ratio of correlations [47]. At this stage, the researcher needs to compare it to a predefined threshold [48, 49]. Hence, if the value of the HTMT is greater than its predefined threshold, the researcher can conclude that there is the non-existence of discriminant validity. This study used the lenient threshold of 0.90 [50] as stated in Table 2.

Quality Production Improvement and System Safety: QPI 16 - CZOTO 10   Materials Research Forum LLC
Materials Research Proceedings 34 (2023) 323-333   https://doi.org/10.21741/9781644902691-38

Table 1. *Composite Reliability [Prepared by the authors]*

Construct	Items	Loadings	AVE	CR	Cronbach's Alpha
Safety Concern	SC1	0.900	0.804	0.966	0.959
	SC2	0.915			
	SC3	0.887			
	SC4	0.887			
	SC5	0.873			
Safety Motivation	SM1	0.913	0.797	0.951	0.936
	SM2	0.958			
	SM3	0.928			
Supervisor Safety Role	SS1	0.881	0.871	0.953	0.926
	SS2	0.939			
	SS3	0.948			
	SS4	0.889			
Safety Compliance	SCB1	0.896	0.699	0.921	0.892
	SCB2	0.920			
	SCB3	0.886			
	SCB4	0.894			
	SCB5	0.884			
	SCB6	0.895			
	SCB7	0.901			
Safety Participation	SPB1	0.824	0.837	0.953	0.935
	SPB2	0.797			
	SPB3	0.779			
	SPB4	0.894			
	SPB5	0.881			

Note: CR=Composite Reliability, AVE=Average Variance Extracted

Table 2. *Discriminant Validity of Constructs [Prepared by the authors]*

Constructs	Safety Compliance	Safety Concern	Safety Motivation	Safety Participation	Supervisor Safety
Safety Compliance					
Safety Concern	**0.768**				
Safety Motivation	0.806	**0.835**			
Safety Participation	0.875	0.831	**0.871**		
Supervisor Safety	0.791	0.798	0.876	**0.842**	

Notes: SCB = Safety Compliance Behavior, SPB = Safety Participation Behavior, SM = Safety Motivation, SC = Safety Concern, SS = Supervisor Safety.

Structural Model. Subsequently, the structural model was tested by conducting bootstrapping of 5,000 resamples to determine the path coefficient values and the t values. Therefore, looking at each structural path, safety leadership has a significant effect on safety behavior ($\beta = 0.609$, $p < 0.05$). Furthermore, supervisor safety roles ($\beta = 0.289$, $p < 0.05$) was also significantly affecting to safety behavior in this research. In addition, the results also revealed that safety leadership has a significant effect on supervisor safety roles ($\beta = 0.830$, $p < 0.05$). The results are presented in table 3.

Table 3. Path Coefficient [Prepared by the authors]

Hypothesis	Standard Beta	Standard Error	T Values	P Values
Supervisor Safety -> Safety Behavior	0.289	0.071	4.108	**0.000**
Safety leadership -> Safety Behavior	0.609	0.066	9.245	**0.000**
Safety leadership -> Supervisor Safety	0.830	0.039	21.232	**0.000**

Notes: $*p = 5$ per cent (based on one tail test with 5,000 bootstrapping).

Moreover, this study found the significant mediation role of supervisor safety roles between safety leadership and safety behavior. The results can be seen in table 4.

Table 4. Indirect Effect (Mediation results) [Prepared by the authors]

Hypothesis	Standard Beta	Standard Error	T Values	P Values
Safety leadership -> Supervisor Safety -> Safety Behavior	0.240	0.060	4.014	**0.000**

Notes: $*p = 5$ per cent (based on one tail test with 5,000 bootstrapping).

Furthermore, The R^2 is 0.725, which carries the meaning that 72.5% of the variance in safety performance was explained by safety leadership (safety motivation, safety concern), and supervisor safety roles.

Discussion
The path coefficient analysis was conducted to test the effect of safety leadership and supervisor safety management on workers safety behavior. The current study findings supported all proposed hypotheses, confirming the roles of safety management by the employer and supervisors has a significant effect on workers safety behavior. These results revealed that high safety management roles by the employer and supervisors affect workers safety performance behavior. The present study findings are consisted with studies of Vinodkumar and Bhasi [2], Yang, Wang, Chang, Guo and Huang [26], Subramaniam, Mohd Shamsudin, Mohd Zin, Sri Ramalu and Hassan [51].

Overall, this study's findings suggest that logistics organisations can employ safety management practices to improve safety performance and lower the accident rates [10]. Additionally, safety management practices such as safety motivation, safety concern and supervisor safety roles not only enhance working conditions but also have a favourable impact on worker attitudes and behaviors about safety, which lowers workplace accidents. Additionally, it is advised that Malaysian logistics company owners and managers implement ongoing safety management procedures at workplace. First, a clear direction of safety policies and goals must be established and communicated to both employees and managers [39].

As this study indicated that safety motivation is the second major component, owner-managers must also instil this drive in their staff. In order to increase safety performance, it is proposed that owner-managers demonstrate concern or caring behavior for the safety of their employees [52]. The findings demonstrated how important owner-managers of logistics companies are to guaranteeing firm performance, particularly safety [53].

Managerial Implications

The findings of this study have several managerial implications. First, the safety management is very important for owner-managers of logistics firms. Second, safety leadership critical dimensions such as safety concern and safety motivation are very important for owner/mangers of logistics firm to reduce the accident ratio and make the safe workplace. Third, by understanding the differences in safety leadership and supervisor safety management and their simultaneous effect on safety performance, the owner/managers of logistics companies can develop effective action plans to enhance their companies' safety performance and reduce the accidents. However, this study findings stated that the role of the owner-manager is vital in long-term safety success and safety concern, and safety motivation inspire workers to succeed in safety performance [32].

Limitations

There are still some limitations that need to be mentioned, despite the fact that this study offers practitioners useful contributions. First, only the chosen Malaysian logistic companies were included in the study's sample. As a result, the results are only loosely generalizable even though they presented significant empirical evidence about the relationship between safety management and safety performance.

Second, because this study is cross-sectional, it is impossible to prove causality between the variables when data are obtained at the same time [54]. Despite a cross-sectional study's inability to determine the direction of causality, this constraint can be somewhat overcome by using a theory to identify and explain the causal linkages between the variables [55]. Finally, since this study is only concerned with three factors—safety leadership, the role of the supervisor in terms of safety, and safety behavior—future studies would also cover other aspects of safety management like training, communication, rules and procedures, and worker involvement.

Conclusion

In conclusion, the result of this study indicated that there is a positive significant relationship between employer and supervisor safety management on workers safety performance in terms of safety compliance and safety participation behavior. The results confirmed that the employer and supervisors need to play their roles in inculcating safety behavior among workers to prevent accidents from happening at the workplace. The safety management roles by employer namely safety concern, safety motivation and safety policy have been proven as crucial variables in determining safety performance behavior. Besides, supervisor safety role has the strongest influence on workers safety performance behavior, hence, supervisor role in managing safety need to be given more focus by the researchers and practitioners in occupational safety and health field. Based on this conclusion, all related parties should apply this research findings for their managerial or academia purposes in benefiting occupational safety and health area.

References

[1] T. Derenda, M. Zanne, M. Zoldy. Automatization in road transport: a review, Prod. Eng. Arch. 20 (2018) 3-7. https://doi.org/10.30657/pea.2018.20.01

[2] M.N. Vinodkumar, M. Bhasi. Safety management practices and safety behavior: Assessing the mediating role of safety knowledge and motivation, Accid. Anal. Prev. 42 (2010) 2082-2093. https://doi.org/10.1016/j.aap.2010.06.021

[3] S.S. Zulkifly. Safety Leadership and its Effect on Safety Knowledge-Attitude-Behavior (KAB) of Malaysia Manufacturing Workers, Solid State Technol. 63 (2020) 218-229.

[4] N.H. Zakaria, N. Mansor, Z. Abdullah. Workplace accident in malaysia: Most Common causes and solutions, Business and Management Review 2 (2012) 75-88.

[5] I. Kamardeen, Strategic safety management information system for building projects in Singapore, Eng. Constr. Archit. Manag. 16 (2009) 8-25. https://doi.org/10.1108/09699980910927868

[6] A. Woźny. Selected problems of managing work safety-case study, Prod. Eng. Arch. 26 (2020) 99-103. https://doi.org/10.30657/pea.2020.26.20

[7] M.S. Mashi, C. Subramaniam, J. Johari. The effect of management commitment to safety, and safety communication and feedback on safety behavior of nurses: the moderating role of consideration of future safety consequences, Int. J. Hum. Resour. Manag. (2018) 2565-2594. https://doi.org/10.1080/09585192.2018.1454491

[8] Malaysia O. Social Security, Annual report, 2022.

[9] C.-S. Lu, C.-S. Yang. Safety leadership and safety behavior in container terminal operations, Safety Sci. 48 (2010) 123-134. https://doi.org/10.1016/j.ssci.2009.05.003

[10] K.C. Shang, C.S. Yang, C.S. Lu. The effect of safety management on perceived safety performance in container stevedoring operations, Int. J. Shipp. Transp. Logist. 3 (2011) 323-341. https://doi.org/10.1504/IJSTL.2011.040801

[11] H. Abbasianjahromi, A. Etemadi. Applying social network analysis to identify the most effective persons according to their potential in causing accidents in construction projects, Int. J. Constr. Manag. 22 (2022) 1065-1078. https://doi.org/10.1080/15623599.2019.1683688

[12] M.J.P. Bussier, H.-Y. Chong. Relationship between safety measures and human error in the construction industry: working at heights, Int. J. Occup. Saf. Ergon. 28 (2022) 162-173. https://doi.org/10.1080/10803548.2020.1760559

[13] A.M. Barakat Abuashour, Z. Hassan. A Conceptual Framework for Enhancing Safety Performance by Impact Cooperation Facilitation, Safety Communication and Work Environment: Jordanian Hospitals, Sains Humanika 11 (2019) 81-89. https://doi.org/10.11113/sh.v11n2-2.1659

[14] N. Ghodrati et al. Unintended consequences of productivity improvement strategies on safety behavior of construction labourers; a step toward the integration of safety and productivity, Buildings 12 (2022) art.317. https://doi.org/10.3390/buildings12030317

[15] Z. Lyubykh et al. Shared transformational leadership and safety behaviors of employees, leaders, and teams: A multilevel investigation, J. Occup. Organ. Psychol. 95 (2022) 431-458. https://doi.org/10.1111/joop.12381

[16] M. Bayram, B. Arpat, Y. Ozkan. Safety priority, safety rules, safety participation and safety behavior: The mediating role of safety training, Int. J. Occup. Saf. Ergon. (2021) 2138-2148. https://doi.org/10.1080/10803548.2021.1959131

[17] B. Fernández-Muñiz, J.M. Montes-Peón, C.J. Vázquez-Ordás. Safety leadership, risk management and safety performance in Spanish firms, Saf. Sci. (2014) 295-307. https://doi.org/10.1016/j.ssci.2014.07.010

[18] A. Mohammadi, M. Tavakolan, Y. Khosravi. Factors influencing safety performance on construction projects: A review, Saf. Sci. 109 (2018) 382-397. https://doi.org/10.1016/j.ssci.2018.06.017

[19] D. Ramos, P. Afonso, M.A. Rodrigues. Integrated management systems as a key facilitator of occupational health and safety risk management: A case study in a medium sized waste management firm, J. Clean. Prod. 262 (2020) art.121346. https://doi.org/10.1016/j.jclepro.2020.121346

[20] S.S. Zulkifly, M.R. Baharudin, N.H. Hasan. Safety leadership and safety knowledge-attitude-behavior (KAB) in Malaysia manufacturing SMEs : A higher order two-stage approach of PLS-SEM, Preprints (2021) 1-17. https://doi.org/10.20944/preprints202106.0527.v1

[21] K.T. Hong, L. Surienty, M.N. Selamat. Safety training and safety behavior in the Malaysian SME, J. Occup. Saf. Health 13 (2016) 55-62.

[22] P. Zhang, N. Li, D. Fang, H. Wu. Supervisor-focused behavior-based safety method for the construction industry: Case study in Hong Kong, J. Constr. Eng. Manag. 143 (2017) art. 05017009. https://doi.org/10.1061/(ASCE)CO.1943-7862.0001294

[23] S.S. Zulkifly et al. Validation of a research instrument for safety leadership and safety knowledge-attitude-behavior (KAB) for Malaysia manufacturing set-up, iRASD J. Manag. 3 (2021) 22-34. https://doi.org/10.52131/jom.2021.0301.0023

[24] J. Mullen, E.K. Kelloway, M. Teed. Employer safety obligations, transformational leadership and their interactive effects on employee safety performance, Saf. Sci. 91 (2017) 405-412. https://doi.org/10.1016/j.ssci.2016.09.007

[25] S. Majid et al. The effect of safety risk management and airport personnel competency on aviation safety performance, Uncertain Supply Chain Manag. 10 (2022) 1509-1522. https://doi.org/10.5267/j.uscm.2022.6.004

[26] C.-C. Yang, Y.-S. Wang, S.-T. Chang, S.-E. Guo, M.-F. Huang. A study on the leadership behavior, safety culture, and safety performance of the healthcare industry, WASET Int. J. Humanit. Soc. Sci. 53 (2009) 1148-1155. https://doi.org/10.5281/zenodo.1062282

[27] M. Imran, A.U. Haque, R. Rębilas. Performance appraisal politics and employees' performance in distinctive economies, Pol. J. Manag. Stud. 18 (2018) 135-150. https://doi.org/10.17512/pjms.2018.18.2.11

[28] S.S. Zulkifly et al. The Impact of Superior Roles in Safety Management on Safety Performance in SME Manufacturing in Malaysia, Glob. Bus. Rev. (2021) 1-16. https://doi.org/10.1177/09721509211049588

[29] B. Md Deros et al. Conformity to occupational safety and health regulations in Malaysian small and medium enterprises, Am. J. Appl. Sci. 11 (2014) 499-504. https://doi.org/10.3844/ajassp.2014.499.504

[30] A.A. Aziz et al. A preliminary study on accident rate in the workplace through occupational safety and health management in electricity service, QUEST J. Res. Bus. Manag. 2(12) (2015) 9-15.

[31] S.S. Zulkifly et al. Workplace safety improvement in sme manufacturing: A government intervention, Int. J. Sci. Technol. 4 (2018) 29-39. https://doi.org/10.20319/mijst.2018.42.2739

[32] Z. Hassan, R. Rahim. The Relationship between Supervisor Safety, Safety Management Practices, and Safety Compliance Behavior among Employees, Sains Humanika 11 (2019) 31-36. http://doi.org/10.11113/sh.v11n2-2.1652

[33] C.M. Tam, I.W.H. Fung. Effectiveness of safety management strategies on safety performance in hong kong, Constr. Manag. Econ. 16 (1998) 49-55. https://doi.org/10.1080/014461998372583

[34] R. Tong et al. Impact of safety management system on safety performance: the mediating role of safety responsibility, Eng. Constr. Archit. Manag. (2020) 3155-3170. https://doi.org/10.1108/ECAM-03-2020-0197

[35] C.S. Lu, K.C. Shang. An empirical investigation of safety climate in container terminal operators, J. Saf. Res. 36 (2005) 297-308. https://doi.org/10.1016/j.jsr.2005.05.002

[36] R.A. Noe. Employee training and development, 8th Edition. McGraw-Hill, New York, 2020. ISBN 978-1260043747

[37] V. Holubová. Integrated safety management systems, Pol. J. Manag. Stud. 14 (2016) 106-118. https://doi.org/10.17512/pjms.2016.14.1.10

[38] C. Subramaniam, F.M. Shamsudin. A.S.I. Alshuaibi. Investigating employee perceptions of workplace safety and safety compliance using pls-Sem among technical employees in Malaysia, J. App. Struct. Equ. Modeling 1 (2017) 44-61.

[39] K.C. Shang, C.-S. Yang, C.-S. Lu. The effect of safety management on perceived safety performance in container stevedoring operations, Int. J. Shipp. Transp. Logist. 3 (2015) 323-341. https://doi.org/10.1504/IJSTL.2011.040801

[40] B. Yanar, M. Lay, P.M. Smith. The Interplay Between Supervisor Safety Support and Occupational Health and Safety Vulnerability on Work Injury, Saf Health Work 10 (2019) 172-179. https://doi.org/10.1016/j.shaw.2018.11.001

[41] S.S. Zulkifly, C. Subramaniam, N.H. Hasan. Examining the influence of safety leadership towards safety behavior in SME manufacturing, Occupational Safety and Health 14 (2017) 17-23.

[42] M.A. Griffin, A. Neal. Perceptions of safety at work: A framework for linking safety climate to safety performance, knowledge, and motivation, J. Occup. Health Psychol. 5 (2000) 347-358. https://doi.org/10.1037/1076-8998.5.3.347

[43] R. Likert. A technique for the measurement of attitudes, Archives of Psychology 22 (1932) 5-55.

[44] J.F. Hair, W.C. Black, B.J. Babin, R.E. Anderson, Multivariate data analysis, 7th Edition, Pearson, New York, 2010. ISBN 978-0138132637

[45] C. Fornell, D.F. Larcker. Evaluating Structural Equation Models with Unobservable Variables and Measurement Error, Journal of Marketing Research 18(1) (1981) 39-50. https://doi.org/10.2307/3151312

[46] D. Gefen, D. Straub, M.-C. Boudreau. Structural equation modeling and regression: Guidelines for research practice, Communications of the Association for Information Systems 4 (2000) art.7. https://doi.org/10.17705/1CAIS.00407

Quality Production Improvement and System Safety: QPI 16 - CZOTO 10   Materials Research Forum LLC
Materials Research Proceedings 34 (2023) 323-333   https://doi.org/10.21741/9781644902691-38

[47] J. Henseler, C.M. Ringle, M. Sarstedt. A new criterion for assessing discriminant validity in variance-based structural equation modeling, J. Acad. Mark. Sci. 43 (2014) 115-135. https://doi.org/10.1007/s11747-014-0403-8

[48] Y.M. Yusoff, H.H. Abdullah, H. Hafinaz, Establishing the green human resource management practices model of SMEs in malaysia, European Academic Research 8(2) (2020) 493-504.

[49] Y.M. Yusoff et al. Linking Green Human Resource Management Practices to Environmental Performance in Hotel Industry, Glob. Bus. Rev. 21 (2020) 663-680. https://doi.org/10.1177/0972150918779294

[50] T. Ramayah, J. Cheah, F. Chuah, H. Ting, M.A. Memon. Partial least squares structural equation modeling (PLS-SEM) using SmartPLS 3.0: An updated guide and practical guide to statistical analysis, Pearson, New York, 2018. ISBN 978-967-349-739-3

[51] C. Subramaniam et al. The influence of safety management practices on safety behavior: A study among manufacturing smes in Malaysia, Int. J. Supply Chain Manag. 5 (2016) 148-160. https://doi.org/10.59160/ijscm.v5i4.1370

[52] J.L. Chua, S.R.A. Wahab. The effects of safety leadership on safety performance in Malaysia, Saudi J. Bus. Manag. Stud. 2 (2017) 12-18. https://doi.org/10.21276/sjbms.2017.2.1.3

[53] S.S. Zulkifly, M.R. Baharudin, M.R. Mahadi, S.N.S. Ismail, N.H. Hasan. The effect of owner- manager's safety leadership and supervisor's safety role on safety performance in Malaysia's manufacturing SMEs, J. Technol. Oper. Manag. 16 (2021) 11-24. https://doi.org/10.32890/jtom2021.16.1.2

[54] U. Sekaran, R. Bougie. Research Methods For Business: A Skill Building Approach, 7th Edition, Wiley, Hoboken, 2016. ISBN 978-1119266846

[55] J.F. Hair Jr., M. Sarstedt, L. Hopkins, V.G. Kuppelwieser, Partial least squares structural equation modeling (PLS-SEM): An emerging tool in business research, Eur. Bus. Rev. 26 (2014) 106-121. https://doi.org/10.1108/EBR-10-2013-0128

Quality Production Improvement and System Safety: QPI 16 - CZOTO 10   Materials Research Forum LLC
Materials Research Proceedings 34 (2023) 334-343   https://doi.org/10.21741/9781644902691-39

The Use of Computer Simulation in the Management of Subcontractors and Outsourced Services

KRYNKE Marek[1, a],

[1]Department of Production Engineering and Safety, Faculty of Management, Czestochowa University of Technology, Al. Armii Krajowej 19b, 42-218 Czestochowa, Poland

[a] marek.krynke@pcz.pl*

Keywords: Simulation, Flexsim, Optimization, Production Management

Abstract. The paper examines cooperation among production companies to fulfill orders beyond plant capacity by selecting subcontractors. The developed model focuses on planning the production process to minimize total production costs by deciding where to produce goods before they are actually produced. The concept utilized a 3D FlexSim simulation environment, specifically the built-in optimization module OptQuest, to address the problem. The paper covers the key steps in creating the simulation model and presents the simulation results.

Introduction

In the case of production companies whose processes depend on cooperation with external entities, it is important to take into account the scope of cooperation, starting from the management of raw materials, materials or semi-finished products, to the adoption of a production plan, the production itself, along with the registration of service, transport and storage costs. A cooperator's production plans should be included in the operational schedule of in-house production [1, 2]. Cooperation in the production process is perceived by many production companies as an opportunity to more effectively adapt to the evolution of production systems [3–5]. The role of cooperation in the production of products is growing, especially in very dynamic markets with variable demand and a short product life cycle [6]. This is because for such markets it is sufficient to apply a deferred production strategy and to transfer the last stage of production from industrial enterprises to distribution companies. [7-9].

In the management of the production process, often a large part of the planning process concerns the selection of a subcontractor in terms of its efficiency and low cost, high quality and value of the delivered finished products, as well as safety. [10-12]. Therefore, it is important to establish a set of procedures/recommendations for pricing, selecting contractors, and measures to monitor and improve supplier relation. One should also not forget about the management processes of own means of production, their availability and own costs of process maintenance [13-18]. The development of modern computer technology and many fields of science, especially in the field of production engineering, made it possible to virtually simulate real production processes [19]. Due to the use of comprehensive IT solutions in the field of modeling and simulation of manufacturing processes, a significant economic benefit is achieved, especially in mass production [20]. Therefore, more and more IT solutions are appearing in the area of tracking, monitoring and visualizing the course of production processes in real time [21,22]. Many IT tools are at the disposal of managers today, incl. Technomatix Plant Simulation, Matlab/Simulink, Enterprise Dynamics, Arena, FlexSim, Vensim, Excel/Solver and others. The effectiveness of the production planning and scheduling processes, in particular at the stage of simulating virtual models of production systems, depends primarily on the mathematical models and optimization algorithms used [23]. In complex production systems, this efficiency is translated primarily into achieving a lower level of production costs, shorter production cycle times with a simultaneous high efficiency of data processing. [24]. It is also important that the digital model reacts quickly to changes [25].

Quality Production Improvement and System Safety: QPI 16 - CZOTO 10   Materials Research Forum LLC
Materials Research Proceedings 34 (2023) 334-343   https://doi.org/10.21741/9781644902691-39

Outsourcing tasks to external subcontractors is becoming increasingly common within enterprises due to the high level of specialization required for technological activities and the need to maintain high-quality standards [26-28].

Typical subcontracting involves technological tasks that either demand high qualifications and expensive machinery or impose a significant environmental burden, necessitating certified safeguards and purification systems [29]. Among the highly specialized activities in this regard, metal processing [30-32] coupled with surface treatment [33], including the application of DLC coatings [34-36], can be mentioned. Additionally, the production of high-pressure components [37,38] requires diagnostic control of structural connections [39,40] and certification.

In this manner, end manufacturers of machinery [41], equipment, or railway rolling stock [42] essentially serve as integrators, responsible for ensuring the sustained reliability and stable performance of the final products throughout their intended lifespan, irrespective of wear and tear [43].

Such an approach requires meticulous optimization of design and production processes, as well as monitoring them both internally and at subcontractors and service providers. Given the complexity of the processes and the numerous factors to consider, essential tools include dimensionality reduction methodologies [44] and design of experiments [45-47], including increasingly popular nonparametric approaches [48-50].

The Essence of Simulation and Optimization

Computer modeling and simulation is primarily used as a decision support method [51]. However, simulation techniques are most often used when analytical solutions are too difficult or time-consuming [52,53]. Simulation modeling is useful in many fields of science. It is used to learn about a given process by replacing it with a simplified system that reflects its selected features. The modeling process is defined by the determination of the mathematical relationship between the output value y, and input value x_1, x_2, x_3, ..., x_n. In the simplest version, it is the formula: $y = f(x_1, x_2, x_3, ..., x_n)$, where the variables: x_1, x_2, x_3, ..., x_n, are input values, while y is the output value of the tested system.

There can be more than one input and output variable, it all depends on the complexity of the system / process. The simulation model can be compared to the so-called a black box that can have n input variables and m output variables [54].

Simulation will help in direct identification of cost reduction areas of efficiency improvement, it can also play an important role in risk analysis [46]. The simulation also indicates scenarios of the development of the situation based on the proposed actions [55]. For example, simulation will not be able to predict specific customer requirements for products and services, but it can be used to assess the impact of demand volatility on the ability to respond to this variability in a production system.

Optimizing a process by simulation means finding the best configuration of the input variables that match the best value [56]. Optimization usually consists of maximizing or minimizing a selected parameter [57]. After building each simulation model, it should be validated, i.e. its suitability for the given application assessed. If it turns out that the model correctly reflects reality, then only then can be proceeded to designing experiments and further data analysis [58, 59].

Methodology – Case Study

The purpose of the paper is to discuss the model of cooperation from the perspective of a group of production companies for the purpose of comprehensive order fulfillment. The study took into account the problem of selecting subcontractors selected to fulfill an order exceeding the production capacity of the plant. The developed model focuses on planning the production process when the goods have not yet been produced. Therefore, at the time of deciding where to produce it, so that the total production costs are as low as possible. In this concept, a 3D FlexSim simulation environment was used to solve the problem – 3D FlexSim with built-in optimization module OptQuest.

As part of the research analysis, the following problem was considered. The production plant must fulfill a production order for the production of 10,000 pcs. of products. The production resources available in the form of available machines are insufficient to fulfill this order, which is why the company wants to hire subcontractors. The x1 production system is owned by the company, while the subsequent systems marked as x2÷x10 are rented, hence the different costs of their use. Unit costs and capacities of individual production systems are summarized in Table 1.

Table 1. *Unit cost and efficiency of individual production systems*

Enterprise	Performance [pcs/hour]	Cost [monetary units/hour]
x1	10÷11	100
x2	15÷17	175
x3	20÷23	200
x4	17÷18	180
x5	14÷15	160
x6	18÷20	178
x7	22÷25	205
x8	16÷18	180
x9	25÷26	210
x10	8÷9	155

The implemented product production process can be presented in the FlexSim environment using the model shown in Fig. 1. In this model, standard objects from the program library were used, which were programmed in accordance with the task conditions. Intuitively, it can be assumed that the function of the production systems of individual plants should be performed by *Processors*. Process time for individual production systems is set according to their capacity. *Source* objects are usually generators of many flow elements. In this model, the source works in *Arrival Sequence* mode. In this particular case, the flow element will symbolize the number of items produced per unit of time. This is the variable that will be used in the optimization process, as well as the place where the optimizer will generate results from subsequent iterations. The OptQuest optimizer from OptTek built into the FlexSim platform will be used to solve the example problem. Its operation is based on neural networks and metaheuristic algorithms [35].

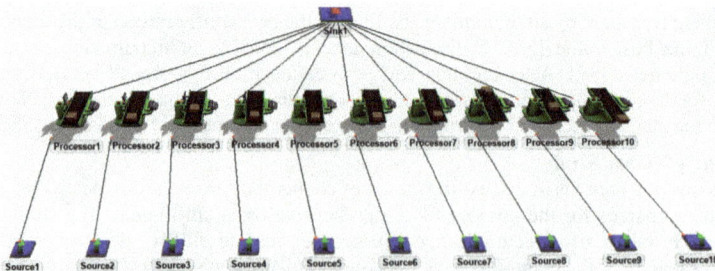

Fig. 1. *Simulation model for the discussed problem (source: own study)*

The objective function is defined as the cost of work of processors reflecting the work of individual plants. The cost is determined by the working time and efficiency of individual production systems. In order to calculate the cost of work for all processors in FlexSim, add the *Financial Analysis* chart to the *Dashboard* and specify the cost for all *Processors*. This parameter must be defined by adding it from the Toolbox library as a *Performance Measure* variable. Assign this objective function as *Financial Analysis - Total* for the previously defined *Financial Analysis* chart (Fig. 2).

The objective function in this task is minimized as the enterprise is interested in lowering the cost. In addition, the total time of task completion is also minimized. The system time from model operation is not listed as a ready-made function in the drop-down list. It is necessary to use the custom code in the optimizer tab for the *Performance Measures* variable. Enter the function in the code editing window [29]:

```
/**Custom Code*/
treenode datanode = parnode(1);
return time();
```

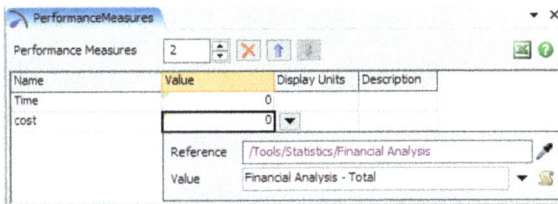

Fig. 2. *Definition of the objective function - the output variable for the total costs and duration of the process (source: own study)*

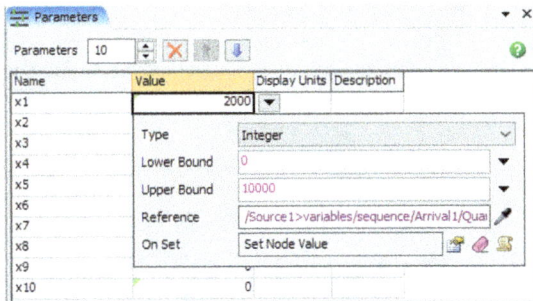

Fig. 3. *Definition of input variables of the objective function (source: own study)*

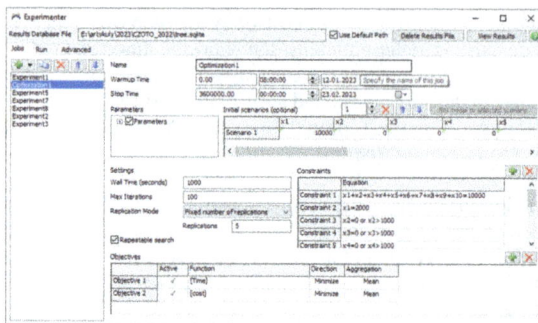

Fig. 4. *FlexSim optimizer settings (source: own study)*

The optimizer needs 10 input parameters, which are the number of flow elements generated by each processor. This is the variable that will be used in the optimization process, as well as the place where the optimizer will generate results from subsequent iterations. This parameter must be defined by adding it from the *Toolbox* library. Then, in the *Value* column, select *Integer* as the type of input variables. The length of the sequence should be set according to the number of products according to the order, i.e. 0÷10000 (Fig. 3).

Boundary conditions and constraints related to the size of the production batch and the production capacity of individual plants should also be specified. Optimizer settings are shown in Figure 4.

The objective function in this task is minimized because the company is interested in reducing the cost. In addition, the total task completion time is also minimized. The optimizer will adjust the values of the flow elements for each *Source* until it finds the optimal value at which the production time and costs are the lowest.

Results Analysis

As a result of the optimizer's work, the best combination of allocation of production orders to individual plants was obtained, while minimizing costs and order execution time. The result of the optimizer's work for 100 iterations, along with the amount of costs and the duration of the production cycle, is shown in Fig. 5.

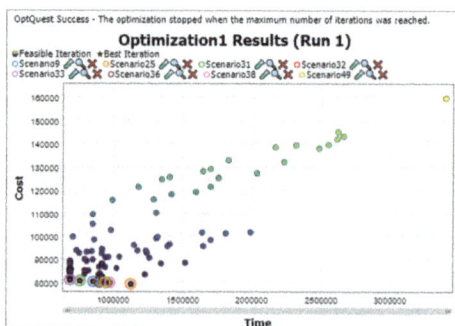

Fig. 5. The result of the optimizer work (source: own study)

Table 2. Summary of the results obtained

Solution	1	2	3	4	5	6	7	8
x1 [pcs]	2000	2000	2000	2000	2000	2000	2000	2000
x2 [pcs]	0	0	0	0	0	0	0	0
x3 [pcs]	0	0	0	0	0	0	0	0
x4 [pcs]	0	0	0	0	0	0	0	0
x5 [pcs]	0	0	0	0	0	0	0	0
x6 [pcs]	0	0	0	0	0	0	0	0
x7 [pcs]	1961	0	2632	1290	3303	1626	1084	1355
x8 [pcs]	0	0	0	0	0	0	0	0
x9 [pcs]	6039	8000	5368	6710	4697	6374	6916	6645
x10 [pcs]	0	0	0	0	0	0	0	0
Time [s]	852906	1129796	758123	947636	687329	900198	976728	938460
Cost [monetary units]	80259	79235	80602	79918	80945	80090	79810	79952

As the best sequence for addressing production orders, the optimizer proposed 8 solutions. In all solutions except 2, all production takes place in 3 plants. The detailed allocation of production orders along with the costs and lead time of a production order are presented in Table 2. Production time depends on many factors, including on the materials used, tools, machines, operator skills, etc. In the quoted calculations, the process time was described by the Uniform distribution function, which will randomly select any numbers from the specified range in accordance with Table 1. The uniform distribution strategy will include numbers with decimal places. Therefore, for the obtained optimal solution, five repetitions (replications) were made and confidence intervals of production costs (Fig. 6a) and execution times (Fig. 6b) were obtained for a single order batch consisting of 10,000 items of products. Graphical interpretation in the form of graphs is shown in Fig. 6.

The presented results were determined as a 95% confidence interval for the studied phenomenon. In practice, this means that there is a 95% probability that the unknown parameter of the population (in this case, the cost and production time will be in the designated numerical range.

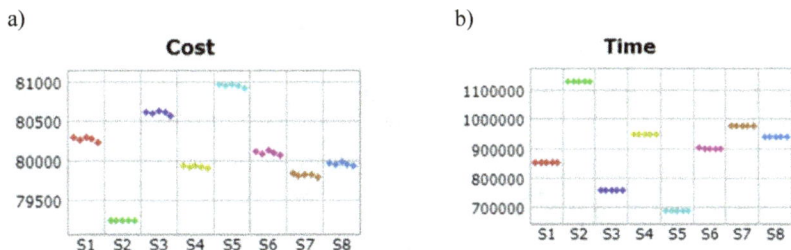

a)

b)

Fig. 6. Confidence intervals for: a) production costs [monetary units], b) execution times [seconds] (source: own study)

Summary
The presented example shows how a discrete event simulation model can improve the planning of the production process. As can be seen in the presented example, looking for savings in the areas

of process engineering allows for a real reduction in production costs, which a large part of companies is not aware of. This factor is of particular importance in the production process supported by external entities, as well as in the logistics sector, where transport, distribution and storage generate high costs. It should also be noted that in addition to costs, production time is also important. It is an indicator of the timely execution of the task, which is particularly important in the case of production in the so-called suction system, i.e. production to order. Such analysis is particularly important in situations where resources and people involved in a given process or operation are shared. Most importantly, however, the simulation shows the impact of plans and schedules on actual performance, and guides managers to choose optimal actions.

References

[1] D. Klimecka-Tatar et al. Sustainable Developement in Logistic – A Strategy for Management in Terms of Green Transport. Manag. Sys. Prod. Eng. 29 (2021) 91-96. https://doi.org/10.2478/mspe-2021-0012

[2] J. Karcz, B. Ślusarczyk. Criteria of quality requirements deciding on choice of the logistic operator from a perspective of his customer and the end recipient of goods, Prod. Eng. Arch. 27 (2021) 58-68. https://doi.org/10.30657/pea.2021.27.8

[3] M. Ingaldi et al. Analysis of problems during implementation of Lean Manufacturing elements, MATEC Web Conf. 183 (2018) art.01004. https://doi.org/10.1051/matecconf/201818301004

[4] R. Ulewicz, R. Kucęba. Identification of problems of implementation of Lean concept in the SME sector, Eng. Manag. Prod. Serv. 8 (2016) 19-25. https://doi.org/10.1515/emj-2016-0002

[5] D. Klimecka-Tatar. Context of production engineering in management model of Value Stream Flow according to manufacturing industry, Prod. Eng. Arch. 21 (2018) 32-35. https://doi.org/10.30657/pea.2018.21.07

[6] K. Knop. Indicating and analysis the interrelation between terms – visual: management, control, inspection and testing, Prod. Eng. Arch. 26 (2020) 110-120. https://doi.org/10.30657/pea.2020.26.22

[7] E. Staniszewska et al. Eco-design processes in the automotive industry, Prod. Eng. Arch. 26 (2020) 131-137. https://doi.org/10.30657/pea.2020.26.25

[8] R. Ulewicz et al. Implementation of Logic Flow in Planning and Production Control, Manag. Prod. Eng. Rev. 7 (2016) 89-94. https://doi.org/10.1515/mper-2016-0010

[9] M. Mazur, H. Momeni. LEAN Production issues in the organization of the company – results, Prod. Eng. Arch. 22(2019) 50-53. https://doi.org/10.30657/pea.2019.22.10

[10] M. Niciejewska et al. Impact of Technical, Organizational and Human Factors on Accident Rate of Small-Sized Enterprises, Manag. Sys. Prod. Eng. 29 (2021) 139-144. https://doi.org/10.2478/mspe-2021-0018

[11] M. Krynke et al. Cost Optimization and Risk Minimization During Teamwork Organization, Manag. Sys. Prod. Eng. 29 (2021) 145-150. https://doi.org/10.2478/mspe-2021-0019

[12] K. Knop. Evaluation of quality of services provided by transport & logistics operator from pharmaceutical industry for improvement purposes, Trans. Res. Procedia 40 (2019) 1080-1087. https://doi.org/10.1016/j.trpro.2019.07.151

[13] M. Ingaldi. A new approach to quality management: conceptual matrix of service attributes, Pol. J. Manag. Stud. 22 (2020) 187-200. https://doi.org/10.17512/pjms.2020.22.2.13

[14] N. Baryshnikova et al. Management approach on food export expansion in the conditions of limited internal demand, Pol. J. Manag. Stud. 21, 2 (2020) 101-114. https://doi.org/10.17512/pjms.2020.21.2.08

[15] K. Knop. Importance of visual management in metal and automotive branch and its influence in building a competitive advantage, Pol. J. Manag. Stud. 22 (2020) 263-278. https://doi.org/10.17512/pjms.2020.22.1.17

[16] R. Ulewicz, M. Blaskova, Sustainable development and knowledge management from the stakeholders' point of view, Pol. J. Manag. Stud. 18 (2018) 363-374. https://doi.org/10.17512/pjms.2018.18.2.29

[17] D. Klimecka-Tatar, M. Ingaldi. Assessment of the Technological Position of a Selected Enterprise in the Metallurgical Industry, Mater. Res. Proc. 17 (2020) 72-78. https://doi.org/10.21741/9781644901038-11

[18] R. Ulewicz. Practical Application of Quality Tools in the Cast Iron Foundry, Manuf. Technol. 14 (2014) 104-111. https://doi.org/10.21062/ujep/x.2014/a/1213-2489/MT/14/1/104

[19] J.M. Garrido. Introduction to Flexsim. In: Object Oriented Simulation: A Modeling and Programming Perspective, J.M. Garrido (Ed.), 31-42, Springer US, 2009. https://doi.org/10.1007/978-1-4419-0516-1

[20] J. Kyncl. Digital Factory Simulation Tools, Manuf. Technol. 16 (2016) 371-375. https://doi.org/10.21062/ujep/x.2016/a/1213-2489/MT/16/2/371

[21] E. Sujová et al. Simulation Models of Production Plants as a Tool for Implementation of the Digital Twin Concept into Production, Manuf. Technol. 20 (2020) 527-533. https://doi.org/10.21062/mft.2020.064

[22] C. Zhuang et al. Digital twin-based smart production management and control framework for the complex product assembly shop-floor, J. Adv. Manuf. Technol. 96 (2018) 1149-1163. https://doi.org/10.1007/s00170-018-1617-6

[23] M. Krynke. Risk Management in the Process of Personnel Allocation to Jobs, System Safety: Human - Technical Facility - Environment 2 (2020) 82-90. https://doi.org/10.2478/czoto-2020-0011

[24] M. Krynke, K. Mielczarek. Applications of linear programming to optimize the cost-benefit criterion in production processes, MATEC Web Conf. 183 (2018) art.04004. https://doi.org/10.1051/matecconf/201818304004

[25] I. Kaczmar. The use of simulation and optimization in managing the manufacturing process – case study, Gospodarka Materiałowa i Logistyka 2016 (4) (2016) 21-28.

[26] [26] S. Borkowski et al. The use of 3x3 matrix to evaluation of ribbed wire manufacturing technology, METAL 2012 – 21st Int. Conf. Metall. Mater. (2012), Ostrava, Tanger 1722-1728.

[27] K. Czerwinska et al. Improving quality control of siluminial castings used in the automotive industry, METAL 2020 – 29th Int. Conf. Metall. Mater. (2020) 1382-1387. https://doi.org/10.37904/metal.2020.3661

[28] A. Pacana et al. Analysis of quality control efficiency in the automotive industry, Transp. Res. Procedia 55 (2021) 691-698. https://doi.org/10.1016/j.trpro.2021.07.037

[29] M. Zenkiewicz et al. Electrostatic separation of binary mixtures of some biodegradable polymers and poly(vinyl chloride) or poly(ethylene terephthalate), Polimery/Polymers 61 (2016) 835-843. https://doi.org/10.14314/polimery.2016.835

[30] D. Siwiec et al. Improving the process of achieving required microstructure and mechanical properties of 38mnvs6 steel, METAL 2020 29th Int. Conf. Metall. Mater. (2020) 591-596. https://doi.org/10.37904/metal.2020.3525

[31] P. Jonšta et al. The effect of rare earth metals alloying on the internal quality of industrially produced heavy steel forgings, Materials 14 (2021) art.5160. https://doi.org/10.3390/ma14185160

[32] T. Lipinski et al. Influence of oxygen content in medium carbon steel on bending fatigue strength, Eng. Rural Develop. 21 (2022) 351-356. https://doi.org/10.22616/ERDev.2022.21.TF116

[33] N. Radek et al. The influence of plasma cutting parameters on the geometric structure of cut surfaces, Mater. Res. Proc. 17 (2020) 132-137. https://doi.org/10.21741/9781644901038-20

[34] N. Radek et al. Microstructure and tribological properties of DLC coatings, Mater. Res. Proc. 17 (2020) 171-176. https://doi.org/10.21741/9781644901038-26

[35] N. Radek et al. Influence of laser texturing on tribological properties of DLC coatings, Prod. Eng. Arch. 27 (2021) 119-123. https://doi.org/10.30657/pea.2021.27.15

[36] N. Radek et al. Operational properties of DLC coatings and their potential application, METAL 2022 – 31st Int. Conf. Metall. Mater. (2022) 531-536. https://doi.org/10.37904/metal.2022.4491

[37] G. Barucca et al. The potential of Λ and Ξ- studies with PANDA at FAIR, Europ. Phys. J. A 57 (2021) art.154 https://doi.org/10.1140/epja/s10050-021-00386-y

[38] M. Domagala et al. The Influence of Oil Contamination on Flow Control Valve Operation, Mater. Res. Proc. 24 (2022) 1-8. https://doi.org/10.21741/9781644902059-1

[39] N. Radek et al. The impact of laser welding parameters on the mechanical properties of the weld, AIP Conf. Proc. 2017 (2018) art.20025. https://doi.org/10.1063/1.5056288

[40] N. Radek et al. Properties of Steel Welded with CO2 Laser, Lecture Notes in Mechanical Engineering (2020) 571-580. https://doi.org/10.1007/978-3-030-33146-7_65

[41] R. Ulewicz, M. Mazur. Economic aspects of robotization of production processes by example of a car semi-trailers manufacturer, Manufacturing Technology 19 (2019) 1054-1059. https://doi.org/10.21062/ujep/408.2019/a/1213-2489/MT/19/6/1054

[42] N. Radek, R. Dwornicka. Fire properties of intumescent coating systems for the rolling stock, Commun. – Sci. Lett. Univ. Zilina 22 (2020) 90-96. https://doi.org/10.26552/com.C.2020.4.90-96

[43] S. Marković et al. Exploitation characteristics of teeth flanks of gears regenerated by three hard-facing procedures, Materials 14 (20210 art. 4203. https://doi.org/10.3390/ma14154203

[44] J. Pietraszek, E. Skrzypczak-Pietraszek. The uncertainty and robustness of the principal component analysis as a tool for the dimensionality reduction. Solid State Phenom. 235 (2015) 1-8. https://doi.org/10.4028/www.scientific.net/SSP.235.1

[45] R. Dwornicka, J. Pietraszek. The outline of the expert system for the design of experiment, Prod. Eng. Arch. 20 (2018) 43-48. https://doi.org/10.30657/pea.2018.20.09

[46] J. Pietraszek et al. Challenges for the DOE methodology related to the introduction of Industry 4.0. Prod. Eng. Arch. 26 (2020) 190-194. https://doi.org/10.30657/pea.2020.26.33

[47] B. Jasiewicz et al. Inter-observer and intra-observer reliability in the radiographic measurements of paediatric forefoot alignment, Foot Ankle Surg. 27 (2021) 371-376. https://doi.org/10.1016/j.fas.2020.04.015

[48] J. Pietraszek. The modified sequential-binary approach for fuzzy operations on correlated assessments, LNAI 7894 (2013) 353-364. https://doi.org/10.1007/978-3-642-38658-9_32

[49] J. Pietraszek et al. Non-parametric assessment of the uncertainty in the analysis of the airfoil blade traces, METAL 2017 – 26th Int. Conf. Metall. Mater. (2017) 1412-1418. ISBN 978-8087294796

[50] J. Pietraszek et al. The non-parametric approach to the quantification of the uncertainty in the design of experiments modelling, UNCECOMP 2017 Proc. 2nd Int. Conf. Uncert. Quant. Comput. Sci. Eng. (2017) 598-604. https://doi.org/10.7712/120217.5395.17225

[51] M. Matuszny. Building decision trees based on production knowledge as support in decision-making process, Prod. Eng. Arch. 26 (2020) 36-40. https://doi.org/10.30657/pea.2020.26.08

[52] M. Beaverstock et al. Applied Simulation: Modeling and Analysis Using FlexSim. BookBaby, Pennsauken Township, 2018. ISBN 978-0983231974

[53] M. Drbúl et al. Simulation Possibilities of 3D Measuring in Progressive Control of Production, Manufacturing Technology 16 (2016) 53-58. https://doi.org/10.21062/ujep/x.2016/a/1213-2489/MT/16/1/53

[54] I. Kaczmar. Komputerowe modelowanie i symulacje procesów logistycznych w środowisku Flexsim. PWN, Warszawa, 2019. ISBN 978-8301205447

[55] S. Setamanit. Evaluation of outsourcing transportation contract using simulation and design of experiment, Pol. J. Manag. Stud. 18 (2018) 300-310. https://doi.org/10.17512/pjms.2018.18.2.24

[56] M. Krynke et al. Analysis of the Problem of Staff Allocation to Work Stations, QPI 2021 Qual. Prod. Improv. (2019) 545-550. https://doi.org/10.2478/cqpi-2019-0073

[57] T.D.C. Le et al. Optimal vehicle route schedules in picking up and delivering cargo containers considering time windows in logistics distribution networks: A case study, Prod. Eng. Arch. 26 (2020) 174-184. https://doi.org/10.30657/pea.2020.26.31

[58] J. Kyncl et al. Tricanter Production Process Optimization by Digital Factory Simulation Tools, Manuf. Technol. 17 (2017) 49-53. https://doi.org/10.21062/ujep/x.2017/a/1213-2489/MT/17/1/49

[59] FlexSim: User manual (2017).

Quality Production Improvement and System Safety: QPI 16 - CZOTO 10   Materials Research Forum LLC
Materials Research Proceedings 34 (2023) 344-353   https://doi.org/10.21741/9781644902691-40

Use of Selected Tools of Quality Improvement in a Company Producing Parts for the Automotive Industry – Case Study

KNOP Krzysztof [a] *

Department of Production Engineering and Safety, Faculty of Management, Czestochowa University of Technology, Al. Armii Krajowej 19b, 42-218 Czestochowa, Poland

[a]krzysztof.knop@wz.pcz.pl

Keywords: Quality Management, Quality Tools, Improvement, Automotive Industry

Abstract. The article is a case study on the use of selected quality management tools in a company from the automotive industry for the purpose of improving the basic production process carried - the assembly process. Quality tools such as Pareto-Lorenz diagram, Ishikawa diagram, and 5WHY method were used. The analyses carried out with the use of quality tools made it possible to identify non-conformities that most often occurred during the assembly process of the tested products, indicate the causes of these non-conformities, and develop suggestions for improvement. The most frequently found non-compliance of products was the disconnection of an element during assembly, due to damage to materials in the warehouse. The root cause turned out to be the lack of control of the state of materials available in the warehouse according to the control plans in force at the plant. The need to introduce changes in the organization of warehouse work, in particular, to increase supervision over the control of the condition of materials and to introduce 100% control of products manufactured in the plant, was indicated as improvement measures.

Introduction

A company that wants to meet market challenges must get to know customers, their needs and expectations, and then, on this basis, develop and manufacture products that will satisfy them [1]. This requires proper quality management. It covers a wide spectrum of issues related to planning, organizing, coordinating and controlling quality assurance activities [2]. Quality management is one of the concepts of broadly understood enterprise management. In the most general sense, it means planning and organizing the enterprise management system in such a way that it covers everything that affects the fulfillment of quality requirements [3]. Quality management is also understood as continuous improvement of products and activities related to their production [4]. In other words, it is the management of resources and processes carried out in the company through the prism of effects directly related to quality [5].

Commitment to quality requires not only a well-designed system that will be implemented in accordance with applicable procedures, but also orientation towards external and, above all, internal customers [6], because they are responsible for the quality of their work, and thus the quality of manufactured products and implemented processes in the company. Thanks to their knowledge, skills and experience, each of the company's employees contributes to the quality of the entire organization [7, 8]. The aim of the article is to present the use of selected quality management tools in a company producing components for the automotive industry in order to improve the main production process in this company – the assembly process. The methodology presented in this article can be similarly applied in other industries. Quality improvement issues are common in the industry [9-11], and correctly identifying the causes of disruptions allows for the reduction of necessary resources [12].

Usually, quality problems originate either in the organizational area or in the material and technological aspect. In the latter case, they are usually related to environmental factors [13,14], excessive tool wear [15,16], or defects in structural joints [17,18]. To counteract these issues, appropriate selection of materials [19-21] can be made, which either directly meet higher requirements or can be enhanced through the application of coatings [22], including electrospark methods [23-25], or modification of the technological characteristics of the coating [26]. Additionally, the morphological features of the coating can be modified [27-29], influencing its friction and hydrodynamic parameters. Separation from undesirable factors [30,31] is also possible. Complex technological methods primarily require a reduction in the number of analyzed factors [32,33], after which optimization and stabilization of the process can be achieved using statistical tools, in classical form (factorial, RSM, Taguchi) [34,35], or nonparametric methods [36]. The result of such actions is more reliable machinery and equipment [37,38], vehicles [39], and BIM elements [40,41]. High-quality solutions also become attractive to military customers, who have very high-quality requirements [42-44].

Methodolodgy

The aim of the work was to present the results of the analysis of nonconformities in the production of axle shafts for drive systems of passenger cars. The article is a case-study from a company producing products for the automotive industry. The nonconformities analysis was carried out using selected quality tools, such as the Pareto-Lorenz diagram, Ishikawa diagram and 5WHY method.

The analyzed company is a manufacturer of products for the automotive industry. The plant mainly produces driveshafts for passenger cars, which were the subject of the nonconformity analysis. During the performance of individual assembly operations of the driveshafts, nonconformities occur, lead to the fact that the assembly process ends with the production of a defective product. In the automotive industry, accuracy and precision are of particular importance. They determine the safety of the driver and other road users. With this in mind, the analyzed company pays special attention to ensuring high quality of all products manufactured in the plant. For this purpose, the company uses appropriate tools and methods of quality improvement. A Pareto-Lorenz diagram was used to identify the most common nonconformities during the assembly operations of the axle shafts for drive systems. The Pareto-Lorenz diagram is a tool that allows data on emerging problems to be recorded and analyzed so that the most significant problem areas are highlighted [45,46].

Ensuring the high quality of driveshafts required by customers requires also identifying the reasons for the nonconformities found. For this purpose, it is necessary to conduct a cause-and-effect analysis using, for example, a Ishikawa diagram. The analysis is based mostly on five main areas of the problem, such as: 1. Man, 2. Machine, 3. Material, 4. Method, 5. Management, within which there are looking for probable causes of the problem [47,48]. In the cause-and-effect analysis in the company, the following six main areas of the problem were used, such as: 1. Man, 2. Machine, 3. Environment, 4. Material, 5. Method, 6. Measurement. The result of this analysis allowed to indicate the main - root cause of nonconformities, which most often occur during the assembly process and, based on the obtained results, enabled the implementation of corrective actions.

Determining the root cause of nonconformities when installing halfshafts to drive systems requires identifying the source of the problem and the reasons why the problem was not previously diagnosed. For this purpose, the 5 Why method was used in the analyzed company. Its primary goal of using is to find the exact, fundamental reason that causes a given problem by asking a sequence of "why" questions [49,50]. It was carried out in the company in three stages, namely: 1. Determining the reasons for the identified nonconformities, 2. Determining the frequency of nonconformities, 3. Determining why the problem had not been noticed earlier.

Quality Production Improvement and System Safety: QPI 16 - CZOTO 10   Materials Research Forum LLC
Materials Research Proceedings 34 (2023) 344-353   https://doi.org/10.21741/9781644902691-40

Results

The results of the nonconformities analysis for the selected driveshafts with the use of Pareto-Lorenz diagram are presented on Fig. 1. The analysis covered a period of three months of the selected year. The analysis covered 112 driveshafts manufactured in the plant. The results constituting the basis for the analysis were collected during the quality control process carried out after the completion of individual assembly operations and the quality control of the finished product. Most of the nonconformities were found during the quality control of components produced after the completion of individual assembly operations. Most of the problems involved disconnecting one of the mounted elements. Others related to damage to one of the assembled elements. Finished product quality control showed that the complete driveshaft is too loud, the vibration level is too high, or the sensor is too loud. During the assembly of driveshafts for drive systems, 88.76% of defective products identified are caused by one type of nonconformity - disconnection of one of the elements during the assembly operation of the subassembly. The remaining five types of diagnosed nonconformities were the cause of only 11.24% of the identified assembly defects. As a result, measures have been taken to reduce nonconformities related to component detachment during assembly operations of the driveshaft subassemblies. This required identifying assembly operations where such nonconformities most often occur, determining their causes and introducing corrective actions that will allow for their elimination or at least limitation. The results of the Pareto-Lorenz analysis showed that in the analyzed plant, during the assembly of half-shafts for drive systems, the most frequently found nonconformity is the disconnection of one of the elements during assembly operations, which results in the creation of a specific subassembly. This disconnection is not caused by an incorrect implementation of the assembly operation, but by a large amount of damage to the materials used for assembly. Therefore, an attempt was made to determine the main causes of this problem. First, using a teamwork and brainstorming technique, all potential causes that could lead to quality issues from the warehouse's point of view were identified. The identified, probable causes of nonconformity were divided into six categories, specifying the reasons resulting from: 1. People's work, 2. Machines used, 3. Work environment, 4. Material used, 5. Method used, 6. Measurements used. Subsequently, all identified potential causes causing the identified nonconformity were assigned to six categories. Assigning all diagnosed causes to individual categories allowed to develop the Ishikawa cause and effect diagram. It was presented in Fig.2. The analysis of the diagram made it possible to identify the main cause of damage to the materials used in assembly operations.

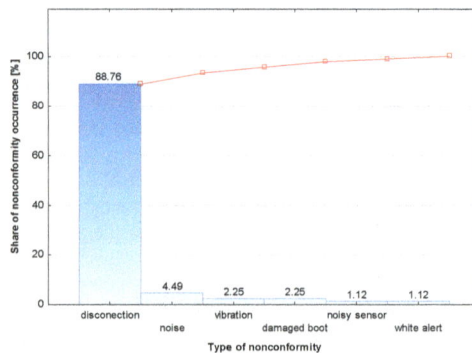

Fig.1. *Pareto-Lorenz diagram for nonconformities found in the process of assembling half shafts to drive systems.*

Quality Production Improvement and System Safety: QPI 16 - CZOTO 10   Materials Research Forum LLC
Materials Research Proceedings 34 (2023) 344-353   https://doi.org/10.21741/9781644902691-40

The conducted Ishikawa cause and effect analysis allow concluding that the damage to the materials used to assemble the axle shafts is mainly due to environmental causes. They relate to the conditions prevailing in warehouses where materials used in assembly operations are collected. It was found that the reasons for these are: improper lighting, and thus poor visibility during transport, high air humidity in the warehouse causing corrosion of materials used in assembly operations, slippery surfaces in the warehouse due to sub-zero temperatures, and high humidity causing rapid wear of wooden containers.

As previously emphasized, damage to the materials used to assemble the axle shafts causes the most common error, which is the detachment of the component during assembly. Therefore, an attempt was made to explain why this nonconformity occurs. Damage to the materials used for assembly is indeed a significant cause, but it should be noticed by employees preparing materials for assembly operations. An in-depth analysis of each of the reasons for disconnecting a component during assembly has shown that the use of such materials during assembly is the responsibility of employees who do not check the condition of materials in warehouses by the control plans in force at the plant, and in the course of performing their duties (picking containers from warehouses) are not focused on their work (conduct phone calls). This results in the fact that very often containers with link shafts are placed in the wrong storage area. This is the main cause of nonconformities during the assembly process of driveshafts for drive systems. They were highlighted in the Fig.2.

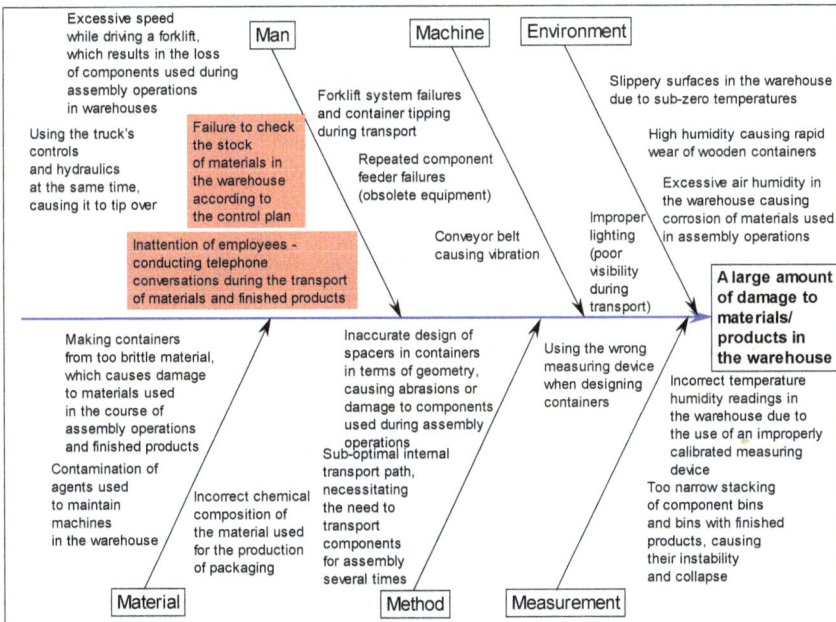

Fig.2. *Ishikawa cause and effect diagram for nonconformities arising during the process of assembling half shafts to drive systems, with an indication of the main causes causing the element to detach during assembly (highlighted in the figure).*

The development and implementation of corrective actions to eliminate the indicated causes of nonconformity during assembly require determining the reasons for the control of the condition of materials in warehouses in accordance with the control plans in force at the plant.

The root cause of the nonconformities during the assembly of driveshafts to drivetrains was identified using the 5 WHY method, based on three stages.

The first stage of the analysis consisted in identifying the causes of nonconformities that were found during the assembly of the driveshaft. At this stage, it was determined what happened in the plant, that the stock of materials in the warehouse was not inspected. Solving this problem required the appointment of a team of employees. It consisted of two warehouse employees, a forklift operator collecting materials from the warehouse for the assembly of axle shafts, and three employees involved in the implementation of individual stages of the assembly process. The team in this composition undertook work on the analysis of nonconformities occurring during assembly. It showed that during assembly operations, the element is most often disconnected. It should be noted that this may be due to a defect in the material used for assembly, and not an error during the assembly operations. The team then began work on determining the possible causes of this problem. There are 18 possible reasons why the material used to assemble the axle shafts are damaged. The condition and quality of deliveries of individual materials as well as the method of their storage and transport to assembly lines were analyzed. The analyzes carried out showed that the main reason for the most frequently found problem, i.e. disconnection of the element, is damage to the materials, which in turn result from the fact that the plant does not carry out inventories of materials by the applicable quality control plan. Carrying out such inventories would avoid the use of inappropriate materials that do not meet quality requirements during assembly.

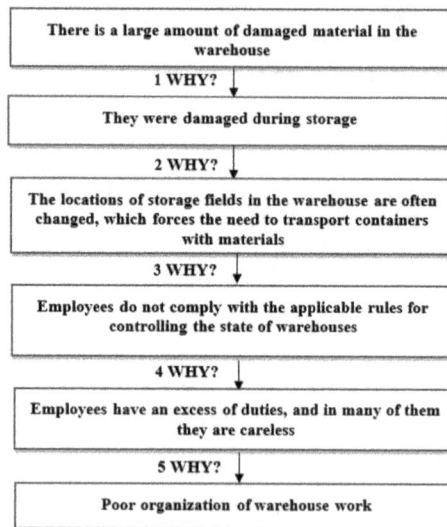

Fig.3. *5WHY analysis for the occurrence of a large amount of damage to the materials used in the assembly process of half shafts for drive systems.*

The second stage of the analysis was to determine the frequency of nonconformities found. The analysis of the quality control cards for the finished product and the quality control cards for the subassemblies made after the completion of the individual stages of the assembly process, carried out by the appointed team of employees - the pressing of the cross, the assembly of the cover and the pressing of the joint, showed that in 79.46% of cases, this nonconformity was found. It was recognized that the problem required an immediate solution, the more so that the disconnection of any element causes the finished product to be defective and cannot be used.

The third stage of the analysis was to determine why this problem had not been identified earlier. After analyzing all possible causes, the appointed team concluded that the control of the stock of materials in the warehouse is not carried out properly, i.e. according to the quantity and quality of materials in force at the plant. This required finding the source of this cause. For this purpose, the 5WHY analysis was carried out. By asking the question "why?" five times the main reason for the occurrence of the most frequently found nonconformities was indicated. This analysis is presented in Fig.3. It showed that the root cause of the identified problem is the poor organization of the warehouse work. This applies both to non-compliance with the applicable rules of quality control of materials in the warehouse, as well as the improper performance of the employees' work.

To eliminate or at least reduce the identified nonconformity, it is proposed to increase supervision over employees. First of all, it is necessary to supervise the control of warehouse stock, especially the quality of materials used for assembly processes. It is proposed that the quantity control of materials and the correct location of containers in the warehouse be controlled before the end of each work shift, i.e. three times a day by appropriately selected employees. These employees should take full responsibility for the implementation of the control plans. It is equally important to introduce supervision over employees employed in the warehouse and transporting materials to assembly lines. There should be a ban on using the plant from private mobile phones. Conversations made by employees during work cause them to be inattentive, which results in making numerous mistakes, especially incorrect placement of containers with materials in the warehouse. Implementation of the proposed changes can significantly contribute to reducing damage to materials in the warehouse, and thus the use of materials that meet quality requirements in assembly processes. This will reduce the number of the most common nonconformities, i.e. detachment of the element during assembly.

Summary
The qualitative analysis carried out with the use of selected quality tools showed that there were many nonconformities during the implementation of assembly processes in the analyzed company. It has been shown that the most common error found is the disconnection of the element during assembly. This makes it impossible to carry out further assembly operations, and in the worst case, it even leads to the production of a defective product, inconsistent with the recipient's requirements. As a result of the cause-and-effect analysis, it turned out that the reason for this nonconformity is damage to the materials used for assembly. The conducted analysis also showed that these damages were most often caused by reasons inherent in the working environment. These were: improper lighting, and thus poor visibility of components during transport, high air humidity in the warehouse causing corrosion of materials used in assembly operations, slippery surfaces in the warehouse caused by sub-zero temperatures, and high humidity causing rapid wear of wooden containers. In-depth analyzes allowed to identify the most important cause of the largest number of nonconformities arising during the assembly process. It was the lack of control of the state of materials available in the warehouse according to the control plans in force at the plant. The lack of these controls results in the use of incorrect materials during assembly, which leads to the production of a defective product. Elimination or at least limitation of identified nonconformities requires the implementation of improvement actions. On the basis of the analyzes carried out, it was concluded that they should include changes in the organization of warehouse work, in particular increasing supervision over the control of the condition of materials. In order to increase

the productivity of the production plant, highly effective quality control should be implemented [51]. This control should be carried out by competent employees, in accordance with the control plans in force at the plant. It was also recognized that the elimination or at least reduction of nonconformities that occur during assembly requires increased supervision over employees both in the warehouse and those responsible for the internal transport of materials and employees performing assembly operations. It is also essential to carry out 100% inspections of the driveshafts produced in-house. Implementation of the proposed actions can significantly contribute to improving the quality of products manufactured by the analyzed plant. One way to improve the assembly process can also be the use of optimization tools, such as the FlexSim discrete event simulation software package [52].

To sum up, quality management in assembly processes refers both to the correct implementation of assembly operations, as well as to the use of appropriate materials, as their incorrect selection or damage during assembly may cause inconsistencies that cause the final product to be defective.

References

[1] D. Hoyle. Quality Management Essentials. 1st Edition, Butterworth-Heinemann, New York, 2007. ISBN 978-0750667869

[2] L. Carpinetti et al. Quality management and improvement: A framework and a business-process reference model, Bus. Proc. Manag. J. 9 (2003) 543-554. https://doi.org/10.1108/14637150310484553

[3] M. Mas-Machuca et al. Quality management: a compulsory requirement to achieve effectiveness, Total Qual. Manag. Bus. Excell. 32 (2018) 1-20. https://doi.org/10.1080/14783363.2018.1548275.

[4] M. Potkany et al. Influence of Quality Management Practices on the Business Performance of Slovak Manufacturing Enterprises, Acta Polytech. Hungarica 17 (202) 161-180. https://doi.org/10.12700/APH.17.9.2020.9.9

[5] M. Potkany et al. Nature and Potential Barriers of Facility Management in Manufacturing Enterprises, Pol. J. Manag. Stud. 23 (2021) 327-340. https://doi.org/10.17512/pjms.2021.23.1.20

[6] G. Smith. The meaning of quality, Total Qual. Manag. 4 (1993) 235-244. https://doi.org/10.1080/09544129300000038

[7] S. Zembski, R. Ulewicz. Managing the Process of Adjusting the Competences of Employees Aged Forty or Over to the Requirements of Industry 4.0, Technical Transactions 119 (2022) art. e2022015. https://doi.org/10.37705/TechTrans/e2022015

[8] Z. Stacho et al. The Organizational Culture as a support of Innovation Processes´ Management: A Case Study, Int. J. Qual. Res. 10 (2016), 769-783. https://doi.org/10.18421/IJQR10.04-08

[9] S. Borkowski et al. The use of 3x3 matrix to evaluation of ribbed wire manufacturing technology, METAL 2012 - 21st Int. Conf. Metallurgy and Materials (2012), Ostrava, Tanger 1722-1728.

[10] K. Czerwinska et al. Improving quality control of siluminial castings used in the automotive industry, METAL 2020 29th Int. Conf. Metall. Mater. (2020) 1382-1387. https://doi.org/10.37904/metal.2020.3661

[11] A. Pacana et al. Analysis of quality control efficiency in the automotive industry, Transp. Res. Procedia 55 (2021) 691-698. https://doi.org/10.1016/j.trpro.2021.07.037

Quality Production Improvement and System Safety: QPI 16 - CZOTO 10   Materials Research Forum LLC
Materials Research Proceedings 34 (2023) 344-353   https://doi.org/10.21741/9781644902691-40

[12] A. Maszke et al. Problems in the implementation of the lean concept at a steel works – Case study, MATEC Web of Conf. 183 (2018) art.01014. https://doi.org/10.1051/matecconf/201818301014

[13] T. Lipinski, J. Pietraszek. Influence of animal slurry on carbon C35 steel with different microstructure at room temperature, Engineering for Rural Development 21 (2022) 344-350. https://doi.org/10.22616/ERDev.2022.21.TF115

[14] T. Lipiński, J. Pietraszek. Corrosion of the S235JR Carbon Steel after Normalizing and Overheating Annealing in 2.5% Sulphuric Acid at Room Temperature, Mater. Res. Proc. 24 (2022) 102-108.

[15] S. Marković et al. Exploitation characteristics of teeth flanks of gears regenerated by three hard-facing procedures, Materials 14 (20210 art. 4203. https://doi.org/10.3390/ma14154203

[16] P. Regulski, K.F. Abramek The application of neural networks for the life-cycle analysis of road and rail rolling stock during the operational phase, Technical Transactions 119 (2022) art. e2022002. https://doi.org/10.37705/TechTrans/e2022002

[17] N. Radek et al. The impact of laser welding parameters on the mechanical properties of the weld, AIP Conf. Proc. 2017 (2018) art.20025. https://doi.org/10.1063/1.5056288

[18] N. Radek et al. Properties of Steel Welded with CO2 Laser, Lecture Notes in Mechanical Engineering (2020) 571-580. https://doi.org/10.1007/978-3-030-33146-7_65

[19] R. Ulewicz et al. Structure and mechanical properties of fine-grained steels, Period. Polytech. Transp. Eng. 41 (2013) 111-115. https://doi.org/10.3311/PPtr.7110

[20] P. Jonšta et al. The effect of rare earth metals alloying on the internal quality of industrially produced heavy steel forgings, Materials 14 (2021) art.5160. https://doi.org/10.3390/ma14185160

[21] T. Lipinski, J. Pietraszek, A. Wach. Influence of oxygen content in medium carbon steel on bending fatigue strength, Engineering for Rural Development 21 (2022) 351-356. https://doi.org/10.22616/ERDev.2022.21.TF116

[22] N. Radek et al. Technology and application of anti-graffiti coating systems for rolling stock, METAL 2019 28th Int. Conf. Metall. Mater. (2019) 1127-1132. ISBN 978-8087294925

[23] N. Radek, K. Bartkowiak. Laser Treatment of Electro-Spark Coatings Deposited in the Carbon Steel Substrate with using Nanostructured WC-Cu Electrodes, Physics Procedia 39 (2012) 295-301. https://doi.org/10.1016/j.phpro.2012.10.041

[24] N. Radek et al. The effect of laser treatment on operational properties of ESD coatings, METAL 2021 30th Ann. Int. Conf. Metall. Mater. (2021) 876-882. https://doi.org/10.37904/metal.2021.4212

[25] N. Radek et al. The impact of laser processing on the performance properties of electro-spark coatings, 14th World Congress in Computational Mechanics and ECCOMAS Congress 1000 (2021) 1-10. https://doi.org/10.23967/wccm-eccomas.2020.336

[26] N. Radek et al. Microstructure and tribological properties of DLC coatings, Mater. Res. Proc. 17 (2020) 171-176. https://doi.org/10.21741/9781644901038-26

[27] N. Radek et al. The WC-Co electrospark alloying coatings modified by laser treatment, Powder Metall. Met. Ceram. 47 (2008) 197-201. https://doi.org/10.1007/s11106-008-9005-7

[28] N. Radek et al. The influence of plasma cutting parameters on the geometric structure of cut surfaces, Mater. Res. Proc. 17 (2020) 132-137. https://doi.org/10.21741/9781644901038-20

[29] P. Kurp, H. Danielewski. Metal expansion joints manufacturing by a mechanically assisted laser forming hybrid method – concept, Technical Transactions 119 (2022) art. e2022008. https://doi.org/10.37705/TechTrans/e2022008

[30] M. Zenkiewicz et al. Electrostatic separation of binary mixtures of some biodegradable polymers and poly(vinyl chloride) or poly(ethylene terephthalate), Polimery/Polymers 61 (2016) 835-843. https://doi.org/10.14314/polimery.2016.835

[31] M. Zenkiewicz et al. Modeling electrostatic separation of mixtures of poly(ϵ-caprolactone) with polyfvinyl chloride) or polyfethylene terephthalate), Przemysl Chemiczny 95 (2016) 1687-1692. https://doi.org/10.15199/62.2016.9.6

[32] J. Pietraszek et al. The principal component analysis of tribological tests of surface layers modified with IF-WS2 nanoparticles, Solid State Phenom. 235 (2015) 9-15. https://doi.org/10.4028/www.scientific.net/SSP.235.9

[33] J. Pietraszek, E. Skrzypczak-Pietraszek. The uncertainty and robustness of the principal component analysis as a tool for the dimensionality reduction. Solid State Phenom. 235 (2015) 1-8. https://doi.org/10.4028/www.scientific.net/SSP.235.1

[34] R. Dwornicka, J. Pietraszek. The outline of the expert system for the design of experiment, Prod. Eng. Arch. 20 (2018) 43-48. https://doi.org/10.30657/pea.2018.20.09

[35] J. Pietraszek et al. Challenges for the DOE methodology related to the introduction of Industry 4.0. Prod. Eng. Arch. 26 (2020) 190-194. https://doi.org/10.30657/pea.2020.26.33

[36] J. Pietraszek. The modified sequential-binary approach for fuzzy operations on correlated assessments, LNAI 7894 (2013) 353-364. https://doi.org/10.1007/978-3-642-38658-9_32

[37] A. Goroshko et al. Construction and practical application of hybrid statistically-determined models of multistage mechanical systems, Mechanika 20 (2014) 489-493. https://doi.org/10.5755/j01.mech.20.5.8221

[38] R. Ulewicz, M. Mazur. Economic aspects of robotization of production processes by example of a car semi-trailers manufacturer, Manufacturing Technology 19 (2019) 1054-1059. https://doi.org/10.21062/ujep/408.2019/a/1213-2489/MT/19/6/1054

[39] N. Radek, R. Dwornicka. Fire properties of intumescent coating systems for the rolling stock, Commun. - Sci. Lett. Univ. Zilina 22 (2020) 90-96. https://doi.org/10.26552/com.C.2020.4.90-96

[40] A. Bakowski et al. Frequency analysis of urban traffic noise, ICCC 2019 20th Int. Carpathian Contr. Conf. (2019) 1660-1670. https://doi.org/10.1109/CarpathianCC.2019.8766012

[41] J.M. Djoković et al. Selection of the Optimal Window Type and Orientation for the Two Cities in Serbia and One in Slovakia, Energies 15 (2022) art.323. https://doi.org/10.3390/en15010323

[42] W. Przybył et al. Virtual Methods of Testing Automatically Generated Camouflage Patterns Created Using Cellular Automata, Mater. Res. Proc. 24 (2022) 66-74. https://doi.org/10.21741/9781644902059-11

[43] N. Radek et al. Operational tests of coating systems in military technology applications, Eksploat. i Niezawodn. 25 (2023) art.12. https://doi.org/10.17531/ein.2023.1.12

[44] W. Przybył et al. Microwave absorption properties of carbonyl iron-based paint coatings for military applications, Def. Technol. 22 (2023) 1-9. https://doi.org/10.1016/j.dt.2022.06.013

[45] D.R. Bamford, R.W. Greatbanks. The use of quality management tools and techniques: a study of application in everyday situations, Int. J. Qual. Rel. Manag. 22 (2005) 376–392. https://doi.org/10.1108/02656710510591219

[46] K. Knop. Analysis and Quality Improvement of the UV Printing Process on Glass Packaging, Qual. Prod. Improv. 3 (2021) 314–325. https://doi.org/10.2478/cqpi-2021-0031

[47] G. Paliska, D. Pavletic, M. Sokovic. Quality tools – systematic use in process industry, J. Achiev. Mater. Manuf. Eng. 25 (2007) 79-82.

[48] A. Pacana, R. Ulewicz. Analysis of causes and effects of implementation of the quality management system compliant with ISO 9001, Pol. J. Manag. Stud. 21 (2020) 283-296. https://doi.org/10.17512/pjms.2020.21.1.21

[49] J. Tarí, V. Sabater. Quality tools and techniques: Are they necessary for quality management? Int. J. Prod. Econ. 92 (2004) 267-280. https://doi.org/10.1016/j.ijpe.2003.10.018

[50] K. Knop, R. Ulewicz. Solving Critical Quality Problems by Detecting and Eliminating their Root Causes – Case-Study from the Automotive Industry, Mater. Res. Proc. 24 (2022) 181-188. https://doi.org/10.21741/9781644902059-27

[51] R. Ulewicz et al. Logistic Controlling Processes and Quality Issues in a Cast Iron Foundry, Mater. Res. Proc. 17 (2020) 65-71. https://doi.org/10.21741/9781644901038-10

[52] M. Krynke. The Use of Computer Simulation Techniques in Production Management, Mater. Res. Proc. 24 (2022) 126-133. https://doi.org/10.21741/9781644902059-19

Quality Production Improvement and System Safety: QPI 16 - CZOTO 10   Materials Research Forum LLC
Materials Research Proceedings 34 (2023) 354-363   https://doi.org/10.21741/9781644902691-41

Geometrical Accuracy of Flexspline Prototypes Made by FDM/MEM Methods

PACANA Jacek[1,a,] DWORNICKA Renata[2,b *] and PACANA Andrzej[3,c]

[1]Rzeszow University of Technology, Rzeszów, Poland; ORCID: 0000-0001-7495-9319

[2]Cracow University of Technology, Kraków, Poland; ORCID: 0000-0002-2979-1614

[3]Rzeszow University of Technology, Rzeszów, Poland; ORCID: 0000-0003-1121-6352

[a]pacanaj@prz.edu.pl, [b *]renata.dwornicka@pk.edu.pl, [c]app@prz.edu.pl

Keywords: 3D Printing, Polymeric Materials, FDM, MEM, Harmonic Drive, Flexspline, Coordinate Measurements

Abstract. This article presents an analysis of the geometrical accuracy of flexspline prototypes that were a component of harmonic drive. To produce the test models, a MEM (Melting and Extrusion Modeling) method was used. This additive method allows the production of elements with very complicated shapes from various polymeric materials. As the test model, a flexspline made of ABS (acrylonitrile-butadiene-styrene) was assumed. Such a special gear with a complicated construction was chosen, because its unique design allows one to obtain clear conclusions from the analysis. To expand the scope of research, models of flexspline with four different construction solutions were made and tested. As a step of the analysis, contact measurements were performed for several flexspline models on the coordinate measuring machine to check their dimensions. The verification of the geometrical accuracy of flexspline models will allow to assess the usefulness of additive methods for the production of prototypes and finished products.

Introduction

Rapid prototyping (RP) methods are a group of technologies currently universally used in many sectors of industry [1,2]. Their advantages are not only used in prototyping processes, but are often the main manufacturing method of machine parts. The quality of products manufactured by additive methods depends mainly on the method itself, but also on the materials used in the production process [3-5] and imposing its quality [6,7]. For most manufactured elements, dimensional accuracy is very important. It allows to ensure the basic functions of the product and ensure proper cooperation with other machine components. The diversity of additive methods used for the production of machine components does not allow for their overall assessment. Therefore, for each of them, it should be conducting tests on prototypes in the initial phase of product design regarding their dimensional accuracy and functionality [8,9]. For machine components operating with load, strength and durability analyses should also be carried out.

Rapid prototyping is a technique that supports both the design and implementation processes of new solutions. This significantly saves energy [10-12] and reduces environmental impact [13]. However, depending on the rapid prototyping technology used, it often involves unsatisfactory accuracy and quality [14], which often requires finishing processing. Such an approach often requires organizational changes [15], the development of new failure mode and effects analysis scenarios [16-18], and the implementation of precise measurement and control techniques [19]. Rapid prototyping has a wide range of applications and often involves the application of special coatings [20,21], including DLC [22-24] or electrospark deposition [25,26], modification of the morphology of the surface layer [27], or welding joints [28]. The presented article introduces a prototype of a flexible cup for a harmonic drive, which due to its significant possible gear ratios

Quality Production Improvement and System Safety: QPI 16 - CZOTO 10   Materials Research Forum LLC
Materials Research Proceedings 34 (2023) 354-363   https://doi.org/10.21741/9781644902691-41

and high precision, can find applications in hydraulic control [29-31] and the military sector [32-34]. Due to the multidimensional dependencies, the implementation of formal statistical techniques [35-37], including expert systems support [38] and nonparametric approaches [39-41], is necessary.

Materials and Methods

In the research, the flexspline of the harmonic drive was included. It is a gear with a thin-walled body with a flat bottom at one end and a toothed wheel rim on the other side of it (Fig.1). The unique sleeve design causes very big problems when making a flexspline by classic reductive machining. In addition, during work, it is deformed by a generator, causing its precisely defined deformation [42]. The small-module toothed wheel rim of the flexspline meshes with the circular spline, transferring the torque to the other gear components. Harmonic drives are characterized by very high kinematic accuracy, so to provide it, the geometrical accuracy of the flexspline itself is also necessary. Unfortunately, a big problem is the relatively low durability of these gears, thus new constructions are constantly being designed. The new shapes of flexspline will reduce stress levels in their body, and so increase the strength and durability of harmonic drives. The use additive methods for the production of such special gears offers new technological potentials. Layered adding of next cross sections of the flexsplines allows them to be made in new shapes that were previously impossible. Additive manufacturing does not require any additional tooling or special tools, and many maintenance operations are automatic.

a)   b)

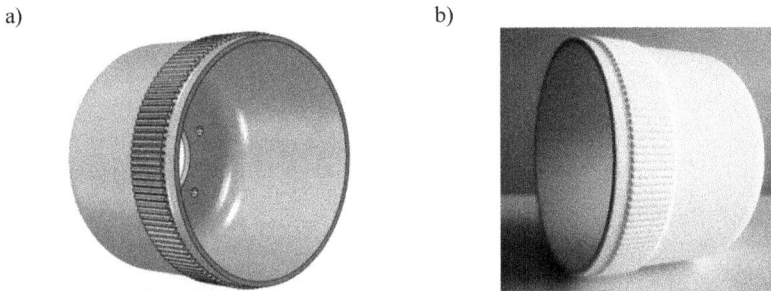

Fig. 1. Standard flexspline of harmonic drive: a) numerical CAD model, b) prototype made by additive method

Prototyping

The MEM method was used to make physical models of the flexspline of harmonic drive. It is a variation of the FDM (Fused Deposition Modeling) method adapted to devices from company TierTime. This method was chosen due to its high availability among additive manufacturing technology and its growing popularity, which guarantees its rapid development [43,44]. The relatively low price of devices used in the FDM method, with many of its advantages, makes it applicable not only for the manufacturing of prototypes but also in the production of finished products. UP Plus 2 and UP mini 2 printers with a single nozzle numerically controlled head were used to produce test models of flexspline.

Because they are printers from the same producer, they have the same dedicated software to prepare models and printing. The UP Studio program (Fig.2a) was therefore used to import models of flexspline as the STL file, prepare the printout and control it. The process of printing a flexspline on the UP mini 2 is shown in Fig.2b.

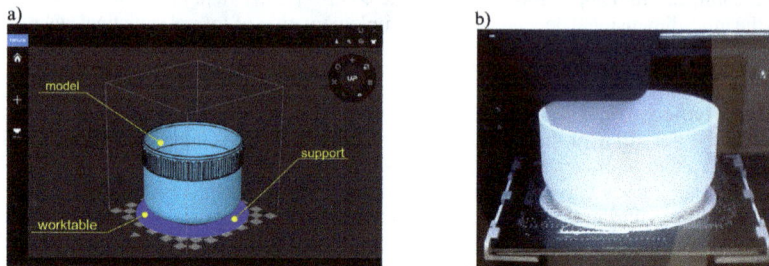

Fig. 2. *Preparation of printing models: a) the UP Studio program with a flexspline model, b) a flexspline model of flexspline, b) model of the flexspline during printing*

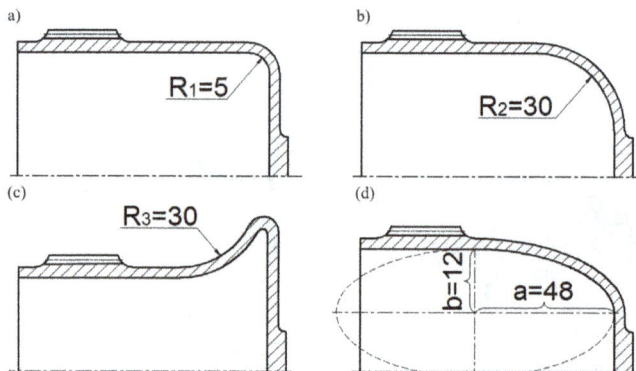

Fig. 3. *Design variants of flexspline models in variants: a) base, b) with enlarged radius, c) with external flange, d) elliptical*

For bench research, several models of flexspline were made, which differed in the shape of the body sleeve. Four variants of construction were selected, together with characteristic geometric parameters, which are presented in Fig.3 with simplified. All wheels were designed based on the same geometric assumptions, and only the dimensions indicated in figure 3were changed. For every of the design variants, models were made while maintaining the same settings of 3D printers and from the same material, which was ABS [45].

Analysis of Prototypes Geometry
After each of the physical models was made, they had to be checked for completeness and correctness of shape. In addition to the general assessment of the condition of the model, the supports had to also be removed and cleaned.

For the prepared physical models of the flexspline, an evaluation of their selected geometrical parameters was performed. Precise control of models produced using the MEM method included two stages:
- initial verification of the main dimensions of the flexspline model.
- checking the accuracy of the surface mapping of the physical model in relation to the CAD output model.

Quality Production Improvement and System Safety: QPI 16 - CZOTO 10   Materials Research Forum LLC
Materials Research Proceedings 34 (2023) 354-363   https://doi.org/10.21741/9781644902691-41

The comparison of the results of the measurements obtained in each step did not show any general differences. However, local divergencies appeared in few cases, resulting from errors characteristic of the applied incremental method. There were sometimes point blobs or gaps of the material on the model, caused by the out-of-control over the melted material in the 3D printing process. A common problem in the manufacturing of models using the FDM/MEM method is the shrinkage of the material that is other than that assumed by the producer. Initial measurements concerned checking the dimensions of the printed models, the wall thickness of the flexspline sleeve, and the diameters of the holes using a caliper. Evaluation of the dimensional accuracy of the printed physical models provided satisfactory results. None of the products had significant differences between the theoretical and the real dimensions. The observed dimensional differences of less than 1% should be taken as a very good result considering the measuring tools used. This proves both the high accuracy of the 3D printers used, and the correct design of the manufacturing process, the proper selection of its parameters, and the use of good materials.

The analysis of geometrical parameters of physical models made by the rapid prototyping method cannot be limited only to general dimensional control. Therefore, the thin-walled models produced by the MEM method were measured precisely using the contact direct method. Measurements were performed using a Roland MDX-40A coordinate milling machine (Fig.4), additionally equipped with a ZSC-1 scanning head [46]. The contact scanning process was chosen due to the high accuracy of the method, and the need to measure models with both large dimensions and detailed toothed wheel rim area. The scan is performed by direct contact of the scanning head tip with the measured material, so it does not require any additional model preparation. Fig.4 shows the process of scanning the flexspline on a coordinate measuring machine. The enlarged fragment (Fig. 4b) shows the tip of the scanning module in relation to the measured teeth with a 0.7mm module.

The scanning process consisted of making measurements in the coordinate system of the CNC machine at the moment of contact of the head with the surface of the model. The scan result is saved in the form of a point cloud, and can also be converted into a triangle mesh, creating a numerical model of the measured part.

Fig. 4. *Measuring a flexspline model on a Roland MDX-40*

The analysis concerned precise linear measurements in maincross sections on models of flexspline. For all tested wheels, a measurement was performed in the plane along their axis, passing through the center of the tooth space. To perform the analysis of dimensions, multiple scans of each of the flexspline physical models were performed, comparing the obtained results. No clear differences were observed between the subsequent measurement results.

Based on the results of the gear measurements, a more accurate analysis of the accuracy of their execution was performed using the specialized GOM Inspect program. It is universally used as inspection software for final quality control on production lines. It allows to download and analyze data from various measuring machines, but also to analyze any surface in the form of a point cloud [47-49]. Measurement results concerning external profiles of flexsplines and also CAD models of flexsplines were imported into the GOM Inspect program. Since the CMM measurement was performed only in the main axis of the flexspline, a cross-section in program was defined on the same plane from the full 3D model. A control cross-section was therefore defined in order to compare the theoretical values with the results measured for the prototype of the flexspline. The GOM Inspect window, with the tool to create a cross section based on a CAD model, is shown in Fig.5.

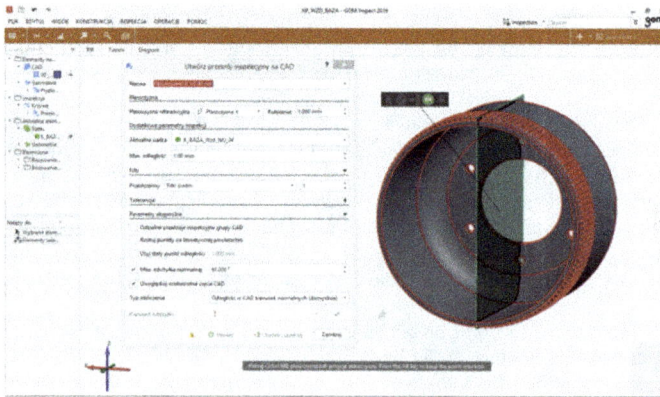

Fig. 5. *View of the GOM Inspect window*
with the active command to define a section on the CAD model

a)   b)

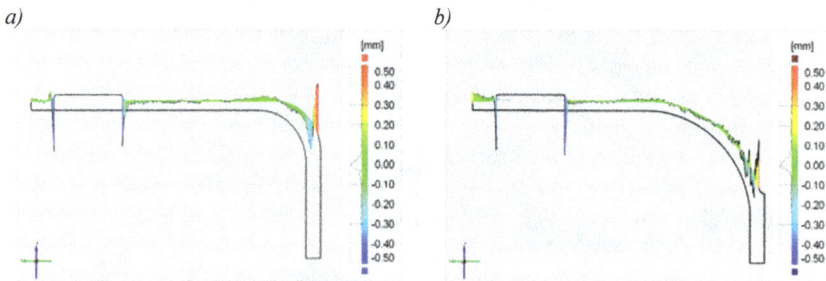

Fig. 6. *Dimensional deviations for the flexspline:*
a) base variant, b) enlarged radius variant

a)   b)

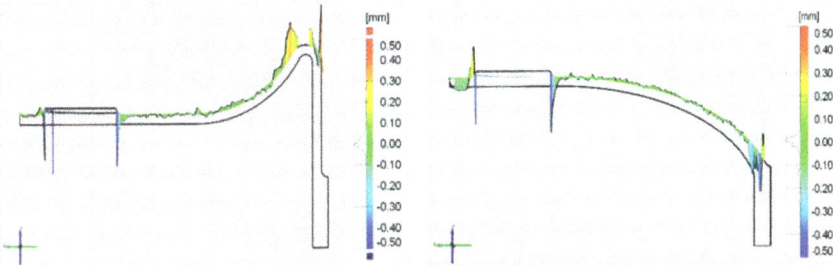

Fig. 7. *Dimensional deviations for the flexspline:*
a) variant with outer flange, b) elliptical variant

The result of the control of the accuracy of the models is presented in the form of deviations between the real and theoretical outline. They can be read for individual points, or in a more representative graphic form as a graph on the created cross-section. Figures 6 and 7 show the deviation diagrams obtained in the GOM Inspect program for the longitudinal cross section of the flexspline of the harmonic drive. It can be seen that the divergencies between the real and theoretical models are not large. Innermost deviations do not exceed 0.05 mm, and these values exist mainly on the body of the flexspline.

The biggest differences are noticeable in the transition radius between the toothed wheel rim and the body sleeve and near the bottom. In these areas, the profile is not parallel to the axis of the wheel, so the layered structure of surfaces strongly affects the values of the measured dimensional deviations. Also, the processing shrinkage of the material, which resulted the shortening of the physical models, has an impact on the divergencies exactly in the width of the gear rim and the total length of the flexspline. Dimensional differences can be also a effect of errors on the external surface of the physical models at the bottom of the flexspline, due to the use of support structures during printing. After removing the supports, small shifts or discontinuities of the material layers often remain in the place of their attachment, which resulted in high values of deviations near the bottom, in the graph regarding the base variant.

The description attached to each of the charts contains a histogram showing the distribution of results for given deviation values. This provides the information that the majority of the profile obtained as a measurement of real models is compatible with the theoretical one, or very close to it. Therefore, the quality of physical models printed using the MEM method from ABS material can be highly evaluated. However, after their production, shrinkages occur and appropriate scaling of CAD models should be applied to avoid dimensional errors in functional prototypes.

Conclusions
The analysis of the dimensional accuracy of the models produced using the FDM/MEM method from ABS polymer confirmed the legitimacy of using 3D printing for the production of functional prototypes of harmonic drives. However, already at the design of the elements, the characteristic material parameters and limitations of additive methods should be implemented. The production of real models from polymer materials should assume the physical characteristics specific to each material. Especially for large models, material shrinkage can significantly affect the dimensions of products. In the case of the MEM method, in which the material is applied in layers, it is also important to position the model relative to the direction of printing. The element should be positioned in such a way that surfaces especially important for strength or appearance are in one layer. In the direction of applying the layers of the model, there are disturbances in the external surfaces, the size of which depends on the resolution of the 3D printer.

Quality Production Improvement and System Safety: QPI 16 - CZOTO 10   Materials Research Forum LLC
Materials Research Proceedings 34 (2023) 354-363   https://doi.org/10.21741/9781644902691-41

The verification showed a relatively high geometrical accuracy of the flexsplines models produced by additive methods. Of course, the dimensional accuracy and external structure of the prototypes provided cannot be compared with the quality of models made of polymer materials by injection molding. Steel models of the flexspline also have much higher dimensional accuracy, quality of external surfaces, and strength. However, only incremental methods allow one to obtain ready-made flexsplines without the use of extra handles, molds, or specialized tools, which greatly reduces production costs. Often, dimensional accuracy is not a main parameter determining the possibilities of using the product. This is the case with the analyzed harmonic drive, which has a unique principle of operation. In this type of gears, there are several or even tens of pairs of teeth that are meshing at the same time. Such a way of working together with the forced deformation of the flexspline causes the initial geometric to be not so important. In the case where the wave gear is not very much loaded, it can be made successfully of polymer materials by additive methods, which is cheaper than other technologies. However, it should be seen that the process of printing wave gear components is much lengthier than, for example, injection into molds. Therefore, additive methods can be used successfully for the production of prototypes and for small production for less demanding finished products. When designing machine elements for production with use one of the rapid prototyping techniques, they should be constructed in such a way as to take full advantage of the benefits, and avoid problems resulting from the limitations of the chosen method.

References

[1] P. Dudek. FDM 3D printing technology in manufacturing composite elements, Arch. Metall. Mater. 58 (2013) 1415-1418. https://doi.org/10.2478/amm-2013-0186

[2] G. Budzik. Possibilities of utilizing 3DP technology for foundry mould making, Arch. Foundry Eng. 7 (2007) 65-68.

[3] A. Pacana, K. Czerwinska, L. Bednarova, Comprehensive improvement of the surface quality of the diesel engine piston, Metalurgija 58 (2019) 329-332.

[4] M. Korzyński, et al. Fatigue strength of chromium coated elements and possibility of its improvement with ball peening, Surface & Coatings Technology 204 (2009) 615-620. https://doi.org/10.1016/j.surfcoat.2009.08.049

[5] A. Pacana et al., Study on improving the quality of stretch film by Shainin method, Przemysl Chemiczny, 93 (2014) 243-245. https://doi.org/10.12916/przemchem.2014.243

[6] A. Gazda et al. Study on improving the quality of stretch film by Taguchi method, Przemysl Chemiczny 92 (2013) 980-982.

[7] A. Pacana et al. Effect of selected factors of the production process of stretch film for its resistance to puncture, Przemysl Chemiczny 93 (2014), 2263-2264. https://doi.org/10.12916/przemchem.2014.2263

[8] J. Pacana, R. Oliwa. Use of rapid prototyping technology in complex plastic structures Part I. Bench testing and numerical calculations of deformations in harmonic drive made from ABS copolymer, Polymers 64 (2019) 56-62. https://doi.org/10.14314/polimery.2019.1.7

[9] J. Pacana, O. Markowska. The analysis of the kinematic accuracy of the actual harmonic drive on a test bench, Advances in Manufacturing Science and Technology 40 (2016) 47-54.

[10] L. Dabek et al. Boiling heat transfer augmentation on surfaces covered with phosphor bronze meshes, MATEC Web of Conf. 168 (2018) art.7001. https://doi.org/10.1051/matecconf/201816807001

[11] M. Szczepaniak et al. T. Use of the maximum power point tracking method in a portable lithium-ion solar battery charger, Energies 15 (2022) art.26. https://doi.org/10.3390/en15010026

[12] S. Maleczek et al. Tests of Acid Batteries for Hybrid Energy Storage and Buffering System—A Technical Approach, Energies 15 (2022) art.3514. https://doi.org/10.3390/en15103514

[13] A. Deja et al. Analysis and assessment of environmental threats in maritime transport, Transp. Res. Procedia 55 (2021) 1073-1080. https://doi.org/10.1016/j.trpro.2021.07.078

[14] K. Knop, R. Ulewicz. Assessment of technology, technological resources and quality in the manufacturing of timber products, Proc. Digitalisation and Circular Economy: Forestry and Forestry Based Industry Implications (2019) 251-256. ISBN 978-9543970421

[15] R. Ulewicz, D. Jelonek, M. Mazur. Implementation of logic flow in planning and production control, Management and Production Engineering Review 7 (2016) 89-94. https://doi.org/10.1515/mper-2016-0010

[16] G. Filo, P. Lempa. Analysis of Neural Network Structure for Implementation of the Prescriptive Maintenance Strategy, Mater. Res. Proc. 24 (2022) 273-280. https://doi.org/10.21741/9781644902059-40

[17] J. Fabis-Domagala, M. Domagala. A Concept of Risk Prioritization in FMEA of Fluid Power Components, Energies 15 (2022) art.6180. https://doi.org/10.3390/en15176180

[18] P. Lempa, G. Filo. Analysis of Neural Network Training Algorithms for Implementation of the Prescriptive Maintenance Strategy, Mater. Res. Proc. 24 (2022) 281-287. https://doi.org/10.21741/9781644902059-41

[19] A. Gadek-Moszczak, P. Matusiewicz. Polish stereology – A historical review, Image Analysis and Stereology 36 (2017) 207-221. https://doi.org/10.5566/ias.1808

[20] N. Radek et al. Technology and application of anti-graffiti coating systems for rolling stock, METAL 2019 28th Int. Conf. Metall. Mater. (2019) 1127-1132. ISBN 978-8087294925

[21] N. Radek, J. Konstanty, J. Pietraszek, Ł.J. Orman, M. Szczepaniak, D. Przestacki. The effect of laser beam processing on the properties of WC-Co coatings deposited on steel. Materials 14 (2021) art. 538. https://doi.org/10.3390/ma14030538

[22] N. Radek et al. Microstructure and tribological properties of DLC coatings, Mater. Res. Proc. 17 (2020) 171-176. https://doi.org/10.21741/9781644901038-26

[23] N. Radek et al. Influence of laser texturing on tribological properties of DLC coatings, Prod. Eng. Arch. 27 (2021) 119-123. https://doi.org/10.30657/pea.2021.27.15

[24] N. Radek et al. Operational properties of DLC coatings and their potential application, METAL 2022 31st Int. Conf. Metall. Mater. (2022) 531-536. https://doi.org/10.37904/metal.2022.4491

[25] N. Radek. Determining the operational properties of steel beaters after electrospark deposition, Eksploatacja i Niezawodnosc 44 (2009) 10-16.

[26] N. Radek et al. The effect of laser treatment on operational properties of ESD coatings, METAL 2021 30th Ann. Int. Conf. Metall. Mater. (2021) 876-882. https://doi.org/10.37904/metal.2021.4212

Quality Production Improvement and System Safety: QPI 16 - CZOTO 10   Materials Research Forum LLC
Materials Research Proceedings 34 (2023) 354-363   https://doi.org/10.21741/9781644902691-41

[27] N. Radek et al. The influence of plasma cutting parameters on the geometric structure of cut surfaces, Mater. Res. Proc. 17 (2020) 132-137. https://doi.org/10.21741/9781644901038-20

[28] N. Radek, J. Pietraszek, A. Goroshko. The impact of laser welding parameters on the mechanical properties of the weld, AIP Conf. Proc. 2017 (2018) art.20025. https://doi.org/10.1063/1.5056288

[29] G. Filo, E. Lisowski, M. Domagała, J. Fabiś-Domagała, H. Momeni. Modelling of pressure pulse generator with the use of a flow control valve and a fuzzy logic controller, AIP Conf. Proc. 2029 (2018) art.20015. https://doi.org/10.1063/1.5066477

[30] G. Barucca et al. The potential of Λ and Ξ- studies with PANDA at FAIR, Europ. Phys. J. A 57 (2021) art.154 https://doi.org/10.1140/epja/s10050-021-00386-y

[31] M. Domagala et al. CFD Estimation of a Resistance Coefficient for an Egg-Shaped Geometric Dome, Appl. Sci. 12 (2022) art.10780. https://doi.org/10.3390/app122110780

[32] W. Przybył et al. Virtual Methods of Testing Automatically Generated Camouflage Patterns Created Using Cellular Automata, Mater. Res. Proc. 24 (2022) 66-74. https://doi.org/10.21741/9781644902059-11

[33] N. Radek et al. Operational tests of coating systems in military technology applications, Eksploat. i Niezawodn. 25 (2023) art.12. https://doi.org/10.17531/ein.2023.1.12

[34] W. Przybył et al. Microwave absorption properties of carbonyl iron-based paint coatings for military applications, Def. Technol. 22 (2023) 1-9. https://doi.org/10.1016/j.dt.2022.06.013

[35] J. Pietraszek, A. Goroshko. The heuristic approach to the selection of experimental design, model and valid pre-processing transformation of DoE outcome, Adv. Mater. Res. 874 (2014) 145-149. https://doi.org/10.4028/www.scientific.net/AMR.874.145

[36] J. Pietraszek, N. Radek, A.V. Goroshko. Challenges for the DOE methodology related to the introduction of Industry 4.0. Prod. Eng. Arch. 26 (2020) 190-194. https://doi.org/10.30657/pea.2020.26.33

[37] B. Jasiewicz et al. Inter-observer and intra-observer reliability in the radiographic measurements of paediatric forefoot alignment, Foot Ankle Surg. 27 (2021) 371-376. https://doi.org/10.1016/j.fas.2020.04.015

[38] R. Dwornicka, J. Pietraszek. The outline of the expert system for the design of experiment, Prod. Eng. Arch. 20 (2018) 43-48. https://doi.org/10.30657/pea.2018.20.09

[39] J. Pietraszek. The modified sequential-binary approach for fuzzy operations on correlated assessments, LNAI 7894 (2013) 353-364. https://doi.org/10.1007/978-3-642-38658-9_32

[40] J. Pietraszek et al. Non-parametric assessment of the uncertainty in the analysis of the airfoil blade traces, METAL 2017 26th Int. Conf. Metall. Mater. (2017) 1412-1418. ISBN 978-8087294796

[41] J. Pietraszek et al. A. Advantages and disadvantages of various uncertainty assessment methods in material and technological predictive models, World Congress in Comput. Mech. and ECCOMAS Congress 1000 (2021) 1-8. https://doi.org/10.23967/wccm-eccomas.2020.053

[42] D. W. Dudley. Harmonic Drive Arrangements, Gear Handbook, McGraw-Hill, New York, 1962.

[43] A. Gądek-Moszczak, N. Radek, S. Wronski, J. Tarasiuk. Application the 3D image analysis techniques for assessment the quality of material surface layer before and after laser treatment, Advanced Materials Research 874 (2014) 133-138.
https://doi.org/10.4028/www.scientific.net/AMR.874.133

[44] C. K. Chua, K. F. Leong. 3D Printing and Additive Manufacturing: Principles and Applications: The 5th Edition of Rapid Prototyping: Principles and Applications, World Scientific, Singapur, 2017. https://doi.org/10.1142/10200

[45] ABS – material data, product card [online]. 2022. [viewed 2022-12-11]. Available from: https://f3dfilament.com/product-category/techniczne/abs-fx

[46] J. Pacana, A. Pacana. The impact of geometry errors on the motion transmission of dual path gearing, Advances in Science and Technology Research Journal 10 (2016) 94-100. https://doi.org/10.12913/22998624/65134

[47] GOM Inspect Software GOM Inspect, Evaluation Software for 3D Point Clouds. [online]. 2022. [viewed 2022-12-11]. Available from: https://www.3dteam.pl/wp-content/uploads/2020/11/GOM-Software.pdf

[48] S. Masood, W. Song. Development of new metal/polymer materials for rapid tooling using fused deposition modeling, Materials Design 25 (2004) 587–94.
https://doi.org/10.1016/j.matdes.2004.02.009

[49] J. Pacana, A. Pacana. Analysis of possibilities of using polymeric materials for testing prototypes of harmonic drive, Materials Research Proceedings 5 (2018) 61-66.
http://dx.doi.org/10.21741/9781945291814-11

Quality Production Improvement and System Safety: QPI 16 - CZOTO 10   Materials Research Forum LLC
Materials Research Proceedings 34 (2023) 364-373   https://doi.org/10.21741/9781644902691-42

Comparing Qualitative Raster Maps

WIECZOREK Małgorzata [1, a] and PRZYBYŁ Wojciech [2, b]

[1] University of Wrocław, Plac Uniwersytecki 1, 50-137 Wrocław, Poland

[2] Military Institute of Engineer Technology, ul. Obornicka 136, 50-961 Wrocław, Poland

[a] malgorzata.wieczorek@uwr.edu.pl, [b] przybyl@witi.wroc.pl

Keywords: Raster Maps, Validation, Filters

Abstract. Spatial diversity of the natural environment can be presented using raster qualitative data. They can be the result of collecting field data or be the result of stochastic modelling of a certain complexity of the environment. One example of such modelling is the determination of similar elements of relief forms on the basis of morphometric variables. The use of unsupervised methods for clustering raster data in modelling can produce different maps. It is necessary to assess the compatibility of the obtained maps in order to assess how significant the differences between them are. The article presents selected stochastic and deterministic methods for assessing the spatial distribution of data. Exemplary methods of measuring variability at the global and local level were discussed.

Introduction

In the era of relatively easy access to numerical data, including raster data, both qualitative and quantitative, the attention of scientists is directed not so much to the ways of obtaining data, but to the ways of using them, analyzing, modelling, and then assessing the results of their validation. Currently, the application of image analysis and classification (i.e. rasters) is used in many fields. The most common methods include the analysis of satellite and meteorological images, morphometric classification, medical image analysis (including tomograms, roentgenograms, ultrasound images, etc.), monitoring of product wear and quality (such as metal products, fabrics, paper, and foodstuffs), as well as analysis of surface images and microscopic images of material structures to monitor tool wear.

Algorithms have already been developed for recognizing objects in satellite images or diagnostic images used in medicine, often using neural networks or fuzzy logic. Comparing two images and assessing whether they are statistically [1] the same when these are modeling results and there is no pattern to relate to is still a challenge, although the topic has already been addressed in the literature [2]. This paper presents two such research situations along with a proposal of measures to assess the similarity of two images. The obtained results and limitations resulting from the measures used were discussed. In the article, the authors focused on exemplary global and local indicators that allow the assessment of differences between two images.

The methodology presented in this article goes beyond its applications solely in the field of topographic maps, as image analysis, particularly for images featuring stochastic fields, can be widely applied in industry and management, significantly impacting product quality enhancement [3-5], as well as modifying management methods [6,7] and related analytics [8,9]. The applications of such analyses encompass various issues, ranging from the assessment of corrosion phenomena [10], both in reference to typical metallic construction materials [11-13] and special alloys [14], to the visual evaluation of separator performance [15], and even to the automatic assessment of surfaces [16], coatings [17-19], and structural welds [20]. Currently, this significantly influences the organization of enterprises that have surpassed the level of Industry 3.0 and are reaching Industry 4.0, impacting the means of automatic control, for instance, in the application of DLC

Quality Production Improvement and System Safety: QPI 16 - CZOTO 10   Materials Research Forum LLC
Materials Research Proceedings 34 (2023) 364-373   https://doi.org/10.21741/9781644902691-42

[21-23] and ESD [24-26] coatings, as well as in the automation of thermal insulation evaluation using infrared cameras [27-29]. It is also a significant tool supporting the detection of machine damage [30,31], particularly hydraulic power equipment [32-34].

Automated methods for processing image maps are also becoming a vital tool in process optimization [35-37], including the application of qualitative and/or subjective factors [38-40], although this typically requires prior strong dimensionality reduction [41,42] to avoid unwanted correlations.

Area of Study

A fragment of the Owl Mountains (a mountain range in the Sudetes in central-southern Poland, located in the Lower Silesian Voivodeship) was selected for the research area with the intervention introduced so as to obtain a flat area. It is an area of 8,725 m in the E-W direction and 6,725 m in the N-S direction (269 rows x 349 columns). Using a digital elevation model (DEM) with a field resolution of 25 m, a k-median clustering with Manhattan metrics method was developed. This grouping was based on morphometric variables calculated from DEM, such as elevation, slope, aspect, curvature.

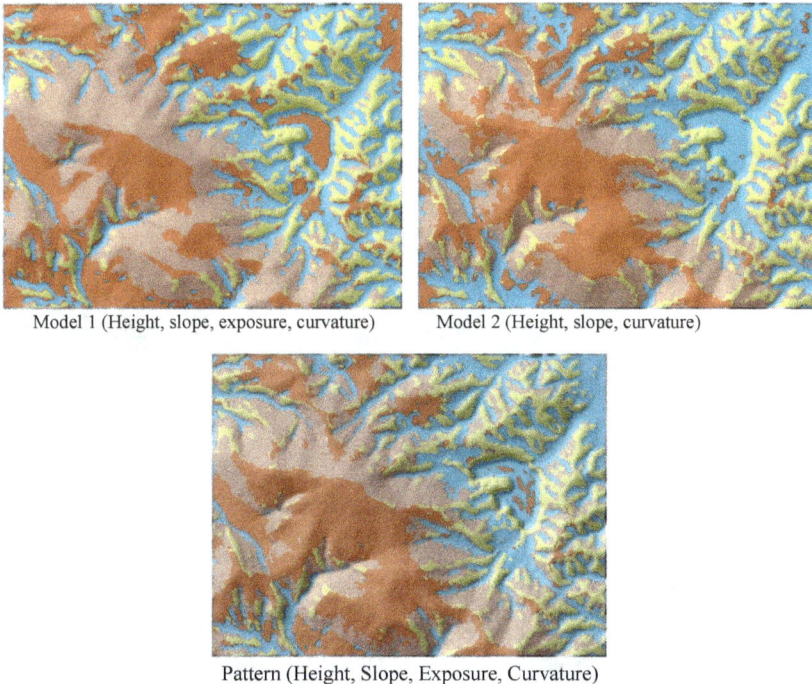

Model 1 (Height, slope, exposure, curvature)   Model 2 (Height, slope, curvature)

Pattern (Height, Slope, Exposure, Curvature)

Fig.1. *Models and pattern of the area subject to classification, divided into classes*

Three images (Fig.1) were selected for the calculations, broken down into 4 classes (clusters) of landform elements: Model 1 – developed on the basis of all 4 morphometric variables, Model 2 – obtained on the basis of height, inclination and curvature, and the third considered as a Pattern - model obtained similarly to model 1, but characterized by the greatest intergroup variability (sum of square between clusters). Exposure is not a morphometric variable that is commonly included

Quality Production Improvement and System Safety: QPI 16 - CZOTO 10   Materials Research Forum LLC
Materials Research Proceedings 34 (2023) 364-373   https://doi.org/10.21741/9781644902691-42

in modelling. Having a tool (apart from the expert's assessment) to assess to what extent the inclusion of an exposure gives a significantly different modelling effect, it would be helpful in assessing the legitimacy of its inclusion in the model. An expert assessment of the advantages and disadvantages of modelling elements of sculpture forms with exposure was presented by Wieczorek and Migoń [43], but advanced similarity measures were not used in this assessment.

Problem to Solve

In this case, we are interested in the answer to the question: How to assess which model fits the pattern better? Although they are not identical, is the scale of similarity large enough to consider them identical models? Where to put this limit of assumed similarity? Obtaining answers to such questions would allow us to decide whether it is worth dividing into classes using 3 or 4 morphometric variables as input.

Another situation in which a measure of statistical significance would be needed in determining the differences between two rasters with qualitative data is the processing of maps by means of filtering. The key question is: After how many filters do we get an image that is too smooth? Fig.2 shows the pattern and its modification by carrying out the filtering procedure 5 times. The agreement between the two images is 92.6% (explained later in the article). The parameters set during filtering (the size of the filtering window and the filtering method), the number of filters and the homogeneity of the image are also important. The smaller information noise, the greater the differences between the original and the final image – however, we leave these considerations for another research study.

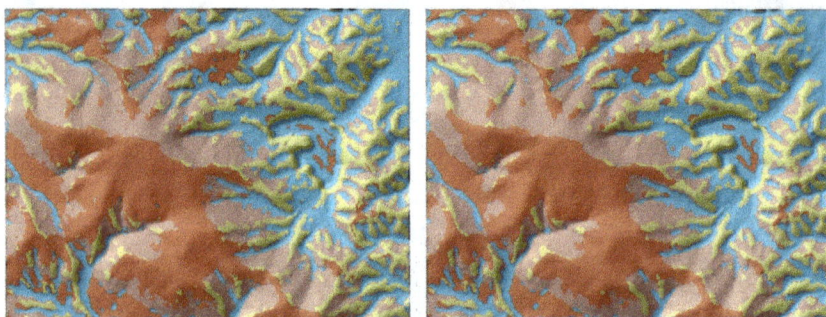

Fig.2. *Pattern and its modification after 5-fold majority filtration (majority in 3x3 window).*

Research Procedure

Map comparison procedures are derived from four spatial data analysis traditions:

1. An accuracy rating that characterizes a match (or mismatch) between a reference map considered to be accurate and one or more of its approximations.
2. Detection of changes that are interpreted as a function of time.
3. Model comparison, where the (predicted or simulated) model results are compared with observed data and/or with other model results (pixel-to-pixel consistency is not expected here).
4. The last tradition is to compare the landscape. The key feature of this approach is that the comparison is undertaken against one or more spatial metrics computed from the maps.

In practice, there are a number of methods used to compare and measure differences between raster images. They can be divided into: objective methods - using image descriptors based on mathematical models, and subjective methods – based on observations made by expert observers.

Quality Production Improvement and System Safety: QPI 16 - CZOTO 10   Materials Research Forum LLC
Materials Research Proceedings 34 (2023) 364-373   https://doi.org/10.21741/9781644902691-42

Three methods were selected in the article: differential, using Cohen's kappa coefficient, and based on the Haralick event matrix (Co-occurrence Matrix). These methods were applied globally – to the entire area and locally – to selected fragments of areas. Different areas and adjacent areas were deliberately selected to test the effectiveness of the measures for different types of terrain.

Global Measures

The simplest approach to finding differences between two raster maps is to subtract one raster from the other and mark the differences. The results of such subtraction are shown in Fig.3, where the differences between the pattern and Model1 are marked in green, and the differences between the pattern and Model2 in yellow. The overall agreement was 62% for Model1 and 76% for Model2, respectively (Table 1). Setting a threshold for the allowable number of differing pixels is a half-way solution, because there is still the question of distributing these differences in a spatial context.

Table 1. Value of differences between the models and the standard based on Cohen's kappa coefficient

Model 1		Model 2	
Cohen's kappa agreement coefficient	agreement	Cohen's kappa agreement coefficient	agreement
0.71	62%	0.82	76%

Table 2. Values of textural properties for Pattern, Model 1 and Model 2

Property	Model 1	Standard	Model 2
Angular Second Moment	0.1913	0.1895	0.1864
Contrast	0.4885	0.3595	0.3797
Correlation	0.8067	0.8563	0.8496
Sum of Squares	1.2638	1.2513	1.262
Inverse Difference Moment	0.9036	0.9104	0.9073
Sum Average	2.9844	2.9951	3.1949
Sum Variance	4.5665	4.6457	4.6683
Sum Entropy	2.3805	2.4063	2.421
Entropy	2.7962	2.7874	2.8102
Difference Variance	0.1228	0.1228	0.121
Difference Entropy	0.8035	0.7724	0.7928
Information Measures of Correlation I	-0.5898	-0.5966	-0.5861
Information Measures of Correlation II	0.9498	0.9513	0.9492

The second proposed global measure is Cohen's Kappa Coefficient [44] – determined according to the equation [1]. This coefficient takes values from -1 to 1. The closer the value is to 1, the greater the agreement between the model and the standard. For the study area, Cohen's kappa coefficient was 0.71 for Model1 and 0.82 for Model2 (Table 1). Based on these two global measures, Model2 is closer to the benchmark, but is this close enough to conclude that the maps are consistent at a certain level of statistical significance?

$$K = \frac{\sum_{i=1}^{n} c_{ji} - \sum_{i=1}^{n} \frac{c_j c_i}{c}}{c_{ji} - \sum_{i=1}^{n} \frac{c_{ji} c_{ii}}{c}}$$

[1]

where c_{ij} – values in the i-row and j-column, dots mean summing up e.g. c_i – summing up over all columns in row i, c – total sum, n – number of classes.

The third method used to determine the differences was the Cartesian distance from the textural properties according to Haralick [45], determined on the basis of the Co-occurrence Matrix. Table 2 contains the values of individual properties for the pattern and models.

Table 3. *The value of differences (taking into account the classes) between the models and the standard based on the Cartesian distance from the textural properties according to Haralick after averaging for angles*

Model	Class 1	Class 2	Class 3	Class 4
Model1	0.173	0.057	0.189	0.071
Model2	0.115	0.003	0.040	0.167
The best fitting	model 2	model 2	model 2	model 1

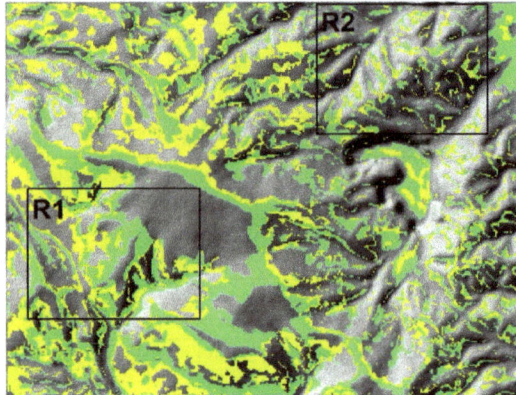

Fig.3. The concept of locality in similarity measures can be understood as a similarity within subclusters or a measure of global similarity calculated for a smaller area (subregion - R1, R2).

Fig.3 shows that greater differences occur in the western part of the study area (R1 rectangle). Difference pixels are larger clusters and there are more of them. In the eastern part of the area, the differences are minor and do not constitute compact patches (R2 rectangle). This experiment shows that differences (Table 3) can be distributed very differently in space, so one global measure of this nature is not sufficient. In addition to the spatial distribution of differences, there is also the issue of their share in a given class, which was already indicated in their research by Pontius et al. [46] and Boots and Csillag [42]. Hence, a local approach seems to be more promising.

Local Measures

In order to determine local measures, 2 sets of sub-areas were designated. One resulted from the simple division of the entire area into equal adjacent sub-areas (Fig.4). The second method of division was to designate two sub-areas that differed according to the field expert (Fig.5).

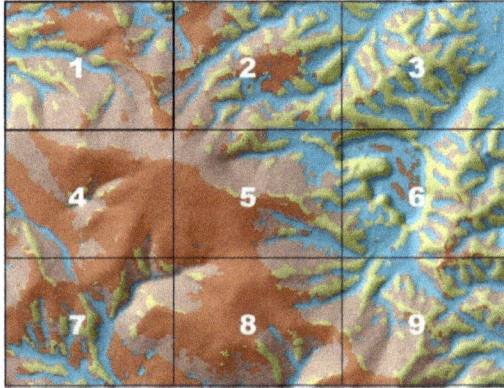

Fig.4. *Scheme of division of the area into equal, adjacent sub-areas*

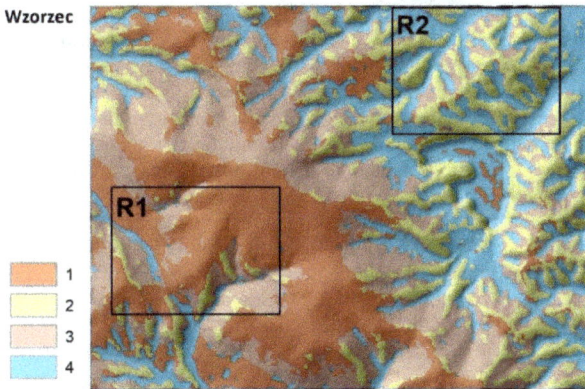

Fig.5. *Designation of two sub-areas with large geomorphological differences*

Therefore, the determination of local measures consisted in determining global measures for 9 modules (tiles being sub-areas of the entire study area) and for two selected frames, for which the largest and smallest differences between the Pattern and Models 1 and 2 were recorded (Table 4). It is worth noting that the Haralick distance does not indicate Model 2 as closer to the standard for all sub-areas, which is the case for Cohen's kappa coefficient. For Cohen's kappa coefficient, a value of 0.7 was adopted as the limit for a satisfactory fit. Although model 2 has a better match to the pattern everywhere, the critical value of 0.7 was exceeded only for 4 out of 9 sub-regions

Quality Production Improvement and System Safety: QPI 16 - CZOTO 10   Materials Research Forum LLC
Materials Research Proceedings 34 (2023) 364-373   https://doi.org/10.21741/9781644902691-42

Table 4. *Summary of differences based on Haralick distance and Cohen's Kappa coefficient for Region 1 (R1) and Region 2 (R2)*

Module	Distances according to Haralick		Cohen's Kappa Coefficient	
	Model 1	Model 2	Model 1	Model 2
1	0.54	0.70	0.59	0.69
2	1.48	0.90	0.63	0.65
3	1.66	0.19	0.64	0.84
4	0.95	1.17	0.41	0.64
5	0.46	0.29	0.65	0.82
6	1.47	0.30	0.65	0.82
7	0.32	0.87	0.50	0.59
8	1.40	0.42	0.34	0.63
9	0.64	0.13	0.66	0.83
R1	1.01	0.84	0.42	0.65
R2	0.92	0.05	0.75	0.83

Conclusions

The number of pixels in which two raster maps differ is very specific information, but not sufficient to consider two maps similar or not, because there is no information about the spatial distribution of these differences. The presented measures and indicators are an attempt to develop a tool to assess the significance of differences between qualitative raster maps in the quantitative and spatial context. Even with expert knowledge, it is difficult to indicate the boundary between similar and dissimilar depiction of a given phenomenon. None of the presented measures is sufficient to unequivocally indicate the critical value allowing two maps to be considered similar. The experiment indicates that you should look for a combination of different indicators. The second approach is to determine similarity measures for subregions and then combine them. However, at this stage it is difficult to indicate the optimal number of subregions. The question remains how to combine the obtained measures for subregions into one summary indicator, which requires further research.

References

[1] Z. Rudnicki. Investigation of some discrimination features of texture images, AGH, Automatyka, 2009, T. 13, z. 3/1, 959-969.

[2] B. Boots, F. Csillag. Categorical maps, comparisons, and confidence, J. Geogr. Syst 8 (2006) 109-118. https://doi.org/10.1007/s10109-006-0018-9

[3] S. Borkowski et al. The use of 3x3 matrix to evaluation of ribbed wire manufacturing technology, METAL 2012 - 21st Int. Conf. Metallurgy and Materials (2012), Ostrava, Tanger 1722-1728.

[4] R. Ulewicz. Outsorcing quality control in the automotive industry, MATEC Web of Conf. 183 (2018) art. 03001. https://doi.org/10.1051/matecconf/201818303001

[5] K. Czerwinska et al. Improving quality control of siluminial castings used in the automotive industry, METAL 2020 29th Int. Conf. Metall. Mater. (2020) 1382-1387. https://doi.org/10.37904/metal.2020.3661

[6] P. Fobel, A. Kuzior. The future (Industry 4.0) is closer than we think. Will it also be ethical? AIP Conference Proceedings 2186 (2019) art.80003. https://doi.org/10.1063/1.5137987

[7] A. Kuzior, J. Zozul'ak. Adaptation of the Idea of Phronesis in Contemporary Approach to Innovation, Manag. Sys. Prod. Eng. 27 (2019) 84-87. https://doi.org/10.1515/mspe-2019-0014

[8] A. Kuzior, A. Kwilinski. Cognitive Technologies and Artificial Intelligence in Social Perception, Manag. Sys. Prod. Eng. 30 (2022) 109-115. https://doi.org/10.2478/mspe-2022-0014

[9] L. Cedro. Model parameter on-line identification with nonlinear parametrization – manipulator model, Technical Transactions 119 (2022) art. e2022007. https://doi.org/10.37705/TechTrans/e2022007

[10] T. Lipinski, J. Pietraszek. Influence of animal slurry on carbon C35 steel with different microstructure at room temperature, Engineering for Rural Development 21 (2022) 344-350. https://doi.org/10.22616/ERDev.2022.21.TF115

[11] A. Szczotok et al. The Impact of the Thickness of the Ceramic Shell Mould on the $(\gamma + \gamma')$ Eutectic in the IN713C Superalloy Airfoil Blade Casting, Arch. Metall. Mater. 62 (2017) 587-593. https://doi.org/10.1515/amm-2017-0087

[12] D. Klimecka-Tatar, M. Ingaldi. Assessment of the technological position of a selected enterprise in the metallurgical industry, Mater. Res. Proc. 17 (2020) 72-78. https://doi.org/10.21741/9781644901038-11

[13] P. Jonšta et al. The effect of rare earth metals alloying on the internal quality of industrially produced heavy steel forgings, Materials 14 (2021) art. 5160. https://doi.org/10.3390/ma14185160

[14] A. Dudek et al. The effect of alloying method on the structure and properties of sintered stainless steel, Arch. Metall. Mater. 62 (2017) 281-287. https://doi.org/10.1515/amm-2017-0042

[15] M. Zenkiewicz et al. Electrostatic separation of binary mixtures of some biodegradable polymers and poly(vinyl chloride) or poly(ethylene terephthalate), Polimery 61 (2016) 835-843. https://doi.org/10.14314/polimery.2016.835

[16] N. Radek et al. The influence of plasma cutting parameters on the geometric structure of cut surfaces, Mater. Res. Proc. 17 (2020) 132-137. https://doi.org/10.21741/9781644901038-20

[17] N. Radek et al. Technology and application of anti-graffiti coating systems for rolling stock, METAL 2019 28th Int. Conf. Metall. Mater. (2019) 1127-1132. ISBN 978-8087294925

[18] N. Radek et al. The effect of laser beam processing on the properties of WC-Co coatings deposited on steel. Materials 14 (2021) art. 538. https://doi.org/10.3390/ma14030538

[19] N. Radek et al. Formation of coatings with technologies using concentrated energy stream, Prod. Eng. Arch. 28 (2022) 117-122. https://doi.org/10.30657/pea.2022.28.13

[20] N. Radek et al. The impact of laser welding parameters on the mechanical properties of the weld, AIP Conf. Proc. 2017 (2018) art.20025. https://doi.org/10.1063/1.5056288

[21] N. Radek et al. Microstructure and tribological properties of DLC coatings, Mater. Res. Proc. 17 (2020) 171-176. https://doi.org/10.21741/9781644901038-26

[22] N. Radek et al. Influence of laser texturing on tribological properties of DLC coatings, Prod. Eng. Arch. 27 (2021) 119-123. https://doi.org/10.30657/pea.2021.27.15

[23] N. Radek et al. Operational properties of DLC coatings and their potential application, METAL 2022 – 31st Int. Conf. Metall. Mater. (2022) 531-536. https://doi.org/10.37904/metal.2022.4491

[24] N. Radek. Determining the operational properties of steel beaters after electrospark deposition, Eksploatacja i Niezawodnosc 44 (2009) 10-16.

[25] N. Radek, J. Konstanty. Cermet ESD coatings modified by laser treatment, Arch. Metall. Mater. 57 (2012) 665-670. https://doi.org/10.2478/v10172-012-0071-y

[26] N. Radek, K. Bartkowiak. Laser Treatment of Electro-Spark Coatings Deposited in the Carbon Steel Substrate with using Nanostructured WC-Cu Electrodes, Physics Procedia 39 (2012) 295-301. https://doi.org/10.1016/j.phpro.2012.10.041

[27] A. Bakowski et al. Frequency analysis of urban traffic noise, ICCC 2019 20th Int. Carpathian Contr. Conf. (2019) 1660-1670. https://doi.org/10.1109/CarpathianCC.2019.8766012

[28] J.M. Djoković et al. Selection of the Optimal Window Type and Orientation for the Two Cities in Serbia and One in Slovakia, Energies 15 (2022) art.323. https://doi.org/10.3390/en15010323

[29] Ł.J. Orman et al. Analysis of Thermal Comfort in Intelligent and Traditional Buildings, Energies 15 (2022) art.6522. https://doi.org/10.3390/en15186522

[30] R. Ulewicz, M. Mazur. Economic aspects of robotization of production processes by example of a car semi-trailers manufacturer, Manuf. Technol. 19 (2019) 1054-1059. https://doi.org/10.21062/ujep/408.2019/a/1213-2489/MT/19/6/1054

[31] S. Blasiak et al. Rapid prototyping of pneumatic directional control valves, Polymers 13 (2021) art.1458. https://doi.org/10.3390/polym13091458

[32] G. Filo et al. Modelling of pressure pulse generator with the use of a flow control valve and a fuzzy logic controller, AIP Conf. Proc. 2029 (2018) art. 20015. https://doi.org/10.1063/1.5066477

[33] E. Lisowski et al. Flow Analysis of a 2URED6C Cartridge Valve, Lecture Notes in Mechanical Engineering 24 (2021) 40-49. https://doi.org/10.1007/978-3-030-59509-8_4

[34] M. Domagala et al. CFD Estimation of a Resistance Coefficient for an Egg-Shaped Geometric Dome, Appl. Sci. 12 (2022) art.10780. https://doi.org/10.3390/app122110780

[35] R. Dwornicka, J. Pietraszek. The outline of the expert system for the design of experiment, Prod. Eng. Arch. 20 (2018) 43-48. https://doi.org/10.30657/pea.2018.20.09

[36] J. Pietraszek et al. Challenges for the DOE methodology related to the introduction of Industry 4.0. Prod. Eng. Arch. 26 (2020) 190-194. https://doi.org/10.30657/pea.2020.26.33

[37] B. Jasiewicz et al. Inter-observer and intra-observer reliability in the radiographic measurements of paediatric forefoot alignment, Foot Ankle Surg. 27 (2021) 371-376. https://doi.org/10.1016/j.fas.2020.04.015

[38] J. Pietraszek. The modified sequential-binary approach for fuzzy operations on correlated assessments, LNAI 7894 (2013) 353-364. https://doi.org/10.1007/978-3-642-38658-9_32

[39] J. Pietraszek et al. Non-parametric assessment of the uncertainty in the analysis of the airfoil blade traces, METAL 2017 – 26th Int. Conf. Metall. Mater. (2017) 1412-1418. ISBN 978-8087294796

[40] J. Pietraszek et al. The non-parametric approach to the quantification of the uncertainty in the design of experiments modelling, UNCECOMP 2017 Proc. 2nd Int. Conf. Uncert. Quant. Comput. Sci. Eng. (2017) 598-604. https://doi.org/10.7712/120217.5395.17225

[41] J. Pietraszek et al. The principal component analysis of tribological tests of surface layers modified with IF-WS2 nanoparticles, Solid State Phenom. 235 (2015) 9-15. https://doi.org/10.4028/www.scientific.net/SSP.235.9

[42] J. Pietraszek, E. Skrzypczak-Pietraszek. The uncertainty and robustness of the principal component analysis as a tool for the dimensionality reduction. Solid State Phenom. 235 (2015) 1-8. https://doi.org/10.4028/www.scientific.net/SSP.235.1

[43] M. Wieczorek, P. Migoń, 2014. Automatic relief classification versus expert and field-based landform classification for the medium-altitude mountain range, the Sudetes, SW Poland, Geomorphology 206 (2014) 133–146. https://doi.org/10.1016/j.geomorph.2013.10.005

[44] P.A. Longley, M.F. Goodchild, D.J. Maguire, D.W. Rhind. Geographical Information Systems: Principles, Techniques, Management and Applications, 2nd Ed., Abridged, Wiley, Hoboken, 2005. ISBN 978-0-471-73545-8

[45] X. Huang, X. Liu, L. Zhang. A Multichannel Gray Level Co-Occurrence Matrix for Multi/Hyperspectral Image Texture Representation, Remote Sens. 6 (2014) 8424-8445. https://doi.org/10.3390/rs6098424

[46] R.G. Pontius Jr., N.R. Malizia. Effect of category aggregation on map comparison, LNCS 3234 (2004) 251-268. https://doi.org/10.1007/978-3-540-30231-5_17

Quality Production Improvement and System Safety: QPI 16 - CZOTO 10   Materials Research Forum LLC
Materials Research Proceedings 34 (2023) 374-379   https://doi.org/10.21741/9781644902691-43

Enhancing Materials Science through Computer Image Analysis and IQA Approaches

PIWOWARCZYK Adam[1,a*] and JASTRZĘBSKA Ilona[2,b]

[1]Cracow University of Technology, Faculty of Mechanical Engineering, al. Jana Pawła II 37, 31-864, Cracow, POLAND

[2]AGH University of Science and Technology in Cracow, Faculty of Materials Science and Ceramics, al. A. Mickiewicza 30, 30-059 Cracow, POLAND

[a]adam.piwowarczyk@pk.edu.pl*, [b]ijastrz@agh.edu.pl*

Keywords: Image Analysis, Materials Science, Microstructure, Image Quality Assessment, Machine Learning

Abstract. Computer image analysis allows for various object detection and classification as well as conducting measurements on microscope images. Technological development, including the application of artificial intelligence solutions, speeds up the classification of large data sets. In the context of image analysis, an essential issue is image quality assessment (IQA). Applying the tools of image analysis, or artificial intelligence, linked with the appropriate IQA approach reveals the opportunity to develop materials with better functional and technological properties more sustainably and effectively.

Introduction

Quality is a universal and significant topic of current decades. It concerns every area of our lives. Daily, we make our decisions based on certain variables, which help us to make the right decision. The process of human analysis begins with creating images using our eyesight, then, the received data are analysed by our brains. The classification is a process that appears during this analysis. Based on classification we select what is right or wrong. This applies to every aspect of our lives, e.g., shopping (choosing the best possible fruit, products, electronic devices or books). The constant development of technology can positively influence all aspects of human life, e.g., the computer became an indispensable assistant at work, which, thanks to its components, allows for improving the work of algorithms in numerous and diverse tasks, permitting to make them efficiently and, usually, correctly (omitting the human factor). Every product, process or system shall represent proper quality, which means that it shall meet the standards of a user. To obtain that, at the stage of preparing the concept, the process shall be divided into components followed by their deep analysis. The quality of the product is verified at the stage of tests, during which extracting as much information as possible shall be required.

The tests and experiments should be planned in detail to get reliable results. In this context, the design of experiment (DoE, *Design of Experiment*) methodology is a valuable approach to planning the work sustainably and effectively [1,2]. It is necessary to establish the goals and select appropriate methods used in the experiment for the verification of results. This allows for obtaining results which can be decisive in the research with the simultaneous minimization of random errors. DoE methodology can be used to analyse the effects of various factors on the product as well as the interaction of input parameters (changing one parameter can influence the value of another). Commonly, it is used to estimate the influence of process input parameters on process output. Applying this methodology, we can determine how strong the influence of individual parameters is on the output, or, we can state that the parameters are interdependent. Also, the DoE provides the tools dedicated to solving the problems caused by multiple factors acting simultaneously. DoE

Quality Production Improvement and System Safety: QPI 16 - CZOTO 10   Materials Research Forum LLC
Materials Research Proceedings 34 (2023) 374-379   https://doi.org/10.21741/9781644902691-43

is not a new term in science, and it has been widely reviewed in scientific publications [3-6]. The application of DoE methodology in image analysis has great potential for the exploration of the material's properties, as well as the development of materials with better properties, in materials science. This work aims to show some aspects of applying the microstructure image analysis to get a deeper insight into possible solutions in this area. Image analysis made on scanning electron microscopy photos can be one step in planning a robust experiment, e.g., optimization of ceramic material of better properties, as SEM analysis is one of the fundamental methods in designing ceramic materials.

Selected Aspects in a Current Image Analysis

The use of image analysis is becoming an increasingly popular tool today. It is used, among others, in materials engineering [7-9], medical engineering [10], security [11], or quality assurance [12]. The application of image analysis gives added value in every field mentioned.

Classical image analysis comprises the relevant sequence of operations. The first stage is image acquisition. Here, problems may arise like noise, shadow or image quality problems during image acquisition. The second stage is image pre-processing. At this point, we can make the image "corrected" and fit it for further work. The next process is the detection of objects in the image and their measurement (quantitative analysis). If everything is done correctly, conclusions can be drawn.

Algorithms developed by humans require much more experience in preparation but allow for better control of factors affecting the final result. Automated image analysis is today a standard method in the quality control of metallic materials, especially in the evaluation of grain size, graphite shape and non-metallic content [13]. Automatically generated solutions based on machine learning (ML) are efficient and fairly accurate classifiers. Machine learning methods belong to the field of artificial intelligence (AI). ML algorithms have been improving the learning and processing of data. It can be compared to the natural human process of learning by experience. These algorithms create a mathematical model based on collected data, called a "training dataset", which produces predictive or decision-making results without directly programming each step of data processing (as typical for convection algorithms). Comparing the application of ML with classical methods, the working model relates a several steps: collection of training data, characteristic feature extraction, development of model and test data [14].

Neural networks are the most important tool in machine learning. They produce versatile and scalable mathematical models which process the information. Neural networks can be applied to large and complex tasks such as image analysis. They vary greatly depending on the network architecture, learning method and activation function. An example of a neural network suitable for image analysis is a group of convolutional networks (CNN, *Convolutional Neural Network*). Their results exceed the human abilities to interpret images, owing to ever-increasing computing power and a large amount of used training data [13].

Image analysis can be defined as a set of methods to facilitate the extraction of information from images. Arranging the functions in the right sequence allows the creation of an algorithm that will be an effective tool. Three main groups of image analysis processing can be distinguished: filtration, morphological operation and binarization. Each of them has its specific tasks [15-17].

The important fact is that every sector of industry uses different types of dedicated software, based on different programming languages. Creating applications can take place at any time as well as any place on earth. We only need an appropriate compiler dedicated to the language in which we create the code. An example of a language that is often used for research, didactics as well as in business solutions is *Python* [18].

The advantage of *Python* is its easy construction which makes it user-friendly and universal. The syntax is very intuitive and, therefore, the code is easy to understand. Considering this fact, writing commands forces the user to use habits, such as making indentations in the code. Finally,

Quality Production Improvement and System Safety: QPI 16 - CZOTO 10   Materials Research Forum LLC
Materials Research Proceedings 34 (2023) 374-379   https://doi.org/10.21741/9781644902691-43

it makes the code readable and clear. Thanks to that, we can quickly create an application, make any corrections and presents a readable code to another researcher, which greatly facilitates the work. Moreover, a large number of libraries can be utilized, presented in Table 1, supporting many existing programmes. *Python* is widely used in areas such as data science, big data or machine learning.

One of the libraries for image processing is *OpenCV* (*Open-Source Computer Vision Library*). It is a free open-source library for commercial and educational use without the need of sharing projects. One of the main goals of the *OpenCV* library is to provide a tool that allows people at different levels of programming and imaging to create both simple programs and advanced projects. *OpenCV* functions include motion analysis and tracking, image processing (filtering, edge detection, corner detection, histograms), face detection system, motion detection, image glueing (creation of views or panorama), machine learning (ML algorithms supporting the above functions are: k-means, deep neural networks, artificial neural networks, decision trees, Bayesian classifier).

Table 1. *List of selected Python language libraries, and their description.*

Name	Description
NumPy	A set of basic tools that allow for advanced mathematical calculations with matrices and sets/vectors
SciPy	A set of tools that allows for many mathematical operations and, most importantly, many numerical methods such as integration, numerical differentiation, algorithms for solving differential equations, linear algebraic algorithms, Fourier transformation and signal processing
Pandas	Toolkit for relational data analysis
Matplotlib	Graphing module
SciKit-Learn	An open-source library of ML algorithms for specific applications, such as image processing, classification, clustering or pattern mining
TensorFlow	A library for machine learning that features high performance and scalability

Image Quality Assessment (IQA) can be classified into three categories, namely the Full Reference method (FR or FR-IQA), Reduced Reference method (RR or RR-IQA) and No Reference method (NR or NR-IQA). Each of them has characteristics and is dedicated to a separate-aim use. The main division is based on the information content of the original image. The requirement of full or partial information is related to FR and RR methods, while the NR method does not require any information. The NR method is the most practical out of the three methods because, in most cases, it is not possible to refer to the pristine original image of the distorted image [19].

NR-IQA helped to create VPMI (*Visual Parameter Measurement Index*) [20]. First of all, in [20] five basic visual parameters were analysed which the human eye uses to assess the quality of the image. In this method, only low-level features that affect the image quality can be examined. The measured parameters included average information entropy, average chromatic level factor, average gradient, average luminance and average bandwidth factor. VPMI is based on the integration of these visual metrics to assess image quality. It is established that the maximum VPMI corresponds to the best quality of the colour image. The obtained results suggest that the proposed method is universal and can be used to evaluate not only the original image but also the distorted image.

Quality Production Improvement and System Safety: QPI 16 - CZOTO 10   Materials Research Forum LLC
Materials Research Proceedings 34 (2023) 374-379   https://doi.org/10.21741/9781644902691-43

Challenges in the Quality of Images

Exploring the properties of the new material is a time-consuming process, utilizing the application of numerous experimental techniques. Image analysis is becoming a powerful tool in this area of research. But, poor image quality, noise or shadow in the analysed image disturb in performing reliable object detection.

Image quality plays a key role during image analysis, and, can be disrupted by multiple problems. In most cases, the original image shall be initially subjected to image processing, before its analysis, to get better object detection and conduct more reliable measurements. Noise is one of the problems, commonly encountered during image analysis. This can be solved by applying dedicated and combined filtering and binarization functions, as shown in Fig.1.

The use of appropriate operations in image processing enables further work on the image free from distortions (which can significantly affect the detection of objects). If we analysed the images from the point of view of how many objects are in the photograph, we would intuitively count only large objects. As shown in Fig.1b, performing measurements on the image with noise produces 5309 detected objects, while analysis conducted on the image after applying an appropriate filter shows only 15 objects, as presented in Fig.1d. So, the image with noise contained almost 354 times more elements than the image after filtration. Thus, the method for image quality assessment is essential during image analysis.

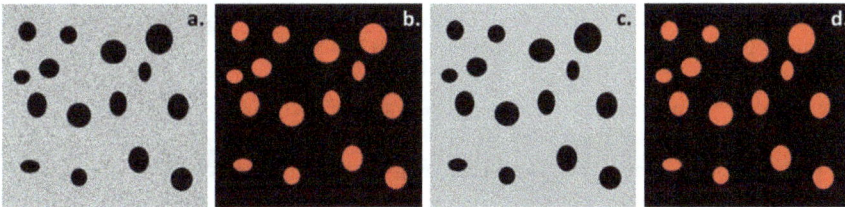

Fig. 1. *Comparison of a. an image with noise, with b. its binary image, c. image after applying the median filter (free from the noise), and d. its binary image.*

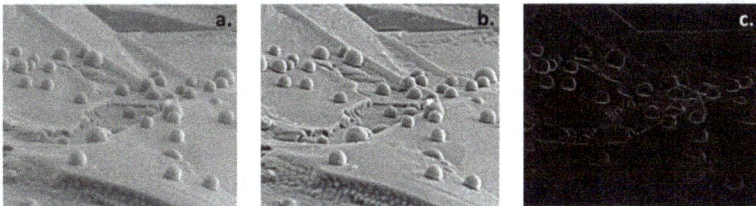

Fig. 2. *Comparison of a. an image with b. image sharpened and c. object edges detected.*

The filtering process is used to extract the information needed for image analysis. An example of such an analysis was attached in Fig.2, which aimed to detect spherical metallic droplets on the surface spinel-structure materials obtained by the arc-plasma melting technique [21,22]. Contrary to point operations, filters are contextual operations. The task of filters is noise reduction, image sharpening or edge detection [17]. The edge detection can be further used to recognize and quantify the metallic doplets on the surface of spinel material.

Following the trends of artificial intelligence applications in the development of image analysis, machine vision will greatly improve the efficiency, quality and reliability of fault finding. Visually

Quality Production Improvement and System Safety: QPI 16 - CZOTO 10   Materials Research Forum LLC
Materials Research Proceedings 34 (2023) 374-379   https://doi.org/10.21741/9781644902691-43

excellent optical illumination platforms and appropriate imaging techniques are prerequisites for high-quality images. Image processing and analysis are a tool to get information from images and deep learning is having a significant impact on the field of image analysis. The use of deep learning will play an increasingly important role in the further development of visual inspection fields [23].

Conclusions

To understand the application of image analysis it is necessary and recommended to get familiar with the methods that have been used for years, and then start experimenting with machine learning (ML) or neural networks (CNN).

Quality control is a very important issue in R&D activities. The Design of the Experiment (DoE) methodology can facilitate the research work, making it more effective and sustainable.

Acknowledgements

This research was supported by The National Centre for Research and Development, Grant no. LIDER/14/0086/L-12/20/NCBR/2021, 2022-2025 (Principal Investigator: I. Jastrzębska).

References

[1] J. Pietraszek, A. Goroshko. The heuristic approach to the selection of experimental design, model and valid pre-processing transformation of DoE outcome, Adv. Mater. Res. 874 (2014) 145-149. https://doi.org/10.4028/www.scientific.net/AMR.874.145

[2] J. Pietraszek, N. Radek, A.V. Goroshko. Challenges for the DOE methodology related to the introduction of Industry 4.0, Prod. Eng. Arch. 26 (2020) 190-194. https://doi.org/10.30657/pea.2020.26.33

[3] R.A. Fisher. The Arrangement of Field Experiments, J. Ministry of Agric. 33 (1926) 503-515. https://doi.org/10.23637/rothamsted.8v61q

[4] K. Hinkelmann, O. Kempthorne. Design and Analysis of Experiments. Vol. 1. Introduction to Experimental Design. Wiley, Hoboken, 2008.

[5] R.H. Myers, D.C. Montgomery, C.M. Anderson-Cook. Response Surface Methodology: Process and Product Optimization Using Designed Experiments. Wiley, Hoboken, 2009.

[6] R.L. Mason, R.F. Gunst, J.L. Hess. Statistical design and analysis of experiments: with applications to engineering and science. Wiley, Hoboken, 1989.

[7] I. Jastrzębska, A. Piwowarczyk. Traditional vs. Automated Computer Image Analysis—A Comparative Assessment of Use for Analysis of Digital SEM Images of High-Temperature Ceramic Material, Materials 16 (2023) art. 812. https://doi.org/10.3390/ma16020812

[8] M. Ludwig et al. Corrosion of magnesia-chromite refractory by PbO-rich copper slags, Corrosion Science 195 (2021) art. 109949. https://doi.org/10.1016/j.corsci.2021.109949

[9] L. E. Vivanco-Benavides et al. Machine learning and materials informatics approaches in the analysis of physical properties of carbon nanotubes: A review, Computational Materials Science 201 (2022) art. 110939. https://doi.org/10.1016/j.commatsci.2021.110939

[10] X. Chen et al. Recent advances and clinical applications of deep learning in medical image analysis. Medical Image Analysis, Volume 79 (2022) art. 102444. https://doi.org/10.1016/j.media.2022.102444

[11] A. Singh et al. Real-time intelligent image processing for security applications. J. Real-Time Image Process. 18 (2021) 1787-1788. https://doi.org/10.1007/s11554-021-01169-w

[12] A. Piwowarczyk, L. Wojnar. Perspektywy rozwoju wykorzystania metod analizy obrazu w zagadnieniach bezpieczeństwa i zapewnienia jakości, in: R.Ulewicz, R.Dwornicka (Eds.)

Quality Production Improvement and System Safety: QPI 16 - CZOTO 10   Materials Research Forum LLC
Materials Research Proceedings 34 (2023) 374-379   https://doi.org/10.21741/9781644902691-43

Praktyczne aspekty zarządzania produkcją i bezpieczeństwem, Wyd. PK, Kraków, 2019, pp.83-91.

[13] A. Piwowarczyk, L. Wojnar. Machine learning versus human-developed algorithms in image analysis of microstructures. QPI 1 (2019) 412-416. https://doi.org/10.2478/cqpi-2019-0056

[14] A. Gądek-Moszczak, R. Filipowska. Efficiency and accuracy of simulated microstructure images generated by machine learning. 14th WCCM-ECCOMAS Congress 2020, January, 11-15, 2021: virtual congress (2021) art. 270129. 10.23967/wccm-eccomas.2020.129

[15] R. Tadeusiewicz, P. Korohoda. Komputerowa analiza i przetwarzanie obrazów, Wydaw. Fundacji Postępu Telekomunikacji, Kraków, 1997.

[16] J.C.F. Russ, B. Neal, The Image Processing Handbook, 7th Ed., CRC Press, Boca Raton, 2016. https://doi.org/10.1201/b18983

[17] L. Wojnar, Analiza obrazu. Jak to działa? Wyd. PK, Kraków, 2020. ISBN 978-8366531529

[18] S. Klosterman, Data Science Projects with Python. Second Edition, Packt Publishing, Birmingham, 2021. ISBN 978-1838551025

[19] J. Yang et al. No-reference image quality assessment focusing on human facial region, Signal Process. Image Commun. 78 (2019) 51-61. https://doi.org/10.1016/j.image.2019.05.011

[20] Y.H. Liu, K.F. Yang, H.M. Yan. No-Reference Image Quality Assessment Method Based on Visual Parameters, J. Electron. Sci. Technol. 17 (2019) 171-184. https://doi.org/10.11989/JEST.1674-862X.70927091

[21] I. Jastrzębska et al. Crystal structure and Mössbauer study of $FeAl_2O_4$. Nukleonika 60 (2015) 47-49. https://doi.org/10.1515/nuka-2015-0012

[22] I. Jastrzębska, J. Szczerba, P. Stoch. Structural and microstructural study on the arc-plasma synthesized (APS) $FeAl_2O_4$-$MgAl_2O_4$ transitional refractory compound, High Temp. Mater. Process. 36 (2017) 299-303. https://doi.org/10.1515/htmp-2015-0252

[23] Z. Ren, F. Fang, N. Yan, Y. Wu. State of the Art in Defect Detection Based on Machine Vision. Int. J. Pr. Eng. Man.-G.T. 9 (2022) 661-691. https://doi.org/10.1007/s40684-021-00343-6

Quality Production Improvement and System Safety: QPI 16 - CZOTO 10   Materials Research Forum LLC
Materials Research Proceedings 34 (2023) 380-389   https://doi.org/10.21741/9781644902691-44

Research on the Effectiveness of a Visual Protection of Military Objects in Field Conditions and in a Virtual Environment

PRZYBYŁ Wojciech[1a*], MAZURCZUK Robert[1b],
SZCZODROWSKA Bogusława[1c] and JANUSZKO Adam[1d]

[1]Military Institute of Engineer Technology, ul. Obornicka 136, 50-961 Wrocław, Poland

[a]przybyl@witi.wroc.pl, [b]mazurczuk.r@witi.wroc.pl, [c]szczodrowska@witi.wroc.pl,
[d]januszko@witi.wroc.pl

Keywords: Military Camouflage, Visual Protection, Field Conditions

Abstract: The article presents two methodologies for testing the effectiveness of camouflaging military equipment. The tested objects were camouflaged with deforming camouflage dedicated to the environment and season. The tests covered the visible spectrum and were carried out both in real conditions - training ground, and in a virtual environment – on a specially prepared calibrated stand, characterized by compliance with real conditions and human perception of vision – presented colors, as well as in terms of the amount of information and detail rendering. These two test methods were compared, attention was drawn to the high correlation of results and to some characteristic properties of both separately. The work confirms the legitimacy of developing and improving virtual research methods as supporting in the design and testing of modern camouflage patterns, especially at the initial stage of pattern design. It can also be useful when assessing masking systems in other ranges of electromagnetic radiation: UV, near infrared, or thermal range.

Introduction

Armed conflicts are an inseparable part of human history, but also the present and probably the future as well. Conducting military operations, both offensive and defensive, is associated with incurring huge costs, primarily social and material. However, one of the cheapest methods of protecting manpower and equipment are effective camouflage systems.

According to the definition [1] camouflage is a type of protection of combat operations, and consists in hiding forces and resources from recognizing the enemy or misleading him about the location of his own troops. Cloaking is therefore a broad issue (Fig.1) and depending on the level of command, it may concern the strategic, operational or direct (tactical) level. In turn, taking into account the means, forces or specifications used, we can distinguish concealment, presence or disinformation.

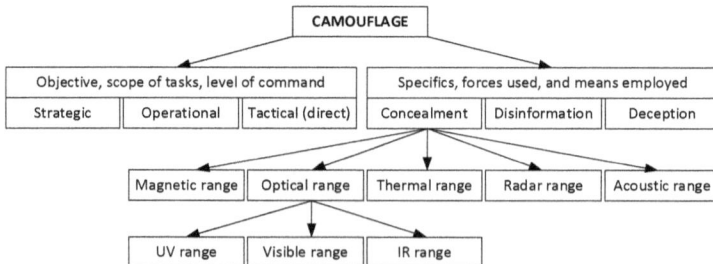

Fig.1. Distribution of masking and its types (source: [2])

Quality Production Improvement and System Safety: QPI 16 - CZOTO 10   Materials Research Forum LLC
Materials Research Proceedings 34 (2023) 380-389   https://doi.org/10.21741/9781644902691-44

New cloaking systems are constantly being developed and introduced, requiring the assessment of the effectiveness of the cloaking. In accordance with the above scheme, this article focuses on the verification of direct (tactical) camouflage in the context of concealment in the optical range, in particular in the visible range (Fig.2).

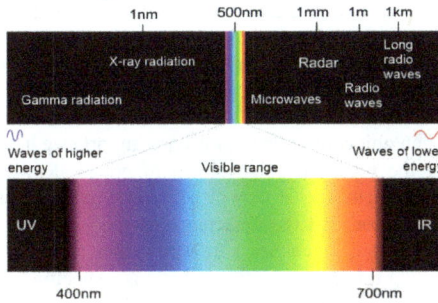

Fig.2. *Scheme of the spectrum of electromagnetic radiation*

Thus, the main task of effective camouflage is to eliminate the unmasking features, i.e. those that make it possible to distinguish one's own objects from the background of the terrain. In this respect, these are the differences in the patterns and colors of the spots forming the pattern of the background and the object, as well as the smoothness / dullness of the surface. For this purpose, camouflage patterns applied to military objects (equipment and equipment, vehicles, buildings, etc.) or uniforms are used [3].

In practice, it is difficult to obtain a universal camouflage that would suit any type of background: forest, desert, urban, winter, etc. Work on adaptive camouflage, which changes itself depending on the background, is carried out in many research centers, but meanwhile, beyond the laboratory level, they didn't come out. Therefore, in practice, camouflages dedicated to the appropriate type of object, background, and even the season of the year are widely used (Fig.3).

Fig.3. *Seasonal camouflage uniforms for Europe*

Within the types of patterns, a mimetic pattern is distinguished, which is to make it similar (mimicry) to the nearest background, and thus make it difficult for the observer to detect the object. However, patterns of this type do not break the silhouette, which makes it easy to recognize the object after detection, and thus to decide on the means of destruction used to destroy the object. Mimicry also does not ensure the continuity of colors and textures in the case of the movement of

Quality Production Improvement and System Safety: QPI 16 - CZOTO 10   Materials Research Forum LLC
Materials Research Proceedings 34 (2023) 380-389   https://doi.org/10.21741/9781644902691-44

the object. Therefore, in camouflage painting, especially on vehicles, deforming camouflage is also used, built on the basis of large spots in relatively high-contrast colors.

Investigation of Camouflage Effectiveness

The study of the effectiveness of masking consists in observing the background (edge) on which the object with camouflage (dedicated to the background) has been placed. A group of observers who should be experienced in recognizing military equipment takes part in the observations. The minimum number of observers is 3. During the tests, the degree of visibility of the object from the given distances is recorded. The degree of visibility is understood as a scale commonly used in reconnaissance:

- Lack (-) – no presence of an object of potential military importance in the observation area;
- Detection (W) – finding the presence of an object of potential military importance in the observation area – finding something that may be of interest to the observer;
- Recognition (R) – determination that the detected object is a specific type of object, e.g. a person, a wheeled armored personnel carrier, a tracked vehicle;
- Identification (I) – specifying that the recognized object is a specific type of object, e.g. a man is a soldier, a wheeled armored personnel carrier is APC "Rosomak", and a tracked vehicle is BPWP "Borsuk".

As part of this article, the study of the effectiveness of masking the type of object known to observers - the HMMWV wheeled vehicle (Fig.4) was carried out.

Fig.4. *HMMWV vehicle*

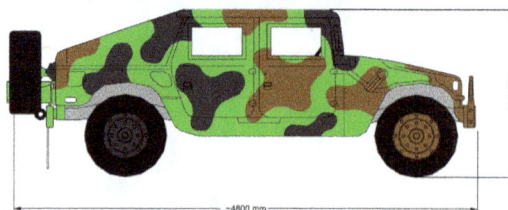

Fig.5. *Silhouette of the HMMWV vehicle in deforming camouflage (designed by WITI)*

For this purpose, a 1:1 scale model (silhouette) was used, made of steel sheet, covered with a camouflage pattern applied with special paints for camouflage painting in green, brown and black colors (Fig.5). Such an exemplary selection of colors and stains can be used in Central and Eastern Europe in the spring-summer-autumn season.

The tests were carried out for the scenery dedicated to this pattern - the background of deciduous, coniferous and mixed forests (woodland) in the spring-autumn period. The study was aimed at determining the effectiveness of masking in natural conditions in the most unfavorable situation for masking, and the most favorable for diagnosis. Therefore, the object was set on the outskirts of forests, on flat terrain, without obstacles to visibility (e.g. other vegetation), in a direction and direction that prevented the sun from blinding observers, and ensured good illumination of the object. It was also noted that the object should be visible in the range of at least ¾ of its height. The tests were carried out only on days without precipitation (fog, rain), at temperatures not causing air waves (vibrations) and with similar insolation conditions (cloudy sky) for each series [4].

The tests were carried out using two methods: in field conditions and in a virtual environment, and then the results were compared. The research methodologies were based on the Polish defense standard NO-80-A200 *"Special paints for camouflage painting. Requirements and test methods"* [5].

Field Trials

The object was placed on the edge of the forest in the selected and marked sector (outer) of observation. The observers did not know where the objects were located. They started the observations from a distance of 1,500 m. Within a maximum of two minutes, they determined the degree of object visibility (absence, detection, recognition) on the test cards. After this time, they approached 100 m closer to the object to re-determine the degree of visibility of the object. This procedure was carried out up to a distance of 500 m (Figure 6).

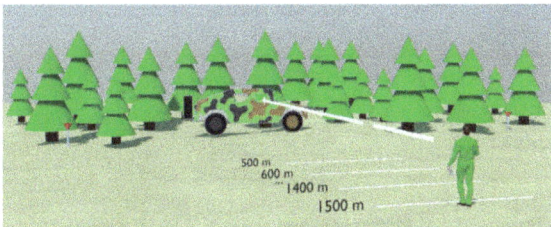

Fig.6. Scheme of conducting field tests

Virtual Environment

During the field tests, the terrain background was also imaged using the Canon EOS 50D Mark III camera (focal length 50 mm, matrix size 36×24 mm, resolution 3744 × 5616 – Table 1) at a scale of 1:10000 and a distance of 500 m (Fig.7).

The photos had to be characterized by appropriate parameters [6], to reflect the actual conditions as much as possible and include descriptive data (metadata) – Table 2

The tests in the virtual environment took place in a specially adapted room with limited access to external light sources and equipped with a multimedia projector, a screen and two positions – an observer and a research supervisor (Fig.8). To ensure the greatest color fidelity, the entire image path has been calibrated [7].

Quality Production Improvement and System Safety: QPI 16 - CZOTO 10   Materials Research Forum LLC
Materials Research Proceedings 34 (2023) 380-389   https://doi.org/10.21741/9781644902691-44

Fig. 7. *An example image taken from a distance of 500 m during field tests*

Table 1. *Display parameters*

Focal length	Matrix dimensions		Resolution		Density	
	width	height	width	height	information	details
[mm]	[mm]	[mm]	[px]	[px]	[px/mm]	[mm/px]
50	36	24	5616	3744	156	0.01

Table 2. *Required display parameters*

Parameter	Minimal value	Remark
Spatial resolution	3600px × 2400px	Ensuring sufficient photo quality – avoiding the visibility of pixels in the presented photos
Total resolution	8 bit/channel	Ensuring sufficient total range
Color space	Adobe RGB	Ensuring a sufficient wide color space – presentation of as many colors as possible
File compression type	Lossless	Ensuring appropriate image quality – avoiding artifacts caused by lossy compression methods
White balance	Performing a white (grey) pattern	Ensure colour fidelity
Sharpness	Good resolution of details	Ensuring sufficient sharpness of the photo - presentation of the details of the photo also when the photo is enlarged
Exposure	Balance and full tonal range	Ensuring a wide tonal range for the examined background – distinguishing details for bright fragments, midtones and dart parts of the image
Location	Designation of the location of the imagery	Providing an indication of where the imagery was taken – either description or by geographical coordinates
Azimuth	Designation of the observed azimuth	Provides indication of the direction of the display
Date and time	Date and time stamp	Unambiguous and precise determination of the date and time of the imaging
Scenery	Possibility of specifying the scenery	Clear marking of the nature of the scenery (coniferous, deciduous, mixed forest, urbanized area, desert, demi-desert, rocky, steppe, etc.)
Season	Possibility to specify the season	Clear indication of the season (spring, summer, autumn, snowless winter, snowy winter, etc.)
Outline distance	Distance designation	Unique term to show the distance from the outskirts (background)
Lens focal	Focal length designation	An unambiguous indication of the focal length for the lens used to image
Color swatch	Image for color swatch	Increase in perceptual color fidelity for camouflage with palette matching with samples

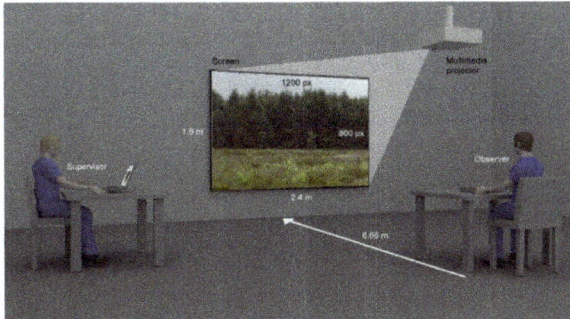

Fig.8. *Scheme of conducting research in a virtual environment*

Table 3.   *Parameters of background fragments for example images 1:1000 (50mm focal length, 36×24mm matrix))*

Distance from the background	Size of the background		Resolution		Density	
	width	height	width	height	Information	details
[m]	[m]	[m]	[px]	[px]	[px/m]	[mm/px]
1000	360	240	5616	3744	7.8	124.2

Table 4.   *Summary of observation results (OK indicates the agreement of the results of both methods, NOK indicates the discrepancy of the results of both methods)*

Trial No	Observer No	Field trials		Virtual trials*	
		distance W (detection)	distance R (recognition)	Detected at 1000 m	Recognized at 1000 m
1	1	1300	800	Yes (OK)	– (OK)
	2	1400	800	Yes (OK)	– (OK)
	3	1300	800	– (NOK)	– (OK)
2	1	1200	600	Yes (OK)	– (OK)
	2	1300	700	– (NOK)	– (OK)
	3	1200	600	Yes (OK)	– (OK)
3	1	1100	700	– (NOK)	– (OK)
	2	1100	600	– (NOK)	– (OK)
	3	1300	700	Yes (OK)	– (OK)
4	1	1000	800	Yes (OK)	– (OK)
	2	1000	800	Yes (OK)	– (OK)
	3	1000	800	Yes (OK)	– (OK)
5	1	600	500	– (OK)	– (OK)
	2	600	500	Yes (NOK)	– (OK)
	3	600	500	– (OK)	– (OK)
6	1	800	600	Yes (NOK)	– (OK)
	2	800	500	Yes (NOK)	– (OK)
	3	800	600	– (OK)	– (OK)
Average		**1022**	**661**	**11×W, 7×Null**	–

Taking into account the optical parameters of the camera [8] (focal length 50 mm, matrix size 36×24 mm, resolution 3744×5616), projector resolution (1200×800 px), screen size (2.4×1.6 m) and perceptual capabilities of the human sense of sight [9], it was established the distance of the observer to the screen at 6.67 m, which corresponded to the observation from 1000 m and provided enough information (pixels) to represent the scenery, while preventing the perception of the pixel nature of digital images displayed on the screen by the projector with a given resolution (Table 3).

The methodology of virtual research was based on the previously mentioned Polish defense standard and recommendations developed during the work of the NATO RTO-AG-SCI-095 working group [10]. In this method, the observers, also within 2 minutes, had to determine the degree of visibility of the object and, in order to verify the correctness, indicate the location of the object in the image. Table 4 contains test results for 6 tests carried out by 3 observers for each test in field conditions and in a virtual environment.

Conclusions

The results obtained by both methods are characterized by a high convergence of results. They were analyzed in terms of two degrees of object visibility – detection (W) and recognition (R). The scope of reconnaissance is a decisive factor, because usually in the course of combat operations, it is on its basis that decisions about further tactical or operational plans are made. Based on the results obtained for all tests and for each test, we can conclude that in the virtual method, an object with a deforming camouflage dedicated to woodland (special paints) was not recognized at a distance of 1000 m by observers using the unarmed eye. The results from field tests confirm this, because the recognition was carried out from a distance of 500 to 800 m (average 661 m), which is a distance of less than 1000 m. This means that the object, while approaching it, was later recognized.

On the other hand, in terms of detection (W), the results are mostly consistent (11 out of 18 trials), but for a few results there are discrepancies (5 out of 18 trials). Analyzing both test environments, it can be seen that during the tests in the real environment, the correctness of the detection cannot be verified, because as it turned out during the tests in the virtual environment, some observers incorrectly indicated the location of the object detection, which could be detected thanks to the laser pointer on the screen (and then the wrong detection is not counted), which was impossible to check during field tests. Another factor affecting the results could have been weather conditions, because despite conducting the research in the conditions recommended for the weather, some of its dynamics – slight wind, operation of the sun, shadows, or a wider range of perception affected the results. The availability of the observers themselves is also not without significance, as during laboratory tests they are not exposed to the inconvenience of field conditions – they make assessments in a comfortable position, without the need to move around, often in difficult training ground, being influenced by weather conditions. Summing up, it should be considered:

- studies in a virtual environment are largely consistent with environmental studies;
- research in a virtual environment, especially at the initial stage of designing and checking the effectiveness of masking, can give reliable results;
- both research methods should be further developed, e.g. in the virtual method, the perceptual compatibilityp with the natural environment should be deepened, and in the field method, an element verifying the correctness of the detected object should be added.

Optical military camouflage has a purely utilitarian character, although many of its elements can find applications beyond the military context. This particularly applies to clothing aspects, where quasi-military attire enjoys significant popularity among youth and survival enthusiasts. It also extends to the automotive and railway industries, as it provides the opportunity to enhance the quality of products [11-13] by skillfully optically masking geometric imperfections. Specifically,

Quality Production Improvement and System Safety: QPI 16 - CZOTO 10   Materials Research Forum LLC
Materials Research Proceedings 34 (2023) 380-389   https://doi.org/10.21741/9781644902691-44

this relates to the surface of sheet metal, especially in the case of railway wagons [14], which require special treatment with fillers prior to applying paint coatings. Surface treatment techniques [15] and the ability to correct it through the application of additional coatings [16-18], combined with masking imperfections, would offer an excellent quality tool for reducing the visibility of non-removable structural joints [19,20]. These techniques may also be necessary at a later stage when the need arises for the renovation of vehicle coatings due to wear [21] and corrosion [22-24]. Conversely, the opposite action, avoiding masking, or even anti-masking, which signals presence, is necessary for certain devices or their components that may be hazardous or whose position should be immediately visible. This may occur when, among other factors, applied special coatings such as electrospark deposition [25-27] or diamond-like carbon coatings [28-30] cause the part to visually blend with the surroundings. The evaluation of camouflage or, conversely, anti-masking, largely relies on subjective assessments based on the observations of an observer panel, often using Likert scales. The multidimensionality of visual masking techniques becomes easier to analyze when the dimensionality of the issue is initially reduced, for example, using one of the linear discrimination methods [31,32]. In subsequent steps, one can proceed based on established design of experiments (DOE) methodologies [33-35], and in specific cases, utilizing nonparametric approaches [36-38].

References

[1] M. Laprus (ed.). Leksykon wiedzy wojskowej, MON, Warszawa, 1979.

[2] Maskowanie wojsk i wojskowej infrastruktury obronnej, DD/3.20, MON/SG WP, Warszawa 2010.

[3] R. Bogacki. Maskowanie kolorem. Mat. III Konf. Naukowo Technicznej, WITI 1997, vol.1, p.108.

[4] Instrukcja o maskowaniu wojsk. Sztab generalny WP, MON, Warszawa, 1977.

[5] Norma Obronna NO-80-A200, Farby specjalne do malowania maskującego. Wymagania i metody badań, MON, Warszawa, 2014.

[6] W. Przybył, I. Plebankiewicz, A. Januszko, C. Śliwiński, W. Malej. Zobrazowania środowiska oraz modele obiektów wojskowych w wirtualnej metodzie oceny skuteczności maskowania, in: Wybrana problematyka w technologiach inżynierii mechanicznej, (2020), Politechnika Świętokrzyska, Kielce, 234-242.

[7] C. Murphy, B. Fraster, F. Bunting. Real World Color Management, 2nd Edition, Peachpit Press Publications, 2004. ISBN 978-0321267221

[8] Canon Inc. Technical datasheet for Canon EOS 50D Mark III.

[9] P. Francuz. Imagia, w kierunku neurokognitywnej teorii obrazu, Wyd. KUL, Lublin, 2013. ISBN 978-83-7702-706-6

[10] NATO RTO, Guidelines for Camouflage Assessment Using Observers (RTO-AG-SCI-095), NATO RTO, 2006.

[11] R. Ulewicz. Outsorcing quality control in the automotive industry, MATEC Web of Conf. 183 (2018) art.03001. https://doi.org/10.1051/matecconf/201818303001

[12] R. Ulewicz, F. Nový. Quality management systems in special processes, Transp. Res. Procedia 40 (2019) 113-118. https://doi.org/10.1016/j.trpro.2019.07.019

[13] A. Pacana et al. Analysis of quality control efficiency in the automotive industry, Transp. Res. Procedia 55 (2021) 691-698. https://doi.org/10.1016/j.trpro.2021.07.037

[14] N. Radek, R. Dwornicka. Fire properties of intumescent coating systems for the rolling stock, Commun. - Sci. Lett. Univ. Zilina 22 (2020) 90-96. https://doi.org/10.26552/com.C.2020.4.90-96

[15] N. Radek et al. The influence of plasma cutting parameters on the geometric structure of cut surfaces, Mater. Res. Proc. 17 (2020) 132-137. https://doi.org/10.21741/9781644901038-20

[16] N. Radek et al. Technology and application of anti-graffiti coating systems for rolling stock, METAL 2019 28th Int. Conf. Metall. Mater. (2019) 1127-1132. ISBN 978-8087294925

[17] N. Radek et al. The effect of laser beam processing on the properties of WC-Co coatings deposited on steel. Materials 14 (2021) art. 538. https://doi.org/10.3390/ma14030538

[18] N. Radek et al. Formation of coatings with technologies using concentrated energy stream, Prod. Eng. Arch. 28 (2022) 117-122. https://doi.org/10.30657/pea.2022.28.13

[19] N. Radek et al. The impact of laser welding parameters on the mechanical properties of the weld, AIP Conf. Proc. 2017 (2018) art.20025. https://doi.org/10.1063/1.5056288

[20] N. Radek et al. Properties of Steel Welded with CO2 Laser, Lecture Notes in Mechanical Engineering (2020) 571-580. https://doi.org/10.1007/978-3-030-33146-7_65

[21] S. Marković et al. Exploitation characteristics of teeth flanks of gears regenerated by three hard-facing procedures, Materials 14 (20210 art. 4203. https://doi.org/10.3390/ma14154203

[22] M. Scendo et al. Purine as an effective corrosion inhibitor for stainless steel in chloride acid solutions, Corr. Rev. 30 (2012) 33-45. https://doi.org/10.1515/CORRREV-2011-0039

[23] T. Lipinski, J. Pietraszek. Influence of animal slurry on carbon C35 steel with different microstructure at room temperature, Engineering for Rural Development 21 (2022) 344-350. https://doi.org/10.22616/ERDev.2022.21.TF115

[24] T. Lipiński, J. Pietraszek. Corrosion of the S235JR Carbon Steel after Normalizing and Overheating Annealing in 2.5% Sulphuric Acid at Room Temperature, Mater. Res. Proc. 24 (2022) 102-108.

[25] N. Radek. Determining the operational properties of steel beaters after electrospark deposition, Eksploatacja i Niezawodnosc 44 (2009) 10-16.

[26] N. Radek, J. Pietraszek, A. Gadek-Moszczak, Ł.J. Orman, A. Szczotok. The morphology and mechanical properties of ESD coatings before and after laser beam machining, Materials 13 (2020) art. 2331. https://doi.org/10.3390/ma13102331

[27] N. Radek et al. The effect of laser treatment on operational properties of ESD coatings, METAL 2021 30th Ann. Int. Conf. Metall. Mater. (2021) 876-882. https://doi.org/10.37904/metal.2021.4212

[28] N. Radek et al. Microstructure and tribological properties of DLC coatings, Mater. Res. Proc. 17 (2020) 171-176. https://doi.org/10.21741/9781644901038-26

[29] N. Radek et al. Influence of laser texturing on tribological properties of DLC coatings, Prod. Eng. Arch. 27 (2021) 119-123. https://doi.org/10.30657/pea.2021.27.15

Quality Production Improvement and System Safety: QPI 16 - CZOTO 10   Materials Research Forum LLC
Materials Research Proceedings 34 (2023) 380-389   https://doi.org/10.21741/9781644902691-44

[30] N. Radek et al. Operational properties of DLC coatings and their potential application, METAL 2022 – 31st Int. Conf. Metall. Mater. (2022) 531-536. https://doi.org/10.37904/metal.2022.4491

[31] J. Pietraszek et al. The principal component analysis of tribological tests of surface layers modified with IF-WS2 nanoparticles, Solid State Phenom. 235 (2015) 9-15. https://doi.org/10.4028/www.scientific.net/SSP.235.9

[32] J. Pietraszek, E. Skrzypczak-Pietraszek. The uncertainty and robustness of the principal component analysis as a tool for the dimensionality reduction. Solid State Phenom. 235 (2015) 1-8. https://doi.org/10.4028/www.scientific.net/SSP.235.1

[33] J. Pietraszek, A. Szczotok, N. Radek. The fixed-effects analysis of the relation between SDAS and carbides for the airfoil blade traces. Archives of Metallurgy and Materials 62 (2017) 235-239. https://doi.org/10.1515/amm-2017-0035

[34] R. Dwornicka, J. Pietraszek. The outline of the expert system for the design of experiment, Prod. Eng. Arch. 20 (2018) 43-48. https://doi.org/10.30657/pea.2018.20.09

[35] J. Pietraszek, N. Radek, A.V. Goroshko. Challenges for the DOE methodology related to the introduction of Industry 4.0. Prod. Eng. Arch. 26 (2020) 190-194. https://doi.org/10.30657/pea.2020.26.33

[36] J. Pietraszek. The modified sequential-binary approach for fuzzy operations on correlated assessments, LNAI 7894 (2013) 353-364. https://doi.org/10.1007/978-3-642-38658-9_32

[37] J. Pietraszek et al. Non-parametric assessment of the uncertainty in the analysis of the airfoil blade traces, METAL 2017 26th Int. Conf. Metall. Mater. (2017) 1412-1418. ISBN 978-8087294796

[38] J. Pietraszek et al. The non-parametric approach to the quantification of the uncertainty in the design of experiments modelling, UNCECOMP 2017 Proc. 2nd Int. Conf. Uncert. Quant. Comput. Sci. Eng. (2017) 598-604. https://doi.org/10.7712/120217.5395.17225

Quality Production Improvement and System Safety: QPI 16 - CZOTO 10   Materials Research Forum LLC
Materials Research Proceedings 34 (2023) 390-399   https://doi.org/10.21741/9781644902691-45

Data Science Challenges of Automated Quality Verification Process in Product Data Catalogues

NIEMIR Maciej [1,2,a] * and MRUGALSKA Beata [2,b]

[1] Łukasiewicz – Poznań Institute of Technology

[2] Poznań University of Technology, Poland

[a]maciej.niemir@pit.lukasiewicz.gov.pl, [b]beata.mrugalska@put.poznan.pl

Keywords: Product Catalogues, Product Data Quality Management, Master Data Synchronization, Machine Learning in Data Quality, GPT-3 in Product Catalogues

Abstract. Product master data are an essential and key component of purchasing processes, ensuring the smooth running of business operations within companies. Unfortunately, due to the lack of a single, complete, worldwide information system storing reference data, managing the data, maintaining its quality, reliability, and timeliness, requires building quality assurance teams for such processes in most companies. There are numerous errors in product data, and identification and correction of them are time-consuming, especially for large data sets that contain many millions of products. These errors are due to the so-called human factor but are also the result of technical errors and limitations of IT systems. Therefore, in the paper, we proposed a number of solutions by category and group that can automate, simplify, and shorten the master data management process. There are also presented examples of data validation using a variety of techniques, rule-based, dictionary-based, and machine learning, that enable mass verification of both images, textual parameters, digital parameters, and classifiers, while indicating the probability of errors in specific attributes as well as in their combination, and in some cases correcting or proposing correct records. The performed tests illustrate the magnitude of problems and potential on a sample dataset.

Introduction

The exchange of digital information is an indispensable part of modern commerce. Good-quality and complete data play a key role [1] [2], as they have a significant impact on the efficiency of business transactions, as well as minimizing the occurrence of errors. Negligence can negatively affect the credibility of companies [3] and generates unnecessary costs [4].

One of the most important areas in terms of information exchange is product data, which is extremely difficult to standardize and maintain due to its diversity in terms of properties and parameters depending on the product group. Work on standardization in various aspects and levels of depth has been ongoing for years. The European Commission is working on regulations, in the interests of consumer safety, the GS1 Organization (GS1 GDM - Global Data Model) is developing models, which indicate the most important elements from the point of view of B2B and B2C cooperation. However, there are no complete and respected standards in this area for all manufactured products worldwide [5]. The problem is exacerbated by the fact that manufacturers do not universally make information about their products available in digital form, so supply chain participants often build product data sets themselves [6], using various available sources or manually creating them from scratch. As a result, commonly used product attributes, such as identification number, name, brand, photo, composition, net content [5], are entered in different, often incorrect ways [7], which can be particularly noticeable for fields that allow the entry of descriptive data or in the form of graphical objects. This, in turn, is very important when shopping on the Internet, where the key to search is not product identifiers, but precisely their names, descriptions, and images [5]. Incorrect data are published on shop websites, search engines, and auction portals, and these are further duplicated and propagated in further databases. As a result of this, manufacturers no longer have control over the actual and reliable description of their products on the Internet, and everyone loses [8].

Alternatively, product data can be synchronized with the catalogues of companies that specialize in obtaining and sharing the correct data, through so-called "data pools". A data pool is a centralized data repository where trading partners (retailers, distributors, or suppliers) can obtain, maintain, and exchange product information in an agreed format. Suppliers can upload data to the data pool, which retailers receive through their data pool. However, not everyone wants to use such a solution, pointing out, among other things, its high cost and lack of flexibility [9]. The solution also does not guarantee the complete assortment needed to populate the database and the set of product data attributes required, and the aggregation of data from different sources again raises the issue of standardization, as can be seen in the results of the study of name similarity in product catalogues [7].

Since all the necessary information is usually found on the label or on an information sheet attached to the product (the manufacturer takes care of these data), as already mentioned, companies often decide to build their own data catalogues, adapting them to their own needs and taking care of their quality on their own, improving the acquired data or creating them from scratch for the IT system. This fact is not surprising given that in e-Commerce, a unique marketing description of a product gives a noticeable competitive advantage ([10] ,[11]), hence the need to combine reliable information from the manufacturer with data created for the shop.

Data collection is an expensive process. Assuming an extremely short period of time - for example, ten minutes to retrieve one product from the warehouse, read the data from the label, verify, and correct the data in the IT system - for a thousand products, the time required to complete the task will be about one month's work for one employee, while with forty thousand, which is the average stock of many B2C companies, it is as much as four years of continuous work. If one were to add to this the need to create one's own product descriptions, the time required to create them would significantly lengthen the entire process, and there is no guarantee that, with such a volume, the data will be entered correctly and that the content of the fields will be consistent and complete. And this is where we touch on the main source of data quality issues. These are:

- error, typos, negligence;
- lack of staff training;
- software errors;
- lack of data standardization.

An important solution to the problem of data entry errors is the implementation of appropriate control procedures [12]. To streamline the process of data control and correction and also to minimize its costs, it is important that this role be taken over as much as possible by the information system. For such purposes, companies use Product Information Management (PIM) software. PIM focusses on the central management of product information to support business processes that involve customer-centric product information [13]. In this classic approach, when developing PIM software, developers and analysts focus on creating advanced data structures, over how information is stored, sometimes on workflows. They are not consulted on technical solutions to improve inter-organizational coordination of master data quality [9]. At the implementation planning stage, the product data found are usually unstructured, erroneous, and reconciling stakeholders and agreeing on the scope of the data is in itself a challenge and requires compromise [14]. Quality control of entries recedes into the background, and the solutions developed generally have only basic data control solutions to:

- Protect the database from critical errors that prevent data storage (limitation of character length or record format);

Quality Production Improvement and System Safety: QPI 16 - CZOTO 10   Materials Research Forum LLC
Materials Research Proceedings 34 (2023) 390-399   https://doi.org/10.21741/9781644902691-45

- Take care of basic data consistency (basic obligatory fields according to the data model, taking care of correct relations in database objects);
- Implement basic data masks (e.g. check digit in the GTIN).

Such a solution usually results in tools that are prone to numerous irregularities on the side of data editing, logic, and field content - the system does not monitor such errors.

An In-Depth Study on the Possibilities of Controlling and Improving Product Data

On the basis of a catalogue with 42 million descriptions of different products collected by 50,000 manufacturers (crowdsourcing), an analysis of the feasibility of additional data validation and quality improvement mechanisms was performed and some of them were investigated. The aim was to find various methods and algorithms that would potentially improve the process of automating data control and quality. In doing so, an innovative assumption was made that validation does not necessarily imply the rejection of the data, but allows for interaction to further validate the data, drawing the user's attention to a potential error. This human-machine interaction provided the opportunity to take a broader view of quality, to include the possibility of prompting, and to validate data that are not 100% certain. Basic industry-independent product attributes [5] were analyzed, enriching them with several additional elements available in the database:

- Basic common attributes: GTIN, Product name, Brand name, Product image, Product classification, Net content with unit;
- Additional common attributes: Product webpage, Marketing description.

Results and Discussion

The preparation of data for analysis should always start with the normalization process in order to standardize the data and prepare them for further verification processes. Skipping this step or reversing the order of operations may result in biases in the performance of other validators, such as the measurement of text field lengths, the examination of duplicates, and the application of dictionary rules. Normalization processes must include, for text fields, conversions to a common character set, and decoding of characters (e.g. HTML entities). For other data, it is necessary to standardize the storage form, e.g., the URL standard (RFC 1738) for web pages and conversions to common numeric formats (text to number conversion, decimal separator conversion). The normalization process itself is usually based on established rules, data converters that do not affect the content of the fields.

Algorithms were run on the database studied, resulting in the correction of up to 9.5% of the products, despite the fact that the system already had basic normalization measures implemented. The analysis showed the need for additional rules, such as the removal of non-printing characters, removal of the transition to a new line in single-line fields, conversion of hard spaces, removal of multiple spaces and tabs, and remnants of underprepared data imports (double quotation marks, quotation marks at the beginning and end of the text, etc.). The prepared, normalized data were further analyzed for the feasibility of using various algorithms and techniques that would be used to build effective data quality validators in two areas:

1. Basic error validation and dependency error validation;
2. Quality validation and suggestions.

For the first area, it was assumed that errors detected would result in the absolute rejection of the data and that the system would not be able to make any changes to the database, since the assumption for this type of validation is to eliminate data that do not comply with standards and logic. The second area includes quality control, which will result in a process in which the user will have to make decisions based on recommendations and/or full communication about the consequences of entering poor quality data.

Quality Production Improvement and System Safety: QPI 16 - CZOTO 10   Materials Research Forum LLC
Materials Research Proceedings 34 (2023) 390-399   https://doi.org/10.21741/9781644902691-45

Validators Based on Defined Patterns and Dictionaries

In order to extend the basic validators implemented in PIM-class information systems, it is useful to review the entire data set by grouping the data of individual attributes to look for specific similar errors to develop common patterns for them. When examining the available dataset, many of such patterns were found that had not been captured by previous validators. For example, the following rule was discovered: If product names do not start with a letter or a number, they are probably incorrect. A test of such a rule by manually verifying 1,000 products confirmed its effectiveness (100%). Likewise, one word meant an error in the product name, as it did not describe the product well enough, taking into account its variants, net content, and brand, and was only a common name describing a type of product, e.g., 'bread.' Another rule is the occurrence of multiple specific characters in the product name (e.g. multiple underscores), which appeared to be 100% indicative of an incorrectly entered product in the database. The number of words in the brand name was also examined, more than four words containing two characters each for the two hundred occurrences tested meant an error in each case. It is worth mentioning that many duplicate names were also found after the data normalization process. This type of validation does not only rely on regular expressions (RegEx) to look for specific patterns in the text. It is also possible to verify the size of the images in terms of their technical parameters such as height, width, and resolution, as well as their compression quality. For web pages, it is possible to verify that the page exists and that it opens correctly. Many validators can also rely on dictionary rules, i.e. specifying prohibited content, e.g. prohibited brand names, prohibited words, and prohibited page content (e.g. page under construction).

The implementation of patterns filled a gap in the validators of the "Basic error validation & dependency error validation" class, giving a large impact on the quality results. For the data set surveyed, 21 million attribute errors were detected in this way. However, these types of validators have significant limitations - the dictionaries need to be updated, e.g. with page templates under construction, with a list of forbidden names, while any validator counting the number of words or regulating the use of specific characters does not guarantee that the data entered will be factually correct, that the product name really is a name and that the image corresponds to it, that the product is correctly classified. Furthermore, it has been noted that when users are informed of a specific problem, e.g. too few characters or too few words, some of them deliberately circumvent the problem by entering additional letters instead of actually correcting the data.

Data Scraping & Structured Data on Websites

Data scraping is the process of extracting specific content from a web page. This makes it possible to obtain and process specific selected extracts from any web page. However, this technique has significant limitations if it is not clear where to look on the page for relevant product attributes. The verification will then be narrowed down to looking for commonalities in product names, brands, and GTINs in page titles or headings to find similarities. However, there is a standard, widely publicized by companies such as Google and Microsoft, that describes how to include metadata on a website so that it is clear where to look for attributes, for example, for product data. The standard is described in detail on Schema.org. The extraction of metadata populated on pages compliant with the schema.org standard for product data can be part of the validation of the product website entered into the PIM system and at the same time part of the validation of other attributes such as name, GTIN, brand, and image. However, it should be noted that schema.org only standardizes metadata, i.e., data about the data, not the content of the data, so the data found are not necessarily a source of reliable information.

The potential for using websites for validation was investigated. From the product database studied, 14447 different domains of websites identified as product pages for 5.8M products were extracted, and then one of the pages in the domain was randomly selected and reviewed. The research found that 15.4% of the domains had the product name stored in the structured data on the sample pages and 2.88% of them had the GTIN number. This gives the number of potential verifications for the products assigned for these domains: 14.66% for names and 5.81% for GTINs in the entire database.

Anomaly Detection

PIM-class information systems for numeric data usually limit values to the range accepted by the database, without paying attention to the constraints imposed by the logic of the field in question. One reason for this is that it is difficult to specify ranges of values when building an information system, especially when they depend on multiple parameters. For example, completely different ranges can be adopted for the net content of a product in the context of different units and for different product categories. However, if the database is filled with a certain number of products, an analysis can be performed on the basis of existing data, and boundary values can be determined with a certain probability of being a potential error. On this basis, the data can be discarded or, safer in this case, a potential data quality problem can be brought to the attention of the user, i.e. a data quality check in terms of 'quality verification and suggestions'. In the product database studied, the net content of the product was analyzed and errors were detected for more than 11% of the products. A deeper analysis of the issue points to the possibility of additional verification methods using machine learning algorithms [15], which makes it possible to verify data not only for extreme data, but also for those that occur within a range if a given numerical field shows such trends. Net content is an excellent example of this type of behavior, as products in certain categories are sold in a standardized packaging or in a certain standardized number of pieces [16].

Semantic Understanding of Textual Attributes

An important area that can significantly improve the validation of text attributes in the context of product data is the area related to natural language processing, in particular, issues related to semantic understanding. Using the so-called transformer architecture [17], on a properly trained language model, it is possible to achieve very good results related to correct text classification, spam detection, grammar verification, spelling [18], text comparison, as well as translation and transformation of text into other text. Data verification capabilities for data imputation [19] are also indicated, allowing attention to deficiencies in critical data for consumer health, such as allergens in food products [16]. Thus, it gives a whole range of possibilities for data correction which are not available to rule-based validators. Unfortunately, the problem in applying such solutions is usually the need to train the model and then properly train the set for appropriate applications, which is a costly process. However, it turns out that this is not the case for all cases and solutions. At present, for some solutions, pre-trained models available under open licenses are sufficient for the tasks, and the fine-tunning can be based on found data, the so-called dirty data.

An analysis was carried out on the studied dataset, on the basis of which the trained BERT ("Bidirectional Encoder Representations from Transformers") model with a classification head learnt from 100k unverified data found in the database (so-called dirty data) was able to successfully assign a product to a classification (in this case the GS1 GPC - "Global Product Classification"). A match of 96% was achieved, while for 4% the algorithm indicated a different category. Based on 200 manually verified samples:

- 44% the algorithm incorrectly suggested a different category.
- 40% the algorithm indicated a better category than the one found
- 8% of the data could not be correctly classified due to an ambiguous product name.
- 8% of the data fit into more than one category and the assignment issue was disputed.

The result of the test tentatively confirms the feasibility and effectiveness of the selected algorithm, even without additional effort to correctly annotate the data. Differences in the result of the algorithm and the user entering the data would be subject to appropriate interaction, an alert with a hint, which fits into the validation class " Quality verification and suggestions".

Quality Production Improvement and System Safety: QPI 16 - CZOTO 10   Materials Research Forum LLC
Materials Research Proceedings 34 (2023) 390-399   https://doi.org/10.21741/9781644902691-45

In addition to the BERT models, which are effective for specialized tasks, but require some training on the relevant data, one can also use the much more extensive Generative Pre-trained Transformer (GPT-3) models [20], which are distinguished by the fact that for some applications they do not need additional training to get good results. The following examples using ChatGPT (GPT-3.5) without any coaching have demonstrated effectiveness in the context of recognizing the correct layout of brand, name and net content in product names, recognizing the correct units of net content, and suggesting the correct product name from the available attributes.

The product names were invented for the study. There was no in-depth research and comparative testing of BERT vs. GPT-3, nor of the effectiveness of the validators on a larger dataset.

Table 1. Examples of data validation solutions using GPT-3.5

Question	Answer
The product has the brand name "Bludoo", a net content of 1kg, and is a washing powder. What name should it have if it starts with a brand name?	"Bludoo Washing Powder 1kg"
Is the full product name 'bludoo' appropriate if the convention of such a name should consist of the brand, the common name, and the net content?	No, the full product name "Bludoo" is not appropriate if the convention for such a name should consist of the brand, the common name, and the net content. In this case, the full product name should be "Bludoo Washing Powder 1kg" as it includes all three elements, the brand name, the common name, and the net content, which adheres to the convention.
Is the product name 'bread' appropriate if it should consist of the brand, name, and net content?	No, the product name 'Bread' is not appropriate if it should consist of the brand, name, and net content. Bread is a common name for a food product and does not provide enough information about the product. The name should include the brand, the specific product name and net content, such as "Brand X Whole Wheat Bread 500g" or "Brand Y Sourdough Bread 1 loaf"
Is the product name "MEGA Men's Shirt 1kg" correct if the name should consist of the brand, common name, and net content or number of pieces?	No, the product name "MEGA Men's Shirt 1kg" is not correct. The weight of 1kg is not appropriate for a shirt as it is not a weight measure that is commonly used for clothing products. Instead, the name should consist of the brand, common name, and number of pieces. A more appropriate name for this product would be "MEGA Men's Shirt - 1 piece" or "MEGA Men's Shirts - Pack of 3"
Is the product name "MEGA Men's Shirt 1pc" correct if the name should consist of the brand, the common name and the net content or the number of pieces sold?	Yes, the product name "MEGA Men's Shirt 1pc" is correct if the name should consist of the brand, the common name, and the net content or the number of pieces sold. It contains the brand name "MEGA", the common name of the product "Men's Shirt", and the number of pieces sold "1pc". This name is informative and adheres to the convention of providing the necessary information about the product.
I already have a product named MEGA Men's Shirt 1pc in the database, and my database should have unique product names. How do I make it stand out?	There are a few ways to make a product with a similar name stand out in a database: Add a unique identifier: You could add a unique identifier such as a SKU or product code to the name of the product, e.g. "MEGA Men's Shirt 1pc - SKU 123456" Include specific details: You could include specific details about the product in the name, such as the color, size, or material, e.g. "MEGA Men's Blue Cotton Shirt 1pc" Add a version number: If you have multiple versions of the same product, you could add a version number to the name, e.g. "MEGA Men's Shirt 1pc v2.0" Use different naming conventions: You could use different naming conventions for different products, e.g. "MEGA Men's Shirt 01" instead of "MEGA Men's Shirt 1pc" Use combination of above: you could use a combination of above methods to make it stand out. It is important to note that the most important thing is to make sure that the new name is unique and clearly different from existing names, while still providing the necessary information about the product.
Is the product name "Bludoo Wshing Powder 1kg" correct?	No, the product name "Bludoo Wshing Powder 1kg" is not correct. It contains a spelling mistake "Wshing" instead of "Washing". The correct name should be "Bludoo Washing Powder 1kg". It is important to ensure that the product names are spelled correctly, as it can lead to confusion and inaccuracies when searching for or identifying the product.
Is the product name "Happido the orange juice 1l" correct?	No, the product name "Happido the orange juice 1l" is not correct. The inclusion of "the" is not necessary and it doesn't add any additional information to the product name. Also, it's not common to include the word "the" when naming a product. The correct name should be "Happido Orange Juice 1l", which is more informative and simple. It's important to use conventions when naming products to make them easily recognizable and searchable.

The product description contains the following information: "High-performance pork ham, steamed, made of joined pieces of meat, with the addition of water. The product contains: pork meat - 76.9%, water, modified starch, salt, thickeners: E407, E415; stabilizers: E331, E452, E450, E451; antioxidant E316, spice extracts (with celery), maltodextrin, citrus fibre, pork collagen protein, flavouring, preservative E250". Is something missing from the description if the product will be sold in the European Union?	The product description you provided contains a lot of information about the ingredients and additives used in the product, but there are a few things that may be missing when the product is sold in the European Union (EU). Allergen information: EU regulations require that products containing certain allergens, such as celery, must be clearly indicated in the ingredients list. Nutritional information: EU regulations require that certain nutritional information, such as energy value, protein, carbohydrates, and fat content, must be provided on the label. Origin: EU regulations require that the country of origin of the ingredients must be indicated on the label. Net weight: EU regulations require that the net weight of the product must be indicated on the label. Storage conditions: EU regulations require that storage conditions must be indicated on the label if they are necessary to ensure the product's safety or quality. It's recommended to check with the relevant authorities in the EU country where the product will be sold to ensure compliance with all the regulations.

The following examples are for illustrative purposes and indicate how the model can be used to validate the data. The following are suggestions for using natural language processing algorithms to create quality validators.

- Detection of whether a product category has been selected correctly. Suggesting the correct category;
- Detection of incorrect product names (spam);
- Detection of similar brands and prevention of duplication. Suggesting the correct brand name;
- Verification of spelling and grammatical errors. Suggesting correct data;
- Suggestion of missing data based on learnt domain knowledge;
- Suggestion of a better product name based on existing attributes;
- Suggestion to change/reformat the data descriptions.

Fig.1. *Answer to the question "Format the description to make it readable for the buyer".*

Enhancing Product Data Validation with Computer Vision

Validation of product data quality is worth enriching with computer vision techniques. With the help of appropriate algorithms, it is possible to verify the correct placement of the product in the photograph, as well as the color and uniformity of the background, which is important for maintaining appropriate standards [5], [6]. When machine learning models are implemented, objects can be recognized and assigned to a category, confirming the consistency of the product name and description with the photo. It can be verified whether the photo shows the packaging of a product, one particular product, or several products, or a product in a scene showing an example of use. You can verify in the case of clothing whether it is the clothing itself or a model presenting it, or you can verify that the photo contains only the company logo. Such validations are of great importance from the point of view of sales in the e-Commerce market [6]. Good-quality product images can be used to conclusively verify the compatibility and consistency of a product's core attributes. With optical character recognition techniques, data can be read from photos, and with natural language processing techniques, data can be compared with attributes stored in text, and errors or gaps in data can be found, which can be crucial, for example, when determining the presence of allergens in food products [16]. The effects of such validators will still not be 100% effective in object recognition or text recognition for a long time, especially when considering a large number of product categories that differ in industry specifics, but they can help improve quality, so it is worth implementing them as suggestions in the "Quality verification and suggestions" class of validators.

An experiment was run on detecting background irregularities in photos on the tested data set to observe the scale of the problem and the need for techniques using image processing. The validator indicated a background color error of 10% of all images submitted.

Conclusions

Maintaining good product data quality is key to minimizing costs in the supply chain. With a large volume of product data, manual data verification is uneconomical and time-consuming, so it makes sense to look for alternatives in the form of suitable automated algorithms implemented in PIM-class systems, especially when the data come from different suppliers or when suppliers themselves enter it into a crowdsourcing model. There are no global standards for most product attributes, which means that data quality problems are not just about simple errors, but are often more complex and widespread. Information systems do not provide such complex and comprehensive algorithms for improving data quality. There are many quality improvement techniques. The process of improving the quality of product data should start with data normalization. This step is crucial as it ensures that all data have a consistent format and structure, making them easier to process and analyze. Simple validation rules and dictionaries can be useful for improving data quality, but they have limitations. These methods only check for basic errors, such as incorrect spelling or formatting, and do not examine the meaning of the content. For a more comprehensive and accurate assessment of data quality, additional validation methods are needed. One such method is the use of natural language processing (NLP) techniques, particularly in the area of transformer-based models. It can help to understand the context and meaning of the data, and detect errors such as inconsistencies or ambiguities in the content. Another valuable tool for improving data quality is computer vision. This technology can be used to validate product images and ensure that they are of the appropriate quality and resolution. This can be particularly important for e-commerce platforms, where product images are the main means of information for customers. Finally, a hybrid combination of multiple validation methods can have a synergistic effect and provide a comprehensive quality assessment of product data. This approach can leverage the strengths of different validation methods and compensate for their limitations, resulting in a more accurate and reliable assessment of data quality.

References

[1] M. Cao, Q. Zhang. Supply chain collaboration: Impact on collaborative advantage and firm performance, J. Oper. Manag. vol. 29 (2011) 163–180. https://doi.org/10.1016/j.jom.2010.12.008

[2] Y. Hole et al. Service marketing and quality strategies, Period. Eng. Nat. Sci. 6 (2018) 182-196. https://doi.org/10.21533/pen.v6i1.291

[3] S.A. Qalati et al. Effects of perceived service quality, website quality, and reputation on purchase intention: The mediating and moderating roles of trust and perceived risk in online shopping, Cogent Bus. Manag. 8 (2021) art.1869363. https://doi.org/10.1080/23311975.2020.1869363

[4] D. Appelbaum et al. Impact of business analytics and enterprise systems on managerial accounting, Int. J. Account. Inf. Syst. 25 (2017) 29-44. https://doi.org/10.1016/j.accinf.2017.03.003

[5] M. Niemir, B. Mrugalska. Basic Product Data in E-Commerce: Specifications and Problems of Data Exchange, Eur. Res. Stud. J. XXIV (2021) 317-329. https://doi.org/10.35808/ersj/2735

[6] M. Niemir, B. Mrugalska. Product Data Quality in e-Commerce: Key Success Factors and Challenges, In: Production Management and Process Control, 36 (2022), AHFE. https://doi.org/10.54941/ahfe1001626

[7] M. Niemir, B. Mrugalska. Identifying the cognitive gap in the causes of product name ambiguity in e-commerce, Logforum 18 (2022) 357-364. https://doi.org/10.17270/J.LOG.2022.738

[8] W.K. Putri, V. Pujani. The influence of system quality, information quality, e-service quality and perceived value on Shopee consumer loyalty in Padang City, Int. Technol. Manag. Rev. 8 (2019) 10-15. https://doi.org/10.2991/itmr.b.190417.002

[9] T. Schäffer, D. Stelzer. Assessing tools for coordinating quality of master data in inter-organizational product information sharing, In: 13th Int. Conf. Wirtschaftsinformatik, February 12-15, 2017, St. Gallen, Switzerland.

[10] T. Wimmer, M. Scholz. Online Product Descriptions–Boost for your Sales? In: 14th Int. Conf. Wirtschaftsinformatik, February 23-27, 2019, Siegen, Germany.

[11] J. Mou et al. Impact of product description and involvement on purchase intention in cross-border e-commerce, Ind. Manag. Data Syst. 120 (2019) 567-586. https://doi.org/10.1108/IMDS-05-2019-0280

[12] A. Haug, J.S. Arlbjørn. Barriers to master data quality, J. Enterp. Inf. Manag. 24 (2011) 288-303. https://doi.org/10.1108/17410391111122862

[13] J. Abraham. Product information management. Theory and practice. Springer, 2014. https://doi.org/10.1007/978-3-319-04885-7

[14] L. Battistello et al. Implementation of product information management systems: Identifying the challenges of the scoping phase, Comput. Ind. 133 (2021) art.103533. https://doi.org/10.1016/j.compind.2021.103533

[15] L. Poon et al. Unsupervised Anomaly Detection in Data Quality Control, In: 2021 IEEE Int. Conf. Big Data, Dec. 2021, 2327-2336. https://doi.org/10.1109/BigData52589.2021.9671672

[16] K. Muszyński, M. Niemir, S. Skwarek. Searching for AI Solutions to Improve the Quality of Master Data Affecting Consumer Safety, In: 22nd Int. Sci. Conf. Business Logistics in Modern Management, October 6-7, 2022, Osijek, Croatia, 121-140. [Online]. Viewed: 10-01-2023. Available: http://blmm-conference.com/wp-content/uploads/BLMM2022_Conference_Proceedings.pdf

Quality Production Improvement and System Safety: QPI 16 - CZOTO 10   Materials Research Forum LLC
Materials Research Proceedings 34 (2023) 390-399   https://doi.org/10.21741/9781644902691-45

[17] J. Devlin et al. BERT: Pre-training of Deep Bidirectional Transformers for Language Understanding, arXiv, 2019, art. 1810.04805. https://doi.org/10.48550/arXiv.1810.04805

[18] S.M. Jayanthi et al. NeuSpell: A Neural Spelling Correction Toolkit, In: Proc. 2020 Conf. Empirical Methods in Natural Language Processing, online, Oct. 2020, 158-164. https://doi.org/10.18653/v1/2020.emnlp-demos.21

[19] W.-C. Lin, C.-F. Tsai. Missing value imputation: a review and analysis of the literature (2006–2017), Artif. Intell. Rev. 53 (2020) 1487-1509. https://doi.org/10.1007/s10462-019-09709-4

[20] B. Ghojogh, A. Ghodsi. Attention Mechanism, Transformers, BERT, and GPT: Tutorial and Survey, OSF Preprints, 17 Dec. 2020. https://doi.org/10.31219/osf.io/m6gcn

Quality Production Improvement and System Safety: QPI 16 - CZOTO 10   Materials Research Forum LLC
Materials Research Proceedings 34 (2023) 400-406   https://doi.org/10.21741/9781644902691-46

Measurements of Vehicle Azimuth Using Acoustic Signals

MADEJ Wiesław

Military Institute of Engineer Technology, Obornicka 136 Street, 50-961 Wrocław, Poland

madej@witi.wroc.pl

Keywords: Acoustic Wave, Microphone Antenna, Propagation, DOA Estimator

Abstract. The location of objects based on the analysis of acoustic signals is widely used in various civil and military systems. Applications that allow the users to determine the location of the sound source are used, for example, in conference rooms for automatic speaker location, enabling unattended camera control, detecting and tracking the movement of objects in area surveillance systems, etc. The paper presents a system for determining the azimuth of a moving motor vehicle developed at the Military Institute of Engineer Technology in cooperation with the Wroclaw University of Technology. The system is designed for the detection, identification and location of armored vehicles and trucks. The use of acoustic signal analysis methods to locate objects allowed for the construction of passive systems that are difficult to detect. This property is particularly important in area surveillance systems and military equipment designs.

Introduction

Technologies based on acoustic signal processing are widely used in various civil- and military systems. The analysis of the acoustic signal parameters enables the identification and localization of their sources. Solutions of this type, those which enable determination of the location of the sound source, are used to detect and track the movement of objects in a military protected areas under surveillance, to detect, identify and locate armored fighting vehicles and trucks, e.g. in anti-tank mines or in systems that locate the shooter's position etc.

The marking of the sound source relative to the measurement system is carried out by determining the direction of the acoustic wave reaching the microphone array. The problem of estimating the direction of arrival of the wave is an issue known as DOA (Direction of Arrival) and is widely described in the literature. Determining the DOA under real conditions is not a simple task due to the influence of the natural environment on the manner the acoustic wave propagates. A pressing problem is the effect of background noise, reflections and interference. A number of algorithms have been developed to solve this issue, among which we can distinguish two basic methods: using estimators of Time Differences of Arrival (TDOA) and estimators of spatial spectrum distribution [1-3]. To a large degree, the efficiency of DOA estimation depends on the fulfilment of the conditions of linearity and isotropic character of the transmission medium and the assumption that the signal source is located at a large distance, so the wave reaching the microphone array can be treated as a plane wave [4].

DOA methods

Two methods of DOA estimation are presented below. They have been analyzed in terms of their applicability in the developed system for tracking the location of motor vehicles.

The TDOA method is based on determining the time (phase) relationships between the signals received from the individual microphones of the array (microphone antenna), the geometry of the antenna and the direction of arrival of the acoustic wave. The spatial distribution of the microphones must be precisely defined so that it is possible to unambiguously determine the appropriate relationships between the received signals. For long distances between the vehicle and the microphone antenna, the vehicle can be treated as a point sound source. In free space, a wave propagates spherically in all directions from a point source. Free space means space in which the influence of other phenomena accompanying wave motion, such as reflections, refractions or wave scattering, is limited to a minimum. In such a situation, with small antenna sizes, we can assume that the incoming wave is a plane wave [2, 5].

In order to ensure the possibility of locating the object for each direction, a study was carried out for a microphone antenna consisting of 5 microphones with the same omnidirectional characteristics, arranged in accordance with Fig.1. This solution guarantees no privileged directions, and thus the same parameters in the entire measuring range (0÷360°).

Microphones marked with indexes 1, 2, 3 and 4 were placed evenly in a circle with a radius R in one plane, while the reference microphone (with index 0) was placed at the center of the array, i.e. at the origin of the reference system.

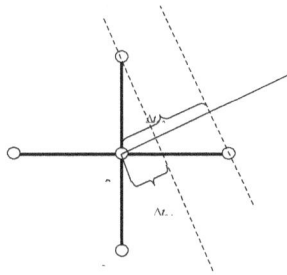

Fig.1. *Phase delay between the antenna microphones*

Determination of the direction of the arrival of the wave is conducted based on measuring the delays of signals received by individual microphones. For the three microphones $(0,1,2)$ and $(0,3,4)$, the following relationships result from the geometrical properties of the described system [7]:

− in the case of microphones $(0,1,2)$:

$$\sin \Phi = \frac{v\Delta t_{0,1}}{R} ; \quad \cos \Phi = \frac{v\Delta t_{0,2}}{R} ; \tag{1}$$

− in the case of microphones $(0,3,4)$:

$$\sin \Phi = \frac{v\Delta t_{0,3}}{R} ; \quad \cos \Phi = \frac{v\Delta t_{0,4}}{R} ; \tag{2}$$

where:

$\Delta t_{i,j}$ - delays between signals picked up by a pair of microphones i, j;
v - the speed of sound;
R - antenna radius;
ϕ - azimuth angle for the sound wave source.

The azimuth angle for the sound wave source was determined based on the knowledge of signal delay from the following dependence:

$$tg\phi = \frac{\Delta t_{0,1}}{\Delta t_{0,2}} \text{ lub } tg\phi = \frac{\Delta t_{0,3}}{\Delta t_{0,4}} \tag{3}$$

To increase the reliability of the result, all pairs of microphones should be analyzed.

There are many methods of beam formatting based on the estimation of the spatial distribution of the spectrum. Most algorithms work well with a single sound source. In a situation where there are many undesirable sound sources and a high level of noise, good results are yielded using the Capon method [2, 3]. This method is often referred to as Minimum Variance Distortionless Response (MVBR) beamforming.

As opposed to noise, the signals of the tracked signal source coming from the individual antenna microphones are correlated. Using this property, it is possible to eliminate the signal noise by estimating the cross-covariance R_{xx} determined from N samples [8]:

$$R_{xx} = \sum_{n=1}^{N} x(t_n) x^T (t_n) = \frac{1}{N} X^H X \qquad (4)$$

$$X = [x(t_1)\ x(t_1)\ ...\ x(t_n)]^T \qquad (5)$$

where:
    t – time;
    X – input signal;
    H – Hermitian transpose symbol (simultaneous transposition of T and complex matrix conjugate).

The direction of the beam is controlled by the appropriate selection of the so-called "control vector". The output signal of the antenna can be written in the form of the relationship [8]:

$$y(t) = w^H x(t) \qquad (6)$$

where:
    t – time;
    x – input signal vector;
    w – weighted vector.

The total power of the antenna output signal is determined by the relationship [8]

$$P(w) = \frac{1}{N}\sum_{n=1}^{N} |y(t_n)|^2 = \frac{1}{N}\sum_{n=1}^{N} w^H x(t_n)\, x^H(t_n) w = w^H R_{xx} w \qquad (7)$$

The MVDR method minimizes the power of the received signal $P(w)$ outside the viewing direction θ provided that:

$$w^H a(\theta) = 1 \qquad (8)$$

where: $a(\theta)$ – steering vector.

The fulfilment of the above requirements is ensured by the weight vector and is determined by the relationship [8]:

$$w = \frac{R_{xx}^{-1} a(\theta)}{a^H(\theta) R_{xx}^{-1} a(\theta)} \qquad (9)$$

Substituting the above dependence into formula (7) we obtain [8]:

$$P(\theta) = \frac{1}{a^H(\theta) R_{xx}^{-1} a(\theta)} \qquad (10)$$

Quality Production Improvement and System Safety: QPI 16 - CZOTO 10   Materials Research Forum LLC
Materials Research Proceedings 34 (2023) 400-406   https://doi.org/10.21741/9781644902691-46

Measurement System

Having analyzed the properties of various DOA methods, expecting a large impact of noise and interference on the operation of the system, the MVDR algorithm was used, which is considered optimal due to the maximization of the signal-to-interference ratio [3].

The measurement system enabling the implementation of the above method of estimating the direction of arrival of the acoustic wave requires the use of solutions that guarantee adequate computing power. This necessity results from the adopted assumption that the system should operate in real-time, and the acceptable level of delay in the results obtained is 0.5 seconds. For this reason, a DSP processor performing all the calculations was used in the system. The process of signal sampling is particularly important for the correct operation of the system. It is necessary to ensure the synchronization of all A/D converters, which eliminates delays between signals at the stage of input signal processing. Errors resulting from the imprecise determination of the sampling moment and phase shifts in individual analogue signal processing paths are impossible to detect and remove in the subsequent stages of signal processing. Fig.2 shows a block diagram of the measuring system. The signals from the microphones are amplified and filtered. The applied low-pass filter is designed to eliminate errors resulting from the phenomenon of aliasing. Aliasing leads to irreversible distortion of the signal due to the penetration of high-frequency harmonics into the low-frequency region due to failure to meet the requirements of the Whittaker–Kotelnikov–Shannon sampling theorem. The operation of the A/C converters is controlled by the synchronization system. The DSP processor processes the received data stream from the A/D converter card and controls the amplification of the amplifiers and the operation of the synchronization system, while the results are sent via the RS485 serial link to the PC.

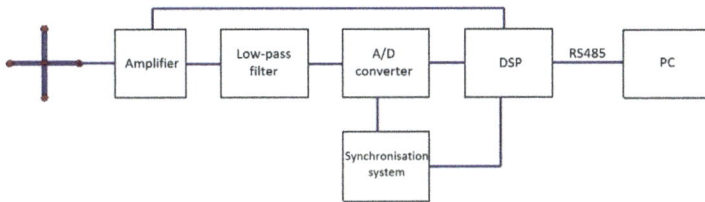

Fig. 2. Measurement system

The antenna used during the measurements is shown in Fig. 3. The distance between the microphones meets the following relationship:

$$D \leq \frac{\lambda}{2} \tag{11}$$

where: D – the distance between the microphones, λ – minimal wavelength.

Fig. 3. *Microphone antenna*

Results

Verification of the system operation was carried out through a series of measurements of the azimuth of the vehicle moving at a constant speed in a straight line over a distance of approx. 200 m. The measurements were carried out for various distances from the antenna of the vehicle's track. Comparison of the obtained results with the theoretically calculated azimuth of the vehicle at a given point on the route made it possible to estimate the obtained measurement accuracy. Theoretical azimuth values were determined from the following relationship:

$$\varphi = arct\left(\cfrac{1}{\cfrac{vt}{d_0} + ctg\left(\cfrac{s_0}{d_0}\right)} \right) \tag{12}$$

where:

φ – azimuth of the vehicle
v – vehicle velocity
d_0 – distance to vehicle route (track)
s_0 – vehicle initial distance (see: Fig. 4).

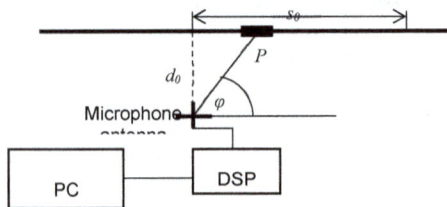

Fig.4. *Arrangement of the system components during the measurements*

Fig. 5 and 6 show: the measured azimuth values (A), the calculated values (Φ) and the measurement error determined as |A-Φ| for two directions of travel for an MTLB tracked armoured fighting vehicle with d_0 distances equal to 10 m and 80 m.

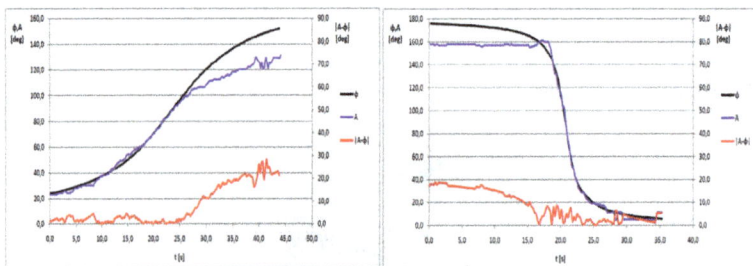

Fig.5. *MTBL vehicle traverses a distance 10 m away*

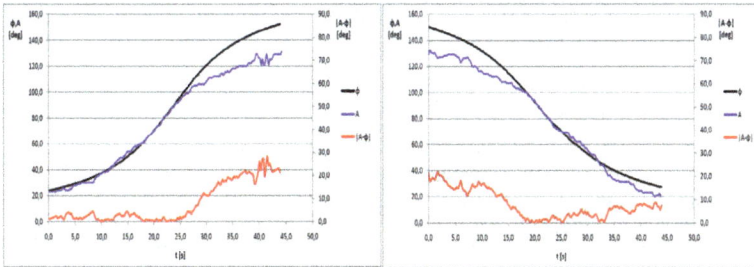

Fig.6. *MTBL vehicle traverses a distance 10 m away*

The obtained results show that the developed system enables the determination of the azimuth of the vehicle with an error under a few degrees. The errors are highly dependent on the distance of the vehicle from the microphone antenna and reach their minimum when the vehicle passes the antenna. This effect is caused by a decrease in the signal level with increasing distance, and thus a stronger influence of noise and unfavourable phenomena related to the propagation of acoustic waves. Furthermore, the observed deviations from the theoretical values for large azimuths result from the fact that in the final part, the vehicle route (track) is curved.

The estimation of the azimuth of the moving vehicle is delayed by 0.25 s. This is a satisfactory result and meets the initial guidelines.

Conclusions

The developed system allows the vehicle azimuth to be measured with satisfactory accuracy in the majority of cases. However, in unfavorable situations, the operation of the system may be strongly dependent on undesirable phenomena related to the propagation of acoustic waves. Important factors of this type include the impact of wind and terrain, with the former being capable of disturbing the direction and speed of sound wave propagation. In the case of large undulations of the terrain, the so-called dead spaces, where the sound, even at a short distance from the sound source, will be weaker or there will be no sound at all. Such a situation may make it impossible to determine the direction with the desired accuracy.

The localization of object direction based on passive acoustic signals emitted by the object is a useful method for spatial orientation, especially when visual techniques are not available. The presented approach seems to have potential applications in medical diagnostics [9] for more precise localization of musculoskeletal injuries and, on a microscopic scale, for detecting surface defects with applied special coatings [10]. Image analysis techniques, including acoustic analysis in this case, as well as statistical methods [11, 12], including non-classical approaches [13, 14], will be helpful in data analysis.

References

[1] J. Dmochowski, J.Benesty, S. Affes. Direction of Arrival Estimation Using the Parameterized Spatial Correlation Matrix, IEEE Transactions on Audio, Speech, and Language Processing 15 (2007) art. 4156192. https://doi.org/10.1109/TASL.2006.889795

[2]V. Kunin et al. Direction of Arrival Estimation and Localization Using Acoustic Sensor Arrays, J. Sensor Technol. 1 (2011) art. 7657. https://doi.org/10.4236/jst.2011.13010

[3] K. Chaudhari, M. Sutaone, P. Bartakke. Comparative Analysis of Planar Acoustic MVDR Beamformer, Int. J. Circuits, Syst. Signal Process. 13 (2019) 732-738.

[4] F. Polak, W. Sikorski, K. Siodła. Location of partial discharges sources using sensor arrays, Przeglad Elektrotechniczny 90(10) (2014) 74-77. https://doi.org/10.12915/pe.2014.10.19

[5] V. Krishnaveni, T. Kesavamurthy, B. Aparna. Beamforming for Direction-of-Arrival (DOA) Estimation-A Survey, Int. J. Comp. App. 61(11) (2013) 4-11. https://doi.org/10.5120/9970-4758

[6] S. Paulose, E. Sebastian, B. Paul. Acoustic Source Localization, Int. J. Adv. Res. Electrical, Electronics and Instr. Eng. 2 (2013) 933-939.

[7] M. Górski, G.Haza, R.Hossa, J. Zarzycki. Opracowanie układów identyfikacji i lokalizacji celu zapalnika kierowanej miny przeciwpancernej, Zakład Teorii Sygnałów Instytut Telekomunikacji, Teleinformatyki i Akustyki Politechniki Wrocławskiej, Wrocław 2010.

[8] Z. Chen, G. Gokeda, Y. Yu. Introduction to Direction-of-Arrival Estimation, Artech House, London, 2010. ISBN 978-1596930896

[9] B. Jasiewicz et al. Inter-observer and intra-observer reliability in the radiographic measurements of paediatric forefoot alignment, Foot Ankle Surg. 27 (2021) 371-376. https://doi.org/10.1016/j.fas.2020.04.015

[10] N. Radek et al. Microstructure and tribological properties of DLC coatings, Mater. Res. Proc. 17 (2020) 171-176. https://doi.org/10.21741/9781644901038-26

[11] J. Pietraszek et al. The parametric RSM model with higher order terms for the meat tumbler machine process, Solid State Phenom. 235 (2015) 37-44. https://doi.org/10.4028/www.scientific.net/SSP.235.37

[12] R. Dwornicka, J. Pietraszek. The outline of the expert system for the design of experiment, Prod. Eng. Arch. 20 (2018) 43-48. https://doi.org/10.30657/pea.2018.20.09

[13] J. Pietraszek. The modified sequential-binary approach for fuzzy operations on correlated assessments, LNAI 7894 (2013) 353-364. https://doi.org/10.1007/978-3-642-38658-9_32

[14] J. Pietraszek et al. Factorial approach to assessment of GPU computational efficiency in surrogate models, Adv. Mater. Res. 874 (2014) 157-162. https://doi.org/10.4028/www.scientific.net/AMR.874.157

Quality Production Improvement and System Safety: QPI 16 - CZOTO 10   Materials Research Forum LLC
Materials Research Proceedings 34 (2023) 407-413   https://doi.org/10.21741/9781644902691-47

Ergonomics of Organizational and Technical Space in the Educational Process of Children in Kindergarten

NICIEJEWSKA Marta [1a]

[1]Czestochowa University of Technology, ul. Armii Krajowej 19B, Poland

[a]marta.niciejewska@pcz.pl

Keywords: Ergonomics, Ergonomic Design, Kindergarten, Occupational Health and Safety, Occupational Safety Management, Technical and Organizational Factors

Abstract. Ergonomics is an interdisciplinary science that applies to various aspects of our lives, extending beyond the workplace. Ergonomics enhances work, education, and leisure activities for lifelong well-being. This article presents the ergonomic principles of the preschool child's world that influence the correct psychophysical development of the child. A quantitative study was conducted in seven non-public kindergartens in the Silesian voivodeship, using direct interviews and elements of overt observation. The study aimed to identify strengths and weaknesses in the ergonomics of kindergartens' organizational and technical spaces, specifically focusing on legal requirements for safe and ergonomic working conditions for children up to the age of 6. The results highlight the challenges faced by designers and kindergarten directors in creating an ergonomic environment for children, revealing numerous difficulties in this area. There is increasing recognition of the importance of ergonomic and safe organizational and technical conditions in early childhood education. The obtained research results are a valuable contribution to a larger study on ergonomics and safety in Polish pre-schools, scheduled for the coming years.

Introduction

Ergonomics is an applied science that is present throughout our lives. Of course, its beginnings, which date back to 1857, concerned primarily work. The name is taken from the Greek: ergon - work, nomos - natural laws. According to the then view, ergonomics as an applied science should define the interdependence between the operation of tools, devices and machines and the protection of the environment in the optimal scope to the possibilities and psychophysical needs of a human being. The goal formulated in this way and the possibilities in relation to humans should be based primarily on the latest achievements in science and technology. Today, after more than 160 years, ergonomics applies to every human activity. We distinguish the ergonomics of work, product and production, but also the ergonomics of living, leisure, the ergonomics of the elderly and the disabled. One of the youngest ergonomics, both at the level of concept and correction, is the ergonomics of the children's world [1-3].

The first stage of a child's education is usually the pre-school period. The architecture and space of the kindergarten building affect the organization of work, as well as aspects related to behavioral needs, including the child's sense of security. Therefore, a well-designed building should also contribute to reducing the sense of stress in children. Equally important is the architecture of the kindergarten building itself - the body, facade solutions, entrance area, development of the facility's surroundings. Particularly important design criteria are the shape of the building and the facade solution, including facade divisions, colors, materials used, and possible graphics emphasizing the character of the building's function. By using appropriate denotative elements, the function of the kindergarten facility is properly recognized and associated in the urban space. The architecture of the building should take into account the aesthetic needs of the main users, i.e. children, i.e. evoke good associations, reactions, evoke positive feelings, and thus encourage attending kindergarten and contacts with peers. Moderation and restraint in the use of colors, graphic forms or details are important in designing the "enclosure" of the kindergarten building, because it is important that the architecture is not infantile, too literally imitating phenomena from the world of nature or fairy tales. The context of the place, the character of the street, housing estate or district in which the kindergarten is located are also important. In any case, however, the function of the kindergarten facility will be characterized by: form, formal and aesthetic features - colors, details, possible graphic elements of the façade, as well as the development of the surroundings, in which the playground and greenery play the dominant role. However, taking into account the needs, possibilities and psychophysical limitations of a preschool child, attention should be paid primarily to the equipment of the kindergarten and the organization of activities. They create an organizational and technical space that is very important in the educational and upbringing process of children in kindergarten. Ergonomics is a very important reference to the entire environment, which consists of e.g. volume of rooms, access to natural lighting, materials that make up floors, walls, equipment, tables, chairs, as well as their adaptation to the anthropometric dimensions of a preschool child [3-12].

Methods

Ergonomics of the child's world is based on general principles of organizational and technical space design based on the selection of appropriate technical equipment and the adjustment of material parameters to the needs of ensuring full safety of use [4-7]. This paper analyzes the organizational and technical space in a kindergarten dedicated to children aged up to 6 years. Children at this age have special needs, because between 3 and 6 years of age is the first critical period of the child's development, in which, in the event of an oversight in the technical design of space, disturbances may occur. Poorly designed space and technical equipment may contribute to an anterior tilt of the pelvis, which results in an enlargement of the lumbar lordosis with simultaneous flattening of the abdomen. Therefore, this is a period in a child's life when the first symptoms of body posture and habits are formed, which very often mark the child's further activity and body positions in relation to sitting on a chair, sitting at a table or on other flat surfaces, e.g. on the floor, carpet, etc. In most European countries, there is a statutory obligation to equip educational institutions, including kindergartens, with ergonomic furniture.

About 20 kindergartens were subject to observation in terms of technical equipment and space as well as the selection of equipment. All kindergartens were located in Poland - 6 of them are private kindergartens, the rest are financed from the state budget (public funding). The qualitative study – a face-to-face interview with elements of open observation – were very similar in terms of the size of the building, the arrangement of individual rooms, the number of children, teachers and equipment.

The face-to-face interview was conducted using a checklist whose questions were based on checklists prepared by the Central Institute for Labor Protection - PIB and the California Childcare Health Program of the University of California San Francisco. The TOL classification was used to develop the checklist. This method assumes that an accident occurs as a result of a component of three types of indirect causes: technical, organizational and human. Consequently, in order to achieve the best results in building a safe and ergonomic environment for people, one should strive to maintain a balance between technical, organizational and human activities. The questions, in accordance with the assumptions of the method, were grouped into technical, organizational and human factors. The technical factors were divided into three areas that were subjected to the study: an educational and leisure room, communication routes and hygienic and sanitary rooms. Each question could be answered with a "yes" or "no" answer. Negative answers ("no") were aimed at identifying weak points that may contribute to the occurrence of accidents in the kindergarten. On the other hand, the answers to "yes" were to indicate the strengths of the surveyed kindergartens in relation to the level of ergonomics of the kindergarten and the safety of children.

Quality Production Improvement and System Safety: QPI 16 - CZOTO 10   Materials Research Forum LLC
Materials Research Proceedings 34 (2023) 407-413   https://doi.org/10.21741/9781644902691-47

Results and Discussion

It is very important to indicate the key aspects that have been observed, these include: selection of construction, selection of materials, lighting, ensuring the proper climate (temperature and ventilation in rooms), ergonomics of movement in space, technical protection (balustrades, covers, etc.), and many others.

All remarks and conclusions were formulated in relation to legal provisions and documents (or standards) which regulate the requirements for technical equipment in kindergartens.

One of the most important issues that should be taken into account when choosing school and kindergarten equipment are legal requirements (on the example of Poland). The main document regulating the selection conditions is § 9 sec. 3 of the Regulation of the Minister of National Education and Sport of December 31, 2002, which reads as follows: "Schools and institutions purchase equipment with appropriate attestations and certificates." The certification obligation covers skeletal products, such as benches, tables and chairs, and box items, i.e. wardrobes. In addition, the obligation also applies to display cases and hangers. Obtaining the certificate is aimed at ensuring the safety of children in kindergartens. In order to obtain the certificate, it is necessary to submit an application by the manufacturer who plans to introduce products to kindergartens. The certification is based on the PN-F-06010-01:1990, PN-EN 1729-1:2016 and PN-EN 1729-2+A1:2016 standards, which are assigned on the basis of the technical documentation of the furniture. Before issuing the certificate, it is also necessary to test all the previously mentioned items of equipment.

Due to the fact that kindergarten furniture is subjected to everyday use, it should be made of high-strength materials that will ensure full safety for users. Particular attention is also paid to the selection (if possible) of natural materials - eg wood, wooden veneers. . However, if there are plastics or materials of chemical origin, they must have an appropriate certificate. An important issue is also the selection of furniture that will be appropriate for the height of students. This has been described in the PN-EN 1729-1:2016-2 standard. It specifies the heights of seats and tables as well as color codes that should be used for individual groups of furniture. Another issue is the color of school furniture. Taking into account the hygiene of children's eyesight, countertops with a matte and smooth surface will work best in a kindergarten. It is also worth choosing products that will be resistant to abrasion, water, grease or impact.

In Polish preschool institutions, more and more attention is paid to safe and ergonomic equipment. All tables and chairs must be conducive to children's development. Ergonomic furniture has all the approvals that confirm that the product will be perfect for schools and kindergartens. It is worth noting that poorly selected equipment of rooms in a kindergarten can significantly reduce the efficiency and cognitive functions of children. Uncomfortable and badly shaped chairs and tables contribute to the development of spinal degeneration, which is difficult to cure. In the case of a young organism, susceptibility to diseases may be even greater. Certificates confirm compliance with all ergonomic principles. A very important role is played by the proper adjustment of the seat and backrest. First of all, in modern preschool institutions, hard wooden chairs are being replaced by soft seats. The aforementioned European Union standard EN 1729 precisely defines how to adjust furniture to the needs of preschool children. Chairs and armchairs must be adapted to the natural curvature of the human spine.

Ergonomics is not only about school furniture. It is equally important to ensure the safety of the youngest, i.e. children in kindergartens. During crazy games, various unforeseen accidents can happen. Frames, doors, free-standing shelves and all furniture with sharp corners are a great danger. Therefore, it is worth planning the entire space in the kindergarten correctly so that children do not hurt themselves. It is a good idea to use special covers, e.g. for door frames. One could be tempted to assess the "occupational" risk for children in the space of such a room. In spaces for children, it is better to limit the amount of glass or mirrors. On the floor there should be a good quality soft carpet that prevents the wards from getting cold. All these elements will create a space ideal for the development of the youngest. There are several more factors influencing good conditions for development.

Good lighting must be installed in kindergartens. In each classroom there should be several lamps with a neutral color of light - 3000 Kelvin. Of course, no lamp must blink as this can be distracting and even cause vision problems. In addition, in rooms intended for small children, it is not recommended to use open luminaires with easy access to bulbs that can break while playing, and instead of popular discharge fluorescent lamps full of heavy metals, it is recommended to choose much safer LED bulbs, which, if broken, do not will emit harmful substances into the environment.

Teachers and babysitters in kindergartens must remember to regularly ventilate the rooms. Crowded kindergartens or classrooms do not give freedom of movement and can lead to numerous stressful situations. Therefore, each student must have their own personal space.

A big problem both in homes and institutions is the failure to adjust the dimensions of furniture to its users. This is due to the financial situation and the lack of awareness of both parents and institutions. In Polish homes, children's furniture is often bought "over the top" and the choice of equipment is determined by the price, not ergonomics. Assigning furniture - tables and chairs to a specific child may be the cheapest and easiest solution, although it is still not a routine activity in many institutions. Ill-fitting furniture worsens the working conditions of internal organs, may result in the emergence or development of posture defects and vision defects.

According to the regulation on technical conditions, if children are expected to stay in buildings without constant supervision, railings should be secured against climbing up and sliding down the railing. In rooms where children may be present, all sources of heat emitting should be shielded - e.g. radiators to prevent them from having direct contact with the hot surface and to prevent burns. According to statistical data, more than 1% of children under the care of educational institutions (hereinafter: "facilities") are involved in accidents in Poland. In 2020/2021, 139 accidents of children in kindergartens were recorded in the Śląskie Voivodeship, including one serious accident. The most common causes of injuries were the child's inattention, other causes, unintentional actions of the child and other people. There have also been intentional actions by others and unintentional strikes. Only 1% of the causes of injuries were deliberate actions of children and poor technical condition of the equipment.

Pursuant to the Education Law, the kindergarten implements the core curriculum for pre-school education. Not all children develop in the same way, therefore the kindergarten should take into account and respect the individual needs of each pupil. The content of education should be selected adequately to the child's level of development, taking into account their interests and way of reasoning. The kindergartens that voluntarily took part in the qualitative study – a face-to-face interview with elements of open observation – were very similar in terms of the size of the building, the arrangement of individual rooms, the number of children, teachers and equipment. Starting the study, the author of this article assumed the fact that no kindergarten is able to guarantee a completely risk-free environment. However, it should strive to minimize the risk by removing causes that may threaten the life or health of children and appropriate planning in order to effectively respond to these threats. By analyzing the entire research process, the weaknesses most frequently indicated by the surveyed teachers were grouped, consisting of various elements of the organizational and technical area. These are primarily:

- tables and chairs of the same size for all children, regardless of their height, without the possibility of adjustment,
- hard tables and chairs, made of wood, not very comfortable to sit for a long time,
- no rounded edges of chests of drawers, tables and window sills,

- no carpet/carpet attached to the floor over the entire surface,
- lack of radiator covers protecting against direct contact with the heating surface and the possibility of moving the knob regulating the radiator heat,
- too poorly lit corridor leading to the toilets for children,
- lack of appropriate, average height of washbasins for children - too low for older children,
- inadequate organization of the "working" space for artistic and technical tasks - lack of appropriate protection of the remaining elements of the equipment against damage and soiling.

The strengths of the organizational and technical space, on the other hand, include the following elements - indicated by the surveyed teachers and observed by the author of this article:

- very ergonomically friendly changing rooms with colorful furniture, individualized both for the characteristics of healthy children and those with various disabilities (e.g. for a child with poor eyesight, the locker was marked with bright colors and the name was written in bold);
- benches in the dressing rooms are stable, have the appropriate width of the seat part, with a suitable place for changeable shoes;
- changing rooms have adequate ventilation and access to natural light.
- passageways and communication routes are sufficiently wide.

Of course, ergonomics is not only the elements included in the organizational and technical area. Ergonomics are also those elements that determine the appropriate psychophysical well-being, i.e. physical and psychosocial factors and environmental conditions in which they spend most of their time during the day. This article, however, focuses primarily on the organizational and technical area of kindergartens, which are the basis for the proper formation of both the body and the soul of a small person. It should be remembered that there are elements of ergonomic space that are the basis for further, appropriate psychophysical development of a child at preschool age.

It is obvious that each child has different perceptual abilities, imagination and perceives the world differently. It also has other anthropometric dimensions. Therefore, both furniture, chairs, tables, as well as devices, toys and equipment that he uses throughout his preschool life, should be adapted to his - already mentioned - anthropometric dimensions, as well as possibilities, needs and psychophysical limitations. It is very important to ensure the freedom to play, develop and rest, provide support to children who develop in a dissonant, slower or accelerated way, support children's independent discovery of the world, including the world of nature, create a variety of natural situations that build the child's sensitivity, support in building a system of values and familiarizing with social norms, supporting body activity through e.g. physical, environmental and music classes, learning about safety, including road safety, learning independence and creating healthy habits, preparing to start learning at school by supporting mechanisms of learning.

Summary

The basis of safety and ergonomics in the kindergarten environment, which should be provided to the child obligatorily, because it has reference to legal provisions, regulations and standards, is the most important. The key aspects in that have been indicated: selection of construction, selection of materials, lighting, ensuring the proper climate (temperature and ventilation in rooms), ergonomics of movement in space, technical protection (balustrades, covers, etc.), and many others. The most common poorly prepared space elements include: tables and chairs of the same size for all children (lack of personal adjustment), heavy tables and chairs, no rounded edges of chests of drawers, lack of radiator covers protecting against direct contact with the heating surface, too poorly lit corridor, lack of appropriate, average height of washbasins for children – too low for older children, inadequate organization of the "working" space for artistic and technical tasks – lack of appropriate protection of the remaining elements of the equipment against damage and soiling. The safety of children in kindergarten is an absolute priority. However, organizational measures alone will not ensure their safety. Technical measures must be employed, including proper materials [13], shaping of surfaces [14] in conjunction with appropriate protective coatings [15, 16], and for highly critical components, special coatings [17, 18]. Reliable and strong welds [19], appropriately processed and secured, are essential elements. The proper location and securing of windows [20] also significantly impact safety. All these elements are numerous and mutually influential, thus requiring an appropriate analysis of their interactions [21]. In this regard, suitable statistical methods [22-24] come to the aid, including non-classical approaches that do not require a priori assumptions about the model's form [25-27].

References

[1] R. Michnik et al. Research on the Stability of the Users of Chair with a Spherical Base. Adv. Intell. Syst. Comput. 623 (2017) 299-307. https://doi.org/10.1007/978-3-319-70063-2_32

[2] A.K. Mańka et al. Ergonomics of the Early Schoolchildren's Position in Terms of the teaching and Learning Process Efectiveness, Edukacja-Technika-Informatyka 3 (2022) 133-140.

[3] K. Goniewicz et al. Management of child injuries in traffic and other accidents: the WHO policy guidelines, Health Problems of Civilization 12(3) (2018)157-162.

[4] J.S. Antonenko et al. Design Features of Furniture and Equipment for Entrance Areas of Kindergartens. IOP Conf. Ser. Mater. Sci. Eng. 463 (2018) art. 042013. https://doi.org/10.1088/1757-899X/463/4/042013

[5] M. Grozdanovic et al. Methodological framework for the ergonomic design of children's playground equipment: A Serbian experience. Work 48 (2014) 273-288. https://doi.org/10.3233/WOR-131661

[6] B. Iliev et al. Comparison of anthropometric dimensions of preschool children and chairs in kindergartens in North Macedonia, Bulgaria and Croatia. Heliyon 9 (2023) art. e14483. https://doi.org/10.1016/j.heliyon.2023.e14483

[7] D. Domljan et al. Equipping building for upbringing and education. Volume I. University of Zagreb, Zagreb, 2015. ISBN 978-9532920314

[8] G.T. Jones, G.J. Macfarlane. Epidemiology of low back pain in children and adolescents, Arch. Dis. Child. 90 (2005) 312-316. https://doi.org/10.1136/adc.2004.056812

[9] L. Liu et al. Ergonomics Design Research on Writing Tools for Children, Procedia Manuf. 3 (2015) 5859-5866. https://doi.org/10.1016/j.promfg.2015.07.890

[10] C.L. Bennett, D. Tien. Ergonomics for Children and Educational Environments - Around the World. In: Proc. the XVth Triennial Congress of the Int. Ergon. Assoc. and the 7th Joint Conf. of Ergon. Soc. of Korea, Seoul, Korea, 2003. ISBN 978.8990838001

[11] J. Charytonowicz, A. Jaglarz. Ergonomic Formation of Hygienic-Sanitary Spaces in Consideration of Health, Safety and Well-Being of Children. Adv. Intell. Syst. Comput. 600 (2018) 64-76. https://doi.org/10.1007/978-3-319-60450-3_7

[12] P.A. Farahani et al. Design of an Ergonomics Assessment Tool for Playroom of Preschool Children. Physical Treatments 6 (2017) 217-226. https://doi.org/10.18869/nrip.ptj.6.4.217

[13] P. Jonšta et al. The effect of rare earth metals alloying on the internal quality of industrially produced heavy steel forgings, Materials 14 (2021) art.5160. https://doi.org/10.3390/ma14185160

Quality Production Improvement and System Safety: QPI 16 - CZOTO 10　　　　Materials Research Forum LLC
Materials Research Proceedings 34 (2023) 407-413　　　　https://doi.org/10.21741/9781644902691-47

[14] N. Radek et al. The influence of plasma cutting parameters on the geometric structure of cut surfaces, Mater. Res. Proc. 17 (2020) 132-137. https://doi.org/10.21741/9781644901038-20

[15] N. Radek et al. Technology and application of anti-graffiti coating systems for rolling stock, METAL 2019 28th Int. Conf. Metall. Mater. (2019) 1127-1132. ISBN 978-8087294925

[16] N. Radek et al. Formation of coatings with technologies using concentrated energy stream, Prod. Eng. Arch. 28 (2022) 117-122. https://doi.org/10.30657/pea.2022.28.13

[17] N. Radek et al. Microstructure and tribological properties of DLC coatings, Mater. Res. Proc. 17 (2020) 171-176. https://doi.org/10.21741/9781644901038-26

[18] N. Radek et al. Influence of laser texturing on tribological properties of DLC coatings, Prod. Eng. Arch. 27 (2021) 119-123. https://doi.org/10.30657/pea.2021.27.15

[19] N. Radek et al. The impact of laser welding parameters on the mechanical properties of the weld, AIP Conf. Proc. 2017 (2018) art.20025. https://doi.org/10.1063/1.5056288

[20] J.M. Djoković et al. Selection of the Optimal Window Type and Orientation for the Two Cities in Serbia and One in Slovakia, Energies 15 (2022) art.323. https://doi.org/10.3390/en15010323

[21] B. Jasiewicz et al. Inter-observer and intra-observer reliability in the radiographic measurements of paediatric forefoot alignment, Foot Ankle Surg. 27 (2021) 371-376. https://doi.org/10.1016/j.fas.2020.04.015

[22] J. Pietraszek et al. The fixed-effects analysis of the relation between SDAS and carbides for the airfoil blade traces. Arch. Metall. Mater. 62 (2017) 235-239. https://doi.org/10.1515/amm-2017-0035

[23] R. Dwornicka, J. Pietraszek. The outline of the expert system for the design of experiment, Prod. Eng. Arch. 20 (2018) 43-48. https://doi.org/10.30657/pea.2018.20.09

[24] J. Pietraszek et al. Challenges for the DOE methodology related to the introduction of Industry 4.0. Prod. Eng. Arch. 26 (2020) 190-194. https://doi.org/10.30657/pea.2020.26.33

[25] J. Pietraszek. The modified sequential-binary approach for fuzzy operations on correlated assessments, LNAI 7894 (2013) 353-364. https://doi.org/10.1007/978-3-642-38658-9_32

[26] J. Pietraszek et al. The fuzzy approach to assessment of ANOVA results, LNAI 9875 (2016) 260-268. https://doi.org/10.1007/978-3-319-45243-2_24

[27] J. Pietraszek et al. The non-parametric approach to the quantification of the uncertainty in the design of experiments modelling, UNCECOMP 2017 Proc. 2nd Int. Conf. Uncert. Quant. Comput. Sci. Eng. (2017) 598-604. https://doi.org/10.7712/120217.5395.17225

Quality Production Improvement and System Safety: QPI 16 - CZOTO 10
Materials Research Proceedings 34 (2023) 414-421

Materials Research Forum LLC
https://doi.org/10.21741/9781644902691-48

Evaluation of Safe and Hygienic Work Conditions in the COVID-19 Era: A Case Study in a Production Company

KIELESIŃSKA Agata[1, a *]

[1]Faculty of Management, Czestochowa University of Technology, Al. Armii Krajowej 19b, 42-218 Czestochowa, Poland

[a]agata.kielesinska@pcz.pl

Keywords: COVID 19, Occupational Risk

Abstract. In recent times, the importance of occupational health and safety has escalated for modern enterprises. This shift can be attributed to the emergence of the COVID-19 virus, which compelled numerous businesses to adopt crisis management strategies and embrace remote work arrangements. It can be argued that the pandemic necessitated employers to implement suitable health and safety measures in order to sustain their operations in the labor market. The objective of this study is to evaluate occupational risks, considering the implications arising from the pandemic, and subsequently assess the provision of safe and hygienic conditions during this period. It is important to note that workplace hygiene requirements can vary based on the company, industry, and job role. Different work environments will entail distinct hygiene demands, and certain occupations or industries may even entail additional risks warranting extra protection. Hence, the assessment focused solely on production positions such as machine operators, fitters, quality inspectors, and warehouse workers.

Introduction

The SARS-CoV-2 virus pandemic has created a number of new challenges for employers and obliged them to take specific actions. With the advent of the COVID-19 pandemic, modern enterprises had to quickly adapt to the prevailing conditions in order to survive on the market. At that time, employers played a key role along with employees involved in occupational health and safety (OSH) activities. [1,2]. Under international law, the obligations to ensure the safety and health of workers are the responsibility of:

- an employer - in relation to employees employed under an employment contract and performing work for the employer on a basis other than an employment relationship (including self-employed), provided that the work is performed at the workplace or in a place indicated by the employer;;
- an entrepreneur - in relation to persons employed by him on a basis other than an employment relationship (including self-employed).

Among the actions aimed at ensuring the safety and health of employees undertaken by the employer, the basic action is to assess the risk at the workplace and apply the necessary preventive measures to reduce this risk (according to the Labor Code). Pursuant to the above provisions, the employer, when assessing occupational risk, is obliged to take into account all factors occurring in the work environment and related to its performance. Due to the COVID-19 epidemic, in addition to the existing threats, a new threat caused by the SARS-CoV-2 coronavirus has appeared in workplaces. Therefore, the employer is obliged to take actions aimed at limiting the risk related to exposure to this biological agent [3-5]. The purpose of the study is to assess occupational risk (taking into account aspects resulting from the pandemic situation), and then to assess safe and hygienic conditions during the pandemic among production employees.

Materials and Methods

Risk assessment is the process of evaluating the risks to the safety and health of workers arising from workplace hazards. It is a regular analysis of all aspects of the job, covering the following areas: what could cause injury or damage, whether hazards can be eliminated, and if not, what preventive or protective measures are or should be in place to control the risk. It should be emphasized that risk assessment is only one part of the overall process used to control risks in the workplace [6]. For this reason, it should be treated as part of good business practice and a means to ensure effective improvement of operational activities [7, 8].

The currently applicable occupational health and safety regulations contain a reference to occupational risk assessment, distinguishing a number of factors subject to assessment. Taking into account the nature of the impact on the affected person, in accordance with the standards, four types of dangerous and harmful factors can be distinguished, i.e.: physical (e.g. noise, vibration, radiation, low and high temperature); chemical (e.g. dust, toxic, irritating, allergenic substances, etc.); biological (e.g. viruses, fungi and microorganisms); psychophysical (e.g. physical and neuro-psychic stress).

Taking into account the work positions of employees of the production division from the group of machine operators, fitters, quality inspectors and warehousemen selected for the purpose of work, the occupational risk assessment before 2020 did not include the threats related to the occurrence of the biological agent, which is the SARS-CoV-2 virus (Table 1).

Table 1. *Occupational risk assessment for production positions considering pandemic situation: machine operator, fitter, quality inspector, and warehouse worker [own study based on company information]*

No.	Hazards	Indicator value			Occupational risk level (R)
		S	E	P	
1	Sound volume too high	5	5	10	250
2	Electric current	15	2	1	30
3	High/low temperatures	3	4	8	96
4	Artificial lighting	3	4	8	96
5	Slippery surfaces	7	6	2	84
6	Sharp tools	3	4	9	108
7	Moving elements of machines and devices	3	6	4	144
8	Toxic substances	7	6	1	42
9	Substances and solutions containing metals and their derivatives	7	6	1	42
10	Aerosols, paints, solvents	7	6	0.5	21
11	Pathogenic microorganisms, bacteria, viruses, fungi	4	5	9	180
11a	**Contact with a person infected with the SARS-CoV-2 virus**	2	5	7	70
11 b	**The spread of the SARS-CoV-2 virus**	7	5	6	210
12	Working in a shift system	3	3	1	9
13	Flexible scope of duties	3	3	1	9
14	High degree of responsibility for the implementation of the entrusted tasks	3	2	1	6

Quality Production Improvement and System Safety: QPI 16 - CZOTO 10   Materials Research Forum LLC
Materials Research Proceedings 34 (2023) 414-421   https://doi.org/10.21741/9781644902691-48

Based on the data presented in Table 1, it can be concluded that despite the increased risk of Covid 19 infection, noise is still the most dangerous factor. So, the employee at the workplace is most exposed to too much noise. It is worth noting that even before the declaration of a pandemic, the risk associated with the presence of biological agents "Pathogenic microorganisms - bacteria, viruses, fungi" was very high. The introduction of the state of pandemic introduced some kind of restrictions (partly obligatory), which allowed for a slight minimization of this threat. However, the introduction of additional items 11a and 11b finally meant that the risk associated with the occurrence of coronavirus (as a biological agent) is high. The research was carried out independently in two departments of the company, among production employees, i.e. machine operators, fitters, quality inspectors, as well as warehouse employees. The main assumption of the study was to determine whether the employer adjusted the way of organizing work and working conditions in such a way as to ensure full safety for employees regardless of the current situation. The survey involved 52% men and 48% women (which is a kind of balance), with the positions of warehousemen and machine operators mostly men, and the positions of assembly and quality inspectors were women. As many as 30% of the respondents are employees aged 31 to 35, with the younger staff coming mainly from department 2. Most of the people participating in the survey (37%) had technical education, a comparably large group of respondents were employees with secondary education (33%). The largest group consisted of respondents with 11-16 years of professional experience (37%) and employees with a slightly shorter work experience of 6-10 years (31%).

Results and Discussion
Moving on to the main part of the questionnaire, 9 aspects were analyzed relating to the disclosure of actual activities and work organization in the company, taking into account the observance of safety rules – both commonly known work safety and those resulting from the restrictions introduced.

Thus, in the first place, respondents were asked to rate the following statement "In the enterprise, safety is the most important thing". The results of the obtained tests are shown in Fig.1.

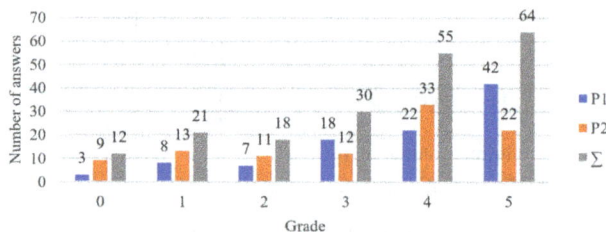

Fig.1. *Comparison of assessments of the statement*
"Safety is the most important thing in the company".

As shown in Fig.1, in the first branch of the company (X), the employees rated the importance of safety in the company very well, as many as 42 respondents gave the statement "Safety is the most important in the company" the highest rating. On the other hand, in the second branch (Y), only 22 employees gave the highest rating. Taking into account the summary assessment, it can be concluded that security aspects play a very important role.

With regard to the data presented in Fig.2 (*Management supports safety activities*), significant discrepancies can be observed in the respondents' assessments. In this case, 45 employees from the first department (X) awarded the highest rating, thus confirming the management's

Quality Production Improvement and System Safety: QPI 16 - CZOTO 10   Materials Research Forum LLC
Materials Research Proceedings 34 (2023) 414-421   https://doi.org/10.21741/9781644902691-48

commitment to creating safe working conditions. On the other hand, in department 2 (Y), the same number of respondents (45) gave a rating of 3 – it is difficult to say, which directly indicates a marginal or complete lack of involvement of the management in the ways of organizing safe and hygienic working conditions. As in the previous question, the summary assessment indicates that the management supports the activities in the OSH scope.

Regarding question 3 "*All employees are involved in general health and safety activities*", employees in both departments are quite in agreement – Fig.3. And the graphs for both departments are quite flat. In department 1 (X) the highest number of ratings was obtained by note 2 (22 respondents), while in department 2 (Y) the highest number of ratings was obtained by note 0 (27 respondents).

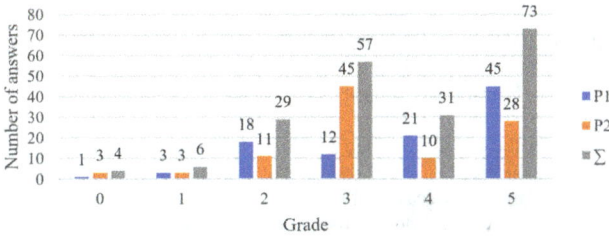

Fig.2. Comparison of assessments of the statement "Management supports safety activities".

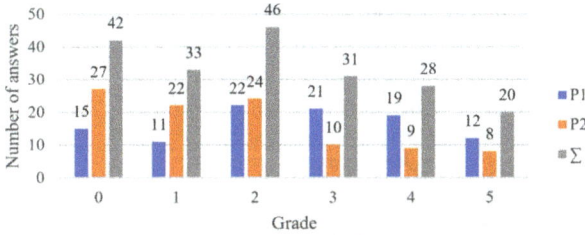

Fig.3. Comparison of assessments of the statement "All employees are involved in general health and safety activities".

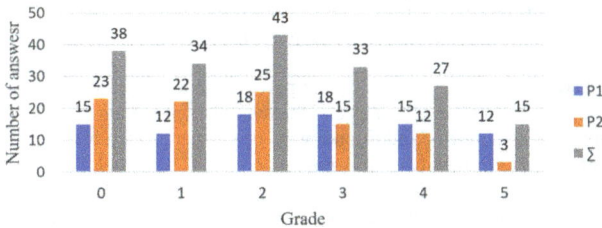

Fig.4. Comparison of assessments of the statement "Employees are constantly informed about changes resulting from the Regulation of the Council of Ministers".

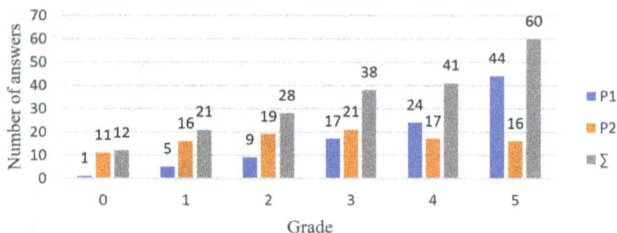

Fig.5. *Comparison of ratings for the statement "Employees comply with all health and safety rules and the sanitary regime against COVID-19".*

Which also informs that the involvement of employees is determined by the interest and actions of the management. This statement is confirmed by the structure of the answers to question 4 "Employees are constantly informed about changes resulting from the Regulation of the Council of Ministers" – Fig.4.

As can be seen from the chart presented in Fig.4 (Employees are constantly informed about changes resulting from the Regulation of the Council of Ministers), employees of department 2 (Y) are not informed or are only occasionally informed about changes in safety regulations in the field of sanitary regime introduced due to pandemic state. In department 2 (Y), almost 75% of respondents gave the lowest scores 0-2. On the other hand, in department 1 (X) information on changes in the organization of work in the company is provided to employees, as many as 22 people confirmed it by awarding a rating of 5. This distribution of ratings also affects the structure of answers to question 5 (Fig.5) - Employees comply with all health and safety rules and sanitary regime against COVID-19.

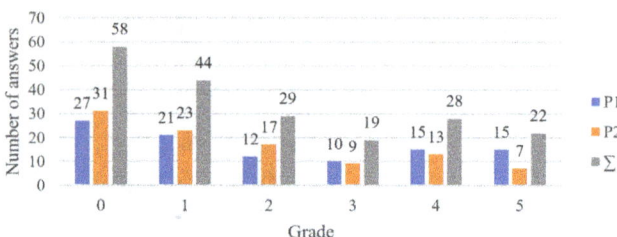

Fig.6. *Comparison of ratings for the statement "Employees have on-the-job training due to the COVID-19 threat before starting work".*

As the data presented in Fig.5 shows, a large percentage of employees from the department1 (X) comply with the health and safety rules and the sanitary regime against COVID-19. As many as 44 respondents gave a score of 5, additionally 24 respondents gave a score of 4 - which is the vast majority. Unfortunately, the situation is not so good in department 2 (Y), where respondents were reluctant to give the highest marks.

However, looking at the summary results, it can be seen that the employees of the tested production unit comply with the health and safety rules, especially in the period of increased risk. Despite such large discrepancies, employees of two departments of the company agree on the necessity and quality of on-the-job training (resulting from the COVID-19 threat) – Fig. 6.

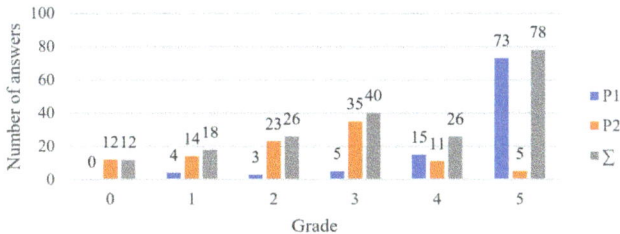

Fig. 7. *Comparison of ratings for the statement "The company keeps a reliable register of dangerous events in relation to the COVID-19 threat".*

The data presented in Fig.6 indicate a high convergence of employees' answers to question 6: *Before starting work, employees receive on-the-job training resulting from the COVID-19 threat.* Unfortunately, the structure of the answers suggests that on-the-job training in terms of the threats caused by COVID-19 is not carried out. Both in unit 1 (X) and in unit 2 (Y) the highest number of ratings was obtained by the score 0 (*I have no opinion*) and 1 (*I completely disagree*). The percentage of positive evaluations (notes 4 and 5) in this case is marginal. Further, significant differences in the structure of answers were noted for question 7: *The company keeps a reliable register of events dangerous to the COVID-19 threat* (Fig. 7).

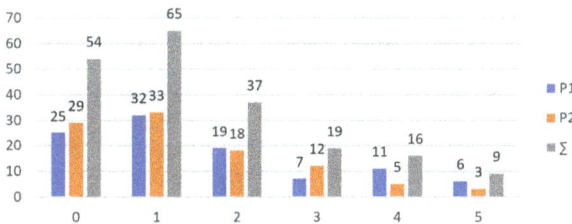

Fig.8. *Comparison of ratings for the statement "Employees (conscientiously) inform about events potentially dangerous to the COVID-19 threat".*

As can be seen (Fig.7), employees of the production division from two departments of the company very differently assessed the statement "*The company keeps a reliable register of dangerous events in relation to the COVID-19 threat*". As many as 73 respondents confirm that a reliable register of events dangerous to the COVID-19 threat is kept in the workplace (employees show very high compliance). On the other hand, employees from department 2 (Y) with their assessments undermine the reliability of the register of dangerous events in the face of the COVID-19 threat. In this context, the structure of ratings obtained for statement 8 seems interesting: Employees (conscientiously) inform about events that are potentially dangerous to the COVID-19 threat – Fig.8. Very high convergence of answers to this question (in both departments) questions the reliability of the records kept.

The correspondence of the answers of employees of the production department in two branches of the company was summarized with question 9 - All employees act in accordance with applicable procedures and instructions. As it results from the data presented in the chart (Fig.9), employees of department 1 (X) confirm (although this confirmation is not unambiguous) that they comply with procedures and instructions (traditional and new ones). On the other hand, employees of department 2 (Y) deny functioning in accordance with applicable rules (instructions and procedures).

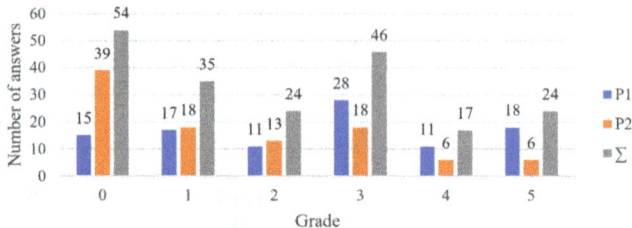

Fig. 9. *Comparison of ratings for the statement „All employees act in accordance with applicable procedures and instructions".*

Summary

The implementation of the tasks assigned to the employee is inscribed in a specific workstation, which is why great emphasis should be placed on the organization of a safe workplace in the area of both the principles of ergonomics and the sciences of organization and management. Despite clear guidelines, rules and regulations, as well as standards related to shaping an ergonomic, but also safe and hygienic workplace, it is sometimes difficult for employers to meet all the needs of employees. This is all the more a big challenge when we have to work in high-risk conditions – a pandemic that is present in the modern professional reality. It turns out that despite the functioning of the same management principles in two departments of the company (the same organizational structure, the same safety policy, the same quality and safety management systems implemented, etc.), employees differently perceive the organization of work in the production division – in one department, the employees rated the importance of safety in the company as very good, while in the second one, the rating is definitely lower. Significant differences in ratings were also observed respondents in relation to the management's involvement in creating safe working conditions. There are also differences in the assessment of whether employees are constantly informed about changes resulting from the Regulation of the Council of Ministers). The results of the conducted research confirmed that employees comply with all health and safety rules and the sanitary regime against COVID-19. But it's definitely different in both departments. Despite such large discrepancies, the employees are aware of the need to participate in training. In addition, there is a very high convergence of answers to this question about reliable register of dangerous events in relation to the COVID-19 threat (in both departments).

The proper implementation of industrial workstations with enhanced safety due to the threat of infectious viruses is a highly challenging issue. The practical implementation of measures such as ventilated isolation suits is not feasible due to restricted freedom of movement. Given the multifaceted nature of the problem, the application of optimization techniques [9-11], including those that do not require a priori model definition [12, 13], can be helpful. It would also be beneficial to employ spatial orientation techniques similar to stereology [14], where appropriate distancing of workers could be considered [15]. Additionally, the implementation of composite-based filtration curtains [16] and virus-killing coatings similar to special coatings [17, 18] could be considered.

References

[1] J. Bartnicka et al. Evaluation of the Effectiveness of Employers and H&S Services in Relation to the COVID-19 System in Polish Manufacturing Companies. Int. J. Environ. Res. Public Health 18 (2021) art.9302. https://doi.org/10.3390/ijerph18179302

[2] N. Baryshnikova et al. Enterprises' strategies transformation in the real sector of the economy in the context of the COVID-19 pandemic, Prod. Eng. Arch. 27 (2021) 8-15. https://doi.org/10.30657/pea.2021.27.2

[3] V. Holubova. Integrated Safety Management Systems, Pol. J. Manag. Stud. 14 (2016) 106-118. https://doi.org/10.17512/pjms.2016.14.1.10.

[4] M.A. Taylor, A.M. Alvero. The Effects of Safety Discrimination Training and Frequent Safety Observations on Safety-Related Behavior, J. Org. Behav. Manag. 32 (2012) 169-193. https://doi.org/10.1080/01608061.2012.698115

[5] N.B. Kurland et al. Business and society in the age of COVID-19: Introduction to the special issue, Bus. Soc. Rev. 127 (2022) 147-157. https://doi.org/10.1111/basr.12265

[6] A. Terje. Risk assessment and risk management: Review of recent advances on their foundation, Eur. J. Oper. Res. 253 (2016) 1-13. https://doi.org/10.1016/j.ejor.2015.12.023

[7] R. Ulewicz et al. Quality and work safety in metal foundry. In: METAL 2020 29th Int. Conf. Metall. Mater., Brno, Czech Republic. Ostrava, Tanger, 1287-1293.. https://doi.org/10.37904/metal.2020.3649

[8] M. Niciejewska, M. Obrecht. Impact of Behavioral Safety (Behavioural-Based Safety – BBS) on the Modification of Dangerous Behaviors in Enterprises, In: System Safety: Human-Technical Facility-Environment 2 (2020) 324-332. https://doi.org/10.2478/9788395720437-040

[9] J. Pietraszek et al. The parametric RSM model with higher order terms for the meat tumbler machine process, Solid State Phenom. 235 (2015) 37-44. https://doi.org/10.4028/www.scientific.net/SSP.235.37

[10] R. Dwornicka, J. Pietraszek. The outline of the expert system for the design of experiment, Prod. Eng. Arch. 20 (2018) 43-48. https://doi.org/10.30657/pea.2018.20.09

[11] J. Pietraszek et al. Challenges for the DOE methodology related to the introduction of Industry 4.0. Prod. Eng. Arch. 26 (2020) 190-194. https://doi.org/10.30657/pea.2020.26.33

[12] J. Pietraszek. The modified sequential-binary approach for fuzzy operations on correlated assessments, LNAI 7894 (2013) 353-364. https://doi.org/10.1007/978-3-642-38658-9_32

[13] J. Pietraszek et al. Factorial approach to assessment of GPU computational efficiency in surrogate models, Adv. Mater. Res. 874 (2014) 157-162. https://doi.org/10.4028/www.scientific.net/AMR.874.157

[14] J. Pietraszek et al. The fixed-effects analysis of the relation between SDAS and carbides for the airfoil blade traces. Arch. Metall. Mater. 62 (2017) 235-239. https://doi.org/10.1515/amm-2017-0035

[15] B. Jasiewicz et al. Inter-observer and intra-observer reliability in the radiographic measurements of paediatric forefoot alignment, Foot Ankle Surg. 27 (2021) 371-376. https://doi.org/10.1016/j.fas.2020.04.015

[16] J. Korzekwa et al. Tribological behaviour of Al2O3/inorganic fullerene-like WS2 composite layer sliding against plastic, Int. J. Surf. Sci. Eng. 10 (2016) 570-584. https://doi.org/10.1504/IJSURFSE.2016.081035

[17] N. Radek et al. Microstructure and tribological properties of DLC coatings, Mater. Res. Proc. 17 (2020) 171-176. https://doi.org/10.21741/9781644901038-26

[18] N. Radek et al. Influence of laser texturing on tribological properties of DLC coatings, Prod. Eng. Arch. 27 (2021) 119-123. https://doi.org/10.30657/pea.2021.27.15

Quality Production Improvement and System Safety: QPI 16 - CZOTO 10   Materials Research Forum LLC
Materials Research Proceedings 34 (2023) 422-429   https://doi.org/10.21741/9781644902691-49

Factors Determining the Perception of OHS by Socially Responsible Entrepreneurs

OLEJNICZAK-SZUSTER Katarzyna[1, a]

[1]The Management Faculty, Czestochowa University of Technology, Armii Krajowej 19B,42-201 Czestochowa, Poland

[a]k.olejniczak-szuster@pcz.pl

Keywords: CSR, OHS, Econometric Models

Abstract. The study identifies the factors which determine the perception of occupational health and safety by socially responsible entrepreneurs. The research task set was carried out on the basis of an analysis of the literature on the subject and the results of surveys conducted in the third quarter of 2021, conducted among 164 entrepreneurs. In the empirical analysis workshop, logit models were estimated, in which determinants belonging to 5 categories were considered. On this basis, using the Gretl software, 8 logit models (4 full and 4 reduced) were built, indicating the relationships between the studied variables.

Introduction

Occupational health and safety (OSH) is a commonly used term for a set of rules for safe and hygienic performance of work, as well as a separate field of knowledge dealing with shaping appropriate working conditions [1]. In recent years, many researchers have focused on OSH issues in various business sectors, due to the fact that they are related to working conditions and the potential impact on business continuity and financial situation of the company [2]. According to Väyrynen et al. [3] most OSH management systems focus on three key pillars: legal compliance, adoption of appropriate standards, and the implementation of good practices. Fisscher [4] is of the opinion that health and safety is considered to be one of the main elements of company ethics, which are closely related to the idea of social responsibility (CSR). This is because a properly implemented and conducted health and safety policy introduces business values to the CSR action programme. In the above context, the identification of factors determining the perception of occupational health and safety by socially responsible entrepreneurs deserves special attention.

Social Responsibility in the Context of Health and Safety

In response to a wide set of interests and expectations of stakeholders, corporate social responsibility (CSR) reflects the extent to which a modern enterprise is actively involved in social initiatives, activities and processes [5, 6]. It is generally accepted that corporations bear three types of responsibility: economic towards their shareholders, social and environmental towards the communities in which they operate [7]. Currently, the concept of CSR is usually assigned three main tasks, i.e. [8]: fulfilling obligations towards various stakeholder groups, responding to social needs and expectations, and using the concept as a company management tool. Considering the role of employees in the overall performance of the organization and its market success, it can be considered that they are its key stakeholders. They not only determine the quality of the product/service that customers receive, but above all they affect efficiency, commitment and job satisfaction. It can be said that they are both drivers of CSR and its beneficiaries [9,10]. In this sense, shaping increasingly safe and hygienic working conditions is becoming increasingly important. The high (safe) standard of workplaces significantly affects the satisfaction of employees, and therefore is the cause of increased work efficiency and higher quality of services. According to M. Pęciłło [11] OSH is an important value of CSR, and the relationship between

Quality Production Improvement and System Safety: QPI 16 - CZOTO 10   Materials Research Forum LLC
Materials Research Proceedings 34 (2023) 422-429   https://doi.org/10.21741/9781644902691-49

OSH and CSR is multifaceted, covering: safety and public health, human resources, work-life balance, basic employee rights, environmental protection, profitability and productivity.

And so, the values of occupational health and safety, which are particularly taken care of by socially responsible entrepreneurs, are manifested mainly in [12]:

- establishing a health and safety system based on employee participation,
- taking into account the specific conditions and needs of various working groups in the assessment and mitigation of risks to safety and health,
- analyzing health and safety problems reported by employees,
- anticipating new risks to safety and health at work,
- adapting workspaces to the psychophysical capabilities of employees, including taking into account the needs of the elderly and people with specific health problems,
- enabling retraining for employees who have suffered accidents,
- identifying psychosocial risks in the workplace and introducing actions to eliminate or reduce them.

Methodological Aspects of the Research

Due to the nature of the data obtained (qualitative-binary data), econometric modeling instruments in the form of binomial models (logit and probit) were used to verify the purpose of the study [13]. In models of this class, the probability of the occurrence of the analyzed phenomenon (explanatory variable) in a given group of respondents (explanatory variable - in comparison with the reference group) is determined. These relationships are reflected in the odds ratio [14]. In other words, models of this class show the relationships that occur between exogenous explanatory variables that describe the characteristics of possible alternatives and the probability of choosing one of the two possible variants (conventionally marked as 0 / 1; yes / no, good / bad). In this sense, this variable takes the value of 1 when the desired event occurs, and the value of 0 when such an event does not occur. In binomial models, it is assumed that the probability is a function of the vector of exogenous variables and the vector of parameters β in the form [15]:

$$P_i = P(y_i = 1) = F x_i \beta \tag{1}$$

where:

$x_i \beta$ - index specifying the i-th observation unit (linear combination of variable and parameter values)

F - is an increasing function of this index

Taking into account the above models of this class, they take the form of [15]:

- logit model:

$$y_i^* = ln\frac{P_i}{1-P_i} = \beta_0 + \sum_{j=1}^{k}\beta_j x_{ij} + u_i \tag{2}$$

where:

y_i^*- what is called a logit

P_i- is determined by the probability of the dependent variable, determined on the basis of the logistic distribution from the following equation:

$$P_i = \frac{P_i}{1-P_i} = e^{y_i^*} = e^{\beta_0 + \sum_{j=1}^{k}\beta_j x_{ij} + u_i}, P_i = \frac{P_i}{1-P_i} = \frac{P_i}{1+e^{-(\beta_0 + \sum_{j=1}^{k}\beta_j x_{ij})}} \tag{3}$$

If:

$y_i^* \to \infty$, then $P_i \to 1$

$y_i^* \to -\infty$, then $P_i \to 0$

$y_i^* = 0$, then $P_i = 0,5$

- probit model:

$$P_i = F\left(\beta_0 + \sum_{j=1}^{k} \beta_j x_{ij}\right) = \int_{-\infty}^{\beta_0 + \sum_{j=1}^{k} \beta_j x_{ij}} \frac{1}{\sqrt{2\pi}} e^{\left(-\frac{t^2}{2}\right)} dt \tag{4}$$

It should be emphasized that in the above models the following relation exists between the β parameters::

$$\beta_{logit} = 1.6\beta_{probit} \tag{5}$$

Considering the above, both the logit and probit models are quite similar. For this reason, the lignite model was chosen when assessing the probability of the occurrence of a given factor. Specific statistical measures are used while assessing the quality of logit models, including:

- Likelihood Ratio Test:

$$LR = -2(I_n\widehat{L_R} - I_n\widehat{L_{UR}}) \tag{6}$$

where:
$\widehat{L_R}$ - the maximum value of the likelihood function log for a model containing only the intercept
$\widehat{L_{UR}}$ - the value of the likelihood function for the full model.

- McFadden's R-squared:

$$R^2_{McFadden} = 1 - \frac{I_n L_{UR}}{I_n L_R}$$
(7)

- Akaike Information Criterion*:

$$AIC = -2lnL(\hat{0}) + 2K$$
(8)

- Bayesian Information Criterion*:

$$BIC = -2lnL(\hat{0}) + K\ln(n) \tag{9}$$

- Hannan-Quinn Information Criterion*:

$$HQC = -2lnL(\hat{0}) + 2K\ln(\ln n)$$
(10)

where * (for AIC, BIC and HQC):
$l(\widehat{0})$ – the log of the likelihood function for the estimated vector
K - number of model parameters,
N – number of observations

- Number of cases of 'correct prediction', by counting the appropriate numbers, on this basis the predicted value of the dependent variable (0) or (1) can be calculated. The cut-off point is 0.5 by default

The input set of independent variables included variables characterizing entrepreneurs (Table 1). As can be seen, the set of explanatory variables (x) includes 5 categories, containing 15 explanatory variables, on the basis of which the research sample can be characterized. Thus, the

Quality Production Improvement and System Safety: QPI 16 - CZOTO 10   Materials Research Forum LLC
Materials Research Proceedings 34 (2023) 422-429   https://doi.org/10.21741/9781644902691-49

sample consists of 51.3% of women and 48.7% of men (reference group). In terms of age, the dominant group of respondents are people aged 25-34 (45.7%), followed by people aged 35-44 (25% of respondents). The reference group in this category are people up to 24 years of age. Most of the respondents were regular employees (81.1% of the respondents), persons holding managerial positions were selected as the reference group (19.9% of the respondents). The respondents are people working mainly in the SME sector, most of them in small companies, i.e. employing from 10 to 49 employees (48.1% of respondents), then in micro-enterprises employing up to 9 employees (28.2% of respondents) and medium-sized (17.38), employing from 50 to 249 employees. In this category, people working in large companies were selected as a reference group. The last category of variables concerned work experience. The dominant group in this category are people with work experience of 1-10 years (53.4%), followed by people working for more than 10 years (30.9% of indications). The smallest group is made up of newly hired people whose work experience does not exceed 12 months - 15.7% of indications (reference group). Taking into account the explained variables (y), they were coded on the basis of respondents' answers to four research problems presented in Table 2.

Table 1. List of explanatory variables (x) [own study]

Independent variable	Group	%	Reference group
GENDER			
F	Female	51.3%	
M	Male	48.7%	*
AGE			
W24	up to 24 y/o	15.8%	*
W2534	25-34 y/o	45.7%	
W3544	35-44 y/o	25%	
W45	over 45 y/o	13.4%	
POSITION			
RW	regular worker	81.1%	
M	managerial	19.9%	*
COMPANY SIZE			
L	Large	15.24%	*
M	Medium	17.38%	
S	Small	48.1%	
MI	Micro	28.2%	
WORK EXPERIENCE			
D12	Under a year	15.7%	*
110	Between 1 and 10 years	53.4%	
D12	Under a year	15.7%	*

Table 2. List of explained variables [own study]

Research area	Percentage structure of responses	
	Yes	No
high (safe) standard of workplaces affects employee satisfaction, thus increasing work efficiency	64.63%	35.36%
the implementation of CSR standards in OHS management has a positive impact on the quality of life of employees	80.48%	19.52%
creating a friendly atmosphere at work, indirectly contributes to counteracting stress at work, including health protection	65.85%	34.15%
actions aimed at improving work-life balance	49.39%	50.61%

Respondents are more likely to report that socially responsible activities have a positive impact on employee satisfaction than allowing employees to choose the forms of working time organization, in particular practices that eliminate the extension of working time beyond the required standard. Respondents equally highly value the creation of a friendly atmosphere, in particular the health of employees and occupational health and safety, which are values that socially responsible entrepreneurs especially care about.

Research Results
Table 3 presents the results of the estimated eight logit models relating to the perception of aspects of selected OHS aspects by socially responsible entrepreneurs.

Table 3. Results of estimation of logit models for the studied variables [own study]

* used observations 1-164

Variable	Coefficient	Standard deviation	z	p-value	Marginal effect
MODEL 1 (pelny)					
const	0.615518	0.206884	2.975	0.0029***	
F	−0.194582	0.123142	−1.580	0.1141	−0.129134
W2534	0.201604	0.121405	1.661	0.0968*	0.129607
W3544	0.0514710	0.144569	0.3560	0.7218	0.0251765
W45	0.112786	0.170513	0.6614	0.5083	0.0670370
RW	−0.0245213	0.148507	−0.1651	0.8689	−0.0182528
M	−0.171972	0.165282	−1.040	0.2981	−0.119822
S	0.0875952	0.115143	0.7607	0.4468	0.0642525
VMI	−0.0765942	0.131822	−0.5810	0.5612	−0.0534703
v18	−0.0809166	0.123327	−0.6561	0.5118	−0.0540757
P10	−0.142286	0.132353	−1.075	0.2824	−0.0935790
MODEL 2 (reduced)					
const	0.403463	0.0966557	4.174	2.99e-05 ***	
W2534	0.208464	0.113508	1.837	0.0663 *	0.135730
MODEL 3 (full)					
const	0.999920	0.141292	7.077	<0.0001	
F	0.0461066	0.0935131	0.4930	0.6220	0.0440357
W2534	−0.356775	0.101922	−3.500	0.0005***	−0.271013
W3544	0.0183353	0.114679	0.1599	0.8730	0.00902923
W45	−0.00851556	0.117795	−0.07229	0.9424	0.00318509

Variable	Coefficient	Standard deviation	z	p-value	Marginal effect
RW	−0.208661	0.0944851	−2.208	0.0272**	−0.170172
M	0.0839683	0.114144	0.7356	0.4620	0.0552888
S	−0.263303	0.0940294	−2.800	0.0051***	−0.201588
VMI	0.0634702	0.0929940	0.6825	0.4949	0.0524804
v18	0.0385344	0.0833341	0.4624	0.6438	0.0552032
P10	0.999920	0.141292	7.077	<0.0001	0.0594076
MODEL 4 (reduced)					
const	1.12281	0.0762324	14.73	<0.0001***	
W2534	−0.366564	0.0909338	−4.031	<0.0001 ***	−0.277178
RW	−0.233255	0.0827381	−2.819	0.0048***	−0.181642
S	−0.241331	0.0919075	−2.626	0.0086***	−0.185119
MODEL 5 (full)					
const	0.483342	0.181885	2.657	0.0079	
F	−0.176431	0.114969	−1.535	0.1249	−0.128261
W2534	0.0821319	0.121381	0.6766	0.4986	0.0535896
W3544	−0.227091	0.189420	−1.199	0.2306	−0.162029
W45	0.0552112	0.146905	0.3758	0.7070	0.0332424
RW	−0.0110799	0.153772	−0.07205	0.9426	−0.0130352
M	−0.0246941	0.112465	−0.2196	0.8262	−0.0161415
S	−0.0931364	0.126634	−0.7355	0.4620	−0.0668347
VMI	0.252865	0.115806	2.184	0.0290	−0.0161415
v18	0.0425340	0.134153	0.3171	0.7512**	0.173735
P10	0.483342	0.181885	2.657	0.0079	0.0274256
MODEL 6 (reduced)					
const	0.403882	0.0990021	4.080	4.51e-05 ***	
110	0.233910	0.111830	2.092	0.0365 **	0.154693
MODEL 7 (full)					
const	0.234774	0.296198	0.7926	0.4280	
F	0.201896	0.167760	1.203	0.2288	0.104847
W2534	−0.142024	0.169960	−0.8356	0.4034	−0.0754381
W3544	−0.0284131	0.192808	−0.1474	0.8828	−0.00636709
W45	−0.0435426	0.251035	−0.1735	0.8623	−0.0189384
RW	0.0261844	0.209216	0.1252	0.9004	0.0156395
M	−0.355439	0.220203	−1.614	0.1065	−0.172035
S	0.0447506	0.163526	0.2737	0.7843	0.0210942
VMI	−0.408033	0.191024	−2.136	0.0327**	−0.198706
v18	−0.0335791	0.167785	−0.2001	0.8414	−0.0210654
P10	0.0867281	0.180267	0.4811	0.6304	0.0456833
MODEL 8 (reduced)					
const	0.248117	0.111277	2.230	0.0258 **	
VMI	−0.363896	0.186943	−1.947	0.0516 *	−0.172808

Explanation: The level of significance of the parameters: *** $\alpha = 0.01$, ** $\alpha = 0.05$, * $\alpha = 0.1$

Looking at the obtained results of logit model estimation, it can be seen that gender is not significant from the point of view of impact on the probability of reporting the analyzed OHS aspects. Taking into account age, statistical significance at the level of α=0.1 can be observed in the case of an increase in employee satisfaction (full and reduced models), and at the level of α=0.01 the implementation of CSR standards in OHS management has a positive impact on the quality of life of employees (full and reduced models). The size of the enterprise turned out to be the factor that most strongly determined the phenomena in model 3 and the reduced model 4, as well as 7 and 8.

Table 4. Data fit measures for the estimation of logit models [own study]

	MODEL 1	MODEL 3	MODEL 5	MODEL 7
FULL MODELS				
Likelihood Ratio Test	9.30318 [0.5036]	29.01 [0.0012]	10.6992 [0.2969]	9.60148 [0.4761]
McFadden's R-squared	-0.060782	0.037806	-0.045099	-0.05523
Hannan-Quinn Information Criterion	235.3793	192.1986	228.0675	250.6691
Number of cases of 'correct prediction'	106 (65.4%)	(77.2%)	109 (67.3%)	96 (59.3%)
	MODEL 2	**MODEL 4**	**MODEL 6**	**MODEL 8**
REDUCED MODELS				
Likelihood Ratio Test	3.31271 [0.0687]	27.1154 [0.0000]	4.38702 [0.0362]	3.99103 [0.0457]
McFadden's R-squared	-0.003225	0.102435	0.001838	-0.000039
Hannan-Quinn Information Criterion	216.2999	172.5291	212.7063	229.8537
Number of cases of 'correct prediction'	106 (64.6%)	127 (77.4%)	108 (65.9%)	93 (56.7%)

The position held is of similar importance, both in models 3 and 4 this factor determined the occurrence of the anbalized phenomenon (creating a friendly atmosphere at work, indirectly contributes to counteracting stress at work, including health protection). In the case of the work experience category, this factor, similarly to the position held, was the strongest determinant of the analyzed variable.

Due to the fact that the data fit measures of the obtained models are similar, the likelihood ratio test, McFadden's R-squared, Hannan-Quinn Information Criterion and Number of cases of 'correct prediction' were selected for the evaluation quality analysis.

Analyzing the measures of data fit to logistic models presented in Table 4, it can be concluded that all the estimated models showed statistical correctness. As a result of the modeling, a high test statistic x^2 was obtained (in all the estimated models). Comparing the number of cases of correct prediction, all models are at a similar level, except that models 4 and 2 are characterized by the highest probability of occurrence.

Conclusion
The work examined selected health and safety issues implemented by socially responsible entrepreneurs. For many employees, one of the most important areas of CSR activities is improving the quality of employees' work, which translates into their greater efficiency. Moreover, the observance of human rights, as well as good relations in the workplace and work in a human-friendly environment also translate into the efficiency and innovation of a modern enterprise, and this also leads to benefits in terms of added value.

References

[1] P. Ziółkowski, Metodyka szkoleń z zakresu bezpieczeństwa i higieny pracy, Wydawnictwo Uczelniane Wyższej Szkoły Gospodarki w Bydgoszczy, Bydgoszcz, 2018. ISBN 978-8365507228

[2] M. Mavroulidis et al. Occupational health and safety of multinational construction companies through evaluation of corporate social responsibility reports, J. Saf. Res. 81 (2022) 45-54. https://doi.org/10.1016/j.jsr.2022.01.005

[3] S. Väyrynen et al. Integrated Occupational Safety and Health Management, Springer, Geneva, 2015. ISBN 978-3319131795

[4] O. Fisscher. Het organiseren van morele competentie – Bedrijfsethiek, Gids voor Personeelmanagement 82 (2003) 12-18.

[5] A. McWilliams, D. Siegel. Corporate Social Responsibility: A Theory of the Firm Perspective. Acad. Manag. Rev. 26 (2001) 117-127. https://doi.org/10.5465/AMR.2001.4011987

[6] S. Cao et al. CSR gap and firm performance: An organizational justice perspective, J. Bus. Res. 158 (2023) art.113692. https://doi.org/10.1016/j.jbusres.2023.113692

[7] Z.Zhang et al. CSR is not a panacea: The influence of CSR on disgust and turnover intention, J. Vocat. Behav. 140 (2023) art.103821. https://doi.org/10.1016/j.jvb.2022.103821

[8] J. Allouche, P. Laroche. The Relationship between Corporate Social Responsibility and Corporate Financial Performance: A Survey, In: J. Allouche (Ed.), Corporate social Responsibility. Vol. 2: Performance and Stakeholders, Palgrave and Macmillan, New York (2006), 4-5.

[9] O. Farooq et al. The multiple pathways through which internal and external corporate social responsibility influence organizational identification and multifoci outcomes: The moderating role of cultural and social orientations, Acad. Manag. J. 60 (2017) 984-985. https://doi.org/10.5465/amj.2014.0849

[10] Y.K. Lee et al. The impact of CSR on relationship quality and relationship outcomes: A perspective of service employees, Int. J. Hospit. Manag. 31 (2012) 745-756. https://doi.org/10.1016/j.ijhm.2011.09.011

[11] M. Pęciłło, Zarządzanie bezpieczeństwem i higieną pracy a społeczna odpowiedzialność biznesu w ujęciu norm SA i ISO, Bezpieczeństwo pracy: Nauka i praktyka 2011(3) (2011) 19-21.

[12] N. Kusiak (Ed.). Przewodnik CSR po bezpiecznym i zrównoważonym środowisku pracy, Ministerstwo Funduszu i Polityki Regionalnej, Warszawa, 2021. ISBN 978-8376107097

[13] M. Gruszczyński. Mikroekonometria: modele i metody analizy danych indywidualnych. Wolters Kluwer, Warszawa, 2010. ISBN 978-8326451843

[14] M. Kunasz, E. Mazur-Wierzbicka. Factors Determining Perception of Mobbing and Sexual Harassment in the Working Environment – Research Results, Marketing i Rynek 11/2017 (2017) 29-36.

[15] D.W. Hosmer, S. Lemeshow. Applied Logistic Regression. Wiley, Hoboken, 2000. ISBN 0-471-35632-8.

[16] T. Kufel. Ekonometria Rozwiązywanie problemów z wykorzystaniem programu GRETL. PWN, Warszawa, 2023. ISBN 978-8301165130.

Quality Production Improvement and System Safety: QPI 16 - CZOTO 10   Materials Research Forum LLC
Materials Research Proceedings 34 (2023) 430-446   https://doi.org/10.21741/9781644902691-50

How to Enhance the Management and Quality of Electronic Publications?

HRABOVSKYI Yevhen[1, a], SZYMCZYK Katarzyna[2, b*]
and LUKIANOVA Viktoriia[3,c]

[1]Department of Computer Systems and Technologies, Simon Kuznets Kharkiv National University of Economics, 9-a Nauky ave. 61166 Kharkiv, Ukraine

[2]Faculty of Management, Czestochowa University of Technology, Dąbrowskiego 69, 42-201 Częstochowa, Poland

[3]Department of Natural Science, Kharkiv National University of Radio Electronics, 14 Nauki ave. 61166 Kharkiv, Ukraine

[a]maxmin903@gmail.com, [b]katarzyna.szymczyk.pcz.pl [c]vlukyanova@ukr.net

Keywords: Managing Page Orientation, Managing Electronic Edition, Digital Publication Management, Automatic Change, Algorithm

Abstract. The electronic publication format has become valuable and easy to access for a modern person. Many readers prefer electronic publications to print ones. Publication houses that want to access more customers and enhance their sales should consider including electronic formats in their product offers. Managing a publishing company requires constant modification and diversification of products. To a large extent, the offer of publications in electronic format has a chance to attract more customers than the traditional printed form. It is primarily about easy access to the searched reading items. The customers can purchase the publication in various electronic formats, e.g., PDF, MOBI, or ePUB format, according to their needs. It does not require going to the library, reading room, stationery store, or waiting for the parcel at the place of residence. The purchased product, paid or free, immediately becomes available after downloading to the appropriate device. It is a very convenient and quick way to get to publications. Therefore, publishing houses should keep up with the times and offer publications in electronic format. However, a number of factors must be taken into account before such a product is created. Often, there is a need to convert the print version of publications into electronic format - a need to convert documents on formats of computer editorial systems in electronic formats. In the paper, the method of changing the orientation of the page of the elevation publication is proposed. Then a set of parameters of the arrangement elements is developed. The algorithm for the automatic change of page orientation of the viewing edition is created. Based on this algorithm, tools have been developed to automatically change the edition page as an appropriate prototype of automation tools (Adobe, USA product) in HTML format using JavaScript programming language (Japan). For the qualitative work of the prototype, the basic rules of the set and layout were analyzed. The designed complex of typical page layouts, in terms of laying elements on the page. An orientation change algorithm with its restrictions has been developed for each typical layout.

Introduction

The purpose of the article is to develop a method for the automatic change in orientation of pages of the address, which will allow one to change the page orientation of the viewing publication with the preservation of the volume of information on the page, in accordance with the rules of dialing and managing. The main problem of this study is as follows. The number of rows per page changes while changing the orientation of the addressing edition, which leads to the inconsistency of the initial content on the page, due to an increase in the incomplete coefficient. But with the task of changing the orientation of the page of the drawing publication in the Folio format, there is a complete matching of the initial layout in terms of content volume on each publication page. The plurality of typical layouts of the page is formulated. On the basis of the analysis of typical layouts, a set of rules for eliminating the disadvantages of an account as a result of changing the orientation of the page of the elevation publication. A system of requirements and restrictions to each rule has been formed.

Under electronic edition, an electronic document (electronic document group) is understood by editorial-publishing elaboration, designed to spread in unchanged form, and has source information. Electronic editions vary with each other by the presence of printed equivalent, thus subdivided into electronic analogs of printed publications and independent editions (digitally born). Electronic editions may be represented by a variety of formats that, in turn, are divided into two groups, if possible, to identify the parameters of the original layout of the printed edition:

a) formats of electronic publications are not capable of playing the separation to the original-layout of the printed publications (fb2, ePUB, MOBI, PRC and others);

b) formats are capable of being identical to the original layout Play the division into printed edition pages (ePDF, Folio).

What happens when changing orientation, it decides specific publishing. It is possible to use both orientations to allow the reader to choose how it is more comfortable with reading a magazine, or restrict it to one of the orientations. You can also apply a tablet turn to enhance communication and as an excuse to create an additional visual event, spectacularly adding additional information or new illustrative blocks. Everything depends on the concept of the publication. A set of elements in both orientations may be identical to the reader who chose only one of them, missing nothing. Also, this set may differ - for surprises or (if the reader is notified and knows what to expect) to describe additional content. The simplest example - one can use a tablet turned in a horizontal position to view a video and a slideshow in full-screen mode.

The main requirement for publications that can reproduce the division into pages in accordance with the original layout of the printed publications is to maintain an absolute identity of each page when converting into an electronic appearance.

However, when the electronic edition is designed for viewing on smartphones and tablet PCs, there is a need to change the edition page orientation. Such editions, as a rule, are generated in two variants: with a vertical and horizontal orientation of the page.

In our times, there are software tools that allow partially to automate the process of changing the orientation of the page of the elevation page. Still, they cannot cover the entire volume of tasks that require this process. Therefore, even after the use of automation of the process of changing the orientation of the page of the elevation of the publication, in any case, it is necessary to manually adjust the accretion, which we understand as a "subtle" recruitment.

In the process of changing the edition page, there are many difficulties associated with the preservation of the identity to the original. The machine tool must be performed by a whole range of operations that require phased decision-making and much time spent.

It should be noted that in changing the edition page, the parameters of the means that use a machine tool to preserve the identity of the original are not regulated. Therefore, in this case, the machine is guided only by its own experience and subjective judgments. As a result, the final product may not correspond to the general technical rules of the set and layout.

That is why forming a complex of models of a "thin" manner is an actual and valuable scientific result. This complex must contain:

1) rules for eliminating disadvantages of recruitment as a result of changing the page orientation of the viewing publication;

2) system of requirements and restrictions for each rule for eliminating disadvantages of recruitment due to changing the orientation of the page of the election edition.

At present, a large number of techniques are devoted to the development of fonts and the formation of rules for the administration of electronic publications [1-5]. As to the need to change the orientation of the page of the elevation of the publication, then it occurs most often when converting the publication into an electronic format. In accordance with various ways of administering electronic publications in specialized literature [6-11], proposed mechanisms for controlling the quality of content and optimizing the interface of multimedia applications. The entire book design system delivers the choice of one type of layout. The design is determined by the content of the book and its type. In turn, the book's contents and its type must be based on the selection of a composite scheme and a page layout.

Materials and Methods

In the context of the stages of development of the method of automatic change orientation of the page of the presentation of the presentation by research methods are the following:

1) investigation of the process of leading editions having two options orientation options:
    - supervise analysis in publishing processes → methods of analysis and synthesis;
    - analysis of layout stages → methods of analysis and synthesis;
2) analysis of the process of changing the page orientation of the viewing publication:
    - analysis of the requirements for publications having two-page orientation options;
    - analysis of existing software tools for changing the page orientation of the address; → methods of analysis and synthesis, empirical method (experiment);
    - analysis of problems that arise in the process of changing the page orientation of the viewing publication → methods of analysis and synthesis, deductions;
    - analysis of the technical rules of the set and the recruitment → methods of analysis and synthesis, deductions;
3) Formation of a complex of models of "thin" making up:
    - forming a plurality of typical layouts of a page → methods of analysis and abstraction;
    - formation of rules for eliminating disadvantages of recruitment as a result of changes in page orientation of page-proof→ methods of analysis and abstraction;
    - formulation of the system of requirements and restrictions for each rule to eliminate the disadvantages of recruitment due to changing the page's orientation of the page-proof. → empirical method (experiment);
4) Development of a prototype of an automated system for a "thin" recruitment:
    - development of a software prototype scenario for changing the page orientation of the election edition → methods of analysis and synthesis, modeling;
    - selection of tools and its justification → method of analysis and synthesis;
    - implementation in the prototype of the formulated complex of models of "thin" making up → induction method;
5) Prototype testing → empirical method (experiment).

Experiments

The Adobe InDesign software is a powerful tool for acting printing products and a leader in modern computer publishing systems, allowing you to realize the most diverse adventures of the author. However, in addition, InDesign has built-in algorithms for converting documents in Folio, Adobe PDF (Interective), Adobe PDF (PRINT), EPS, EPUB, Flash, IDML, JPEG, XML. So, the extraordinary functionality of InDesign caused the choice of this system to realize the problems of this study. InDesign CS6 has a developed functionality to create editions with the "elastic belronty", which is extremely necessary at changing the elevation's orientation for further preservation in the Folio format. However, in addition to explicit advantages, InDesign also has several shortcomings at this stage. First, to obtain a qualitative result of changing the page

Quality Production Improvement and System Safety: QPI 16 - CZOTO 10   Materials Research Forum LLC
Materials Research Proceedings 34 (2023) 430-446   https://doi.org/10.21741/9781644902691-50

orientation, it is necessary to pre-conduct several elastic recruitment settings that require too much time and skills, namely:
- the task of the rules of the "elastic recruitment";
- tying text fragments;
- reconciliation of repetitive content and others.

InDesign allows you to deal with the following rules for "elastic recruitment":
- scaling;
- re-centering;
- holding based on guidelines;
- making-up based on objects.

Zoom is most effective when adapting a page to a layout with the same aspect ratio and orientation. Repeated centering is appropriate when changing the content for a similar device and orientation with a task of larger size. Helping based on guidelines is suitable for simpler pages consisting mainly of text and multiple images. By managing objects based on objects, you can configure the method of correcting each object for a new page. However, after changing the page orientation options, the result of the coup, as a rule, requires further editing, because InDesign only performs work on adding pages and copying content. However, the program has not implemented intelligent content placement algorithms on the pages of the publication; the application places materials in the same places where they are on the source pages. As a result, as a result of changing page orientation, the layout loses its identity to the original layout. Therefore, further placement of text and graphic frames, in terms of the rules of the set and laying, and an aesthetic point of view, is necessary. Secondly, an existing algorithm for changing the page orientation of the viewing publication does not allow the implementation of the methods of moisture and grooves of the text to achieve competent placement of content on a layout with a modified page orientation. So, after changing the page's orientation through InDesign, the transactions of the ink and grooves of the text also have to be carried out manually, which we understand as a "thin making-up". Thirdly, in the process of changing the edition page, the parameters of the tools that use the machine tool to preserve the identity of the original are not regulated. Therefore, in this case, the machine is guided only by its own experience and subjective judgments. As a result, the final product may not correspond to the general technical rules of the set and making-up.

At the stage of changing the publication page's page orientation, several problems require solutions. The main ones are:
- mismatch of the number of lines of the initial layout and layout with modified page orientation;
- violation of the aspect ratio of text and graphic frames to page parties.

Therefore, existing problems cause contradictions, which is that changing the orientation of the page of the elevation edition changes the number of rows on the page. This leads to the inconsistency of the initial content on the page, due to an increase in the incomplete coefficient. But at the same time, changing the orientation of the page of the drawing publication in the Folio format is the full correspondence of the initial layout in terms of content volume on each publication page.

Results

To achieve the goal, it is necessary to form a list of rules for eliminating the disadvantages of recruiting as a result of changing the orientation of the page of the elevation publication. To do this, you need to analyze how the volume of content changes on the page depending on the change in the parameters of the layer elements. Fig. 1 presents a general algorithm for designing printed publications. In this scheme, special attention should be paid to the block "Calculation of the number of lines during design." This calculation is carried out in accordance with a mathematical model designed for a text block. As a rule, any band consists of several blocks: graphic, text, block, occupied by the formula, table, etc.

Quality Production Improvement and System Safety: QPI 16 - CZOTO 10   Materials Research Forum LLC
Materials Research Proceedings 34 (2023) 430-446   https://doi.org/10.21741/9781644902691-50

The main text can be divided into parts, subdivisions, and chapters in any combination. Paragraphs separate every new semantic part. The division of the text under paragraphs facilitates the text of the reader. Therefore, the basis of developing a mathematical model is the principle of separating text into paragraphs.

According to the separation of text on paragraphs, the area occupied by the text may be expressed through the area of all paragraphs of the text:

$$S_{\text{общ}} = \sum_{i=1}^{N} A_i,$$ (1)

where A_i – the area occupied by i paragraph;
N – the number of paragraphs in the text.

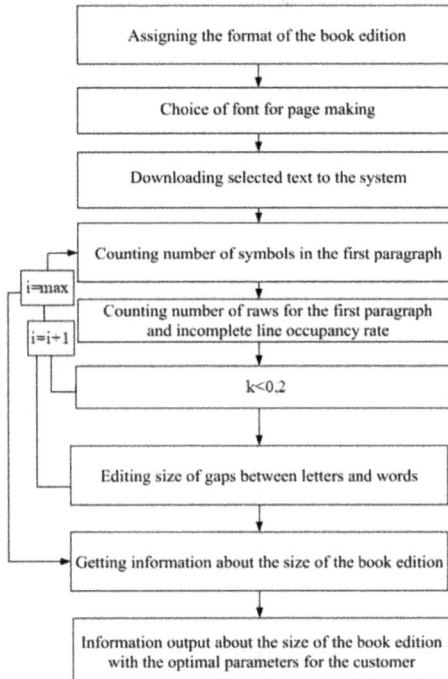

<table>
<tr><td>Assigning the format of the book edition</td></tr>
<tr><td>Choice of font for page making</td></tr>
<tr><td>Downloading selected text to the system</td></tr>
<tr><td>Counting number of symbols in the first paragraph</td></tr>
<tr><td>Counting number of raws for the first paragraph and incomplete line occupancy rate</td></tr>
<tr><td>k<0.2</td></tr>
<tr><td>Editing size of gaps between letters and words</td></tr>
<tr><td>Getting information about the size of the book edition</td></tr>
<tr><td>Information output about the size of the book edition with the optimal parameters for the customer</td></tr>
</table>

i=max

i=i+1

Fig.1. Algorithm for calculating the number of rows when designing.

Considering the area occupied by one paragraph, it should be noted that for the convenience of perception, the beginning of each paragraph must be allocated. The most common way of allocating is the problem of paragraph indentation at the beginning of the first line. In addition, each paragraph includes a basic text with a certain width of signs, letters, and distances between words. As a rule, the number of characters in the future edition is known. Before adjusting, one also specifies the format of the future edition, a font headset, and a keel size for the selected headset. The number of rows N in the paragraph may be presented as a function of text parameters, that is:

$N=f(F,k,k_0,A,\varphi,\gamma,\omega)$,

where F – edition format;
k – the value of points;
k_0 – base point size;
A –area of paragraph;
φ – the value of inter-exploin discharge;
γ – the value of the distance between words;
ω – the width of the signs.

The following formula can calculate the number of lines in the edition:

$$N = \frac{k}{k_0 \cdot F} \cdot \left[A + \left(\sum_{j=1}^{m} \sum_{i=1}^{n} a_{ij} + g \cdot a_k \right) + \sum_{j}^{m} \sum_{i=1}^{n-1} \eta_{ij} + \sum_{j=1}^{m-1} \beta_j + q \right],$$

(2)

where F – edition format, mm;
A – paragraph indentation, mm;
k –signs point size, points;
k_0 – base point size, points;
a_{ij} – width i sign in j word, mm;
g – the number of transfers
a_k – the width of the transfer sign, mm;
n – number of signs in a word;
m – the number of words on one row of strips;
q – the value of the magnitude of the gaps that fall in the end of the lines, mm;
η_{ij} i β_j – width of letters and external distances, mm.

The calculation was carried out for 15 headsets. The arbitrary text of historical literature was taken. The formula determined the errors:

$$n = \frac{|N_v - N|}{N_v} \cdot 100\%,$$

(3)

where N – number of rows obtained when calculating;
N_v – number of rows obtained when a page making-up.
The calculation results differ slightly from the layer results, but the error is minimal. For the main headsets used in the development of editions, the error is not more than 2% (Table 1). However, in certain circumstances, the simulation of the text block may occur. In the first place, it concerns incomplete lines. If the string is filled in less than 10%, then you will have to perform the position of the lines manually. To do this, you can change the width of the sign or reduce the distance between characters or words.

Quality Production Improvement and System Safety: QPI 16 - CZOTO 10   Materials Research Forum LLC
Materials Research Proceedings 34 (2023) 430-446   https://doi.org/10.21741/9781644902691-50

Table 1. *The experimental value of the calculation error*

Font	9	10	11	12	13	14
Times New Roman	-0,29	-0,39	0,48	-0,11	0,30	0,93
Petersburg	-1,19	-0,60	-1,08	0,00	0,00	1,03
Arial	-1,57	0,00	-0,97	0,20	0,18	0,77
Tahoma	-0,92	-0,36	-0,86	0,20	-0,55	0,94
Myriad Pro	-0,99	-1,02	0,12	-0,11	0,88	0,36
Univers	-1,03	-0,23	-0,63	-0,29	0,63	0,83
Miniature	-0,28	0,25	-1,02	-0,21	0,87	1,08
RodeoLight	0,00	1,67	0,43	-0,10	0,82	1,27
a_AntiqueTrady	-1,26	-1,02	-0,62	0,00	-0,44	0,57
a_BodoniNova	-1,57	-0,12	-0,43	0,10	-0,91	0,09
Academy	-0,47	-1,55	-1,80	-0,12	-0,55	0,10
AGGalleon	-0,42	0,63	-0,57	0,31	0,10	1,08
Baltica	0,00	-0,43	-0,49	0,00	0,16	0,38
Verdana	-0,58	-0,84	-0,10	0,09	0,57	1,20
Journal	0,00	-0,23	0,00	0,38	0,18	1,14

Investigation of the impact of the parameters of the publication on the formation of a petroleum set when returned in two stages. The impact of the settlement options for the passage or extension is determined in the first stage. The coefficient of sealing of the PN rows characterizes the degree of moiety and exhibition. The second stage is characterized by the coefficient of filling the text block area – P_S

$$\Delta P_N = (P_N - 1) \cdot 100\ \%, \tag{4}$$

$$\Delta P_S = (P_S - 1) \cdot 100\ \%. \tag{5}$$

The P_N and P_S coefficients are convenient to determine not an absolute value, but their percentage to $P_N = 1$ and $P_S = 1$ with the output parameters of the set. The positive area of the values of these coefficients corresponds to the value of lines, which leads to an increase in the text block area, and the negative - the moiety of the lines, which leads to a decrease in the square of the text block. Investigation of the influence of distances between words was originally analyzed, and then experimental verification of the results obtained in the text and grooves of lines in the text was carried out. In this case, it is essential to introduce such a publication parameter as a coefficient of the fullness of the k incoming row.

When returning to any publication, it is inevitable for incomplete lines to appear in the text. Such lines are filled in less than 20% and worsen the ease of reading text, leaving additional spaces on the page. If you do not monitor incomplete lines and do not remove them, then there will be a lot of short lines that increase their volume. To combat similar lines, there are several tools, the most important of which is to change the value of external distances. It is a powerful tool for combating hanging lines in aggregate with others. Reducing each degree of each of the parameters can completely get rid of incomplete lines. It will significantly reduce the number of rows in the edition, and when changing the orientation of the elevation, it will allow us to capture the required content on the page. To assess the impact of the size of the external distance in the text of historical literature on the number of lines that need to be driven or expelled, we will use a mathematical model. The residual part of the calculation of the number of rows in the paragraph can determine the fill factor K of the incomplete line. That is, if the calculation of the number of lines was, for

Quality Production Improvement and System Safety: QPI 16 - CZOTO 10   Materials Research Forum LLC
Materials Research Proceedings 34 (2023) 430-446   https://doi.org/10.21741/9781644902691-50

example, 5.155 lines, then the filling coefficient will be the residue value: k = 0.155. According to the proposed version of the reading, the entire process can be divided into the following steps:
- calculation of the number of characters in each paragraph;
- definition of the fill factor K in an incomplete line;
- in the case of a small value of the filling coefficient, the string yield, reducing the distance between the letters to the maximum permissible value n_{max} ;
If the result does not suit, use a distance decrease tool between words β;
- summing up the number of lines and receiving the results of the publication.
In addition to the output patterns of the impact of changing the values of the parameters of the distance between the letters and between the words on the effectiveness of the abomination and grooves of the text, the algorithm for changing the orientation of the page of the drawing publication was formulated.

This algorithm is based on the coefficient of proportionality of the k_{np}, which displays the proportional ratio of the parties of the element elements:

$$k_{pr} = \frac{W}{H},$$

(6)

where k_{pr} – coefficient of proportionality of the parties;
W – freum width;
H – freum height.

Therefore, the scheme of operation of the algorithm for changing the page orientation of the page is as follows:
calculation of the coefficient of proportionality for page parameters;
creation if a document with a new page orientation;
calculation of the new size of the elements of the page using the coefficient of proportionality of the k_{np} (multiplication of the coefficient in case of a decrease in the height of the page and division in the event of an increase in the width of the page, and vice versa).
As for the typesetting and forcing of the text, the following algorithm was displayed. If necessary, to manage the text, use the coefficient of k_{fit}.

$$k_{fit} = \frac{V_{par}}{V_{last\ line}},$$

(7)

where k_{fit} – typesetting paragraph coefficient;
V_{par} – a paragraph volume;
$V_{last\ line}$ – volume of the last line of the paragraph.

Quality Production Improvement and System Safety: QPI 16 - CZOTO 10   Materials Research Forum LLC
Materials Research Proceedings 34 (2023) 430-446   https://doi.org/10.21741/9781644902691-50

Table 2. *The formation of rules for eliminating the disadvantages of recruitment as a result of changing the orientation of the page of the page-proof*

№	Case	Operations
1	Page, general parameters	1) Creation of a document using a coefficient of proportionality
2	Text in one column	1) Creation of a document using a coefficient of proportionality; 2) Typesetting of text by the algorithm of the typesetting coefficient
3	Text in several columns	1) Typesetting of text by the algorithm of the typesetting coefficient; 2) If necessary - an increase in the number of columns
4	Strip illustration	1) Horizontal version request; 2)Placing an image on the center of the page, fitting at the height of the page; 3)Proportional image wrapping in graphic frame
5	Open illustration on page width	1) Horizontal version request; 2) Checking the image to the width stock, if any, then create a graphic frame with an output height and width of the size band + proportional image wrapping; 3) Typesetting – forcing of the text
6	Open illustration by page height	1) Request a horizontal version of the image; 2) Creating a graphic frame with an output width and height by size of a band set + proportional image wrapping; 3) Typesetting – forcing of the text
7	Closed illustration	1) Placing an image with an attachment to the edge of the set of a set, with the preservation of the indentation ratio from the top and bottom of the page field, as well as the flow parameters; 2) Typesetting – forcing of the text
8	Deaf illustration	Placing an image with an attachment to the edge of the set of a set, with the preservation of the indentation ratio from the top and bottom of the page field, as well as the flow parameters
9	Text in a few columns + illustration in destination	1) Request a horizontal version of the image; 2) Sllustration on the width of the column with a decrease in height, then fit the contents of the graphic frame + typesetting of the text; 3)Typesetting of the text + proportional reduction of illustration
10	Illustrations of proper geometric forms	Prohibition to change the aspect ratio, placement of frames in the necessary coordinates
11	Several vertical images	Placing illustrations to left – right
12	Several horizontal images	Proportional placement of an image with an attachment to the edge band, with the preservation of the ratio of indentation from the top and bottom of the page field, as well as the flow parameters

For each paragraph, it is necessary to calculate the coefficient of k_{fit}, after which, to achieve maximum efficiency, it is precisely the paragraph in which the coefficient of k_{fit} is the largest. Based on the results of empirical studies, it has been proved that using this coefficient of kfit ink gives significant energy efficiency compared to using the rule of the shortest line on the page. In

Quality Production Improvement and System Safety: QPI 16 - CZOTO 10   Materials Research Forum LLC
Materials Research Proceedings 34 (2023) 430-446   https://doi.org/10.21741/9781644902691-50

accordance with the formed set of page layouts, the rules of possible elimination of the disadvantages of an account as a result of changes in the orientation of the page of the elevation edition (Table 2) were formed.

Thus, based on the results of empirical studies, a number of cases of an acute, as well as options for maintaining the volume of content on the page, were considered. As a result, rules were formulated to change the orientation of the elevation of specific layout cases. The following system of requirements and restrictions allows for avoiding errors and discrepancies in the original layout as a result of changing the orientation of the page of the elevation publication.

As for the publication's text, before changing the orientation of the page of the elevation edition, it is necessary to draw attention to the unilcurrent prepositions and dashes. After all, after changing the orientation of the page, the text will be performed. The specified redemption can lead to the emergence of hanging prepositions and transferring the dash to the following line as the first character. It is not allowed relative to the rules of a set and a manner. For this reason, you should also pay attention to fixed transfer in words.

During typesetting and forcing text, the following restrictions for distillation agents were formed. When grossly typesetting, one uses the tracking tool, which can be reduced by a maximum of up to 30 thousand estimates of conditional units of em and increased according to +30 thousand conditional units of em.

When uniformly typesetting, the text uses several means, such as the distance between words and symbols and scaling characters.

At the same time, based on analysis of the technical rules of the set and a manner, as well as expert opinion of professional machines, the parameters of the uniform moisture of the text may vary in the following limits.

The distance between words can be changed from 80% to 110%, with a step of 5%. The distance between characters is allowed to change from -3% to 3%. Scaling characters can be changed to the opposite side to 95%.

If it is not possible to dig or drive the text using the above limits, an additional increase or decrease is technically impossible; then, in the first place, one should increase the distance:

- after a dot at the end of the sentence;
- after exclamation marks and questions;
- after a semicolon and a colon.

When using tracking, it is recommended to consider several features of a person's perception of a typographic text. For example, the text typed in large font, looks better if the letters in words are more closely (than using a standard interval). It is especially noticeable when the word is typed in full letters. The degree of the necessary correction of interlist passes depends on the goat and the headset. Some headsets require more tangible tracking, while others can do practically without such. Particularly useful tracking in situations when there is a need for a dense set of some parts of the text, for example, in separate graphs of the table. The interlink change is used only when it was impossible to dig or drive the text above the specified methods. As for illustrations, their main requirement is to preserve the type of completion. Changing the type of illustrations on the page is possible only if none of the other content saving on the page did not help. Also, the change in the type of illustrations maybe if this does not contradict the basic design of the publication and the rule of the uniformity of the recruitment throughout the publication. As for the placement of illustrations in text frames, there are several options for completion. First, it is important to check whether the graphic frame will not remain a significant part of the image when it comes to illustrations to a frame with preservation of proportions. Otherwise, you need to perform a manual fitting image in an edition with a modified orientation. Or it is necessary to go to the second way of completing the illustration on the page, namely, an image options request for the corresponding page orientation.

Quality Production Improvement and System Safety: QPI 16 - CZOTO 10   Materials Research Forum LLC
Materials Research Proceedings 34 (2023) 430-446   https://doi.org/10.21741/9781644902691-50

And suppose the illustration has equal width and height (square, circle). In that case, it is prohibited to change the ratio between the width and height by the coefficient of proportionality when changing the orientation of the page of the election edition. Thus, a list of requirements and restrictions on each rule for changing the orientation of the page of the election edition were formed. These requirements and regulations apply to each element of the recruitment and describes the order and variant of actions when using an alternative layout.

Discussion of the results

Based on the developed complex of typical layouts, the support of the pages was developed by the next scenario of the prototype of the means of automatic change in the orientation of the page of the elevation edition. Each type of layout of the page When changing the orientation, there are a number of problems inherent in this type of layout. These problems require certain rules for eliminating the disadvantages of recruitment due to changes in the orientation of the page of the elevation publication. As a result, it was decided to build a system of automation on the principle of individual modules and functions that will meet the necessary layout types. To solve each problem, the decades of recruitment as a result of changing the orientation of the page of the elevation publication may be used several rules. Therefore, creating multiple modes with a specific set of parameters is necessary. Among such parameters, the user program can select the most suitable for a specific type of original edition layout. Thus, the concept of a means of automation of the process of changing the orientation of the page of the elevation publication includes a certain sequence of actions that can be divided by steps:

- choosing the user of a specific mode program (set of settings);
- analysis of the original layout of the publication;
- formation of a script from individual modules and functions required for the layout of type types involved;
- launching the script to change the page orientation of the viewing edition;
- demonstration of the result of the script;
- output information about problem pages to pay attention to.

In addition, it is necessary to reproduce the developed knowledge base in itself:
- complex of typical page layouts;
- list of problems that may arise at the stage of changing the page orientation of the address;
- list of rules necessary to solve problems in the process of changing the page orientation of the viewing publication;
- the system of requirements and restrictions for each rule for solving problems in the process of changing the page orientation of the viewing publication.

Thus, a software product has been developed for practical and scientific value because it will help machine tools avoid many errors when changing the page of the elevation edition. To implement an effective user interaction with the software product, creating a convenient, intuitive interface executed on the usability principles is necessary.

To solve the task of automatic change in the orientation of the eligible publication based on the developed scenario, the use of JavaScript language will be optimal.

Software products such as Folio Producer and Folio Builder are used to create folio products.

The Folio Producer Instrument Instrument Sets Overlay Creator Palette, Mandatory External InDesign Module and Desktop Viewer application.

The Windows Forms or Web application can be used as a prototype interface.

The first stage of the script is the user selection of a program of a specific mode (set of settings). So, the system of recommendations and restrictions was implemented as three modes of prototype operating modes that can be selected from the drop-down list.

Each mode is implemented in the prototype as a separate module, changing each parameter in the form of separate functions.

The «Defalt» mode is used in the rough municipality and exhibition of the text when using the tracking tool that can be reduced by a maximum of up to 30 thousand estimates of conditional em units and increased in accordance with +30 thousand conditional units of em.

Images in this mode are fitted in proportion to graphic frames.

«Advanced» mode is used with a uniform position of the text when using multiple means of moisture and grooves of the text, such as the distance between the words, the distance between symbols, and scaling characters. At the same time, based on analysis of the technical rules of the set and a manner, as well as expert opinion of professional machines, the parameters of the uniform moisture of the text may vary in the following limits.

The distance between words can be changed from 80% to 110%, with a step of 5%. The distance between characters is allowed to change from -3% to 3%. Scaling characters can be changed to the opposite side to 95%.

The third "Custom" mode suggests that the user performs the selection of settings to change the page orientation of the viewing publication. The choice can be done according to the following parameters:
- distance between words;
- distance between characters;
scale characters;
- trekking;
- allow/prohibit altering;
- allow/prohibit change number of columns per page;
- format image freight in graphic frames;
- choosing a directory with image variants for modified orientation.

In addition, users can familiarize themselves with the parameters of modes, basic rules, and restrictions for changing the page orientation for each of the typical layouts by passing "Help".

The next stage of the prototype is the analysis of the original layout of the publication. This process occurs due to reading the values of the styles of the layer elements (Fig. 2 and Fig. 3).

```
for(pStylesIndex=2;pStylesIndex<myDocument.paragraphStyles.length;
pStylesIndex++){
  horDocument.paragraphStyles[pStylesIndex].appliedFont =myDocument.paragraphStyles[pStylesIndex].appliedFont;
  horDocument.paragraphStyles[pStylesIndex].capitalization =myDocument.paragraphStyles[pStylesIndex].capitalization;
  horDocument.paragraphStyles[pStylesIndex].fontStyle=myDocument.paragraphStyles[pStylesIndex].fontStyle;
```

Fig. 2. *Reading parameter values for text frames.*

```
horDocument.paragraphStyles[pStylesIndex].pointSize=myDocument.paragraphStyles[pStylesIndex].pointSize;
horDocument.paragraphStyles[pStylesIndex].spaceBefore=myDocument.paragraphStyles[pStylesIndex].spaceBefore;
horDocument.paragraphStyles[pStylesIndex].tracking=myDocument.paragraphStyles[pStylesIndex].tracking;
horDocument.paragraphStyles[pStylesIndex].underline=myDocument.paragraphStyles[pStylesIndex].underline;
horDocument.paragraphStyles[pStylesIndex].leading=myDocument.paragraphStyles[pStylesIndex].leading;
horDocument.paragraphStyles[pStylesIndex].justification=myDocument.paragraphStyles[pStylesIndex].justification;
horDocument.paragraphStyles[pStylesIndex].dropCapCharacters=myDocument.paragraphStyles[pStylesIndex].dropCapCharacters;
horDocument.paragraphStyles[pStylesIndex].dropCapLines=myDocument.paragraphStyles[pStylesIndex].dropCapLines;
horDocument.paragraphStyles[pStylesIndex].firstLineIndent=myDocument.paragraphStyles[pStylesIndex].firstLineIndent;
```

Fig. 3. *Reading parameter values for text frames.*

Regarding the formed coefficients of the k_{pr} proportionality and the k_{fit}, then in the prototype they are implemented as follows (Fig.4 and Fig. 5).

```
Koef_K[pCounter]=Sum_chars/last_chars;
    }
for(pCounter=0;pCounter<horDocument.textFrames
[frameCounter].paragraphs.length;pCounter++)

 var max=Koef_K[0];
  var max_counter=0;
  canchange=false;
```

Fig. 4. *Implementation in the prototype of the coefficient of typesetting k$_{fit}$.*

```
for(i=1;i<horDocument.textFrames[frameCounter].paragraphs.length;i++)
   {
    if(Koef_K[i]>max && ParagraphChanged[i]==false)
    {
   max=Koef_K[i];
   max_counter=i;
   canchange=true;
    }
   }
   if(max_counter==0 && ParagraphChanged[0]==false)
   canchange=true;
   if(canchange==false)
   break;
```

Fig. 5. *Implementation in the prototype of the algorithm on the basis
of the coefficient of typesetting k$_{fit}$.*

The alternative orientation page is also formed by the coefficient of proportionality (Fig. 6).

```
ratio=myWidth/myHeight;
myPages=myDocument.pages;
var rectangle=0;
for(pageIndex=0;pageIndex<myDocument.pages.length;pageIndex++) {
 myPages[pageIndex]=myDocument.pages[pageIndex];
 with (horDocument.pages[pageIndex].marginPreferences)  {
 var columnCount=myColumnCount;
 var columnGutter=myColumnGutter;
 var top=myMarginTop*ratio;
 var left= myMarginLeft/ratio;
 var bottom=myMarginBottom*ratio;
 var right=myMarginRight/ratio; }
```

Fig. 6. *Implementation in the prototype of page forming on the basis
of the coefficient of proportionality k$_{pr}$.*

Each parameter of the administering elements is checked for existence, and in the event that the style does not contain information about some parameter, the value of this parameter is taken from the base style. Similarly reads information about character style and tables. Placing illustrations occurs according to the following algorithm (Fig 7, Fig. 8).

```
for(gFrameIndex=0;gFrameIndex<myPages[pageIndex].
rectangles.length;gFrameIndex++)
  {
  try
  {
  filename =File(myDocument.pages[pageIndex].rectangles[gFrameIndex].graphics[0].itemLink.filePath);
  image=true;
  }
  catch(error)
  {
  image=false;
  }
myGraphicFrames[gFrameIndex]=myPages[pageIndex].rectangles[gFrameIndex];
  var horY=myGraphicFrames[gFrameIndex].geometricBounds[0]*ratio;
  var horX=myGraphicFrames[gFrameIndex].geometricBounds[1]/ratio;
  var horY2=myGraphicFrames[gFrameIndex].geometricBounds[2]*ratio;
  var horX2=myGraphicFrames[gFrameIndex].geometricBounds[3]/ratio;
  myGFrameWidth=horX2 -horX;
  myGFrameHeight=horY2 - horY;
  imageFrame=horDocument.pages[pageIndex].rectangles.add();
  imageFrame.geometricBounds=[horY, horX, horY2, horX2];
```

Fig. 7. *Algorithm for reading and placing illustrations.*

```
imageFrame.fillColor=myDocument.pages[pageIndex].rectangles
[gFrameIndex].fillColor.name;
   }// end  for(gFrameIndex=0;gFrameIndex<myPages[pageIndex].rectangles.
length; gFrameIndex++)
```

Fig. 8. *Read algorithm and tasks of frames.*

Thus, we get information about the color model and its data on each color paint or sample.

So, after analyzing the elements of the recruitment, the script is formed from individual modules and functions necessary for the layout. As a result of the prototype, one obtains an InDesign document with a modified page orientation (Fig. 9–11).

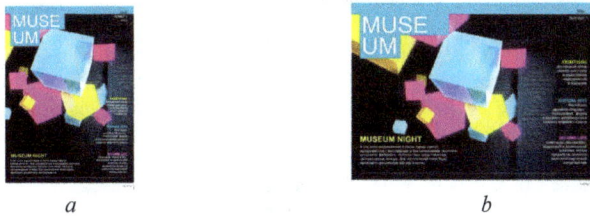

a   b

Fig. 9. *The result of the automated change of the page of the InDesign document with the vertical orientation of the page: a - in the horizontal orientation of the page; b - using the prototype created.*

Materials Research Forum LLC
https://doi.org/10.21741/9781644902691-50

Fig. 10. The result of the automated change of the page of the InDesign document with the vertical orientation of the page: a - in the horizontal orientation of the page; b - using the prototype created.

Fig. 11. The result of the automated change of the page of the InDesign document with the vertical orientation of the page: a - in the horizontal orientation of the page; b - using the prototype created.

Thus, the resulting layout corresponds to the initial content. For a prototype, an interface for interacting with the user has been developed for the auto-edition orientation of the page. Within the framework of this interface, a logo for a software product was also created, a name was given, and a well-designed generic concept of prototype design.

Summary

Some empirical research within the framework of this article allowed analyzing most cases. As a result, the rules for eliminating the disadvantages of recruitment as a result of changing the orientation of the page of the elevation publication in specific cases of a layout. In addition, a certain regularity of the possibility of moisture and grooves of the text was found at certain values of the parameters of the moisture and exhibition, the coefficient of proportionality of the kpr and the coefficient of kfit text is formulated. In addition, scientific research was analyzed to identify the boundary limits of the possibility of changing the parameters that affect the scope of text on the page. The analysis of technical requirements for the original - publication layout was also carried out. As a result, a system of requirements and restrictions on the rules for eliminating the disadvantages of recruitment due to changing the orientation of the page of the elevation publication.

The resulting scientific result can be used at the stage of administering electronic editions in a folio format for a tablet PC, namely, in the process of changing the orientation of the elevation page. As a result, compliance with the final layout of the technical requirements of the set and a recruitment is provided.

The resulting scientific result can be used at the stage of administering electronic editions in a folio format for a tablet PC, namely, in the process of changing the orientation of the elevation page. As a result, compliance with the final layout of the technical requirements of the set and a recruitment is provided:

a) formulated separate modules and functions used for a particular type of page layout;
b) the designed set of parameters of the elements, which fully describes the following elements of the academic as a page, paragraph, symbol, table, and object;
c) the three prototype operating modes are developed, which are characterized by sets of parameters to automatically change the orientation of the page of the elevation edition;
d) an algorithm for automatic change of page orientation of the viewing of the publication is developed;
e) developed toolkits to change the page orientation of the viewing publication automatically;
f) a formulated knowledge base containing a set of rules and restrictions to eliminate the disadvantages of automatic change of page orientation of the viewing publication;
g) The logo for the software product is developed, and the general design concept of the interface is formulated.

The developed system of automatic change in the orientation of the page of the elevation publication allows to:

a) read the necessary information about the style and formatting of the text and its content from the InDesign document;
b) change the page orientation to preserve the identity of the original layout;
c) provide information about problem pages to pay attention to after the orientation change.

The scientific result of the study is to formulate the method of automatic change in the orientation of the page of the drawing publication. The practical effect is the realization of the developed method in the prototype of the automated system for a «thin layer».

References

[1] E. Gardiner. The Electronic Book, Oxford University Press, Oxford, 2010.

[2] N. Hood. Quality in MOOCs : Surveying the Terrain, Commonwealth of Learning, Burnaby, 2016. ISBN 978-1894975803

[3] Y. Hrabovskyi, N. Brynza, O. Vilkhivska. Development of information visualization methods for use in multimedia applications, EUREKA: Physics and Engineering 1 (2020) 3-17. https://doi.org/10.21303/2461-4262.2020.001103

[4] Y. Hrabovskyi, V. Fedorchenko. Development of the optimization model of the interface of multimedia edition', EUREKA: Physics and Engineering 3 (2019) 3-12. https://doi.org/10.21303/2461-4262.2019.00902

[5] Y. Hrabovskyi, O. Yevsyeyev. Development of methodological principles of support-preservation engineering work, Technology audit and production reserves 2/2 (40) (2018) 43-49. https://doi.org/10.15587/2312-8372.2018.127776

[6] E. Lupton, Thinking with Type: A Critical Guide for Designers, Writers, Editors, & Students, Princeton Architectural Press, New York, 2007. ISBN 978-1568989693

[7] R. Martin. Twenty challenges for innovation studies, Sci. Public Policy 43(3) (2016) 432-450. https://doi.org/10.1093/scipol/scv077

[8] D.E. Mitchell, R.K. Ream (Eds.). Professional responsibility – The Fundamental Issue in Education and Health Care Reform, Springer International Publishing, Switzerland, 2014. https://doi.org/10.1007/978-3-319-02603-9

Quality Production Improvement and System Safety: QPI 16 - CZOTO 10   Materials Research Forum LLC
Materials Research Proceedings 34 (2023) 430-446   https://doi.org/10.21741/9781644902691-50

[9] I. Safonov et al. Adaptive Image Processing Algorithms for Printing, Springer, Singapore, 2018. https://doi.org/10.1007/978-981-10-6931-4

[10] M. Saylor. The Mobile Wave: How Mobile Intelligence Will Change Everything, Vanguard Press, New York, 2012. ISBN 978-1593157203

[11] O.M. Vultur, S.G. Pentiuc, V. Lupu. Real-time gestural interface for navigation in virtual environment, 13th Int. Conf. Development and Application Systems (DAS), Suceava, Romania, May 19-21, 2016, 303-307. https://doi.org/10.1109/DAAS.2016.7492592

Quality Production Improvement and System Safety: QPI 16 - CZOTO 10   Materials Research Forum LLC
Materials Research Proceedings 34 (2023) 447-452   https://doi.org/10.21741/9781644902691-51

Job Satisfaction in the Healthcare Sector

BSOUL-KOPOWSKA Magdalena

Czestochowa University of Technology, Al.Armii Krajowej 19b, 42-218 Czestochowa, Poland

m.bsoul_kopowska@pcz.pl   ORCID: 0000-0002-6167-6827

Keywords: Job Satisfaction, Satisfaction Determinants, Reasons for Job Dissatisfaction

Abstract. The objective of this article is to present research regarding the level of job satisfaction of physicians employed in healthcare institutions in the Silesian Voivodeship. The basis for the analysis is a survey carried out between August and October 2022, with a sample of 184 medical workers. The study was carried out by means of a diagnostic survey, using a questionnaire technique. Random purposive sampling was used. Regular surveys will make it possible to improve the stability of employment in the healthcare sector and can provide a basis for corrective measures to be taken by managers of healthcare facilities in order to increase the sense of job satisfaction among doctors.

Introduction

Among all the professions that a person can perform, there is a group of professions that serve others in a special way. Professions with a social mission are those in which the desire to help others and to care for others should appear as the primary motivation (in their current state and their development as well as future well-being). This is often associated with a significant level of responsibility for the people to whom the professional activity is directed. Medical professions can be included in the group of professions defined above.

The COVID-19 pandemic has clearly underlined the significant shortage of medical personnel in Poland. It has been pointed out for years that we have the lowest rate of doctors per 1000 inhabitants in the European Union, and one of the lowest among OECD countries – 2415. The above result is far from the average level of this indicator for the European Union, i.e. 3.8. It is therefore necessary to act proactively to increase the supply of health professionals and optimise the use of human resources in the healthcare sector. In order for this to be possible, it is necessary to monitor the physician' sense of job satisfaction. Systematic research will make it possible to improve the stability of employment in the healthcare sector. This article attempts to explore the level of job satisfaction in the medical community. For this purpose, an empirical study was carried out, described in the next section of this paper.

Job Satisfaction in Theoretical Terms

Work is an essential part of everyone's life, it provides the opportunity for growth, provides necessary financial resources and can contribute to better health and well-being, but at the same time it can be a source of frustration and even illness. It is also an important factor that needs to be monitored in research on the quality of life. The interest regarding employee satisfaction has been around in management sciences for several decades, while undergoing a significant evolution. Starting in the 1940s, it has been one of the most intensively studied phenomena in management and organisational psychology [1].

However, in order to properly understand the term satisfaction, it seems necessary to define the concept. This definition, or rather the lack of it, has caused many problems for the theorists of organisation and management sciences in the past. As the issue has been understood in varied ways by different researchers, this has led to confusion when trying to compare the results of the various studies conducted on the subject over several decades. Job satisfaction is most often defined as a person's positive attitudes and feelings towards the work environment and their job duties [2]. Job satisfaction also means a subjective feeling of satisfaction with one's job, working conditions and salary. It is usually identified with the quality of work, while its assessment takes into account elements such as the employee's career and individual status, health and well-being at work, personal growth, work-life relationships, physical and mental effort, and interpersonal relationships in the work environment [3]. M. Juchnowicz defines job satisfaction as: "a higher level of satisfaction, requiring that work provides intellectual challenges, a sense of success, enjoyment of professional growth and self-fulfilment, as well as full identification with the job and/or organisation. Achieving a sense of job satisfaction additionally requires the impact of internal factors such as learning opportunities, assigned responsibility, recognition from superiors" [4]. In turn, A. Czerw and A. Borkowska state that job satisfaction "is understood as general satisfaction with one's job or activity chosen as a future occupation" [5].

Job satisfaction depends on the balance between what a person invests in the work (e.g. time, commitment) and what he or she receives in return (promotion, salary, development opportunities, colleague relationships). A lack of satisfaction can therefore be said to exist when a large investment on the part of the employee is accompanied by a small profit [6]. Satisfaction also depends on the extent to which the work meets the employee's needs and expectations.

Satisfaction research shows that a lack of job satisfaction results in higher absence rates [7], higher employee turnover, reduced loyalty towards the organisation [8], reduced quality of services provided by the company [9], etc. Employees who are dissatisfied with their jobs are much more likely to display anti-social behaviour, such as a tendency to steal, for instance. Those feeling a high level of satisfaction, on the other hand, are sociable, willing to help their clients [9]. They also show more commitment and identify more with the organisation's activities [10]. Job satisfaction directly contributes to the quality of the work and its results. High job satisfaction makes employees more committed to their tasks and even identify more strongly with the organisation where they are employed and are less likely to leave it. This makes them more productive while increasing the level of services they provide [11]. Therefore, satisfaction research has begun to draw the attention to intrinsic personality traits or cognitive judgements as factors responsible for the emergence of job satisfaction or dissatisfaction [12] [13] especially when it comes to developing and adjusting motivation systems in organisations [14]. Research is also conducted into internal and external factors affecting employee satisfaction and job satisfaction. According to research conducted, the following factors most often impact the level of satisfaction achieved: salary, opportunities for promotion, career development prospects, the value of the tasks performed, job security, stress, work standards, fair treatment of all employees in terms of salary, interpersonal relations [15]. Thanks to numerous studies, researchers are now in possession of broad data on the basis of which they try to provide reliable information on the importance and strength of the influence of individual variables on people's perceived level of life satisfaction. For the purposes of this article, job satisfaction will be understood as the difference between what a person expects and what they experience at work [16].

Research Methodology
The objective of the study was to determine the factors shaping job satisfaction among physicians employed in healthcare institutions in the Silesian Voivodeship and to find out their level of job satisfaction. The study was carried out by means of a diagnostic survey, using a questionnaire technique. Random purposive sampling was used. The questionnaire consisted of 18 closed-ended single- or multiple-choice questions regarding the issues of job satisfaction and motivation assessment. The assessment was made by awarding points on a Likert scale where 1 means 'strongly disagree' and 5 means 'strongly agree'. Using an odd scale allows the respondent to maintain a neutral position. The survey was conducted over a three-month period (August - October) in 2022. After eliminating the incorrectly completed questionnaires, 184 respondents

Quality Production Improvement and System Safety: QPI 16 - CZOTO 10   Materials Research Forum LLC
Materials Research Proceedings 34 (2023) 447-452   https://doi.org/10.21741/9781644902691-51

were statistically analysed. The impact factors were determined based on an analysis of the available literature. In the questionnaire used for the research, motivating factors were listed, taking into account the following division: factors defining basic arrangements (goals, competences, working conditions and financial benefits), team relations, relationships with the direct supervisor (appreciation, communication) and individual motivators (growth, achievements, impact). Due to the number of respondents, the presented results cannot be considered to be representative for the general group of physicians the Silesian Voivodeship, but despite the above, they constitute an interesting empirical material. The research has a pilot character.

Research Results

A total of 184 physicians working in the profession took part in the questionnaire. Of the study group, 107 were women and 77 were men. 74 are employees of hospital wards and clinics. The remaining respondents are employees of both public and non-public open healthcare facilities.

The questionnaire asked physicians about the factors t impacting job satisfaction. Respondents were asked to tick 19 factors that influence their sense of job satisfaction, ranking them from most important to least important. Figure 1 shows the job satisfaction indicators in order from the highest values indicating the level of satisfaction to the lowest values.

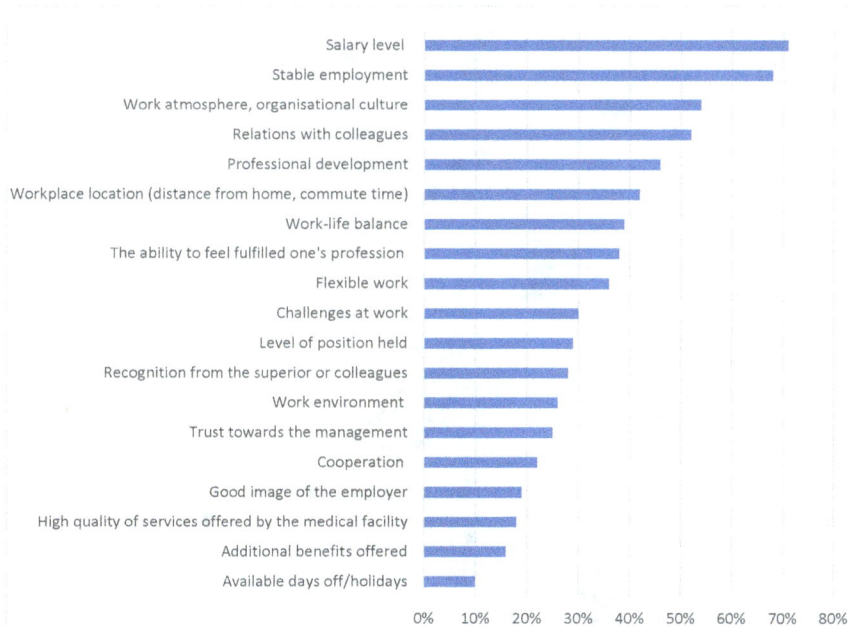

Fig. 1. *Factors shaping the most perceived level of job satisfaction [own work].*

According to the physicians taking part in the questionnaire, the most important factor influencing job satisfaction is the level of remuneration (71%). Perhaps this is due to the fact that most of them are employed on a medical contract basis which makes the livelihood of the physicians and their families entirely dependent on the size of the contract and their actual ability to provide work and also whether they will be guaranteed renewal of the contract and its terms,

Quality Production Improvement and System Safety: QPI 16 - CZOTO 10   Materials Research Forum LLC
Materials Research Proceedings 34 (2023) 447-452   https://doi.org/10.21741/9781644902691-51

especially pay, in subsequent periods. The next most important determinants shaping the sense of satisfaction were employment stability (68%), work atmosphere and organisational culture (54%), relations with colleagues 52% and professional growth opportunities (46%).

On the other hand, factors such as the location of the workplace (42%), work-life balance (39%), opportunity for fulfilment (36%), challenges at work (30%), position held (29%), and recognition from supervisor or colleagues (28%) came next.

After verifying the previously obtained results with the data on factors that impact job satisfaction, and which are related to the rules in terms of cooperation with the supervisor and colleagues, it turned out for the surveyed physicians that the possibility of freedom of action has a significant impact on their sense of job satisfaction: indications (38%) and the possibility of mutual decision-making (29%). This is probably due to the high level of confidence of supervisors in the professional competence of physicians. Another question related to the impact of recognition from colleagues on the sense of job satisfaction. This factor was considered important by 28% of the physicians surveyed.

On the other hand, the least important factors for the surveyed group of respondents are the working environment (26%), trust in management 25%, cooperation (22%), good image of the employer (19%), high quality of services offered by the organisation (18%), the offer of additional benefits (16%) or the availability of days off/holidays (10%).

Grouping the surveyed factors into individual categories: factors related to the economic aspect of work and interpersonal relationships are of greatest importance.

In a further part of the study, physicians were asked how they rated their sense of job satisfaction. Table 1 presents the responses.

Table 1. Assessing one's own job satisfaction [own study]

Assessment	N=184	%
Very high	18	10%
High	72	39%
Average	71	39%
Low	20	11%
Very low	3	1%

Fully positive job satisfaction is rated highly by almost half of the respondents - 90 physicians (49%) of all respondents. Almost as many – 71 (39%) - describe their level of satisfaction as average. The remaining group of physicians surveyed – 23 (12%) – indicate a low or even very low level of satisfaction.

These results differ when compared to a study conducted by the Centre for Studies, Analysis and Information of the Supreme Medical Council in 2022. According to the aforementioned report regarding a study on the medical community, 63% of physicians have a positive opinion on their professional work (22%) – an average of one in seven (14%) – have a bad opinion. Physicians who were significantly more satisfied with their work and rated it higher, in all the dimensions covered by the survey, were those who work outside Poland [17]. The discrepancy between the author's results and the data in the Report is probably due to the large difference in the size of the sample and the participation in the survey of people living outside Poland.

In the group of physicians taking part in the study, the lowest level of satisfaction was related to the impact of work on private and professional life. Interestingly, the survey found that low and average levels of the sense of job satisfaction were more common among younger physicians, aged up to 45 years, than among older physicians, aged over 45 years. The greatest differences in the group of physicians when it comes to their age were noted for satisfaction with their income. Here,

Quality Production Improvement and System Safety: QPI 16 - CZOTO 10   Materials Research Forum LLC
Materials Research Proceedings 34 (2023) 447-452   https://doi.org/10.21741/9781644902691-51

too, there was more dissatisfaction among doctors aged up to 45 years. Similarly, in all the other issues examined, the level of job satisfaction was lower than in the group of physicians under 45, in the senior physicians group.

A physician is a profession in which the desire to help, to care for others, should occur as an essential motivation. In the study conducted, 36% of respondents identified the opportunity for fulfilment in their profession as an important factor shaping their sense of job satisfaction. Salary, professional stability, relationships with colleagues or professional growth are more important to the physicians covered by the study. It is likely that ensuring a sense of satisfaction in terms of salary, a sense of professional stability, good workplace relations and suitable conditions for professional growth all contribute to fulfilment in the medical profession and the physician's attitude towards their profession.

Conclusions

For every human being, work is a process closely linked to their existence. As work is part of an individual's life, it can affect the other areas of their life. This impact can be positive or become a source of psychological discomfort and frustration. Such feelings can have a significant impact on one's private life [18]. This aspect is particularly important when it comes to professions that are exposed to stress due to their specific nature. A physician is such a profession. This is why a sense of satisfaction with work performance and opportunities for growth are so important in a physician's work. This is because it has an impact on the quality of the service provided, as well as on the direct contact between the physician and the patient [19].

It is now widely accepted that work should always be a source of satisfaction. In relation to the medical community, this issue is particularly important nowadays, as the level of professional satisfaction achieved by medics can have a significant impact not only on the results and effectiveness of the physician-employee's work, but above all on the role it can play in shaping the level of care provided to patients. Thus becoming an important component in the process of building patient satisfaction.

The results of the study indicate that:
- satisfaction of the surveyed physicians in the Silesian region is moderate (49%);
- salary was identified as the satisfaction factor with the highest rating;
- the least important factors fin terms of satisfaction were considered to be the employer's image (19%), the quality of services offered by the organisation (18%), the offer of additional benefits (16%) or the available days off/holidays (10%);
- almost half of the respondents said they were satisfied with their job, but only 10% rated the level of satisfaction as high;
- physicians with more seniority rated their job satisfaction higher.

The results of the study can provide a basis for corrective measures to be taken by managers of healthcare facilities in order to raise the level of job satisfaction among physicians and thus ensure greater stability of employment in the healthcare sector.

References

[1] J.F. Carey Jr., I.A. Berg, A.C. Van Dusen. Reliability of ratings of employee satisfaction based on written interview record, J. App. Psychol. 35 (1951) 252-255. https://doi.org/10.1037/h0063073

[2] D.S. Staples, C.A. Higgins. A Study of the Impact of Factor Importance Weightings on Job Satisfaction Measures, J. Bus. Psychol. 13 (1998) 211-232. https://doi.org/10.1023/a:1022907023046

Quality Production Improvement and System Safety: QPI 16 - CZOTO 10   Materials Research Forum LLC
Materials Research Proceedings 34 (2023) 447-452   https://doi.org/10.21741/9781644902691-51

[3] A. Brown, C. Forde, D. Spencer, A. Charlwood. Changes in HRM and job satisfaction, 1998–2004: Evidence from the Workplace Employment Relations Survey. Hum. Resour. Manag. J. 18 (2008) 237-256. https://doi.org/10.1111/j.1748-8583.2008.00069.x

[4] M. Juchnowicz. Management through Engagement. Concepts. Controversies. Applications. PWE, Warszawa, 2010. ISBN 978-83-208-1884-0

[5] A. Czerw, A. Borkowska. The Role of Optimism and Emotional Intelligence in Achieving Job Satisfaction : Implications for Employee Development Programs, Współczesne Zarządzanie 1 (2010) 69-82.

[6] D. P. Schulz, S. E. Schulz, Psychology and work today. Routledge, 2013. ISBN 978-1292021683

[7] D. J. Dwyer, D. C. Ganster. The Effects of Job Demands and Control on Employee Attendance and Satisfaction. J. Organ. Behav. 12 (1991) 595-608. https://doi.org/10.1002/job.4030120704

[8] L.A. Witt, M.N. Beokermen. Satisfaction with initial work assignment and organizational commitment. J. App. Soc. Psychol. 21 (1991) 1783-1792. https://doi.org/10.1111/j.1559-1816.1991.tb00504.x

[9] N. Malhotra, A. Mukherjee. The relative influence of organizational commitment and job satisfaction on service quality of customer-contact employees in banking call centres. J. Serv. Mark. 18 (2004) 162-174. https://doi.org/10.1108/08876040410536477

[10] R.J. Aldag, T.M. Stearns. Management. College Division, South-Western Pub. Co, Cincinnati, 1991. ISBN 978-053880562

[11] J.L. Heskett et al. Putting the service-profit chain to work. Harv. Bus. Rev. 72 (1994) 118-129.

[12] A. Furnham. Personality and intelligence at work. Exploring and explaining individual differences at work. Routledge, 2008. ISBN 978-1841695860

[13] A. Zalewska. Dwa światy. Emocjonalne i poznawcze oceny jakości życia i ich uwarunkowania u osób o wysokiej i niskiej reaktywności. Academica SWPS Publishing House, Warszawa, 2003. ISBN 978-8389281005

[14] B. Bojewska. Organisational Culture and the Enterprise's Strategy. Handel Wewnętrzny 2005(4-5) (2005) 7-13.

[15] G. Gruszczyńska-Malec, M. Rutkowska. Od satysfakcji do motywacji. Personel i Zarządzanie 2005(10) (2005) 58-63.

[16] P.J.D. Drenth, Hk. Thierry, Ch.J. de Wolff. What is organizational psychology? in: Handbook of Work and Organizational Psychology, 2nd Ed. Vol. I, Hove Psychology Press, East Sussex, 1998, 1-9.

[17] Health policy: nearly 2/3 of doctors assess their professional work as good. [online]. 2023. [viewed: 2023-01-31]. Available from: https://politykazdrowotna.com/artykul/nil-blisko-23-lekarzy-dobrze-ocenia-swoja-prace-zawodowa/837191

[18] A. Andruszkiewicz. Behavioural types at work and impact on mental health of nurses, Nursing Problems 18(2) (2010) 91-96.

[19] A. Lubrańska. Psychologia pracy. Podstawowe pojęcia i zagadnienia, Difin Publishing House, Warszawa, 2008. ISBN 978-8380854345

Quality Production Improvement and System Safety: QPI 16 - CZOTO 10   Materials Research Forum LLC
Materials Research Proceedings 34 (2023) 453-459   https://doi.org/10.21741/9781644902691-52

Matching Computational Tools to User Competence Levels in Education of Engineering Data Processing

RADEK Mateusz[1,a], PIETRASZEK Anna[2,b], KOZIEŃ Adam[1,c]
RADEK Katarzyna[3,d] and PIETRASZEK Jacek[4,e] *

[1]Jagiellonian University, ul.Gołębia 24, 31-007 Kraków, Poland

[2]Zofia Nałkowska Seventh High School, ul.Skarbińskiego 5, 30-071 Kraków, Poland

[3]Marshal's Office of the Świętokrzyskie Voivodeship, Al. IX Wiekow Kielc 3, 25-516 Kielce, Poland

[4]Cracow University of Technology, Al. Jana Pawła II 37, 31-864 Kraków, Poland

[a]mateusz6676@wp.pl, [b]anna.g.pietraszek@gmail.com, [c]a.kozien@uj.edu.pl, [d]kradek@onet.eu, [e]jacek.pietraszek@pk.edu.pl

Keywords: Educational Technology, STEM Education, Engineering Education, IT Tools, COVID-19

Abstract. Computational data processing often poses significant challenges for individuals not professionally specialized in this area. In particular, the problem lies in finding suitable computational tools that, on the one hand, have the necessary functionalities and, on the other hand, are easy to use for users with intermediate or low-level competencies. During the academic year 2022/2023, the Faculty of Mechanical Engineering at the Cracow University of Technology conducted classes on data processing for three diverse groups: the third- and fourth-year students of a public high school with a math-physics-computer science profile, undergraduate students in safety engineering, and undergraduate students in applied informatics. The needs regarding the functionalities of the computational tools and the competencies in their usage were compared, both in the assessment of the instructors and the participants themselves. The article presents comparisons of computational tools in terms of various aspects, as well as guidelines for their application for each target group.

Introduction

For many years, there have been ongoing challenges in matching computational tools to users' competency levels, particularly within the realm of education. This issue extends across various educational levels, including secondary and primary schools, as well as higher education. Initially, the development of computational tools focused primarily on achieving the necessary functionalities and hardware capabilities, such as performance and memory size. User convenience and ease of use were secondary considerations, assuming a relatively high level of user competency in technical matters. However, the landscape began to shift with the widespread adoption of graphical user interfaces (GUI) on personal computers in the early 1990s. The introduction of GUIs made computational tools more accessible to individuals with moderate IT skills. The intuitive nature of GUIs allowed users with limited technical expertise to navigate and utilize computational tools more comfortably. A significant turning point in the usability of computing tools came with the release of the first iPhone in 2007. This revolutionary device featured a touch-based user interface, marking a significant departure from traditional computer interfaces. Touch-based graphical interfaces were already available on both personal computers and phones before, but it was Apple that truly focused on the ergonomics of the interface and ease of navigation, enabling average individuals without computer skills to effectively utilize all the features of a smartphone. The iPhone's intuitive touch controls and simplified user experience

Quality Production Improvement and System Safety: QPI 16 - CZOTO 10   Materials Research Forum LLC
Materials Research Proceedings 34 (2023) 453-459   https://doi.org/10.21741/9781644902691-52

appealed to a wide range of users, including those without technical backgrounds or advanced computing skills. This mass-market appeal pushed software developers to prioritize the ergonomics and user-friendliness of their applications, accommodating consumer users who may lack technical competencies.

The demand for computational tools that catered to users with varying levels of technical proficiency grew substantially. Developers recognized the need to bridge the gap between complex computational capabilities and user-friendly interfaces, making it easier for individuals with average or limited technical skills to harness the power of computational tools effectively. This shift in focus led to the development of user-centric design principles, where the emphasis shifted from raw functionality to intuitive interfaces and streamlined user experiences. The goal became to create computational tools that offered the necessary functionalities while being accessible to a broader audience, including those without extensive technical expertise. As a result, developers have increasingly emphasized user-friendliness, leading to the development of intuitive interfaces and ergonomic applications that empower a wider range of users to leverage the benefits of computational tools.

Target Groups of Students
During the winter semester of the academic year 2022/2023, a series of classes were conducted with three distinct groups of students. The first group consisted of first-cycle (Bachelor of Engineering) engineering students pursuing degrees in mechanical disciplines. The classes delivered to this group focused on imparting fundamental computational and statistical methods essential to their field of study. The second group comprised first-cycle (Bachelor of Engineering) engineering students specializing in applied computer science. In addition to covering computational and statistical methods, these classes delved into the wider realm of data processing, encompassing both quantitative and qualitative approaches. Students were equipped with the necessary knowledge and skills to effectively manipulate and analyze data within their field. The third group comprised students in grades 3 and 4 from a public high school, following the math-physics-inf profile. As part of a collaborative initiative between the university and the school, these students were enrolled in computer science classes. The curriculum designed for this group aimed to provide a comprehensive understanding of computer science, encompassing both theoretical concepts and practical applications [1]. Topics covered included the foundations of object-oriented programming languages, as well as the fundamental principles of data processing. The diversification of student groups allowed for tailored instruction that catered to the specific needs and academic backgrounds of each cohort.

Utilized Software Tools
The software utilized in the conducted classes can be categorized based on four key aspects:

a) *Mode of Operation*: The first aspect is the mode of operation, which distinguishes between numerical processing and symbolic processing. Numerical processing involves calculations and operations performed on numerical data, while symbolic processing involves the manipulation and analysis of symbolic expressions and equations.

b) *User Interface*: The second aspect is the user interface, which can be either text-based or graphical. A text-based interface relies on command-line input and output, where users interact with the software through text commands. On the other hand, a graphical interface provides a visual environment with menus, buttons, and interactive elements for users to interact with the software.

c) *Task Formulation*: The third aspect pertains to the formulation of tasks within the software. It involves the methods by which users can express and specify their computational tasks. This can range from selecting pre-defined procedures or functions from a menu, using mathematical

Quality Production Improvement and System Safety: QPI 16 - CZOTO 10   Materials Research Forum LLC
Materials Research Proceedings 34 (2023) 453-459   https://doi.org/10.21741/9781644902691-52

notation to formulate tasks, or employing programming approaches to write algorithms for task execution.

d) *Processing Location*: The fourth aspect considers the location of processing. It distinguishes between local software applications that run on users' computers and web-based applications that are accessed through servers over the internet.

In the classes with mechanical engineering students, two software environments were employed. PTC Mathcad® (numerical/GUI/math-like/local) was utilized for numerical calculations, providing a user-friendly environment for performing engineering calculations. Maple® (numerical-symbolic/ GUI/programming-like/local), on the other hand, was used for symbolic calculations, allowing students to manipulate and solve mathematical expressions symbolically. For exercises from industrial statistics, two different software systems were utilized. Minitab® (numerical/GUI/ programming-like/local) was used for students specializing in Automation and Robotics, providing comprehensive statistical analysis tools. Statistica® (numerical/GUI/programming-like/local), on the other hand, was employed for students from other mechanical engineering subdisciplines, offering a wide range of statistical analysis capabilities.

In the Applied Computer Science program, the initial classes followed a similar pattern. PTC Mathcad® was utilized for numerical calculations, providing a convenient environment for performing various computations. Maple®, known for its symbolic processing capabilities, enabled students to perform symbolic calculations and manipulate mathematical expressions. However, as the curriculum progressed to machine learning classes, the focus shifted towards intensive programming in Python (numerical/text-mode/programming-like/local). Students leveraged various Python libraries and frameworks to delve into machine learning algorithms and applications.

Regarding the high school students' classes, it was known that they had prior exposure to algorithmics based on the C++ programming language. Leveraging their existing knowledge, the curriculum smoothly transitioned to more advanced operations using the C# language within the Microsoft Visual Studio environment (numerical/GUI/programming-like/local). Additionally, the network-accessible Wolfram Alpha environment (numerical-symbolic/GUI/programming-like/web-based) was utilized, providing students with a powerful computational platform for tackling complex tasks and exploring mathematical concepts.

Results

Grades of usefulness were assessed in two ways: through ratings assigned to students for their completed assignments and through interviews conducted with both students and instructors. Two aspects were evaluated: subjective ease of use (student ratings), actual effectiveness in the application (student work ratings), and functionality range (instructor ratings). Additionally, students' free-form feedback on the encountered software was also collected.

For groups of mechanical engineering students, the following results were obtained:

a) Mathcad® – students found it easy to learn with a convenient task input and to result in a retrieval process resembling a pen-and-paper approach. Instructors rated its functionality range as moderate but found it highly effective for instruction.

b) Maple® – students reported it to be difficult to learn, with a convenient task input in calculator-style mode and a somewhat complex programming mode. Instructors found it to have a wide functionality range but moderate instructional effectiveness.

c) Minitab® – students found it moderately difficult to learn, appreciating its convenient menu organization in the form of complete industrial procedures. Instructors rated its functionality range as moderate but found it highly effective for instruction.

d) Statistica® – students found it easy to learn, but noted that its highly dispersed menu structure made it challenging to locate necessary procedures. Instructors found it to have a wide functionality range but moderate instructional effectiveness.

For Applied Computer Science students, the following results were obtained:

a) Mathcad® – students found it easy to learn but noted the complex task formulation and the atypical way of formulating script tasks. Instructors rated its functionality range as moderate, with high instructional effectiveness for basic mathematical analysis.
b) Maple® – students found it easy to learn, appreciating its simple programming mode. Instructors found it to have a wide functionality range but moderate instructional effectiveness due to students' relatively weak preparation in higher mathematics.
c) Python – students found it easy to learn, appreciating its rich libraries, but mentioned the challenge of implementing specific solutions. Instructors rated its functionality range as wide but noted students' tendency to rely on libraries without a deep understanding of the implemented solutions.

For high school students, the following results were obtained:

a) C# and Visual Studio – students found it difficult initially until they became familiar with the workflow in the environment. They found it easier to design and execute applications compared to the relatively outdated Borland C++ Builder used in school. Instructors noted its wide functionality range but significant overhead. It demonstrated good instructional effectiveness, although the environment was relatively unstable compared to Visual Studio Code.
b) Wolfram Alpha® – students found it easy to learn but encountered significant inconsistency in commands, as it sometimes accepted natural language while at other times required formal Wolfram Language syntax. They appreciated its highly clear and rich presentation of results. Instructors rated its functionality range as moderate due to limitations in formulating complex tasks and the inability to run scripts. The system's behavior varied due to continuous development, but it demonstrated high instructional effectiveness.

Discussion

Before delving into the analysis of the obtained results, it is disheartening to acknowledge that there has been a noticeable decline in mathematical competencies among all three groups compared to previous years [2]. This decline can be attributed to two factors that have had a significant impact on the educational landscape. The first factor, which can be characterized as accidental and one-time, is the prolonged period of remote learning necessitated by the Covid-19 pandemic [3]. Particularly during the initial period from March 2020 to June 2020, the abrupt transition to remote education caused a virtual collapse of the entire educational process, spanning from primary schools to higher education institutions [4]. The sudden shift to online platforms posed numerous challenges for both students and teachers, including technological barriers, limited access to resources, and difficulties in maintaining engagement and motivation. The lack of face-to-face interaction and personalized guidance further hindered the learning experience. While efforts were made to establish more systematic remote teaching approaches, especially after the Ministry of Education and Science facilitated uniform access to the Microsoft Teams system from September/October 2020 onwards, it remained an imperfect substitute for traditional classroom instruction. The limitations of remote learning, combined with the disruptions caused by the pandemic, undoubtedly impacted students' mathematical abilities. The second, and unfortunately more persistent, factor contributing to the decline in mathematical competencies is the long-term economic deterioration of the education system. This deterioration is manifested through the systematic reduction of real wages for teachers. As a consequence, the earnings of primary and secondary school teachers, as well as university assistants, have been diminished to the level of minimum wages reserved for completely unskilled workers. Such a situation not only undermines the attractiveness of the teaching profession but also leads to a negative selection of individuals pursuing careers in education. Highly qualified individuals are discouraged from entering the teaching profession due to the uncompetitive compensation offered. As a result, the education system suffers from a scarcity of qualified personnel, leading to overcrowded classrooms, inadequate teacher-student ratios, and compromised instructional quality. The dire financial conditions and the diminishing prestige of the teaching profession have also led to a significant exodus of educators to other industries that offer better financial incentives. This brain drain further exacerbates the quality of education and contributes to the erosion of educational standards.

The consequences of these factors are far-reaching. The scarcity of qualified teachers and the subsequent staff shortages have resulted in canceled classes and the inability to fully implement the prescribed curriculum. Students and learners are deprived of essential knowledge and skills that are fundamental to their academic and personal development. The declining quality of education has a detrimental impact on students' mathematical competencies, hindering their ability to grasp advanced concepts, solve complex problems, and think critically. Moving on to the discussion of the results, it can be concluded that it is advisable to continue using the Mathcad® and Minitab® software programs in the classes for undergraduate students majoring in Mechanical Engineering. However, it is not beneficial to continue the classes using Maple® and Statistica®. The current level of mathematical preparation among students in these study programs is too weak for them to fully benefit from the advantages associated with teaching symbolic processing of mathematical formulas in Maple® and advanced statistical analysis in Statistica®. Therefore, it remains effective to focus solely on providing a purely utilitarian preparation in basic engineering calculations processed by Mathcad® and narrowly focused, professionally-oriented statistical procedures offered by the Minitab® software.

When it comes to students majoring in Applied Computer Science, it is important to address a significant disparity between their perceived level of competence and the observations made by the course instructors. While students may believe they possess strong mathematical skills, the external observations paint a different picture, revealing that their proficiency primarily lies in practical programming skills and utilizing existing libraries. However, their utilization of these libraries often lacks depth and sophistication, as their understanding of the underlying methods and algorithms is relatively weak. Therefore, it is crucial to continue the classes using both Mathcad® and Maple®, with a specific focus on consistently stimulating the enhancement of their mathematical competencies. These software programs offer valuable tools for symbolic mathematical processing and advanced analytical capabilities, which can greatly benefit students in their applied computer science studies. By emphasizing the importance of mathematical foundations and providing targeted exercises and assignments, instructors can help bridge the gap between the student's current skill level and the desired proficiency. Additionally, in the case of classes involving Python, it is essential to emphasize acquiring a solid theoretical understanding of the methods and algorithms available in the libraries. Students should be encouraged to delve beyond the surface-level implementation and gain a deeper comprehension of the underlying mathematical principles. This will enable them to make informed decisions when selecting appropriate methods, understanding their limitations, and interpreting the results accurately. Furthermore, it is important to address the tendency among students to rely heavily on existing libraries without fully grasping the underlying concepts. While libraries can be powerful tools for rapid development and prototyping, it is crucial to emphasize the significance of understanding the theoretical foundations and assumptions behind the implemented methods. Instructors should encourage students to explore the documentation, engage in independent research, and actively seek to deepen their understanding of the mathematical principles involved. By doing so, students

Quality Production Improvement and System Safety: QPI 16 - CZOTO 10   Materials Research Forum LLC
Materials Research Proceedings 34 (2023) 453-459   https://doi.org/10.21741/9781644902691-52

can avoid the pitfalls of blindly applying methods and relying on potentially erroneous results. In the case of high school students, a significant stratification of competencies has been observed, with some students demonstrating exceptional but narrowly focused programming, mathematical, and physics skills while others possess average competencies. By observing their interaction with software programs and the outcomes of their assignments, it is deemed beneficial to continue the classes using Visual Studio and Wolfram Alpha. However, it should be noted that despite its many advantages, particularly its widespread accessibility, Wolfram Alpha serves as a substitute for a comprehensive numerical-symbolic processing system and should be regarded as an initial step towards exploring complete environments such as Mathematica, Maple, Maxima, or Octave. The observed stratification of competencies in high school classrooms highlights the need for tailored instruction that caters to both highly skilled individuals and those with average competencies. By utilizing Visual Studio, students can enhance their programming skills and gain practical experience in software development. This platform offers a versatile and widely used environment that fosters hands-on learning and encourages the exploration of various programming languages. Additionally, incorporating Wolfram Alpha into the curriculum provides students with a powerful computational tool that can assist them in solving complex mathematical and scientific problems. Its accessibility and extensive database make it a valuable resource for quick calculations, data analysis, and obtaining solutions to various mathematical equations. However, it is important to emphasize that Wolfram Alpha should be viewed as an introductory tool, serving as a stepping stone toward more comprehensive numerical-symbolic processing systems. To ensure a well-rounded education, it is advisable to introduce students to complete environments like Mathematica, Maple, Maxima, or Octave. These software programs offer a broader range of mathematical and computational capabilities, enabling students to explore advanced topics, conduct in-depth analyses, and gain a deeper understanding of mathematical concepts. By gradually transitioning students from Wolfram Alpha to these comprehensive systems, they can develop a more comprehensive skill set and better prepared for higher education or professional endeavors in fields such as mathematics, engineering, or scientific research.

Conclusions
The article addresses the issue of aligning computer-based tools for numerical and symbolic engineering computations with the competency levels of three distinct groups of students. The suitability of both fundamental tools (Mathcad®, Minitab®, Visual Studio) and advanced ones (Maple®, Statistica®, Wolfram Alpha®) is evaluated, with the results strongly dependent on the characteristics of the target groups. Concerning students in mechanical engineering, the use of advanced tools is deemed inappropriate due to their relatively weak mathematical background. Instead, a comprehensive understanding and application of basic tools within narrowly defined, vocationally oriented methods are desired. For Applied Computer Science students, it is advisable to continue using existing IT tools while placing a strong emphasis on developing their mathematical competencies and gaining knowledge of the theoretical foundations of the methods employed through available programming libraries. In the case of high school students, it is advantageous to utilize existing tools while considering the varying levels of competency within classes. Providing tasks and projects of different difficulty levels is recommended to both stimulate the interest of highly advanced individuals and foster the growth of average students. This approach aims to avoid detrimental frustration and instill a sense of possibility for all learners. Overall, aligning the selection of computational and symbolic computing tools with the competency levels of different student groups is pivotal for effective teaching and learning outcomes. By tailoring the choice of tools and instructional strategies, educators can enhance student engagement, promote skill development, and cultivate a supportive learning environment for all students. The conclusions drawn from the above findings will be useful for conducting

Quality Production Improvement and System Safety: QPI 16 - CZOTO 10   Materials Research Forum LLC
Materials Research Proceedings 34 (2023) 453-459   https://doi.org/10.21741/9781644902691-52

educational activities in other, more conceptually challenging subjects such as medical statistics [5] and surface layer analysis [6].

References

[1] R. Marshall et al. Teaching STEM to K-12 Students: Undergraduate Students Engaged in Engineering Pedagogical Development in a COVID-Persistent Learning Environment. ASEE Annual Conference and Exposition, Los Angeles, 2021, art.#32886.

[2] M.W. Kier, L.L. Johnson. Exploring How Secondary STEM Teachers and Undergraduate Mentors Adapt Digital Technologies to Promote Culturally Relevant Education during COVID-19. Edu. Sci. 12 (2022) art.48. https://doi.org/10.3390/educsci12010048

[3] T. Dhurumraj et al. Broadening educational pathways to STEM education through online teaching and learning during COVID-19: Teachers' perspectives. J. Balt. Sci. Educ. 19 (2020) 1055-1067. https://doi.org/10.33225/JBSE/20.19.1055

[4] D.S. Wright et al. I will survive: Teachers reflect on motivations to remain in education amidst a global pandemic. J. Res. Sci. Teach. (2022) (early view). https://doi.org/10.1002/tea.21831

[5] B. Jasiewicz et al. Inter-observer and intra-observer reliability in the radiographic measurements of paediatric forefoot alignment, Foot Ankle Surg. 27 (2021) 371-376. https://doi.org/10.1016/j.fas.2020.04.015

[6] N. Radek et al. Microstructure and tribological properties of DLC coatings, Mater. Res. Proc. 17 (2020) 171-176. https://doi.org/10.21741/9781644901038-26

Quality Production Improvement and System Safety: QPI 16 - CZOTO 10   Materials Research Forum LLC
Materials Research Proceedings 34 (2023) 460-467   https://doi.org/10.21741/9781644902691-53

Multi-Criterial Quality of Vintage Audio Equipment in User's Assessment

SROKA Mariusz[1, a *]

[1]Czestochowa University of Technology, Dąbrowskiego 69, 42-201 Częstochowa, Poland

[a]mariusz.sroka@pcz.pl

Keywords: Sound, Quality, Audio Vintage, Electroacoustics

Abstract. The article presents key parameters for users of vintage audio electroacoustic equipment, which determine their quality. The research part presents the results relating to: knowledge of the concept of vintage audio, technical knowledge about electroacoustic equipment and the respondents' opinions on the quality of contemporary audio systems. The key features of your vintage audio equipment, important from the user's point of view, have also been distinguished. A diagnosis was made of whether the respondents attach importance to the price of the equipment and whether they would be willing to pay more for the quality of sound and workmanship. The article ends with a discussion of the research results and a summary.

Introduction

Sound has accompanied man since the beginning of his existence. These were the sounds of nature and the sounds of communication between tribesmen. The sound was used to warn, express emotions, feelings and later became one of the elements of art. Initially very primitive art, which over time turned into music. Primitive sound forms played accidentally on natural musical instruments can serve as an example. The development of human civilization has allowed the construction of many musical instruments, which have survived to modern times in an unchanged form. Vocal singing was also of great importance. Artists practicing this art enjoyed great respect and recognition in every epoch.

The nineteenth century AD brought the technical possibilities of recording sound on various carriers and reproducing it and duplicating it on constructed devices. Initially, these were very primitive devices, looking from today's perspective, but the very desire to record, duplicate and reproduce voice, singing or music at different times and places was important.

The invention of the vacuum tube and later the transistor gave the designers further opportunities to create more and more perfect devices for recording and reproducing sound. Other important inventions (it is impossible to list them all) that contributed to the development of audio systems include:
- invention of the radio;
- stereophonic (quadraphonic);
- computerization (sound processors);
- multi-channel systems.

The purpose of this study was to identify the key parameters of audio systems for users. It should be noted that the comparison and evaluation of the parameters was made without the use of specialized equipment (nor was it based on the results of studies using such equipment). It was purposeful and intentional. Quality in the assessment of a single user is subjective and only the results from the entire surveyed group of evaluators allowed for the formulation of objective (general) conclusions, which were included in the chapters on the discussion of the results and the summary.

The bibliography presented in the article is only intended to encourage the potential reader to explore the issues related to electroacoustics and, possibly, to arouse his interest in this direction.

Quality Production Improvement and System Safety: QPI 16 - CZOTO 10   Materials Research Forum LLC
Materials Research Proceedings 34 (2023) 460-467   https://doi.org/10.21741/9781644902691-53

According to the author of the article, each scientific study, paper or diploma thesis should contain definitions of the terms used and an explanation of their meaning. In a classic study, one should refer to the available bibliography, quote other authors and use the available literature. This article has omitted this form. This is due to the lack of space for such a presentation and also due to the very extensive range of information in this field. The presented list of literature sources is only a substitute of knowledge in the field of electroacoustics [1-4].

Quality Criteria for Audio Equipment

Each person hears subjectively and has their own tastes and unique experiences in this respect. This may be, for example, a preference towards emphasizing low tones or quite the opposite - towards high tones. Quite a large population of people prefer to boost the bass and treble while lowering the midrange. Some prefer to listen the music with low level volume, while others prefer the opposite - the more decibels the better. Some prefer to listen to music through headphones, which is unthinkable for fans of speaker systems.

An important element of the analysis of the quality of audio equipment, which many forget or omit completely, are the acoustic properties (parameters) of the room in which the listening session takes place. As an example, we should mention famous concert halls and world philharmonics. Even the best audio equipment will not sound good in a room that does not meet the basic acoustic parameters. At this point, the question should be asked, how to systematize the parameters of audio equipment so that it is possible to assess its quality? The answer seems very simple. You should compare the technical parameters and the equipment that has the best has the highest quality. Well, it turns out that's not the case at all. In the research part of the article, an attempt will be made to identify these key quality parameters (which may or may not result from technical parameters) that users are guided by when choosing music sets.

For the purposes of this study, the following technical quality criteria (measurable and expressed in appropriate units) were adopted: power, frequency response, total harmonic distortion, signal-to-noise ratio. Other criteria adopted (non-technical) are: musical experience, music scene, quality of workmanship, materials used, external appearance and price.

Research Methodology and Purpose

The respondents of the survey were users of vintage audio equipment from Poland. The condition for participation in the study was the possession and use of vintage audio equipment on a daily basis for at least five years, in order to be able to give authoritative answers. Participation in the study was voluntary and preceded by the verbal consent of the respondent. 30 questionnaires were carried out, of which, after checking the correctness of filling in, consistency and logic of the answers provided, all questionnaires were classified for further analysis. The survey contained 14 questions. All questions in the questionnaire were closed and single-choice questions.

The aim of the conducted research was:

- determining whether the respondents are familiar with the concept of vintage audio;
- obtaining knowledge whether the surveyed group has technical knowledge of audio equipment;
- isolation of the key features (assessment criteria) of the owned vintage audio equipment that are important from the user's point of view;
- diagnosing whether respondents attach importance to the price of equipment and whether they would be willing to pay more for the quality of sound and workmanship;
- getting to know the opinions of the respondents on the quality of currently produced audio systems.

The test results were presented graphically or summarized in tables and analyzed in the chapters below.

Quality Production Improvement and System Safety: QPI 16 - CZOTO 10   Materials Research Forum LLC
Materials Research Proceedings 34 (2023) 460-467   https://doi.org/10.21741/9781644902691-53

Presentation of Research Results

The first survey question from the record part concerned the separation of age groups of respondents. Age groups have been established for this particular study and are not related to commonly accepted age groups for sociological research. It was mainly about determining the generation that most often uses vintage audio equipment. The dominant group is the age range from 41 to 55 years, i.e. as much as 55%. The smallest age group is between 26 and 40 years old. The results for this part of the survey are summarized in Fig.1.

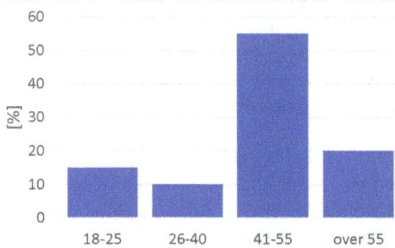

Fig.1. *Age groups of respondents: 15%, 10%, 55% and 20%, respectively [own study].*

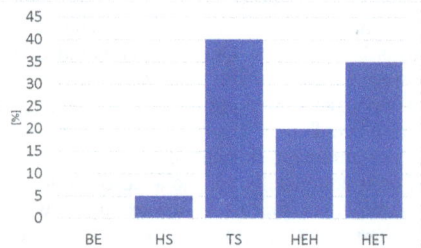

Fig.2. *Education of the respondents: BE – basic education 0%; HS – humanities secondary 5%; TS – technical secondary 40%; HEH – higher education in humanities 20%; HET – higher technical education 35% [own study].*

The second question from the record concerned the education of the participants in the study. The purpose of this question was to find out the type and level of education of the respondents participating in the study. Definitely, because as many as 75% of the respondents have technical education. None of the respondents completed primary education. One quarter of the respondents have secondary or higher education in the humanities. The results for this part of the survey are summarized in Fig.2.

The first survey question in the research part of the questionnaire was aimed at finding out the gradation of knowledge of the concept of audio vintage in the individual self-assessment of the respondents. The task of the respondents was to answer the question: how do you assess your knowledge of the concept of vintage audio? Half of the respondents assess their knowledge of the concept of vintage audio as good, and 30% as very good. Only one fifth considered their knowledge in this area to be at a sufficient level. The results for this survey question are summarized in Fig.3.

Quality Production Improvement and System Safety: QPI 16 - CZOTO 10   Materials Research Forum LLC
Materials Research Proceedings 34 (2023) 460-467   https://doi.org/10.21741/9781644902691-53

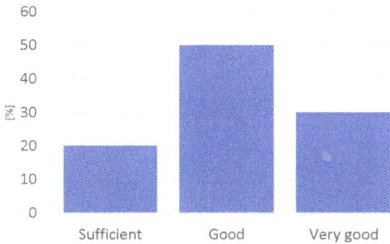

Fig.3. *Knowledge of the concept of vintage audio in the self-assessment of the respondents: 20%, 50%, 30%, respectively [own study].*

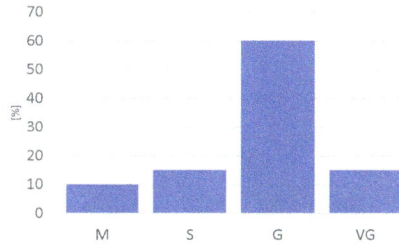

Fig.4. *The level of individual technical knowledge in the field of electroacoustics in the self-assessment of the respondents: M – mediocre 10%; S – sufficient 15%; G – good 60%; VG – very good 15% [own study].*

The second survey question related to the technical knowledge of the respondents. The respondents' task was to determine the level of their own technical knowledge in the field of electroacoustics, regardless of their education. The purpose of this question was to find out the respondents' opinions on their individual, strictly technical, knowledge of issues related to electroacoustics. Three quarters of the respondents describe their technical knowledge in the field of electroacoustics as good or very good. Only 10% of the respondents indicated a mediocre level, and 15% a sufficient one. The results for this survey question are summarized in Fig.4.

The next question concerned the respondents' knowledge of the technical parameters of the used vintage audio equipment. The respondents were asked to determine the level of knowledge about the typical technical parameters of their audio equipment. Specifically, it was about whether these parameters are known to them and can accurately indicate them by giving their values in generally used measurement units. More than half believe that they know the basic parameters of their vintage audio equipment. Five percent do not know this data and 30 know only some of it. Only ten percent declare knowledge of all the technical parameters of the audio equipment used. The results for this survey question are summarized in Fig.5.

The fourth survey question was aimed at finding out the respondents' opinions on the importance and meaning they attach to the catalog parameters of audio systems provided by manufacturers. Sixty-five percent of those surveyed say it matters little to them and twenty - very little. Only five percent of those surveyed believe that this is very important information for them. The results for this survey question are summarized in a Fig.6.

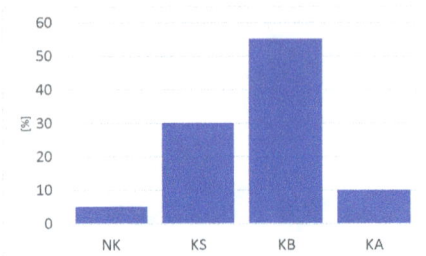

Fig.5. *Knowledge of the technical parameters of your vintage audio equipment: NK – don't know 5%; KS – know some 30%; KB – know basics 55%; KA – know all 10% [own study]*

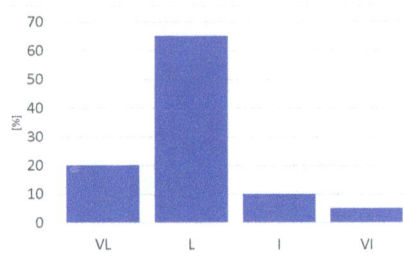

Fig.6. *Importance of technical parameters of vintage audio equipment in the opinion of respondents: VL – very little 20%; L – little 65%; I – important 10% ; VI – very important 5% [own study]*

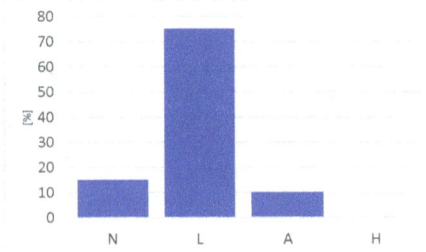

Fig.7. *Failure rate of vintage audio equipment in the opinion of respondents: N – negligible 15%; L – low 75%; A – average 10%; H – high 0% [own study].*

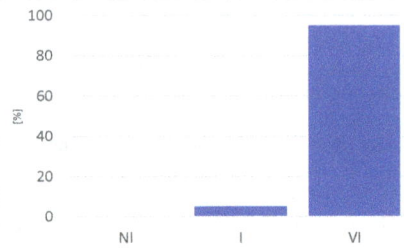

Fig.8. *The importance of the sound quality of vintage audio equipment in the opinion of the respondents: NI – not important 0%; I – important 5%; VI – very important 95% [own study].*

An important question in the survey was the question about the respondents' opinions on the failure rate of their vintage audio equipment. This indirectly gives us an answer about the quality of audio systems produced in the 1970s, 1980s and 1990s. Ninety percent of the respondents indicate a negligible and very low failure rate of their vintage audio equipment. None of the respondents marked the answer with a high failure rate. The results for this survey question are summarized in Fig.7.

The next survey question concerned the importance of the sound quality of vintage audio equipment owned by the respondents. Respondents were asked to indicate how important this parameter was for them. None of the respondents indicated that the sound of audio equipment is unimportant to them. Ninety-five percent of those surveyed answer this question with a "very important" answer. The results for this survey question are summarized in Fig.8.

The next survey question concerned the respondents' opinions regarding the quality of their vintage audio equipment. The quality of workmanship was understood in terms of the electronic components used and the finishing materials used. Three-quarters of respondents describe the quality of workmanship as very important for them from the point of use, and twenty percent as important. Only five percent of the respondents describe the quality of workmanship as unimportant. The results for this survey question are summarized in Fig.9.

The next survey question referred to the importance of aesthetic values that users attach to their equipment. One fourth of the respondents believe that aesthetic values are not important to them. Sixty percent indicate moderate to important. Only fifteen percent of the respondents indicated very important. The results for this survey question are summarized in Fig.10.

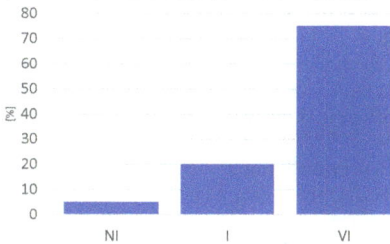

Fig.9. The importance of the quality of vintage audio equipment in the opinion of the respondents: NI – not important 5%; I – important 20%; VI – very important 75% [own study].

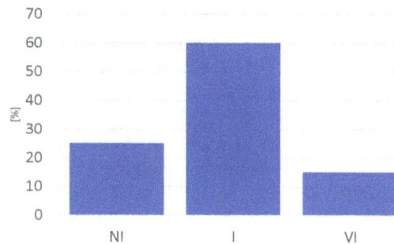

Fig.10. The importance of the aesthetic qualities of vintage audio equipment in the opinion of the respondents: NI – not important 25%; I – important 60%; VI – very important 15% [own study].

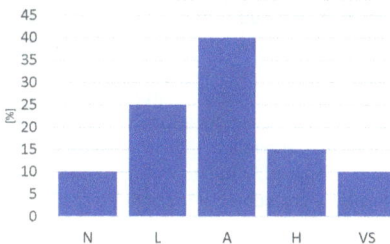

Fig.11. Significance of the price of vintage audio equipment in the opinion of the respondents: N – negligible 10%; L – low 25%; A – average 40%; H – high 15%; VS – very significant 10% [own study].

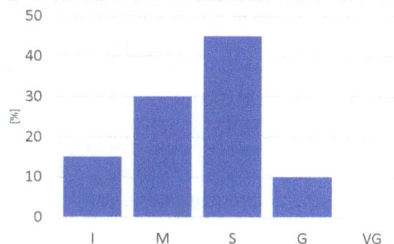

Fig.12. Quality of currently produced audio sets in the opinion of the respondents: I – insufficient 15%; M – mediocre 30%; S – sufficient 45%; G – good 10%; VG – very good 0% [own study].

Another survey question concerned the importance of price when buying vintage audio equipment. Of course, we are talking about the price on the secondary market for obvious reasons. The question also allowed us to determine whether buyers are willing to pay more knowing that they are paying for sound quality and build quality. For one tenth of the respondents, the importance of the purchase price is insignificant. The same number of respondents indicate this parameter as very significant. For sixty-five percent of the respondents, this parameter is of low or medium importance. The results for this survey question are summarized in Fig.11.

The next question from the questionnaire referred to the respondents' opinions on the quality of currently produced audio systems and sets. The approach to assessing quality in this case included subjective criteria such as sound quality, quality of materials used and perception of aesthetics. The results for this survey question are summarized in Figure 12.

Quality Production Improvement and System Safety: QPI 16 - CZOTO 10   Materials Research Forum LLC
Materials Research Proceedings 34 (2023) 460-467   https://doi.org/10.21741/9781644902691-53

Discussion of the Research Results

The largest age group were respondents over 41 (75%), and the least numerous (10%) aged 26 to 40. This is the result of changes that took place in the Polish economy after 1990. It can be said that then our country was flooded with a wave of dubious quality equipment imported from Western countries. Another factor that affects the small share of this age group is that they are professionally active people, bring up children and most likely lack the time (and probably also the financial resources) to pursue their own interests.

The education of the respondents has little influence on their interests towards electroacoustics. Nevertheless, the vast majority of respondents (75%) have technical education. The study did not specify the type of this technical education, and therefore it is difficult to say whether it is related to electroacoustics at least to a small extent.

The respondents know the concept of vintage audio very well. Eighty percent of the respondents assess their knowledge in this area as good or very good. Only twenty percent indicated sufficient. Certainly, this result is influenced by the popularity of the Internet, discussion clubs, forums, the availability of literature, online auctions and the very fact of owning and being interested in vintage audio equipment.

The level of individual technical knowledge in the field of electroacoustics in the self-assessment of the respondents is very good. Seventy-five percent of the respondents declare their level of knowledge in the survey as good or very good. Only twenty-five percent rate their knowledge as moderate or sufficient. The factors that affect this state are practically the same as for the knowledge of the concept of vintage audio.

Users' knowledge of the technical parameters of their vintage audio equipment is a very interesting phenomenon. Only ten percent declare that they know all the parameters of their equipment. The rest know only some or basic knowledge or do not know them at all (5%). Do users not attach so much importance to the parameters provided by manufacturers? Or maybe they just don't remember them or don't even try to remember them? These questions were answered in the survey in the next question. For eighty-five percent of the respondents, the importance of these parameters is low (65%) or very low (20%). It is hard to disagree with such an opinion. After all, we choose the equipment individually and our impressions are subjective and only the listening test is important.

In the opinion of the respondents, the failure rate of vintage audio equipment is very good. Fifteen percent of respondents assess it as negligible and seventy-five percent as small. This gives grounds to claim that vintage audio equipment is reliable. From the author's own experience, degradable elements (over time) may be a problem. Without going into details, this applies to belts and drive wheels (sprockets) made of plastic and the upper suspensions of the speakers (does not apply to suspensions made of canvas).

The sound quality of vintage audio equipment is very important to users. It is hard to be surprised and argue in this regard. After all, users strive for the best listening and the best possible acoustic experience. The same applies to the quality of the equipment, in particular the adjustment elements and the finishing materials used. This also affects the durability and reliability, which allows for long-term and trouble-free use of the equipment.

The aesthetic qualities of vintage audio equipment are of moderate importance in the opinion of the respondents. Of course, if we are dealing with nicely preserved specimens, the joy is double. It's getting harder and harder to find these on the secondary market.

The respondents also referred to the importance of the price of vintage audio equipment. It turns out that they are able to pay more for the quality of sound and workmanship as well as aesthetics than it would result from the economic calculation. This attitude results, among others, from the fact that the offer on the secondary market is limited.

Quality Production Improvement and System Safety: QPI 16 - CZOTO 10   Materials Research Forum LLC
Materials Research Proceedings 34 (2023) 460-467   https://doi.org/10.21741/9781644902691-53

The respondents assessed the quality of currently produced audio sets very critically. In their opinion, this assessment is as follows: unsatisfactory - 15%, mediocre - 30%, sufficient - 45%. No one indicated very good quality. Is it because of sentiment? To some extent, probably yes, but not entirely. This is evidenced by the research results presented earlier, relating to the expectations of users.

Summary

The issues that the author undertook to explore in the work are difficult. The first difficulty stems from the subjectivity of the criteria for evaluating audio sets and their differences from other consumer technical devices. Audio devices are perceived and perceived very individually (no two people hear and feel the same). Therefore, the subjectivity of the conducted research significantly limits their precision.

Conducting repeatable listening tests (especially in amateur conditions) is practically impossible. The perception of aesthetics, on the other hand, goes beyond the limits of quantification. Adoption of any scale in this respect will always result in a lack of appropriate accuracy.

Another difficulty that would seem to go unnoticed is the adoption of quantitative technical criteria. After all, you can compare selected catalog technical parameters and on this basis determine which audio set has better quality. It should only be noted that the differences between the parameters that actually determine the sound quality between individual devices are negligible. They are so small that the human ear cannot tell the difference. Therefore, such a comparison means nothing. A listening test is still necessary for verification, which invariably remains a subjective test.

It is worth emphasizing the fact that a small group of respondents took part in the study, among others because the population of users of vintage audio equipment is generally very small. So, are the test results representative? In that case, the fundamental question inevitably arises: what was the purpose of this article? According to the author, electroacoustics is a very developing and entertaining hobby. It provides entertainment and unforgettable experiences while listening to music. And what is very important, it shapes certain attitudes, teaches respect, patience, striving for perfection and deepens the broadly understood technical culture.

References

[1] M. Feszczuk, Wzmacniacze elektroakustyczne, WKiŁ, Warszawa, 1978.

[2] R. Makarewicz, Dźwięk w środowisku, OWN, Poznan, 1994. ISBN 83-85481060.

[3] E. Brixen. Audio Metering: Measurements, Standards and Practice (Audio Engineering Society Presents) 3rd Edition. ISBN 978-1138909113

[4] A. Witort. Dźwięk i technika Hi-Fi, NOT-SIGMA, Warszawa, 1988. ISBN 83-85001204

Quality Production Improvement and System Safety: QPI 16 - CZOTO 10   Materials Research Forum LLC
Materials Research Proceedings 34 (2023) 468-476   https://doi.org/10.21741/9781644902691-54

Technological Tools in Businesses' Communication with Generation Z

KOROMBEL Anna[1,a] * and ŁAWIŃSKA Olga[1,b]

[1]Czestochowa University of Technology, ul. Dąbrowskiego 69, 42-201 Częstochowa, Poland

[a]anna.korombel@pcz.pl, [b]olga.lawinska@pcz.pl

Keywords: Social Media, Technology, Enterprise, Management, Generation Z

Abstract. The aim of the paper is to assess the effect of businesses' activities involving the use of technological communication tools on Gen Zers' purchasing decisions. The study of Gen Zers' behaviours as a response to businesses' social media activities is part of broader research conducted by the authors among students in Poland and Great Britain in 2020 and 2021. The study used the method of a survey, and, as part of it, the CAWI technique. Descriptive statistics measures were used to analyse the research material. The findings demonstrate that businesses' activities involving the use of technological communication tools have a positive impact on Gen Zers' purchasing decisions. The study also examined the relationship between the analysed activities undertaken by businesses and the respondent's gender.

Introduction

Generation Z, i.e. people born in 1995 (and later) [1] up to 2009, is currently entering the job market [2]. Accounting for around 25% of the world population [3], Gen Zers have the purchasing power that makes them important partners for businesses today. This is also confirmed by economic forecasts – it is estimated that the share of Gen Zers in the workforce will increase fast: from 10% in 2019 to 30% in 2030, while their income will grow almost sevenfold: from around USD 460 billion in 2019 to USD 3.2 trillion in 2030. What is more, Generation Z consumption spending is estimated to increase more than sixfold, from USD 467 billion in 2019 to USD 3.0 trillion in 2030, which accounts for 11% of total spending in the economy [4]. These figures mean that businesses should strive to gain understanding of Gen Zers' behaviours and, subsequently, build and tighten connections with them, as this can bring them measurable benefits such as sustained improvement in business performance [5], increased profitability [6], increased sales revenue [7] and, consequently, maintenance of competitive edge [8]. Businesses have to include Generation Z as part of their business strategy by building lasting relationships with its members if they want to avoid losing market share [9]. If they ignore the current and future impact and power of this Generation, they will face failure [10]. This is because one of the most important goals of a manufacturing company, one that determines its market success, is satisfaction of customer needs and expectations [11].

As Gen Zers are the first generation born into the digital world, which they often perceive as equally important as the real world, businesses should seek and develop contacts with this Generation on social media, defined as "a group of Internet-based applications that build on the ideological and technological foundations of Web 2.0 and that allow the creation and exchange of user-generated content" [12]. Technological tools enable customers to express their opinions and participate in discussions or various events organised by businesses. The content published on social media not only generates needs in young consumers, it also induces them to make unplanned purchases and constitutes an important source of inspiration at the stage of searching for options of need satisfaction. Moreover, social media constitutes a valuable source of information about products and venues where people can express their opinions and share purchasing experiences [13]. Identification of specific customer segments on social media allows businesses to provide personalised content to specific customers based on demographic patterns and shared interests [14]. Meanwhile, generated content, designed to create engagement with brand community on social media, impacts the amount of spending on purchases [15]. In order to build effective relationships with customers on social media, businesses have to engage in constant interaction with potential customers [16] leading to the development of new products. Business success is determined not only by the speed of decision-making, but a solid information support for that process [18].

The aim of the paper is to assess the effect of businesses' activities involving the use of technological communication tools on Gen Zers' purchasing decisions. The authors have noted the need to intensify research in this domain. In order to fill this gap, the authors indicated the research problem by formulating the following research questions:

Q1. Do social media activities undertaken by businesses impact Gen Zers' purchasing decisions?

Q2. Is there a relationship between the gender of Gen Z member and the impact of businesses' social media activities on Generation Z purchasing decisions?

Q3. How do the research findings differ between the respondents from Poland and those from Great Britain?

Methods

The research examining Gen Z attitudes towards brands on social media is part of wider research carried out by the authors among students in Poland and Great Britain in 2020 and 2021. It contained both qualitative and quantitative aspects, and in both cases, the authors employed a method of indirect measurement - survey research - the technique of a survey, and the research tool of a survey questionnaire. The set of variables used in the research was selected based on critical analysis of the literature [19, 20]. There is no consensus in the literature about the starting birth year of Generation Z. The authors adopted the year 1995, which is most often found in the literature, as the cut-off year for the generation under study. In the research, the independent variable was the studied group of respondents, not the entire Generation Z population. Treating the latter as an independent variable would be problematic due to the difficulty with specifying the exact age range for Generation Z. Using Gen Z as a heuristic is useful, according to the authors, as generational profiling is now used as description in popular media and culture. The main research was preceded by a pilot study conducted in 2018 with the aim of identifying and eliminating any potential errors in the survey questionnaire. A total of 157 students participated in the research carried out in Poland, with responses of 151 students (126 females and 25 males) qualified for the analysis. A total of 150 students participated in the research carried out in Great Britain, with responses of 150 people (80 females and 70 males) qualified for further research. In Poland, the data was collected via an online survey questionnaire available on the Webankieta.pl platform, while in Great Britain this process was commissioned to an external entity specialising in surveys. In the research conducted in Poland and in Great Britain, the sampling was non-probabilistic. Using non-probabilistic selection, the authors also applied statistical inference to identify relationships in the groups under study, which is not possible using descriptive statistics. It should be noted that the group of the Polish students qualified for the research was overrepresented by females, impacting the gender structure of the respondents. Based on the results of the research, the number and frequency of the responses provided by respondents to the survey questions were calculated. The authors realise that the sampling technique employed in the research does not enable estimation of an error that occurs when the regularities observed in the sample are generalised to the whole population. In order to identify relationships existing in the groups under study, statistical inference was used, for which a certain significance level was adopted – which is not possible in the case of descriptive statistics. Statistical inference was conducted with ex ante significance level at $\alpha = 0.05$, with p value calculated for each test. By comparing the p value with the level of statistical significance, the authors determined whether

Quality Production Improvement and System Safety: QPI 16 - CZOTO 10   Materials Research Forum LLC
Materials Research Proceedings 34 (2023) 468-476   https://doi.org/10.21741/9781644902691-54

there was sufficient proof to reject H_0 against H_1 $(p < \alpha)$ or not $(p \geq \alpha)$. Statistica v.13 was used to conduct the analyses. The authors are aware of the limitations of survey research, which include superficial understanding of the phenomena in question and the possibility of inaccurate answers provided by respondents. Another critique of the presented research findings may concern the fact that the research was conducted on a small group. Small samples may raise methodological issues (e.g., generalisation is difficult), however, when proper statistical tests are applied, they can be useful in inference [21].

Results

The authors endeavoured to determine whether the activities undertaken by businesses on social media, and if so, which ones, impacted purchasing decisions of Generation Z respondents in the period 2020 to 2021. Figure 1 presents the indicators of structure for the individual activities.

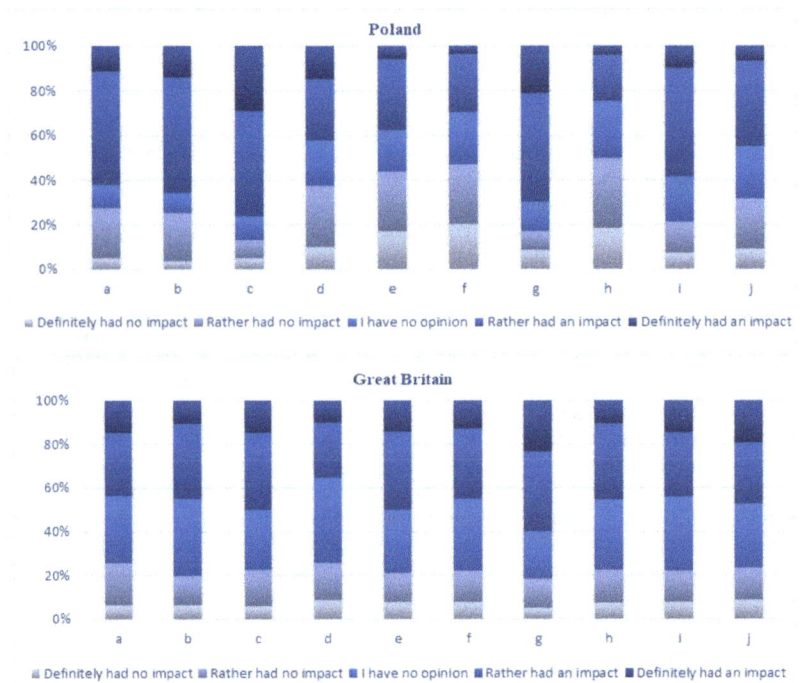

Fig. 1. Assessment of businesses' social media activities impacting purchasing decisions of Generation Z respondents – research findings in Poland and Great Britain in 2020/2021. (a. Publishing information about what is new in the offer, b. Presenting an application of a specific product/service, c. Publishing information about promotion, d. Presenting a test carried out by an expert, e. Recommendation from a known person, f. Participation in a competition, g. Receipt of a discount coupon, h. Making the wall available for asking questions, i. Positive actions by a brand/company that you have liked in particular, j. Social responsibility, social campaigns)

Analysis of the data presented in Fig. 1 shows that the activities undertaken by businesses on social media have had an impact on purchasing decisions of Generation Z respondents. The activities that had a strong impact on purchasing decisions of the respondents both in Poland and Great Britain were: *publishing information about special offers and offering discount coupons.*

Next, the authors conducted statistical analysis and determined the impact of businesses' social media activities on respondents' purchasing decisions depending on the gender (Tables 1 and 2).

Table 1. *Results of Mann-Whitney U test (adjusted for continuity) with respect to the relationship between the effect of businesses' social media activities impacting purchasing decisions and the gender of a Gen Z respondent – results of the research conducted in Poland in 2020/2021*

Variables	Sum of Ranks Female	Sum of Ranks Male	U	Z	p
Publishing information about what is new in the offer & Gender	988750	1588.50	1263.50	1.69	0.091
Presenting an application of a specific product/service & Gender	9787.00	1689.00	1364.00	1.14	0.252
Publishing information about promotion & Gender	10009.00	1467.00	1142.00	2.32	**0.020**
Presenting a test carried out by an expert & Gender	9685.50	1790.50	1465.50	0.56	0.574
Recommendation from a known person & Gender	9628.00	1848.00	1523.00	0.27	0.790
Participation in a competition & Gender	9715.00	1761.00	1436.00	0.72	0.474
Receipt of a discount coupon & Gender	9982.50	1493.50	1168.50	2.17	**0.029**
Making the wall available for asking questions & Gender	9850.00	1626.00	1301.00	1.41	0.157
Positive actions by a brand/company that you have liked in particular & Gender	9930.00	1546.00	1221.00	1.89	0.058
Social responsibility, social campaigns & Gender	10140.00	1336.00	1011.00	2.94	**0.003**

Mann-Whitney U test (adjusted for continuity) (Vector) for the variable: Gender. Marked results are significant with p<0.05000.

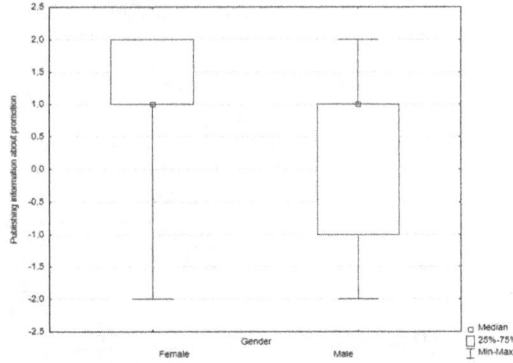

Fig. 2. *The impact of businesses publishing information about special offers on social media on Generation Z respondents' purchasing decisions by gender, in Poland in 2020/2021.*

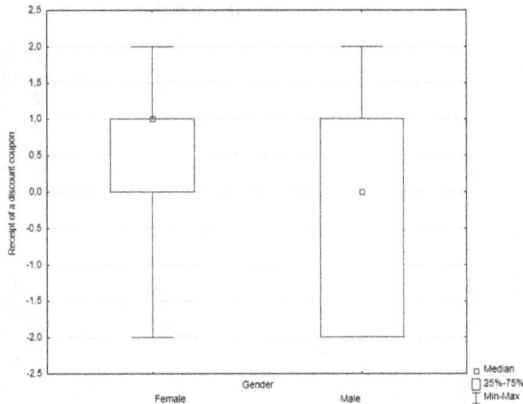

Fig. 3. *The impact of businesses offering discount coupons on social media on Generation Z respondents' purchasing decisions by gender, in Poland in 2020/2021.*

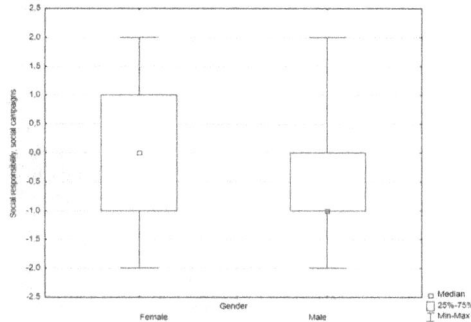

Fig. 4. *The impact of businesses' social responsibility, social campaigns run on social media on Generation Z respondents' purchasing decisions by gender, in Poland in 2020/2021.*

Analysis of the data collected in Poland revealed that gender had a significant impact on the differences between the analysed variables in the following cases:
a) publishing information about special offers ($p = 0.02$),
b) offering discount coupons ($p = 0.03$),
c) social responsibility, social campaigns ($p = 0.003$).
In the case of the aforementioned variables, these activities had a significantly stronger impact on purchasing decisions of females compared to males, which is also confirmed by graphical interpretations (Figures 2, 3 and 4).

Three of the analysed social media activities undertaken by businesses had a different impact on Generation Z respondents' purchasing decisions depending on the respondent's gender.

In the case of the research conducted in Great Britain, no statistically significant correlations were found between the analysed variables ($p>0.05$). The respondent's gender was not significantly correlated with the dependent variables – social media activities undertaken by businesses to impact Gen Zers' decisions to make a purchase.

Table 2. *Results of Mann-Whitney U test (adjusted for continuity) with respect to the relationship between the effect of businesses' social media activities impacting purchasing decisions and the gender of a Generation Z respondent – results of the research conducted in Great Britain in 2020/2021*

Variables	Sum of Ranks Female	Sum of Ranks Male	U	Z	p
Publishing information about what is new in the offer & Gender	6061.00	5264.00	2779.00	-0.08	0.9364
Presenting an application of a specific product/service & Gender	6236.00	5089.00	2604.00	-0.77	0.4408
Publishing information about promotion & Gender	6124.50	5200.50	2715.50	-0.33	0.7425
Presenting a test carried out by an expert & Gender	5617.00	5708.00	2377.00	1.66	0.0969
Recommendation from a known person & Gender	5724.50	5600.50	2484.50	1.23	0.2171
Participation in a competition & Gender	6223.50	5101.50	2616.50	-0.72	0.4735
Receipt of a discount coupon & Gender	6130.50	5194.50	2709.50	-0.35	0.7246
Making the wall available for asking questions & Gender	6112.00	5213.00	2728.00	-0.28	0.7789
Positive actions by a brand/company that you have liked in particular & Gender	6177.00	5148.00	2663.00	-0.53	0.5937
Social responsibility, social campaigns & Gender	6115.50	5209.50	2724.50	-0.29	0.7710

Mann-Whitney U test (adjusted for continuity) (Vector) for the variable: Gender. Marked results are significant with $p < 0.05000$.

Summary

Social media activities undertaken by businesses have an impact on Gen Zers' decisions to make purchases. Both in Poland and Great Britain, publishing information about special offers and offering discount coupons were found to have the biggest impact on Gen Zers' purchasing decisions. With respect to the relationship between the respondent's gender and the effect of businesses' activities, statistically significant correlations between the analysed variables were only found in the research conducted in Poland. Three activities in this group had a significantly stronger impact on purchasing decisions of females compared to males: publishing information about special offers, offering discount coupons, and social responsibility, social campaigns. Among respondents residing in Great Britain, no statistically significant correlations were found between the analysed variables.

The answers found to the research questions will allow entrepreneurs to expand their knowledge of Generation Z behaviours. This knowledge can be used not just for building connections with members of this generation, but for undertaking effective social media activities that will impact their purchasing decisions. Activities that are more tailored to Generation Z expectations will translate into increased turnover for businesses and, consequently, a stronger market position.

The present research has many limitations, which can be seen as directions of future studies. The first limitation is related to the sample size and female over-representation in the the survey conducted in Poland, which excludes the possibility of generalising the findings to the whole Generation Z. Another limitation arises from the research tool applied, in particular from the fact that it entails superficial understanding of the phenomena in question and the possibility of inaccurate answers provided by respondents. Also, not including all potential benefits of building customer-business relationship is definitely a limitation of the research.

Despite the aforementioned limitations, the research can be considered as valuable, as it provides information on Gen Z social media behaviours and expectations regarding customer-business relationships. Future research could be undertaken in other countries to find out whether the same patterns exist among Generation Z customers. Further research should concentrate on identifying the reasons for low engagement in the customer-business relationship on social media among Gen Zers in Poland.

References

[1] I.C. Kamenidou et al. Segmenting the Generation Z cohort university students based on sustainable food consumption behavior: A preliminary study, Sustainability 11 (2019) art.837. https://doi.org/10.3390/su11030837

[2] M. McCrindle. The ABC of XYZ: Understanding the global generations, McCrindle Research Pty Ltd., Sydney, 2014. ISBN 978-0992483906

[3] H. Patel, E. Morrison. Generation Z: Step aside Millennials, Barclays Research Highlights: Sustainable & Thematic Investing. [online] 2019 [Viewed: 31-01-2023] Available from: https://www.cib.barclays/content/dam/ barclaysmicrosites/ibpublic/documents/our-insights/gen-z/Leaflet%20Generation_Z.pdf

[4] Oxford Economics. Gen Z's role in shaping the digital economy. [online] 2021 [Viewed: 31-01-2023] Available from: https://www.oxfordeconomics.com/resource/gen-z-role-in-shaping-the-digital-economy

[5] J.C. Narver, S.F. Slater. The effect of a market orientation on business profitability, J. Market. 54 (1990) 20-35. https://doi.org/10.1177/002224299005400403

[6] S.-H. Chuang, H.-N. Lin. The roles of infrastructure capability and customer orientation in enhancing customer-information quality in CRM systems: Empirical evidence from Taiwan, Int. J. Inf. Manag. 33 (2013) 271-281. https://doi.org/10.1016/j.ijinfomgt.2012.12.003

[7] Z. Soltani et al. The impact of the customer relationship management on the organization performance, J. High Technol. Manag. Res. 29 (2018) 237-246. https://doi.org/10.1016/j.hitech.2018.10.001

[8] W. Reinartz, M. Krafft, W. D. Hoyer. The Customer Relationship Management Process: Its Measurement and Impact on Performance, J. Market. Res. 41 (2004) 293-305. https://doi.org/10.1509/jmkr.41.3.293.35991

[9] H. Schroth. Are you ready for gen Z in the workplace? California Manag. Rev. 61 (2019) 5-18. https://doi.org/10.1177/0008125619841006

[10] S. Sladek, A. Grabinger. Gen Z. The first generation of the 21st Century has arrived! XYZ University [online] 2014 [Viewed: 31-01-2023] Available from: https://www.xyzuniversity. com/wp-content/uploads/2018/08/GenZ_Final-dl1.pdf

[11] K. Knop, R. Ulewicz. Solving Critical Quality Problems by Detecting and Eliminating Their Root Causes – Case-Study from the Automotive Industry, Mater. Res. Proc. 24 (2022) 181-188. https://doi.org/10.21741/9781644902059-27

[12] A. Kaplan, M. Haenlein. Users of the world, unite! The challenges and opportunities of social media, Bus. Horiz. 53 (2010) 59-68. https://doi.org/10.1016/j.bushor.2009.09.003

[13] M. Stachowiak-Krzyżan. The Use of Social Media by Young Consumers in Purchasing Processes, Marketing of Scientific and Research Organizations 31 (2019) 83-108. http://doi.org/10.2478/minib-2019-0005

[14] K. K. Kapoor et al. Advances in Social Media Research: Past, Present and Future, Inf. Sys. Front. 20 (2018) 531-558. https://doi.org/10.1007/s10796-017-9810-y

[15] K. Y. Goh, C. S. Heng, Z. Lin. Social media brand community and consumer behavior: Quantifying the relative impact of user-and marketer-generated content, Inf. Sys. Res. 24 (2013) 88–107. https://doi.org/10.1287/isre.1120.0469

[16] A. Susarla, J. H. Oh, Y. Tan. Influentials, Imitables, or Susceptibles? Virality and word-of-mouth conversations in online social networks, J. Manag. Inf. Sys. 33 (2016) 139-170. https://doi.org/10.1080/07421222.2016.1172454

[17] Á. García-Crespo et al. SEMO: A framework for customer social networks analysis based on semantics, J. Inf. Technol. 25 (2010) 178-188. https://doi.org/10.1057/jit.2010.1

[18] M. Krynke, D. Klimecka-Tatar. The use of Computer Simulation Techniques in Production Management, Mater. Res. Proc. 24 (2022) 126-133. https://doi.org/10.21741/9781644902059-19

[19] J. Gummerus et al. Customer engagement in a Facebook brand community, Manag. Res. Rev. 35 (2012) 857-877. https://doi.org/10.1108/01409171211256578

[20] B. Gregor, T. Kubiak. The assessment of activities conducted by companies in social media in light of research concerning their users, Market. Sci. Res. Org. 4(14) (2014) 36-50. https://doi.org/10.14611/minib.14.04.2014.03

[21] N. Nachar. The Mann-Whitney U: A test for assessing whether two independent samples come from the same distribution, Tutor. Quant. Methods Psychol. 4 (2008) 13-20. https://doi.org/10.20982/tqmp.04.1.p013

Quality Production Improvement and System Safety: QPI 16 - CZOTO 10   Materials Research Forum LLC
Materials Research Proceedings 34 (2023) 477-485   https://doi.org/10.21741/9781644902691-55

Work Safety Factors in the Public Administration of the Post-Covid Period

ZADROS Katarzyna

Częstochowa University of Technology Faculty of Management, ul. Gen. H. Dąbrowskiego 69
42-201 Częstochowa Poland

katarzyna.zadros@pcz.pl

Keywords: Post-Covid Period, Public Administration, Management, Safety Factors, Worker's

Abstract. The article has a theoretical and practical character. It is an introduction to further research on issues related to the impact of the Covid-19 pandemic on the working conditions of employees of public administration institutions during the period of partial control of the health threat caused by the SARS-CoV-2 virus. The article presents issues related to occupational safety management in selected public administration offices and the opinions of employees employed in them on this subject. The focus was on the problems related to providing employees with safe working conditions while these institutions are recovering from the crisis related to the limited access to public services and returning to traditional forms of functioning.

Introduction

Polish labor law guarantees safe, hygienic working conditions and health protection in the working environment for every employee. Managing safety in the working environment is therefore the primary responsibility of every manager. It is determined by a number of environmental, psychosocial, legal and organizational factors, related to individual risk and the social security system. The concept of safety management in the workplace can be understood as the totality of activities planned, organized and undertaken to achieve and maintain a state of safety [1]. The basis for these activities is the fastest possible identification of the threat and its minimization or removal, followed by control and introduction of improvements to increase the level of security.

One of the functional areas of the state in which providing employees with a safe working environment is of particular importance is public administration (central and local government). Its task is to manage public affairs and mobilize human and material resources for the implementation of public goals and tasks set before it, on the basis of applicable legal regulations and in the forms specified by law [2, 3, 4].

For decades, pathogenic microorganisms have practically not been included among the risk factors in the working environment of public administration officials in Poland and in most countries of the world. Only the outbreak of the Covid-19 pandemic radically changed the situation and caused the general closure of access to public offices for citizens who need to settle official matters. Instead of a standard visit to a specific office, the client was obliged to make an appointment by phone or e-mail for a strictly defined time, settle matters only by phone or e-mail, use ePUAP, inboxes placed at the entrance to the facilities or traditional mail [5]. Such a procedure significantly extended the time of dealing with official matters and made it difficult to explain even minor irregularities. At the same time, it meant the rapid development of e-administration necessary to meet the needs of citizens. It also limited the contact of employees with customers, and thus increased the level of their safety in the work environment.

Remote work [6] was also supposed to improve safety and reduce virus emissions, consisting in the performance of tasks entrusted to the employee by the employer outside the place of permanent performance at a specific time and to the extent specified in the employment contract.

The place of its provision may be the home or any other place where it is possible to control the employee [7]. For public administration, with specific exceptions to the law, remote work as an obligation was introduced in Poland on November 3, 2020 [5]. Thanks to this, despite the rapidly spreading pandemic, the continuity of operation of public administration offices was ensured. Office workers, practically overnight, they had to start performing tasks in a new way, often using their own computer equipment, as employers did not always provide them with work tools. According to Art. 3 of the Act of March 2, 2020 on special solutions related to the prevention, counteraction and combating of Covid-19, other infectious diseases and crisis situations caused by them, the employer was also not obliged to inspect such a workplace for compliance with health and safety requirements [8].

Only the production and dissemination of a vaccine protecting against the severe course of the disease and its complications changed the situation. The Polish state, in accordance with the recommendations of the World Health Organization (WHO), introduced the possibility of free vaccination for all citizens. This allowed for a gradual return to standard working conditions, also in public administration. Another factor that gradually began to reduce the risk of the disease was that more and more people acquired immunity as a result of vaccination or infection. In this way, the number of people posing an epidemiological threat gradually began to decrease. The decrease in the incidence and emission of the virus was also favored by climatic conditions and warm seasons.

However, the question arises whether employees employed in public administration institutions assess their current place and working conditions as safe and whether their superiors manage work safety in such a way as to build and strengthen this sense of security. Another problem that will be sought for an answer to is the question of whether the return to the traditional form and working time suits employees, or whether they would prefer to continue to be able to work remotely. These issues will be analyzed and assessed later in the article, based on a survey conducted among employees of selected public administration institutions in Poland.

Material and Method
The method by which the t-study was carried out as a marketing research method, slightly modified by the author, used in the study of service quality, called the Customer Satisfaction Index (CSI), classically understood as an external stakeholder [9]. The indicator is built on the basis of a weighted assessment, and its result consists of the assessment of individual elements and the weights assigned to them [10]. It is calculated based on the following formulas:

$$\text{CSI} = \sum_{i=1}^{n} w_i o_i \qquad \text{(template no. 1)}$$

where:

i, 1 ... n - elements of employee involvement
w_i - the weight of the employee involvement element
o_i - assessment of the employee involvement element

$$\text{CSI}_{max} = \sum_{i=n}^{n} w_{io_{i\,max}} \qquad \text{(template no. 2)}$$

$$\text{CSI\%} = \frac{\text{WZP}}{\text{WZP}_{max}} \times 100\% \qquad \text{(template no. 3)}$$

Converting the numerical value of the indicator to a percentage makes it much easier to analyze the results of the study (Table 1), therefore it was decided to convert it.

The study was conducted in October 2022 in selected 6 offices of large cities (UM) – Gdynia, Częstochowa, Sosnowiec, Toruń, Gliwice and Rzeszów. Such a selection resulted, inter alia, from the fact that the author lives in one of these cities and that in those cities she could obtain consent to provide employees with the questionnaire electronically.

Quality Production Improvement and System Safety: QPI 16 - CZOTO 10   Materials Research Forum LLC
Materials Research Proceedings 34 (2023) 477-485-   https://doi.org/10.21741/9781644902691-55

Table 1. *Evaluation criteria of CSI in percent used in the study [9]*

CSI% value criteria	Rating
0-40	*very bad* – employee completely dissatisfied
41-60	*bad* – employee dissatisfied
61-75	*average* – there are problems with the level of employee satisfaction
76-90	*good* – no problems were found with the employee satisfaction level
91-100	*very good* – highly satisfied employee

Table 2. *Population and number of employees of the offices in the examined cities (data as of December 31, 2021) [Own studies based on BIP information]*

City	Population	Employment at UM	Number of inhabitants per 1 employee
Czestochowa	207 467	1017	204
Gdynia	246 348	1001	246
Gliwice	166 703	632	264
Rzeszów	198 476	1417	140
Sosnowiec	195 978	957	173
Toruń	180 832	432	419

An original questionnaire was used for the study, and the analysis of the responses was based on the weighted assessment principle. The respondents – customer service employees in the surveyed offices – assessed the level of satisfaction and the importance of individual elements included in a given area of the survey on a five-point Likert scale [10], where a rating scale was used for satisfaction: 1 - very dissatisfied; 2 - rather dissatisfied; 3 - indifferent; 4 - rather satisfied; 5 - very satisfied. However, to determine the significance (rank) of a given element for the respondent: 1 - it does not matter; 2 - little importance; 3 - indifferent; 4 - important; 5 - very important.

This scale made it possible to carry out a statistical analysis in accordance with the CSI procedure [11] to determine the level of satisfaction and the significance of individual factors for the examined factors. The substantive questions were supplemented with records, which made it possible to characterize the respondents. The selection of respondents was random, and the sample was relatively small, which makes it impossible to state whether the survey results were representative. However, taking into account that in all surveyed offices the answers were similar, and the factors differentiating them were the same features: gender and age of the survey participants, it seems that representativeness was achieved.

The sample of 217 people were employees of city offices dealing with direct customer service (Table 3). As already mentioned, the study was conducted in city offices in six large cities: Gdynia (Gd) - 49 respondents, Częstochowa (Cz) - 48 respondents, Sosnowiec (S) - 29 participants, Toruń (T) - 31 employees, Gliwice (Gl) - 38 people and Rzeszów (Rz) - 22 participants in the sample.

A statistical participant of the survey will most likely be a woman, usually between 36 and 45 or 26 to 35 years old and with higher education, who work in the office for 6 to 10 years.

Table 3. *Demographic characteristics of the respondents [Own study based on the survey]*

characteristic		City						Σ (%)
		Gd	Cz	S	T	Gl	Rz	
sex	women	47	45	25	29	34	22	202 (93.1)
	men	2	2	3	2	3	1	11 (6.9)
Σ		49	48	29	31	38	22	217 (100)
age	up to 25 years	2	1	2	2	1	2	10 (4.6)
	26-35 years	18	19	15	13	21	9	95 (43.8.)
	36-45 years	23	21	9	13	12	6	84 (38.7)
	46-55 years old	5	6	3	3	4	3	24 (11.1)
	56-65 years	1	1	-	-	-	2	4 (1.8)
Σ		49	48	29	31	38	22	217 (100)
education	lo	2	1	-	1	-	1	5 (2.3)
	lz / vol.	5	5	9	7	4	2	32 (14.7)
	I ° studies	22	23	15	11	15	10	96 (44.3)
	2nd degree studies	20	19	5	12	19	9	84 (38.7)
Σ		49	48	29	31	38	22	217 (100)
work experience in the office	1-2 years	2	1	1	2	1	-	7 (3.2)
	3-5 years	7	9	4	6	6	3	35 (16.1)
	6-10 years	25	24	15	13	17	12	106 (48.8)
	11-15 years	13	11	6	7	11	4	52 (23.9)
	16-20 years	1	-	3	2	2	1	9 (4.1)
	> 20 years	1	3	-	1	1	2	8 (3.7)
Σ		49	48	29	31	38	22	217 (100)

Analysis of the Test Results

The analysis of the results of the survey will begin with questions that were used to determine whether the respondents worked remotely during the lockdown, and if so, how long they worked (Table 4).

Table 4. *Frequency of remote work of residents [Own study based on the survey]*

remote work	city						Σ (%)
	Gd	Vol	S	T	Gl	Rz	
4-5 days a week	1	1	2	2	3	2	11 (5.1)
2-3 days a week	16	12	6	8	11	5	58 (26.8)
1 day a week	17	21	9	11	10	7	75 (34.6)
several days a month	7	5	4	6	7	4	33 (15.1)
single days	4	4	5	1	3	2	19 (8.8)
at all	4	5	3	3	4	2	21 (9.6)
Σ	49	48	29	31	38	22	217 (100)

Only a few respondents did not work remotely at all during the pandemic - 21 indications, i.e. less than 10% of the total. People who performed their work in a remote system once or 2-3 times a week dominated. A very small group were people who worked remotely most of the week, a total of 11 people said that this was how they performed their professional duties.

During the survey, all respondents stated that they had returned to the traditional way of working. This means that they work in the office every day and receive customers traditionally, face to face. Therefore, the question arises whether the employer manages their safety at work in such a way that they do not feel the threat of coronavirus infection SARS-CoV-2 and how it ensures this safety.

Table 5. *The importance for employees of solutions increasing work safety during the pandemic in offices [own study based on research]*

no.	Factor	Rank (C_i)					
		1	2	3	4	5	$\overline{x}_{century}$
1.	temperature monitoring	0	9	199	8	1	3
2.	electronic registration of movement in the facility	1	42	158	12	4	2.89
3.	automatic disinfectant dispensers	17	35	155	7	3	2.74
4.	automatic soap dispensers	4	19	120	66	12	2.72
5.	organization of remote work	6	22	140	45	3	3.07
6.	appointment of clients by the hour	26	102	75	11	3	*2.4*
7.	customer registration over the phone	22	92	82	17	4	2.49
8.	online customer appointments	13	24	110	51	19	3.19
9.	limiting direct contact with customers	7	22	156	28	4	3
10.	customer service via e -PUAP	21	31	135	18	12	2.86
11.	customer service via e-mailbox	9	55	86	49	18	3.06
12.	flexibility of working time	4	12	42	121	38	3.82
13.	adapting working conditions to the needs of the employee	5	14	25	119	54	**3.98**
14.	work at home	7	12	26	124	48	3.92

Table 6. *The level of employees' satisfaction with the solutions improving work safety adopted during the pandemic [own study based on research]*

no.	Factor	Satisfaction level (W_i)					
		1	2	3	4	5	$\overline{x}_{century}$
1.	temperature monitoring	75	94	40	6	2	*1.92*
2.	electronic registration of movement in the facility	7	22	154	29	5	2.91
3.	automatic disinfectant dispensers	16	49	125	21	6	2.78
4.	automatic soap dispensers	14	22	138	26	17	3.04
6.	organization of remote work	23	42	122	25	5	2.75
7.	appointment of clients by the hour	7	31	75	79	25	3.39
8.	telephoning clients	16	25	100	62	14	3.16
9.	online customer appointments	19	27	89	23	9	2.86
10.	limiting direct contact with customers	9	94	69	36	9	2.74
11.	customer service via e -PUAP	36	69	82	25	5	2.51
12.	customer service via e-mailbox	45	105	37	23	7	2.27
13.	flexibility of working time	26	34	61	69	27	3.09
14.	adapting working conditions to the needs of the employee	11	15	21	124	46	***3.86***
15.	work at home	14	25	26	111	41	3.61

The weighted averages for the individual issues asked in the survey were calculated first. This made it possible to determine the importance for employees of the implementation of factors increasing safety in the management of the work environment (Table 5) and the level of satisfaction with the solutions applied by managers (Table 6).

Verifying the information on the importance of individual factors that were asked in the survey, it was found that for the respondents, the most important was again the ability to adjust working conditions to the needs of the employee ($\bar{x}_w = 3.98$), while the respondents considered arranging customers for a specific hour to be the least important ($\bar{x}_w = 2.4$).

The survey also assessed the distribution of responses in terms of employee satisfaction with the use of individual solutions that were introduced and still function despite the lifting of the lockdown, due to the need to ensure safe working conditions. The least satisfactory solution for the respondents was the introduction of automatic temperature monitoring ($\bar{x}_w = 1.92$), while the best assessed was the possibility for managers to adjust working conditions to the needs of the employee ($\bar{x}_w = 3.86$). In the case of all factors, the weighted average rating was 3.1, which means that the level of satisfaction of respondents with working conditions during the period of recovering from the pandemic was average. The conducted analysis also allowed for the calculation of the CSI index and determination of its percentage value (Table 7). The overall CSI value was 3.115, which translates to 62.3% as a percentage.

Table 7. CSI max and CSI % [own study based on research]

Factor	CSI value				
	n=167				
	W_i	C_i	W_i	W_i*C_i	$W_i*C_{i}max$
1.	1.92	3	0.044	0.132	0.22
2.	2.91	2.89	0.067	0.194	0.335
3.	2.78	2.74	0.064	0.175	0.32
4.	3.04	2.72	0.07	0.19	0.35
5.	2.75	3.07	0.063	0.193	0.315
6.	3.39	2.4	0.078	0.187	0.39
7.	3.16	2.49	0.073	0.182	0.365
8.	2.86	3.19	0.066	0.21	0.33
9.	2.74	3	0.063	0.19	0.315
10.	2.51	2.86	0.058	0.166	0.29
11.	2.27	3.06	0.052	0.159	0.26
12.	3.09	3.82	0.071	0.271	0.355
13.	3.86	3.98	0.09	0.358	0.45
14.	3.61	3.92	0.083	0.325	0.415
Σ	43.39		-	3.115	5 CSI$_{max}$
CSI		—		CSI **3.115**	*62.3%*

Comparing this value with the data presented in Table 1, it was found that the general level of satisfaction with the solutions adopted as part of managing work environment safety in the offices of the surveyed cities was at an average level.

The last part of the analysis shows the distribution of respondents' answers regarding their opinion on the continuation of remote work despite the lifting of the lockdown (Table 8).

Quality Production Improvement and System Safety: QPI 16 - CZOTO 10   Materials Research Forum LLC
Materials Research Proceedings 34 (2023) 477-485-   https://doi.org/10.21741/9781644902691-55

Table 8. *Respondents' attitude to remote work after lifting the lockdown [own study based on research]*

Ready to work remotely	City						Σ (%)
	Gd	**Vol**	**S**	**T**	**Gl**	**Rz**	
4-5 days a week	4	3	2	1	3	2	15 (6.9)
2-3 days a week	11	9	6	7	6	5	44 (20.3)
1 day a week	15	10	9	9	8	6	57 (26.3)
several times a month	8	5	3	3	4	4	27 (12.4)
single days	4	9	4	3	6	2	28 (12.9)
at all	7	12	5	8	11	3	46 (21.2)
Σ	49	48	29	31	38	22	217 (100)

The answers on this issue are strongly differentiated, and the fewest people would like to work remotely every day or most of the week - 15 responses. Almost 30 people are ready to provide work in this way in exceptional situations, on single working days or when the need arises. More than a quarter of respondents prefer remote work on one day of the week, and slightly more than 20% 2-3 times a week. This means that employees in the workplace do not feel threatened, but they need to function in a standard way in employment. Probably, working in a stationary mode allows employees to meet the needs of affiliation and contact with other people much better. This is important because these needs have been identified in numerous studies as the most unmet due to the lockdown introduced to limit the spread of the pandemic.

To sum up, it should be stated that the results of the study presented above concern only a fragment of reality. It is therefore advisable to continue research on safety management in the work environment, taking into account the changes in the development of the Covid- 19 pandemic. It should also be noted that the pandemic situation, although it currently seems limited, is such a new phenomenon that scientific research on various aspects of the functioning of public institutions in the conditions resulting from it must be continued. Although there are more and more scientific reports on this issue in the country in the world [13, 14,15,16, et. al.], similarly to the study presented above, studies with a limited scope still dominate. However, there are no comprehensive studies concerning all areas of professional activity and various groups of employees, as well as various areas of public and economic life.

During the lockdown, the number of conducted and presented studies was also significantly limited due to significant difficulties in communication between researchers and office employees. Finally, it is difficult to say how the situation with the spread of the SARS-CoV-2 virus will develop in subsequent periods and, therefore, what solutions will be introduced to protect employees against infection and provide them with optimal working conditions.

All these factors mean that, on the one hand, conducting research on this issue will be of great importance for science [17, 18, 19], and on the other hand, it is difficult to predict how much will be able to run them.

Summary
Research on the management of occupational safety of employees in the conditions of the Covid-19 pandemic is of significant importance in all public and economic entities, especially those that had to completely rebuild their current way of functioning in a short time. This group included public administration offices and their employees. Due to the importance of the problem and its impact on the possibility of using the solutions applied during the pandemic in standard conditions of public administration in the future, undertaking research in this area should be considered fully justified.

The study presented in the article was an attempt to join this trend of scientific analysis. As part of it, the author verified the possibility of using the CSI method to study job satisfaction of officials and, indirectly, the quality of the work environment safety management process in municipal offices of selected cities.

The results of the study, presented in the article, allow us to conclude that it is possible to use this method in relation to internal clients, such as employees. With its help, you can also research and analyze various issues related to the functioning of people in employment and various factors affecting work safety, as well as the level of employee satisfaction with solutions introduced by managers related to managing this safety.

References

[1] Bezpieczeństwo i ochrona zdrowia osób pracujących w czasie epidemii Covid-19. Centralny Instytut Ochrony Pracy – Państwowy Instytut Badawczy. Warszawa 2020. [online] Viewed: 31-01-2023. Available from: https://www.pip.gov.pl/pl/f/v/222228/koronawirus-zalecenia%20ogolne%202020%2005%2019.pdf

[2] S. Jipson, V. Paul. Public administration: theory and practice. University of Calicut, Calicut 2011.

[3] H. Izdebski, M. Kulesza. Administracja publiczna. Zagadnienia ogólne. Liber, Warszawa 2004. ISBN 83-72060762

[4] M. Raczyńska, K. Krukowski. Zarządzanie w administracji publicznej. Od idealnej biurokracji do zarządzania procesowego. UJ, Kraków 2020. ISBN 978-8365688682

[5] Funkcjonowanie urzędów administracji publicznej w okresie epidemii Covid-19. No 173/2021/P/21/094/LSZ. NIK, Warszawa, 2021. [online] Viewed: 31-01-2023. Available from: https://www.nik.gov.pl/plik/id,25613,vp,28386.pdf

[6] M. Panasiuk, E. Berlińska. Praca zdalna w dobie pandemii COVID-19 – perspektywa pracownika, In: P. Walentynowicz, A. Sałek-Imińska (Eds.), Zarządzanie i rynek pracy w warunkach pandemii COVID-19, Bernardinum, Pelpin 2021, 101-112. ISBN 978-8381277600

[7] A. Jeran. Telecommuting (Telework) as a Source of Problems to Realise the Functions of Work. Opuscula Socialogica 2 (2016) 49-61.

[8] J. Szczepański, Ł. Zamęcki. Praca zdalna w administracji publicznej w czasie pandemii Covid-19 – raport z badań, Instytut Nauki o Polityce, Warszawa 2020. [online] Viewed: 31-01-2023. Available from: https://depot.ceon.pl/handle/123456789/20946

[9] J. Woźniak, D. Zimon. Application of CSI Method to Research Satisfaction of Consumer in Selected Sales Network, Modern Management Rev. 21 (2016) 219-228.

[10] J. Woźniak. Evaluation of Usefulness of SERVQUAL and CSI Methods in Context of Logistics Customer Service Research, Research on Enterprise in Modern Economy – theory and practice 2 (2017) 237-248.

[11] S. Minta, M. Cempiel. Consumer satisfaction survey of a traditional product on the example of "oscypek" cheese, Annals of the Polish Association of Agricultural and Agrobusiness Economists 19(6) (2017) 176-181.

[12] B. Olbrych. Organization of Direct Interview with Survey Questionnaire to Conduct Research into Quality of Services, Acta Universitatis Lodziensis. Folia Oeconomica 227 (2009) 137-149.

[13] Z. Allam, D.S. Jones. On the Coronavirus (COVID-19) Outbreak and the Smart City Network: Universal Data Sharing Standards Coupled with Artificial Intelligence (AI) to Benefit Urban Health Monitoring and Management. *Healthcare* 8 (2020) art.46. https://doi.org/10.3390/healthcare8010046

[14] S. Sooryaa Muruga Thambiran. How COVID accelerated smart city development, GCN (2020). [online] Viewed: 31-01-2023. Available from: https://gcn.com/state-local/2020/10/how-covid-accelerated-smart-city-development/315780/

[15] K. Pepłowska. Biurowość w czasach pandemii. Wpływ epidemii COVID-19 na informatyzację biurowości w jednostkach administracji samorządowej. Archeion 123 (2021) 281-309. https://doi.org/10.4467/26581264ARC.21.008.14488

[16] A. I. Syamila, G. Nurika. Health and safety practices during COVID-19: How to ensure workplace environment safety and health, J. Hum. Care 2 (2012) 253-263. http://doi.org/10.32883/hcj.v6i2.1203

[17] A. Dolot. Wpływ pandemii COVID-19 na pracę zdalną – perspektywa pracownika. E-mentor 1(83) (2022). https://doi.org/10.15219/em83.1456

[18] M. Fukowska, T. Koweszko. Analiza stanu psychicznego i satysfakcji z pracy personelu medycznego w okresie pandemii COVID-19. Psychiatria 19(1) (2022) 79-88. https://doi.org/10.5603/PSYCH.a2021.0043

[19] V. Thinh-Van et al. The COVID-19 pandemic: Workplace safety management practices, job insecurity, and employees' organizational citizenship behavior, Saf. Sci. 145 (2022) art.105027. https://doi.org/10.1016/j.ssci.2021.105527

Keyword Index

About the Editors

Robert Ulewicz is a professor at the Częstochowa University of Technology (Poland), where he serves as the head of the Department of Production Engineering and Safety at the Faculty of Management. His primary area of interest lies in quality and safety issues, as well as the application of advanced technologies in the context of Industry 4.0 transformation in small and medium-sized enterprises. He is the founder and editor-in-chief of the scientific journal "Production Engineering Archives" (ISSN: 2353-5156; e-ISSN: 2353-7779), which serves as a significant source of knowledge for the academic and industrial communities. His numerous scientific publications have contributed to the deepening of knowledge in the field of production engineering. His research contributes to the understanding and enhancement of safety and quality standards across various sectors of the economy. In 2022, Professor Robert Ulewicz achieved remarkable recognition by being ranked among the TOP 2% most influential scientists in the world. This confirms his significant contribution to the development of science and his inspiring influence on both the industry and the academic community. The results obtained are shaping the transformation in the field of manufacturing engineering, opening up new horizons for future technologies and economic development.

Norbert Radek is a professor at the Kielce University of Technology (Poland), where he is the director of the Centre for Laser Technologies of Metals at the Faculty of Mechatronics and Mechanical Engineering. His main research interests focus on the surface engineering and the technological aspects of special metallic coatings, especially those modified with a laser beam, as well as special paint coatings, including railway and military ones. His numerous publications deal with these issues in industrial and military context. He cooperates with numerous enterprises and scientific institutes, civil and military.

Jacek Pietraszek is a professor at the Cracow University of Technology (Poland), where he holds the position of the Department Chair for Applied Computer Science at the Faculty of Mechanical Engineering. His primary and profound area of expertise resides in industrial statistics and data science, with a particular focus on the experimental design (DOE). He serves as the editor-in-chief of the scientific journal "Technical Transactions" (ISSN: 0011-4561; e-ISSN: 2353-737X). His prolific body of work, including numerous publications, encompasses a wide spectrum of data analysis within the vary domains of industry, ranging from production engineering, mechanical engineering, through materials engineering, and even reaching to phytochemistry. His research, cooperation with numerous industrial and consulting companies contribute to improving the quality of research conducted in various industries. His active participation within the Polish Committee for Standardization further underscores his commitment to elevating research and industrial standards. In 2020, he was ranked among the TOP 2% of the most influential scientists in the world.